SET THEORY

STUDIES IN LOGIC

AND

THE FOUNDATIONS OF MATHEMATICS

VOLUME 86

NORTH-HOLLAND PUBLISHING COMPANY
AMSTERDAM • NEW YORK • OXFORD

SET THEORY

WITH AN INTRODUCTION TO DESCRIPTIVE SET THEORY

K. KURATOWSKI
and
A. MOSTOWSKI

1976

NORTH-HOLLAND PUBLISHING COMPANY
AMSTERDAM • NEW YORK • OXFORD
PWN—POLISH SCIENTIFIC PUBLISHERS
WARSZAWA

Library of Congress Catalog Card Number: *74 83731*
North-Holland ISBN: *0 7204 0470 3*

This book is a translation of the original Polish *Teoria mnogości* published by PWN — Polish Scientific Publishers, 1966 in the series "Monografie Matematyczne"

First edition 1968
Second, completely revised edition 1976

Publishers:

PWN — Polish Scientific Publishers — Warszawa
North-Holland Publishing Company
Amsterdam • New York • Oxford

SOLE DISTRIBUTOR FOR THE U.S.A. AND CANADA:
ELSEVIER/NORTH HOLLAND INC.
52 VANDERBILT AVENUE
NEW YORK, N.Y. 10017

PRINTED IN POLAND
(D.R.P.)

PREFACE TO THE FIRST EDITION

The creation of set theory can be traced back to the work of XIXth century mathematicians who tried to find a firm foundation for calculus. While the early contributors to the subject (Bolzano, Du Bois Reymond, Dedekind) were concerned with sets of numbers or of functions, the proper founder of set theory, Georg Cantor, made a decisive step and started an investigation of sets with arbitrary elements. The series of articles published by him in the years 1871–1883 contains an almost modern exposition of the theory of cardinals and of ordered and well-ordered sets. That step toward generalizations which Cantor made was a difficult one was witnessed by various contradictions (antinomies of set theory) discovered in set theory by various authors around 1900. The crisis created by these antinomies was overcome by Zermelo who formulated in 1904–1908 the first system of axioms of set theory. His axioms were sufficient to obtain all mathematically important results of set theory and at the same time did not allow the reconstruction of any known antinomy. Close ties between set theory and philosophy of mathematics date back to discussions concerning the nature of antinomies and the axiomatization of set theory. The fundamental problems of philosophy of mathematics such as the meaning of existence in mathematics, axiomatics versus description of reality, the need of consistency proofs and means admissible in such proofs were never better illustrated than in these discussions.

After an initial period of distrust the newly created set theory made a triumphal inroad in all fields of mathematics. Its influence on mathematics of the present century is clearly visible in the choice of modern problems and in the way these problems are solved. Applications of set theory are thus immense. But set theory developed also problems of its own. These problems and their solutions represent what is known as abstract set theory. Its achievements are rather modest in comparison to the applications of set theoretical methods in other branches of mathematics, some of which owe their very existence to set theory. Still, abstract set theory is a well-established part of mathematics and the knowledge of its basic notions is required from every mathematician.

Recent years saw a stormy advance in foundations of set theory. After breaking through discoveries of Gödel in 1940 who showed relative consistency of various set-theoretical hypotheses the recent works of Cohen allowed him and his successors to solve most problems of independence of these hypotheses while at the same time the works of Tarski showed how deeply can we delve in the domain of inaccessible cardinals whose magnitude surpasses all imagination. These recent works will certainly influence the future thinking on the philosophical foundations of mathematics.

The present book arose from a mimeographed text of Kuratowski from 1921 and from an enlarged edition prepared jointly by the two authors in 1951. As a glance on the list of contents will show, we intended to present the basic results of abstract set theory in the traditional order which goes back still to Cantor: algebra of sets, theory of cardinals, ordering and well-ordering of sets. We lay more stress on applications than it is usually done in texts of abstract set theory. The main field in which we illustrate set-theoretical methods is general topology. We also included a chapter on Borel, analytical and projective sets. The exposition is based on axioms which are essentially the ones of Zermelo–Fraenkel. We tried to present the proofs of all theorems even of the very trivial ones in such a way that the reader feels convinced that they are entirely based on the axioms. This accounts for some pedantry in notation and in the actual writing of several formulae which could be dispensed with if we did not wish to put the finger on axioms which we use in proofs. In some examples we use notions which are commonly known but which were not defined in our book by means of the primitive terms of our system. These examples are marked by the sign #.

In order to illustrate the role of the axiom of choice we marked by a small circle ° all theorems in which this axiom is used. There is in the book a brief account of the continuum hypothesis and a chapter on inaccessible cardinals. These topics deserve a more thorough presentation which however we could not include because of lack of space.

Also the last chapter which deals with the descriptive set theory is meant to be just an introduction to the subject.

Several colleagues helped us with the preparation of the text. Dr M. Mączyński translated the main part of the book and Mr R. Kowalsky collaborated with him in this difficult task. Professor J. Łoś wrote a penetrating appraisal of the manuscript of the 1951 edition as well as of the present one. His remarks and criticism allowed us to eliminate many errors and inaccuracies. Mr W. Marek and Mr K. Wiśniewski read the manuscript and the galley proofs and helped us in improving our text. To all these persons we express our deep gratitude.

KAZIMIERZ KURATOWSKI
ANDRZEJ MOSTOWSKI

PREFACE TO THE SECOND EDITION

The second edition of our *Set Theory* differs essentially from the first—which was translated from the Polish edition (by Professor M. Mączyński)—by the extension of its content. Our aim was to introduce the reader to some chapters of set theory which actually seem to be especially attractive and are cultivated by a large and still growing number of mathematicians.

The major changes introduced in the new edition are:

1. We wrote a new chapter on trees containing also a short introduction to the partition calculus and we completely rewrote Chapter 9 of the old edition dealing with inaccessible cardinals.

2. We introduced four chapters on descriptive set theory which replace Chapter 10 of the old edition. These four chapters (which were written by K. Kuratowski) contain

a. A short survey of the theory of Borel sets and Borel-measurable functions, preceded by a fairly general theory of L-measurable functions (where L is an arbitrary σ-lattice).

b. An insight into the theory of Souslin (analytic) sets and—more generally—of projective sets.

c. Some results on measurable selectors, mostly found within the last few years.

Some results presented in Chapters 11–14 are new as far as we know, whereas the first 10 chapters contain only results which are known from the literature.

We consistently tried to remain within the framework of the classical set theory. For this reason we did not include into our book any of the exciting recent results in whose proofs one uses model theoretical methods or notions borrowed from advanced parts of mathematical logic. See Mostowski [1].

We welcome this opportunity to express our gratitude—in addition to persons mentioned in the Preface to the first edition—to our younger colleagues, J. Kaniewski, W. Marek, R. Pol, and P. Zbierski who read the manuscript, engaged in numerous discussions and provided many suggestions and corrections.

It is also our pleasure to express our thanks to Dr B. S. Niven from the White Agricultural Research Institute, South Australia, for correcting our English and to Mrs D. Wojciechowska for her help in preparing our manuscript.

Finally our thanks go to the North-Holland Publishing Company, as well as to the Polish Scientific Publishers and personally to Mrs Z. Osek, Mr W. Muszyński and Mr J. Panz for their assistance in the publication of this book.

KAZIMIERZ KURATOWSKI
ANDRZEJ MOSTOWSKI

CONTENTS

Preface to the first edition v
Preface to the second edition viii

CHAPTER I. Algebra of sets

§ 1. Propositional calculus 1
§ 2. Sets and operations on sets 4
§ 3. Inclusion. Empty set 7
§ 4. Laws of union, intersection, and subtraction 10
§ 5. Properties of symmetric difference 13
§ 6. The set 1, complement 18
§ 7. Constituents 21
§ 8. Applications of the algebra of sets to topology 27
§ 9. Boolean algebras 33
§ 10. Lattices 42

CHAPTER II. Axioms of set theory. Relations. Functions

§ 1. Set theoretical formulas. Quantifiers 46
§ 2. Axioms of set theory 52
§ 3. Some simple consequences of the axioms 58
§ 4. Cartesian products. Relations 62
§ 5. Equivalence relations. Partitions 66
§ 6. Functions 69
§ 7. Images and inverse images 74
§ 8. Functions consistent with a given equivalence relation. Factor Boolean algebras 78
§ 9. Order relations 80
§ 10. Relational systems, their isomorphisms and types 85

CHAPTER III. Natural numbers. Finite and infinite sets

§ 1. Natural numbers 89
§ 2. Definitions by induction 93
§ 3. The mapping J of the set $N \times N$ onto N and related mappings . . 98
§ 4. Finite and infinite sets 102

CHAPTER IV. Generalized union, intersection and Cartesian product

§ 1. Set-valued functions. Generalized union and intersection 107
§ 2. Operations on infinite sequences of sets 117
§ 3. Families of sets closed under given operations 121
§ 4. σ-additive and δ-multiplicative families of sets 124
§ 5. Reduction and separation properties 127
§ 6. Generalized cartesian products 129
§ 7. Cartesian products of topological spaces 133
§ 8. The Tychonoff theorem 137
§ 9. Reduced direct products 140
§ 10. Infinite operations in lattices and in Boolean algebras 145
§ 11. Extensions of ordered sets to complete lattices 152
§ 12. Representation theory for distributive lattices 158

CHAPTER V. Theory of cardinal numbers

§ 1. Equipollence. Cardinal numbers 164
§ 2. Countable sets 169
§ 3. The hierarchy of cardinal numbers 174
§ 4. The arithmetic of cardinal numbers 178
§ 5. Inequalities between cardinal numbers. The Cantor–Bernstein theorem and its generalizations 181
§ 6. Properties of the cardinals $\mathfrak{a}$ and $\mathfrak{c}$ 188
§ 7. The generalized sum of cardinal numbers 191
§ 8. The generalized product of cardinal numbers 195

CHAPTER VI. Linearly ordered sets

§ 1. Introduction 201
§ 2. Dense, scattered, and continuous sets 205
§ 3. Order types ω, η, and λ 210
§ 4. Arithmetic of order types 217
§ 5. Lexicographical ordering 220

CHAPTER VII. Well-ordered sets

§ 1. Definitions. Principle of transfinite induction 224
§ 2. Ordinal numbers 228
§ 3. Transfinite sequences 230
§ 4. Definitions by transfinite induction 233
§ 5. Ordinal arithmetic 239
§ 6. Ordinal exponentiation 245
§ 7. Expansions of ordinal numbers for an arbitrary base 248
§ 8. The well-ordering theorem 254
§ 9. Von Neumann's method of elimination of ordinal numbers . . . 262

CHAPTER VIII. Alephs and related topics

§ 1. Ordinal numbers of power $\mathfrak{a}$ 267
§ 2. The cardinal $\aleph(\mathfrak{m})$. Hartogs' aleph 270
§ 3. Initial ordinals 272
§ 4. Alephs and their arithmetic 275
§ 5. The exponentiation of alephs 280
§ 6. The exponential hierarchy of cardinal numbers 284
§ 7. The continuum hypothesis 290
§ 8. The number of prime ideals in the algebra $P(A)$ 296
§ 9. $\mathfrak{m}$-disjoint sets 300
§ 10. Families of disjoint open sets 302
§ 11. Equivalence of certain statements about cardinal numbers with the axiom of choice 308

CHAPTER IX. Trees and partitions

§ 1. Trees 315
§ 2. The lexicographical ordering of zero-one sequences. η_ξ sets 319
§ 3. König's infinity lemma 326
§ 4. Aronszajn's trees 329
§ 5. Souslin trees 332
§ 6. Some partition theorems 336

CHAPTER X. Inaccessible cardinals

§ 1. Normal functions and stationary sets 342
§ 2. Weakly and strongly inaccessible cardinals 348
§ 3. A digression on models of Σ°[TR] 352
§ 4. Higher types of inaccessible numbers 356
§ 5. Weakly compact cardinals 360
§ 6. Measurable cardinals 366
§ 7. Measurable cardinals and reduced products 375

Incroduction to descriptive set theory

CHAPTER XI. Auxiliary notions

§ 1. The notion of a metric space. Various fundamental topological notions 386
§ 2. Exponential topology. Compact-open topology 392
§ 3. Complete and Polish spaces 396
§ 4. $\boldsymbol{L}$-measurable mappings 400
§ 5. The operation $\mathscr{A}$ 409
§ 6. The Lusin sieve 412

CHAPTER XII. Borel sets. $\boldsymbol{B}$-measurable functions. Baire property

§ 1. Elementary properties of Borel subsets of a metric space 415

§ 2. Ambiguous Borel sets 417
§ 3. Borel-measurable functions 419
§ 4. ***B***-measurable complex and product functions 421
§ 5. Universal functions for Borel classes 423
§ 6. Borel subsets of Polish spaces 426
§ 7. Further properties of Borel sets 427
§ 8. Baire property 428

CHAPTER XIII. Souslin spaces. Projective sets

§ 1. Souslin spaces. Fundamental properties 434
§ 2. Applications of countable order types to Souslin spaces 444
§ 3. Coanalytic sets (*CA*-sets) 447
§ 4. The σ-algebra $\bar{\boldsymbol{S}}$ generated by Souslin sets and the $\bar{\boldsymbol{S}}$-measurable mappings 451
§ 5. The ***PCA***-sets and sets of higher projective classes 455

CHAPTER XIV. Measurable selectors

§ 1. The general selector theorem 458
§ 2. Selectors for measurable partitions of Polish spaces 463
§ 3. Selectors for point-inverses of continuous mappings. 466

Bibliography 476

List of important symbols 496

Subject index 502

CHAPTER I

ALGEBRA OF SETS

§ 1. Propositional calculus

Mathematical reasoning in set theory may be presented in a very clear form by making use of logical symbols and by basing arguments on the laws of logic formulated in terms of such symbols. In this section we shall present some basic principles of logic in order to refer to them later in this chapter and in the remainder of the book.

We shall designate arbitrary sentences by the letters $p, q, r, \ldots$ We assume that all of the sentences to be considered are either true or false. Since we consider only sentences of mathematics, we shall be dealing with sentences for which the above assumption is applicable.

From two arbitrary sentences, p and q, we can form a new sentence by applying to p and to q any one of the connectives:

and, or, if ... then ..., if and only if.

The sentence *p and q* we write in symbols $p \wedge q$. The sentence $p \wedge q$ is called the *conjunction* or the logical *product* of the sentences p and q which are the *components* of the conjunction. The conjunction $p \wedge q$ is true when both components are true. On the other hand, if any one of the components is false then the conjunction is false.

The sentence *p or q*, which we write symbolically $p \vee q$, is called the *disjunction* or the logical *sum* of the sentences p and q (the components of the disjunction). The disjunction is true if either of the components is true and is false only when both components are false.

The sentence *if p then q* is called the *implication* of q by p, where p is called the *antecedent* and q the *consequent* of the implication. Instead of writing *if* p *then* q we write $p \to q$. An implication is false if the consequent is false and the antecedent true. In all other cases the implication is true.

If the implication $p \to q$ is true we say that q *follows* from p; if we know that the sentence p is true we may conclude that the sentence q is also true.

In ordinary language the sense of the expression "if ..., then ..." does not entirely coincide with the meaning given above. However, in mathematics the use of such a definition as we have given is useful.

The sentence p *if and only if* q is called the *equivalence* of the two component sentences p and q and is written $p \equiv q$. This sentence is true provided p and q have the same logical value; that is, either both are true or both are false. If p is true and q false, or if p is false and q true, then the equivalence $p \equiv q$ is false.

The equivalence $p \equiv q$ can also be defined by the conjunction

$$(p \to q) \wedge (q \to p).$$

The sentence *it is not true that* p we call the *negation* of p and we write $\neg p$. The negation $\neg p$ is true when p is false and false when p is true. Hence $\neg p$ has the logical value opposite to that of p.

We shall denote an arbitrary true sentence by V and an arbitrary false sentence by F; for instance, we may choose for V the sentence $2 \cdot 2 = 4$, and for F the sentence $2 \cdot 2 = 5$.

Using the symbols F and V, we can write the definitions of truth and falsity for conjunction, disjunction, implication, equivalence and negation in the form of the following true equivalences:

(1) $F \wedge F \equiv F, \quad F \wedge V \equiv F, \quad V \wedge F \equiv F, \quad V \wedge V \equiv V,$

(2) $F \vee F \equiv F, \quad F \vee V \equiv V, \quad V \vee F \equiv V, \quad V \vee V \equiv V,$

(3) $(F \to F) \equiv V, \quad (F \to V) \equiv V, \quad (V \to F) \equiv F, \quad (V \to V) \equiv V,$

(4) $(F \equiv F) \equiv V, \quad (F \equiv V) \equiv F, \quad (V \equiv F) \equiv F, \quad (V \equiv V) \equiv V,$

(5) $\neg F \equiv V, \quad \neg V \equiv F.$

Logical *laws* or *tautologies* are those expressions built up from the letters $p, q, r, \ldots$ and the connectives $\wedge$, $\vee$, $\to$, $\equiv$, $\neg$ which have the property that no matter how we replace the letters $p, q, r, \ldots$ by arbitrary sentences (true or false) the entire expression itself is always true.

The truth or falsity of a sentence built up by means of connectives from the sentences $p, q, r, \ldots$ does not depend upon the meaning of the sentences $p, q, r, \ldots$ but only upon their logical values. Thus we can test whether an expression is a logical law by applying the following method: in place of the letters $p, q, r, \ldots$ we substitute the values F and V in every possible manner. Then using equations (1)–(5) we calculate the logical value of the expression for each one of these substitutions. If this value is always true, then the expression is a tautology.

Example. The expression $(p \wedge q) \rightarrow (p \vee r)$ is a tautology. It contains three variables p, q and r. Thus we must make a total of eight substitutions, since for each variable we may substitute either F or V. If, for example, for each letter we subtitute F, then we obtain $(F \wedge F) \rightarrow (F \vee F)$, and by (1) and (2) we obtain $F \rightarrow F$, namely V. Similarly, the value of the expression $(p \wedge q) \rightarrow (p \vee r)$ is true in each of the remaining seven cases.

Below we give several of the most important logical laws together with names for them. Checking that they are indeed logical laws is an exercise which may be left to the reader.

$(p \vee q) \equiv (q \vee p)$ *law of commutativity of disjunction,*

$[(p \vee q) \vee r] \equiv [p \vee (q \vee r)]$ *law of associativity of disjunction,*

$(p \wedge q) \equiv (q \wedge p)$ *law of commutativity of conjunction,*

$[p \wedge (q \wedge r)] \equiv [(p \wedge q) \wedge r]$ *law of associativity of conjunction,*

$[p \wedge (q \vee r)] \equiv [(p \wedge q) \vee (p \wedge r)]$ *first distributive law,*

$[p \vee (q \wedge r)] \equiv [(p \vee q) \wedge (p \vee r)]$ *second distributive law,*

$(p \vee p) \equiv p, \quad (p \wedge p) \equiv p$ *laws of tautology,*

$(p \wedge F) \equiv F, \quad (p \wedge V) \equiv p$
$(p \vee F) \equiv p, \quad (p \vee V) \equiv V$ *laws of absorption.*

In these laws the far reaching analogy between propositional calculus and ordinary arithmetic is made apparent. The major differences occur in the second distributive law and in the laws of tautology and absorption. In particular, the laws of tautology show that in the propositional calculus with logical addition and multiplication we need use neither coefficients nor exponents.

$[(p \to q) \wedge (q \to r)] \to (p \to r)$	*law of the hypothetical syllogism,*
$(p \vee \neg p) \equiv V$	*law of excluded middle,*
$(p \wedge \neg p) \equiv F$	*law of contradiction,*
$p \equiv \neg\neg p$	*law of double negation,*
$\neg(p \vee q) \equiv (\neg p \wedge \neg q)$ $\neg(p \wedge q) \equiv (\neg p \vee \neg q)$	*de Morgan's laws,*
$(p \to q) \equiv (\neg q \to \neg p)$	*law of contraposition,*
$(p \to q) \equiv (\neg p \vee q),$	
$F \to p, \quad p \to p, \quad p \to V.$	

Throughout this book whenever we shall write an expression using logical symbols, we shall tacitly state that the expression is true. Remarks either preceding or following such an expression will always refer to a proof of its validity.

§2. Sets and operations on sets

The basic notion of set theory is the concept of *set*. This basic concept is, in turn, a product of historical evolution. Originally the theory of sets made use of an intuitive concept of set, characteristic of the so-called "naive" set theory. At that time the word "set" had the same imprecisely defined meaning as in everyday language. Such, in particular, was the concept of set held by Cantor,[1]) the creator of set theory.

Such a view was untenable, as in certain cases the intuitive concept proved to be unreliable. In Chapter II, §2 we shall deal with the antinomies of set theory, i.e. with the apparent contradictions which appeared at a certain stage in the development of the theory and

[1]) Georg Cantor (1845–1918) to whom we owe the creation of set theory was a German mathematician, professor at the University of Halle. He published his basic papers on set theory in "Mathematische Annalen" during the years 1879–1893. These papers were reprinted in Cantor [7]; this volume contains also a biography of Cantor written by E. Zermelo.

were due to the vagueness of intuition associated with the concept of set in certain more complicated cases. In the course of the polemic which arose over the antinomies it became obvious that different mathematicians had different concept of sets. As a result it became impossible to base set theory on intuition.

In the present book we shall present set theory as an axiomatic system. In geometry we do not examine directly the meaning of the terms "point", "line", "plane" or other "primitive terms", but from a well-defined system of axioms we deduce all the theorems of geometry without resorting to the intuitive meaning of the primitive terms. Similarly, we shall base set theory on a system of axioms from which we shall obtain theorems by deduction. Although the axioms have their source in the intuitive concept of sets, the use of the axiomatic method ensures that the intuitive content of the word "set" plays no part in proofs of theorems or in definitions of set theoretical concepts.

Sometimes we shall illustrate set theory with examples furnished by other branches of mathematics. This illustrative material involving axioms not belonging to the axiom system of set theory will be distinguished by the sign # placed at the beginning and at the end of the text.

The primitive notions of set theory are "set" and the relation "to be an element of". Instead of *x is a set* we shall write $Z(x)$, and instead of *x is an element of y* we shall write $x \in y$.[1]) The negation of the formula $x \in y$ will be written as x non $\in y$, or $x \notin y$ or $\neg(x \in y)$. To simplify the notation we shall use capital letters to denote sets; thus if a formula involves a capital letter, say A, then it is tacitly assumed that A is a set. Later on we shall introduce yet another primitive notion: $x\mathrm{TR}y$ (*x is the relational type of y*). We shall discuss it in Chapter II.

For the present we assume four axioms:

I. AXIOM OF EXTENSIONALITY: *If the sets A and B have the same elements then they are identical.*

[1]) The symbol $\in$ is derived from the Greek letter *epsilon*. The use of this letter for the elementhood relation was introduced by Peano [2] who selected it as the abbreviation of the Greek word "to be" (ἐστί). Many other mathematical and logical symbols also originated with Peano.

A.[1]) Axiom of union: *For any sets A and B there exists a set which contains all the elements of A and all the elements of B and which does not contain any other elements.*

B.[1]) Axiom of difference: *For any sets A and B there exists a set which contains only those elements of A which are not elements of B.*

C.[1]) Axiom of existence: *There exists at least one set.*

The axiom of extensionality can be rewritten in the following form:

$$\textit{if, for every } x, \quad x \in A \equiv x \in B, \quad \textit{then} \quad A = B,$$

where the equality sign between the two symbols indicates that they denote the same object.

It follows from axioms I and A that for any sets A and B there exists exactly one set satisfying the conditions of axiom A. In fact, if there were two such sets C_1 and C_2, then they would contain the same elements (namely those which belong either to A or to B) and, by axiom I, $C_1 = C_2$.

The unique set satisfying the conditions of axiom A is called the *union* of two sets A and B and is denoted by $A \cup B$. Thus for any x and for any sets A and B we have the equivalence

(1) $$x \in A \cup B \equiv (x \in A) \vee (x \in B).$$

Similarly, from axioms I and B, it follows that for any sets A and B there exists exactly one set whose elements are all the objects belonging to A and not belonging to B. This set, called the *difference* of A and B, is denoted by $A-B$. For any x and for arbitrary sets A and B we have

(2) $$x \in A-B \equiv (x \in A) \wedge (x \notin B).$$

By means of de Morgan's law and the law of double negation (§1, p. 4) it follows that

(3) $$\neg(x \in A-B) \equiv \neg(x \in A) \vee (x \in B),$$

i.e. x is not an element of $A-B$ if x is not an element of A or x is an element of B.

[1]) In Chapter II these axioms will be replaced by more general ones.

Using the operations $\cup$ and $-$, we can define two other operations on sets.

The *intersection* $A \cap B$ of A and B we define by

$$A \cap B = A - (A - B).$$

From the definition of difference we have for any x

$$x \in A \cap B \equiv (x \in A) \wedge \neg(x \in A - B),$$

from which, by means of (3) and the first distributive law (see p. 3), it follows that

$$\begin{aligned} x \in A \cap B &\equiv (x \in A) \wedge [\neg(x \in A) \vee (x \in B)] \\ &\equiv [(x \in A) \wedge \neg(x \in A)] \vee [(x \in A) \wedge (x \in B)] \\ &\equiv F \vee [(x \in A) \wedge (x \in B)] \equiv [(x \in A) \wedge (x \in B)], \end{aligned}$$

and finally

(4) $$x \in A \cap B \equiv (x \in A) \wedge (x \in B).$$

Hence the intersection of two sets is the common part of the factors; the elements of the intersection are those objects which belong to both factors.

The *symmetric difference* of two sets A and B is defined as

(5) $$A \dot{-} B = (A - B) \cup (B - A).$$

The elements of the set $A \dot{-} B$ are those objects which belong to A and not to B together with those objects which belong to B and not to A.

Exercises

1. Define the operations $\cup$, $\cap$, $-$ by means of: (a) $\dot{-}$, $\cap$, (b) $\dot{-}$, $\cup$, (c) $-$, $\dot{-}$.

2. Show that it is not possible to define either the sum by means of the intersection and the difference, or the difference by means of the sum and the intersection

§3. Inclusion. Empty set

A set A is said to be a *subset* of a set B provided every element of the set A is also an element of the set B. In this case we write $A \subset B$ or $B \supset A$ and we say that A is *included* in B. The relation $\subset$ is called the *inclusion relation*.

The following equivalence results from this definition

(1) $$\{\textit{for every } x \ (x \in A \rightarrow x \in B)\} \equiv A \subset B.$$

Clearly, from $A = B$ it follows that $A \subset B$, but not conversely. If $A \subset B$ and $A \neq B$ we say that A is a *proper* subset of B. If A is a subset of B and B is a subset of A then $A = B$, i.e.

$$(A \subset B) \wedge (B \subset A) \rightarrow (A = B).$$

To prove this we notice that from the left-hand side of the implication we have for every x

$$x \in A \rightarrow x \in B \quad \text{and} \quad x \in B \rightarrow x \in A,$$

from which we obtain the equivalence $x \in A \equiv x \in B$, and thus $A = B$ by axiom I.

It is easy to show that if A is a subset of B and B is a subset of C, then A is a subset of C:

(2) $$(A \subset B) \wedge (B \subset C) \rightarrow (A \subset C),$$

i.e. the inclusion relation is *transitive*.

The union of two sets contains both components; the intersection of two sets is contained in each component:

(3) $$A \subset A \cup B, \quad B \subset A \cup B,$$

(4) $$A \cap B \subset A, \quad A \cap B \subset B.$$

In fact, from $p \rightarrow (p \vee q)$ it follows that for every x

$$x \in A \rightarrow [(x \in A) \vee (x \in B)],$$

from which, by 2 (1),[1] $x \in A \rightarrow x \in (A \cup B)$, and by (1) we obtain $A \subset A \cup B$. The proof of the second formula of (3) is similar, the proof of (4) follows from the law $(p \wedge q) \rightarrow p$.

From 2 (2) it follows that

$$A - B \subset A.$$

Thus the difference of two sets is contained in the minuend.

The inclusion relation can be defined by means of the identity relation and one of the operations $\cup$ or $\cap$. Namely, the following equivalences hold

(5) $$(A \subset B) \equiv (A \cup B = B) \equiv (A \cap B = A).$$

[1]) 2 (1) denotes formula (1) in §2.

In fact, if $A \subset B$ then for every x, $x \in A \to x \in B$; thus, by means of the law $(p \to q) \to [(p \vee q) \to q]$,

$$[(x \in A) \vee (x \in B)] \to (x \in B)$$

which proves that $A \cup B \subset B$. On the other hand, $B \subset A \cup B$ and hence $A \cup B = B$.

Conversely, if $A \cup B = B$, then by (3) $A \subset B$.

The second part of equivalence (5) can be proved in a similar manner.

It follows from axiom B that if there exists at least one set A then there also exists the set $A - A$ which contains no element. There exists only one such set. In fact, if there were two such sets Z_1 and Z_2, then (for every x) we would have the equivalence

$$x \in Z_1 \equiv x \in Z_2 .$$

This equivalence holds since both components are false. Thus, from axiom I, $Z_1 = Z_2$.

This unique set which contains no element is called the *empty set* and is denoted by $\emptyset$. Thus for every x

$$x \notin \emptyset,$$

i.e.

$$(x \in \emptyset) \equiv F.$$

The implication $x \in \emptyset \to x \in A$ holds for every x since the antecedent of the implication is false. Thus

$$\emptyset \subset A,$$

i.e. the empty set is a subset of every set.

Formula 2 (1) implies that

$$x \in (A \cup \emptyset) \equiv (x \in A) \vee (x \in \emptyset) \equiv (x \in A) \vee F \equiv x \in A,$$

because $p \vee F \equiv p$. From this we infer

$$A \cup \emptyset = A,$$

and from $\neg F \equiv V$

$$A - \emptyset = A .$$

The identity $A \cap B = \emptyset$ indicates that the sets A and B have no common element, or—in other words—they are *disjoint*.

The equation $B - A = \emptyset$ is equivalent to $B \subset A$.

The role played by the empty set in set theory is analogous to that played by the number zero in algebra. Without the set $\emptyset$ it would not always be possible to perform the operations of intersection and subtraction and the calculus of sets would be considerably more complicated.

§4. Laws of union, intersection, and subtraction

The operations of union, intersection, and subtraction on sets have many properties in common with operations on numbers: namely, union with addition, intersection with multiplication, and subtraction with subtraction. In this section we shall mention the most important of these properties. We shall also prove several theorems indicating the difference between the algebra of sets and arithmetic.

The commutative laws:

$$A \cup B = B \cup A, \qquad A \cap B = B \cap A. \tag{1}$$

These laws follow directly from the commutative laws for disjunction and conjunction.

The associative laws:

$$A \cup (B \cup C) = (A \cup B) \cup C, \qquad A \cap (B \cap C) = (A \cap B) \cap C. \tag{2}$$

Again, these laws are direct consequences of the associative laws for disjunction and conjunction.

Formulas (1) allow us to permute the components of any union or intersection of a finite number of sets without changing the results. Similarly, formulas (2) allow us to group the components of such a finite union or intersection in an arbitrary manner. For example:

$$\begin{aligned} A \cup \{B \cup [C \cup (D \cup E)]\} &= [A \cup (D \cup C)] \cup (B \cup E) \\ &= (E \cup C) \cup [B \cup (A \cup D)]. \end{aligned}$$

In other words, we may eliminate parentheses when performing the operation of union (or intersection) on a finite number of sets.

The distributive laws:

$$\begin{aligned} A \cap (B \cup C) &= (A \cap B) \cup (A \cap C), \\ A \cup (B \cap C) &= (A \cup B) \cap (A \cup C). \end{aligned} \tag{3}$$

The proofs follow from the distributive laws for conjunction over disjunction and disjunction over conjunction, given in §1.

The first distributive law is completely analogous to the corresponding distributive law in arithmetic. Similarly, as in arithmetic, from this law it follows that in order to intersect two unions we may intersect each component of the first union with each component of the second union and take the union of those intersections:

$$\begin{aligned}&(A\cup B\cup \ldots \cup H)\cap(X\cup Y\cup \ldots \cup T)\\&\quad=(A\cap X)\cup(A\cap Y)\cup \ldots \cup(A\cap T)\cup(B\cap X)\cup(B\cap Y)\cup\\&\qquad\ldots \cup(B\cap T)\cup \ldots \cup(H\cap X)\cup(H\cap Y)\cup \ldots \cup(H\cap T).\end{aligned}$$

The second distributive law has no counterpart in arithmetic.

The laws of tautology:

(4) $$A\cup A=A,\qquad A\cap A=A.$$

The proof is immediate from the laws of tautology $(p\vee p)\equiv p$ and $(p\wedge p)\equiv p$.

We shall prove several laws of subtraction.

(5) $$A\cup(B-A)=A\cup B.$$

PROOF. By means of (1) and (2) of § 2 we have

$$x\in[A\cup(B-A)]\equiv(x\in A)\vee[(x\in B)\wedge\neg(x\in A)],$$

from which, by the distributive law for disjunction over conjunction,

$$\begin{aligned}x\in[A\cup(B-A)]&\equiv[(x\in A)\vee(x\in B)]\wedge[(x\in A)\vee\neg(x\in A)]\\&\equiv(x\in A)\vee(x\in B),\end{aligned}$$

since $(x\in A)\vee\neg(x\in A)\equiv V$, and V may be omitted as a component of a conjunction. Thus

$$x\in[A\cup(B-A)]\equiv x\in(A\cup B),$$

which proves (5).

From (5) we conclude that the operation of forming difference of sets is not the inverse of the operation of forming their union. For example, if A is the set of even numbers and B the set of numbers divisible by 3 then the set $A\cup(B-A)$ is different from B, for it contains all even numbers.

On the other hand, if $A \subset B$, we have by (5) and 3 (5)

$$A \cup (B-A) = B,$$

as in arithmetic.

(6) $$A-B = A-(A \cap B).$$

PROOF.

$$\begin{aligned} x \in A-(A \cap B) &\equiv (x \in A) \wedge \neg(x \in A \cap B) \equiv (x \in A) \wedge \neg[(x \in A) \wedge (x \in B)] \\ &\equiv (x \in A) \wedge [\neg(x \in A) \vee \neg(x \in B)] \\ &\equiv [(x \in A) \wedge \neg(x \in A)] \vee [(x \in A) \wedge \neg(x \in B)] \\ &\equiv F \vee [(x \in A) \wedge \neg(x \in B)] \equiv [(x \in A) \wedge \neg(x \in B)] \\ &\equiv x \in A-B. \end{aligned}$$

The distributive law for intersection over subtraction has in the algebra of sets the following form

(7) $$A \cap (B-C) = (A \cap B)-C.$$

This law follows from the equivalence

$$\begin{aligned} x \in A \cap (B-C) &\equiv [(x \in A) \wedge (x \in B) \wedge \neg(x \in C)] \\ &\equiv [(x \in A \cap B) \wedge \neg(x \in C)] \\ &\equiv x \in (A \cap B)-C. \end{aligned}$$

From (7) it follows that

$$A \cap (B-A) = (A \cap B)-A = (B \cap A)-A = B \cap (A-A) = B \cap \emptyset = \emptyset.$$

Thus

$$A \cap (B-A) = \emptyset.$$

De Morgan's laws for the calculus of sets take the following form

(8) $$\begin{aligned} A-(B \cap C) &= (A-B) \cup (A-C), \\ A-(B \cup C) &= (A-B) \cap (A-C). \end{aligned}$$

In the proofs we make use of de Morgan's laws for the propositional calculus.

The following identities are given without proof.

(9) $$(A \cup B)-C = (A-C) \cup (B-C),$$

(10) $$A-(B-C) = (A-B) \cup (A \cap C),$$

(11) $$A-(B \cup C) = (A-B)-C.$$

The following formulas illustrate the analogy between the inclusion relation and the "less than" relation in arithmetic:

(12) $$(A \subset B) \wedge (C \subset D) \rightarrow (A \cup C \subset B \cup D),$$

(13) $$(A \subset B) \wedge (C \subset D) \rightarrow (A \cap C \subset B \cap D),$$

(14) $$(A \subset B) \wedge (C \subset D) \rightarrow (A - D \subset B - C).$$

From (14) it follows as an easy consequence that

(15) $$(C \subset D) \rightarrow (A - D \subset A - C),$$

which is the counterpart of the arithmetic theorem:

$$x \leqslant y \rightarrow z - y \leqslant z - x.$$

Exercises

1. Prove the formula

$$N(A \cup B) = N(A) + N(B) - N(A \cap B),$$

where $N(X)$ denotes the number of elements of the set X (under the assumption that X is finite).

Hint: Express $N(A-B)$ in terms of $N(A)$ and $N(A \cap B)$.

2. Generalize the result of Exercise 1 in the following way

$$N(A_1 \cup A_2 \cup \ldots \cup A_n) = \sum_i N(A_i) - \sum_{i,j} N(A_i \cap A_j) + \sum_{i,j,k} N(A_i \cap A_j \cap A_k) - \ldots,$$

where the indices of the summations take as values the numbers from 1 to n, and they are different from each other.

3. Applying the result of Exercise 2 show that the number of integers less than n and prime to n is given by the formula

$$n\left(1 - \frac{1}{p_1}\right)\left(1 - \frac{1}{p_2}\right) \ldots \left(1 - \frac{1}{p_r}\right),$$

where $p_1, p_2, \ldots, p_r$ denote all different prime factors of n.

§5. Properties of symmetric difference[1])

The symmetric difference $A \dot{-} B$ was defined in §2, p. 7 by the formula

(0) $$A \dot{-} B = (A - B) \cup (B - A).$$

[1]) The properties of symmetric difference were investigated very extensively by Stone [1]; see also Hausdorff [2].

The operation $\dot{-}$ is commutative and associative:

(1) $$A \dot{-} B = B \dot{-} A,$$

(2) $$A \dot{-} (B \dot{-} C) = (A \dot{-} B) \dot{-} C.$$

Formula (1) follows directly from (0).

To prove (2) we transform the left-hand and right-hand sides of (2) by means of (0):

$$\begin{aligned} A \dot{-} (B \dot{-} C) &= A \dot{-} [(B-C) \cup (C-B)] \\ &= \{A - [(B-C) \cup (C-B)]\} \cup \{[(B-C) \cup (C-B)] - A\}. \end{aligned}$$

Using (8), (9), (10), and (11) of § 4, we obtain

$$\begin{aligned} &A \dot{-} (B \dot{-} C) \\ &= \{[A-(B-C)] \cap [A-(C-B)]\} \cup [(B-C)-A] \cup [(C-B)-A] \\ &= \{[(A-B) \cup (A \cap C)] \cap [(A-C) \cup (A \cap B)]\} \cup [B-(C \cup A)] \cup \\ &\quad \cup [C-(B \cup A)] = [(A-B) \cap (A-C)] \cup [(A-B) \cap B] \cup \\ &\quad \cup [(A-C) \cap C] \cup (A \cap B \cap C) \cup [(B-(C \cup A)] \cup [C-(B \cup A)] \\ &= [A-(B \cup C)] \cup [B-(C \cup A)] \cup [C-(A \cup B)] \cup (A \cap B \cap C). \end{aligned}$$

Thus the set $A \dot{-} (B \dot{-} C)$ contains the elements common to all the sets A, B, and C as well as the elements belonging to exactly one of them.

To transform the right-hand side of (2) it is not necessary to repeat the computation. It suffices to notice that by means of (1)

$$(A \dot{-} B) \dot{-} C = C \dot{-} (A \dot{-} B),$$

from which (substituting in the formula for $A \dot{-} (B \dot{-} C)$ the letters C, A, B for A, B, C, respectively) we obtain

$$\begin{aligned} &(A \dot{-} B) \dot{-} C \\ &\qquad = [C-(A \cup B)] \cup [A-(B \cup C)] \cup [B-(C \cup A)] \cup (C \cap A \cap B) \\ &\qquad = [A-(B \cup C)] \cup [B-(C \cup A)] \cup [C-(A \cup B)] \cup (A \cap B \cap C). \end{aligned}$$

Thus the associativity of the operation has been proved. It follows from (1) and (2) that we may eliminate parentheses when performing the operation $\dot{-}$ on a finite number of sets.

The operation of intersection is distributive over $\dot{-}$, that is,

(3) $$A \cap (B \dot{-} C) = (A \cap B) \dot{-} (A \cap C).$$

In fact, it follows from (6) and (7) of § 4 that

$$\begin{aligned} A \cap (B \dot{-} C) &= A \cap [(B-C) \cup (C-B)] \\ &= [(A \cap B)-C] \cup [(A \cap C)-B] = [B \cap (A-C)] \cup [C \cap (A-B)] \\ &= \{B \cap [A-(A \cap C)]\} \cup \{C \cap [A-(A \cap B)]\} \\ &= [(A \cap B)-(A \cap C)] \cup [(A \cap C)-(A \cap B)] \\ &= (A \cap B) \dot{-} (A \cap C). \end{aligned}$$

The empty set behaves as a zero element for the operation $\dot{-}$, that is,

(4) $$A \dot{-} \emptyset = A.$$

In fact, $(A-\emptyset) \cup (\emptyset - A) = A \cup \emptyset = A$.

The theorems which we have proved so far do not indicate any essential difference between the operations $\dot{-}$ and $\cup$. However, a difference can be seen in the following theorems.

(5) $$A \dot{-} A = \emptyset.$$

In fact, $A \dot{-} A = (A-A) \cup (A-A) = \emptyset$.

The operation of union has no inverse operation. In particular, we have seen that the operation of subtraction is not an inverse of the union operation. However, there does exist an operation inverse to the operation $\dot{-}$: for any sets A and C there exists exactly one set B such that $A \dot{-} B = C$, namely $B = A \dot{-} C$. In other words:

(6) $$A \dot{-} (A \dot{-} C) = C,$$

(7) $$A \dot{-} B = C \rightarrow B = A \dot{-} C.$$

In fact, (2), (4) and (5) imply

$$A \dot{-} (A \dot{-} C) = (A \dot{-} A) \dot{-} C = \emptyset \dot{-} C = C \dot{-} \emptyset = C,$$

which proves (6). If $A \dot{-} B = C$ then $A \dot{-} (A \dot{-} B) = A \dot{-} C$ and hence $B = A \dot{-} C$ by means of (6).

Thus (6) and (7) indicate that the operation $\dot{-}$ does have an inverse: the operation $\dot{-}$ itself.

In algebra and number theory we investigate systems of objects usually called *numbers* with two operations + and $\cdot$ (called *addition* and *multiplication*). These operations are always performable on those objects and satisfy the following conditions:

(i) $$x+y = y+x,$$

(ii) $$x+(y+z) = (x+y)+z,$$

(iii) there exists a number 0 such that $x+0 = x$,

(iv) for arbitrary x and y there exists exactly one number $z = x-y$ (the *difference*) such that $y+z = x$,

(v) $$x \cdot y = y \cdot x,$$

(vi) $$x \cdot (y \cdot z) = (x \cdot y) \cdot z,$$

(vii) $$x \cdot (y+z) = (x \cdot y)+(x \cdot z).$$

Such systems are called *rings* (more exactly: *commutative rings*). If there exists a number 1 such that for every x

(viii) $$x \cdot 1 = x,$$

then we say that *the ring has a unit element.*

The algebraic computations in rings are performed exactly as in arithmetic. For, in proving arithmetic properties involving addition, subtraction and multiplication, we make use only of the fact that numbers form a commutative ring with unit.

Formulas (1)–(7) show that sets form a ring (without unit) if by "addition" we understand the operation $\dot{-}$ and by "multiplication" the operation $\cap$. A peculiarity of this ring is that the operation "subtraction" coincides with the operation "addition" and, moreover, the "square" of every element is equal to that element.

Using $\dot{-}$ and $\cap$ as the basic operations, calculations in the algebra of sets are performed as in ordinary arithmetic. Moreover, we may omit all exponents and reduce all coefficients modulo 2 (i.e., $2kA = 0$ and $(2k+1)A = A$).

This result is significant because the operations $\cup$ and $-$ can be expressed in terms of $\dot{-}$ and $\cap$. Owing to this fact the entire algebra of sets treated above may be represented as the arithmetic of the ring of sets. In fact, it can easily be verified that:

(8) $$A \cup B = A \dot{-} B \dot{-} (A \cap B),$$

(9) $$A - B = A \dot{-} (A \cap B).$$

Formulas (8) and (4) imply the following theorem:

(10) *if A and B are disjoint, then* $A \cup B = A \dot{-} B$.

The role which the symmetric difference plays in applications is illustrated by the following example.

Let X be a set and $\boldsymbol{I}$ a non-empty family of subsets of X; that is, $\boldsymbol{I}$ is a set whose elements are subsets of X. Suppose that

$$\text{(11)} \qquad \begin{gathered} (Y \subset Z) \wedge (Z \in \boldsymbol{I}) \to (Y \in \boldsymbol{I}), \\ (Y \in \boldsymbol{I}) \wedge (Z \in \boldsymbol{I}) \to (Y \cup Z \in \boldsymbol{I}). \end{gathered}$$

A family of sets satisfying these conditions is called an *ideal*. We say that two subsets A, B of X are *congruent modulo* $\boldsymbol{I}$ if $A \dot{-} B \in \boldsymbol{I}$ and we denote this fact by $A \doteq B \pmod{\boldsymbol{I}}$ or by $A \doteq B$ if the ideal $\boldsymbol{I}$ is fixed.

Since $\varnothing \in \boldsymbol{I}$, it follows from (5) that $A \doteq A$, i.e. the relation $\doteq$ is reflexive. (1) implies that $(A \doteq B) \to (B \doteq A)$, i.e. the relation $\doteq$ is symmetric.

Finally, the identity $A \dot{-} B = (A \dot{-} C) \dot{-} (B \dot{-} C)$ implies that $A \dot{-} B \subset (A \dot{-} C) \cup (B \dot{-} C)$, because the symmetric difference of two sets is contained in their union. By means of (11) we infer that

$$(A \doteq B) \wedge (B \doteq C) \to (A \doteq C),$$

i.e. the relation $\doteq$ is transitive.

Replacing the sign $=$ by the sign $\doteq$ in the previous definitions, we obtain new notions. For example, two sets A and B are said to be *disjoint modulo* $\boldsymbol{I}$ provided $A \cap B \doteq \varnothing$ (see p. 9); we say that A *is included in* B *modulo* $\boldsymbol{I}$ if $A - B \doteq \varnothing$, etc.

A notion dual to ideal is that of a filter. We say that a non-empty family $\boldsymbol{F}$ consisting of subsets of X is a *filter* if it satisfies the conditions

$$\begin{gathered} (Y \subset Z) \wedge (Y \in \boldsymbol{F}) \to (Z \in \boldsymbol{F}), \\ (Y \in \boldsymbol{F}) \wedge (Z \in \boldsymbol{F}) \to (Y \cap Z \in \boldsymbol{F}). \end{gathered}$$

Congruence modulo a filter $\boldsymbol{F}$ is defined in the dual way to that of the congruence modulo an ideal: we say that $A \doteq B \pmod{\boldsymbol{F}}$ if $(X - A) \cup B \in \boldsymbol{F}$ and $(X - B) \cup A \in \boldsymbol{F}$ or equivalently

$$[X - (A \cup B)] \cup (A \cap B) \in \boldsymbol{F}.$$

Exercises

1. Show that the set $A_1 \dot{-} A_2 \dot{-} \ldots \dot{-} A_n$ contains those and only those elements which belong to an odd number of sets A_i $(i = 1, 2, \ldots, n)$.

2. For A finite let $N(A)$ denote the number of elements of A. Prove that if the sets $A_1, A_2, \ldots, A_n$ are finite then

$$N(A_1 \dot{-} A_2 \dot{-} \ldots \dot{-} A_n)$$
$$= \sum_i N(A_i) - 2\sum_{i,j} N(A_i \cap A_j) + 4 \sum_{i,j,k} N(A_i \cap A_j \cap A_k) -$$
$$-8 \sum_{i,j,k,l} N(A_i \cap A_j \cap A_k \cap A_l) + \ldots$$

3. Show that

$$(A_1 \cup A_2 \cup \ldots \cup A_n) \dot{-} (B_1 \cup B_2 \cup \ldots \cup B_n) \subset (A_1 \dot{-} B_1) \cup \ldots \cup (A_n \dot{-} B_n),$$
$$(A_1 \cap A_2 \cap \ldots \cap A_n) \dot{-} (B_1 \cap B_2 \cap \ldots \cap B_n) \subset (A_1 \dot{-} B_1) \cup \ldots \cup (A_n \dot{-} B_n)$$

[Hausdorff].

4. Show that for any ideal $\boldsymbol{I}$ the condition $A \doteq B$ implies

$$A \cup C \doteq B \cup C, \quad A \cap C \doteq B \cap C, \quad A - C \doteq B - C, \quad C - A \doteq C - B.$$

5. For any real number t denote by $[t]$ the largest integer $\leqslant t$. Let A_t be the set of rational numbers of the form $[nt]/n$, $n = 1, 2, \ldots$ Prove that if $\boldsymbol{I}$ is the ideal composed of all finite subsets of the set of rational numbers, then $\neg(A_x \doteq A_y \pmod{\boldsymbol{I}})$ and A_x is disjoint (modulo $\boldsymbol{I}$) from A_y for all irrational numbers $x, y > 0$, $x \neq y$.

6. Let $\boldsymbol{I}$ be an ideal. Then $A \doteq B \bmod \boldsymbol{I}$ iff A is of the form $A = (B-P) \cup Q$ where $P \in \boldsymbol{I}$ and $Q \in \boldsymbol{I}$.

7. Let $\boldsymbol{I}$ be an ideal. Write briefly $A \doteq B$ instead of $A \doteq B \bmod \boldsymbol{I}$. Show that if $A_1 \doteq B_1$ and $A_2 \doteq B_2$, then

$$(A_1 \cup A_2) \doteq (B_1 \cup B_2), \quad (A_1 \cap A_2) \doteq (B_1 \cap B_2), \quad (A_1 - A_2) \doteq (B_1 - B_2).$$

§6. The set 1, complement

In many applications of set theory we consider only sets contained in a given fixed set. For instance, in geometry we deal with sets of points in a given space, and in arithmetic with sets of numbers.

In this section $A, B, \ldots$ will denote sets contained in a certain fixed set which will be referred to either as the space or the *universe* and will be denoted by 1. Thus for every A

$$A \subset 1,$$

from which it follows that

(1) $$A \cap 1 = A, \quad A \cup 1 = 1.$$

The set $1-A$ is called the *complement* of A and is denoted by A^c or $-A$:

$$-A = A^c = 1-A.$$

Clearly,

(2) $$A \cap -A = \emptyset, \quad A \cup -A = 1.$$

Since $--A = 1-(1-A)$, we obtain by (10) of § 4 the following *law of double complementation*

(3) $$--A = A.$$

Setting $A = 1$ in de Morgan's laws (see 4 (8)) and substituting A and B for B and C, we obtain

(4) $$-(A \cap B) = -A \cup -B, \quad -(A \cup B) = -A \cap -B.$$

Thus *the complement of the intersection of two sets is equal to the union of their complements and the complement of the union of two sets is equal to the intersection of their complements.*

It is worth noting that the formulas which we obtained by introducing the notion of complementation are analogous to those of the propositional calculus discussed in §1. To obtain the laws of the propositional calculus (see pp. 2–4) it suffices to substitute in (1)–(4) the equivalence sign for the sign of identity and to interpret the letters $A, B, \ldots$ as propositional variables and the symbols $\cup$, $\cap$, $-$, $\emptyset$, 1 as disjunction, conjunction, negation, the false sentence and the true sentence, respectively. Conversely, theorems of the algebra of sets can be obtained from the corresponding laws of the propositional calculus simply by changing the meaning of symbols. From this point of view calculations on sets contained in a fixed set 1 can be simplified by using the operations $\cup$, $\cap$, $-$.

Subtraction can be defined by means of the operation $-$ and one of the operations $\cup$ or $\cap$. In fact, we have

$$A-B = A \cap (1-B) = A \cap -B$$

and

$$A-B = A \cap -B = -(-A \cup B).$$

The inclusion relation between two sets can be expressed by the identity

$$(5) \qquad (A \subset B) \equiv (A \cap -B = \emptyset).$$

For assuming $A \subset B$ and multiplying both sides of the inclusion by $-B$, we obtain $A \cap -B \subset B \cap -B$ and since $B \cap -B = \emptyset$, we have $A \cap -B = \emptyset$. Conversely, if $A \cap -B = \emptyset$, then

$$A = A \cap 1 = A \cap (B \cup -B)$$

$$= (A \cap B) \cup (A \cap -B) = (A \cap B) \cup \emptyset = A \cap B \subset B.$$

Since $(A = B) \equiv (A \subset B) \wedge (B \subset A)$, it follows from (5) that

$$(A = B) \equiv (A \cap -B = \emptyset) \wedge (B \cap -A = \emptyset),$$

and, since the condition $(X = \emptyset) \wedge (Y = \emptyset)$ is equivalent to $X \cup Y = \emptyset$,

$$(6) \qquad (A = B) \equiv [(A \cap -B) \cup (B \cap -A) = \emptyset] \equiv (A \dot{-} B = \emptyset).$$

It follows directly from (5) that

$$(7) \qquad (A \subset B) \equiv (-B \subset -A).$$

(compare with the law of contraposition, p. 4).

The system of all sets contained in 1 forms a ring where the operation $\dot{-}$ is understood as addition and $\cap$ as multiplication. This ring differs from the ring of sets considered in § 5 in that it has a unit element. The unit is namely the set 1. In fact, formula (1) states that the set 1 satisfies condition (viii) of § 5 characterizing the unit element of a ring.

Hence calculations in the algebra of sets are formally like those in the algebra of numbers.

Exercises

1. The quotient of two sets is defined as follows $A:B = A \cup -B$. Find formulas for $A:(B \cup C)$ and for $A:(B \cap C)$ (counterpart of de Morgan's laws). Compute $A \cap (B:C)$.

2. Prove that for each filter F the congruence modulo F is a reflexive, symmetric and transitive relation and that $A \doteq B \pmod F$ implies $A \cup C \doteq B \cup C$, $A \cap C \doteq B \cap C$ and $X - A \doteq X - B$.

§7. Constituents

In this section we shall consider sets which can be obtained from n arbitrary sets by applying the operations of union, intersection, and difference. We shall show that the total number of such sets is finite and that they can be represented in a certain definite form (*normal form*).

Let $A_1, A_2, \ldots, A_n$ be arbitrary subsets of the space 1. Throughout this section these subsets will remain fixed.

Let

$$A_i^1 = 1 - A_i, \qquad A_i^0 = A_i \qquad \text{for } i = 1, 2, \ldots, n.$$

Each set of the form

$$A_1^{i_1} \cap A_2^{i_2} \cap \ldots \cap A_n^{i_n} \qquad (i_k = 0 \text{ or } i_k = 1 \text{ for } k = 1, 2, \ldots, n)$$

will be called a *constituent*.

The total number of distinct constituents is at most 2^n, because each of the superscripts i_k may have either one of the values 0 and 1. The number of constituents may be less than 2^n; for instance, if $n = 2$ and $A_1 = 1 - A_2$, then there are only three constituents:

$$\emptyset = A_1^0 \cap A_2^0 = A_1^1 \cap A_2^1, \qquad A_1 = A_1^0 \cap A_2^1, \qquad A_2 = A_1^1 \cap A_2^0.$$

Distinct constituents are always disjoint.

In fact, if

$$S_1 = A_1^{i_1} \cap A_2^{i_2} \cap \ldots \cap A_n^{i_n} \quad \text{and} \quad S_2 = A_1^{j_1} \cap A_2^{j_2} \cap \ldots \cap A_n^{j_n}$$

and if for at least one $k \leqslant n$, $i_k \neq j_k$, for instance $i_k = 0$ and $j_k = 1$, then $A_k^{i_k} \cap A_k^{j_k} = \emptyset$. Hence $S_1 \cap S_2 = \emptyset$.

The union of all constituents is the space 1.

It suffices to notice that

$$1 = (A_1^0 \cup A_1^1) \cap (A_2^0 \cup A_2^1) \cap \ldots \cap (A_n^0 \cup A_n^1).$$

By applying the distributive law of intersection with respect to union on the right-hand side of the equation we obtain the union of all the constituents.

The set A_i is the union of all the constituents which contain the component A_i^0.

If $S_1, S_2, \ldots S_h$ are all the constituents, then

$$1 = S_1 \cup S_2 \cup \ldots \cup S_h.$$

Therefore

$$A_i = (A_i \cap S_1) \cup (A_i \cap S_2) \cup \ldots \cup (A_i \cap S_h).$$

If S_p contains the component A_i^1, then $A_i \cap S_p = \emptyset$ because $A_i \cap A_i^1 = A_i \cap (1 - A_i) = \emptyset$. On the other hand, if S_p contains the component A_i^0, then $A_i \cap S_p = S_p$. Thus A_i is the union of those constituents which contain the component A_i^0. Q.E.D.

We shall now prove the following

THEOREM 1: *Each non-empty set obtained from the sets $A_1, A_2, \ldots, A_n$ by applying the operations of union, intersection and subtraction is the union of a certain number of constituents.*

PROOF. The theorem is true for the sets $A_1, A_2, \ldots, A_n$. It suffices to show that if X and Y are unions of a certain number of constituents then the sets $X \cup Y$, $X \cap Y$, $X - Y$ can also be represented as the union of constituents (provided $X \cup Y$, $X \cap Y$, $X - Y$ are non-empty).

Assume that X and Y can be represented as unions of constituents:

$$X = S_1 \cup S_2 \cup \ldots \cup S_k, \qquad Y = \overline{S}_1 \cup \overline{S}_2 \cup \ldots \cup \overline{S}_l.$$

It follows that

$$X \cup Y = (S_1 \cup \ldots \cup S_k) \cup (\overline{S}_1 \cup \ldots \cup \overline{S}_l).$$

Thus $X \cup Y$ is a union of constituents.

From the distributive law for intersection with respect to union, it follows that

$$X \cap Y = (S_1 \cap \overline{S}_1) \cup (S_1 \cap \overline{S}_2) \cup \ldots \cup (S_1 \cap \overline{S}_l) \cup \ldots$$
$$\ldots \cup (S_i \cap \overline{S}_j) \cup \ldots \cup (S_k \cap \overline{S}_l).$$

$S_i \cap \overline{S}_j = \emptyset$ if $S_i \neq \overline{S}_j$; otherwise $S_i \cap \overline{S}_j = S_i$. Thus $X \cap Y$ is a union of constituents

$$X \cap Y = S_{i_1} \cup S_{i_2} \cup \ldots \cup S_{i_p},$$

or else is empty.

If among the constituents $S_{i_1}, S_{i_2}, \ldots, S_{i_p}$ occur all of the constituents $S_1, S_2, \ldots, S_k$, then

$$X - Y = X - (X \cap Y) \subset (S_1 \cup \ldots \cup S_k) - (S_1 \cup \ldots \cup S_k) = \emptyset.$$

Otherwise, let $S_{j_1}, S_{j_2}, \ldots, S_{j_q}$ be those constituents among $S_1, S_2, \ldots$ $\ldots, S_k$ which do not occur among the constituents $S_{i_1}, S_{i_2}, \ldots, S_{i_p}$.

We have

$$\begin{aligned} X-Y &= X-(X\cap Y) \\ &= [(S_{i_1}\cup \ldots \cup S_{i_p})\cup(S_{j_1}\cup \ldots \cup S_{j_q})]-(S_{i_1}\cup \ldots \cup S_{i_p}) \\ &= (S_{j_1}\cup \ldots \cup S_{j_q})-(S_{i_1}\cup \ldots \cup S_{i_p}) \\ &= (S_{j_1}\cup \ldots \cup S_{j_q})-[(S_{j_1}\cup \ldots \cup S_{j_q})\cap(S_{i_1}\cup \ldots \cup S_{i_p})] \\ &= S_{j_1}\cup \ldots \cup S_{j_q}, \end{aligned}$$

because

$$(S_{j_1}\cup \ldots \cup S_{j_q})\cap(S_{i_1}\cup \ldots \cup S_{i_p}) = \emptyset.$$

Thus $X\cup Y$, $X\cap Y$, and $X-Y$ are representable as unions of constituents. Q.E.D.

THEOREM 2: *From n sets by applying the operations of union, intersection, and subtraction at most* 2^{2^n} *sets can be constructed.*

In fact, each such set, with the exception of the empty set, is a union of constituents. Because the number of constituents cannot be greater than 2^n, the number of distinct unions constructed from some (non-zero) number of constituents cannot be greater than $2^{2^n}-1$.

Of particular importance is the case where all of the constituents are different from $\emptyset$. In this case, we say that the sets $A_1, \ldots, A_n$ are *independent.*

THEOREM 3: *If the sets* $A_1, \ldots, A_n$ *are independent, then the number of distinct constituents equals* 2^n.

PROOF. If

$$(0) \qquad S = A_1^{i_1}\cap \ldots \cap A_n^{i_n} = A_1^{j_1}\cap \ldots \cap A_n^{j_n}$$

and not all of the equations $i_1 = j_1, \ldots, i_n = j_n$ hold, then $S = \emptyset$. In fact, if for example, $i_p = 1$ and $j_p = 0$, then intersecting both sides of the last equation in (0) with A_p^1 we obtain $S = \emptyset$. Thus if the sets $A_1, \ldots, A_n$ are independent then equation (0) holds if and only if $i_1 = j_1, \ldots,\ i_n = j_n$. Q.E.D.

Example. Let the set D_m consist of sequences $(z_1, \ldots, z_n)$ such that each z_i equals either 0 or 1 but $z_m = 0$. The sets $D_1, \ldots, D_n$ are independent. In fact, $D_m^{i_m}$ consists of those sequences $(z_1, \ldots, z_n)$ for which $z_m = i_m$. Thus $(i_1, \ldots, i_n) \in D_1^{i_1} \cap \ldots \cap D_n^{i_n}$.

We shall apply the concept of constituents to a discussion of the following *problem of elimination.*[1]) We introduce the abbreviations

$$\Gamma_n^0(A) \equiv \{A \text{ contains at least } n \text{ elements}\},$$

$$\Gamma_n^1(A) \equiv \{A \text{ contains exactly } n \text{ elements}\}.$$

Let $i_1, \ldots, i_n$, $j_1, \ldots, j_n$ be sequences of the numbers 0 and 1. Let $p_1, \ldots, p_n$, $q_1, \ldots, q_n$ be sequences of non-negative integers. We are interested in finding necessary and sufficient conditions for the existence of a set X satisfying the conjunction of the following conditions:

(i)
$$\begin{gathered}\Gamma_{p_1}^{i_1}(X \cap A_1),\ \Gamma_{p_2}^{i_2}(X \cap A_2),\ \ldots,\ \Gamma_{p_n}^{i_n}(X \cap A_n),\\ \Gamma_{q_1}^{j_1}(-X \cap A_1),\ \Gamma_{q_2}^{j_2}(-X \cap A_2),\ \ldots,\ \Gamma_{q_n}^{j_n}(-X \cap A_n).\end{gathered}$$

We assume at first that $n = 1$. Writing i, j, p, q, A instead of i_1, j_1, p_1, q_1, A_1, we obtain the solution:

(ii)
$$[(i = j = 1) \wedge \Gamma_{p+q}^1(A)] \vee \Gamma_{p+q}^0(A).$$

In fact, if there exists a set X satisfying (i) and $i = j = 1$, then A is the union of two sets containing respectively p and q elements, and in this case A contains exactly $p+q$ elements. If $i = 0 \vee j = 0$ then A is the union of two sets, one of which contains at least p elements and the other at least q elements. Therefore A contains at least $p+q$ elements. Conversely, if condition (ii) is satisfied, then it suffices to choose as X any subset of A containing p elements.

Assume that $n > 1$ and $A_1, \ldots, A_n$ are pairwise disjoint. If there exists a set X satisfying (i), then writing $X_s = X$, $s = 1, 2, \ldots, n$, we conclude that

(iii)
$$\Gamma_{p_s}^{i_s}(X_s \cap A_s) \wedge \Gamma_{q_s}^{j_s}(-X_s \cap A_s) \quad \text{for} \quad s = 1, 2, \ldots, n,$$

and by virtue of (ii)

(iv)
$$[(i_s = j_s = 1) \wedge \Gamma_{p_s+q_s}^1(A_s)] \vee \Gamma_{p_s+q_s}^0(A_s), \quad s = 1, 2, \ldots, n.$$

[1]) Problems of elimination were extensively studied in the so-called algebra of logic late in the 19th century; see e.g. Schröder [1]. The particular problem discussed here was formulated and solved by Skolem [1]; it found applications to some problems of mathematical logic; see Feferman–Vaught [1].

Conversely, if (iv) holds then for every s $(1 \leqslant s \leqslant n)$ there exists a set X_s satisfying (iii). Let

$$X = [(X_1 \cap A_1) \cup (X_2 \cap A_2) \cup \ldots \cup (X_n \cap A_n)] \cup \cup (-A_1 \cap -A_2 \cap \ldots \cap -A_n).$$

Therefore

$$-X = [(-X_1 \cup -A_1) \cap (-X_2 \cup -A_2) \cap \ldots \ldots \cap (-X_n \cup -A_n)] \cap (A_1 \cup \ldots \cup A_n).$$

Since the sets A_i are disjoint, we have $X \cap A_s = X_s \cap A_s$ and $-X \cap \cap A_s = -X_s \cap A_s$. By applying (iii) we obtain (i).

Next, we assume that for all r, s $(1 \leqslant r, s \leqslant n)$ either $A_r = A_s$ or $A_r \cap A_s = \emptyset$. We shall designate conditions (i) by $W_1, W_2, \ldots, W_n$, $V_1, V_2, \ldots, V_n$. We shall show that if $A_r = A_s$ then $W_r \rightarrow W_s$, or $W_s \rightarrow W_r$, or else $W_r \wedge W_s \equiv F$. Indeed, if $i_r = i_s = 0$, then $W_s \rightarrow W_r$ if $p_r \leqslant p_s$, and $W_r \rightarrow W_s$ if $p_s < p_r$. If $i_r = 1$ and $i_s = 0$, then $W_r \rightarrow W_s$ in case $p_r \geqslant p_s$, and in case $p_r < p_s$, $W_r \wedge W_s \equiv F$. Finally, if $i_r = i_s = 1$ then $W_r \rightarrow W_s$ for $p_s = p_r$ and otherwise $W_r \wedge W_s = F$. Similarly it can be shown that either $V_r \rightarrow V_s$, or $V_s \rightarrow V_r$, or $V_r \wedge V_s \equiv F$. We conclude that either the conjunction of (i) is false or else we may omit from (i) certain components and obtain an equivalent conjunction in which none of the sets A_s occurs more than once. Thus this case is reduced to the preceding case.

Now we shall reduce the general case to the case in which the sets A_s are either identical or disjoint. For this purpose we note that if $M \cap N = \emptyset$, then

$$\Gamma_p^0(M \cup N) \equiv \Gamma_p^0(M) \vee [\Gamma_{p-1}^1(M) \wedge \Gamma_1^0(N)] \vee \vee [\Gamma_{p-2}^1(M) \wedge \Gamma_2^0(N)] \vee \ldots \vee [\Gamma_0^1(M) \wedge \Gamma_p^0(N)],$$

$$\Gamma_p^1(M \cup N) \equiv [\Gamma_p^1(M) \wedge \Gamma_0^1(N)] \vee [\Gamma_{p-1}^1(M) \wedge \Gamma_1^1(N)] \vee \ldots \ldots \vee [\Gamma_0^1(M) \wedge \Gamma_p^1(N)].$$

By induction, if the sets $S_1, \ldots, S_h$ are pairwise disjoint, then $\Gamma_p^i(S_1 \cup \cup \ldots \cup S_h)$ can be expressed equivalently as a disjunction of conjunctions, where each conjunction has the form

$$\Gamma_{p_1}^{j_1}(S_1) \wedge \ldots \wedge \Gamma_{p_h}^{j_h}(S_h).$$

Represent the sets A_s as unions of constituents; then according to the above remark, each of the conditions (i) can be expressed as a disjunction of conjunctions each of which has the form

$$\Gamma_{v_1}^{u_1}(X \cap S_1) \wedge \ldots \wedge \Gamma_{v_h}^{u_h}(X \cap S_h),$$

or respectively,

$$\Gamma_{z_1}^{w_1}(-X \cap S_1) \wedge \ldots \wedge \Gamma_{z_h}^{w_h}(-X \cap S_h).$$

Applying the distributive law for conjunction over disjunction, we express the conjunction of conditions (i) as a disjunction, each of which is a conjunction whose components have either the form $\Gamma_f^e(X \cap S_g)$ or the form $\Gamma_f^e(-X \cap S_g)$. Sets occurring in each such conjunction are either identical or disjoint. Thus the general case is reduced to the preceding one.

Example. We shall find necessary and sufficient conditions for the existence of a set X satisfying the conditions

$$X \cap A \cap B \neq \emptyset, \qquad -X \cap A \cap B \neq \emptyset,$$

$$X \cap A \not\subset B, \qquad X \cap B \not\subset A.$$

These conditions can be expressed equivalently as the conjunction of the following six conditions:

$$\Gamma_1^0(X \cap A \cap B), \quad \Gamma_1^0(X \cap A \cap -B), \quad \Gamma_1^0(X \cap -A \cap B),$$

$$\Gamma_1^0(-X \cap A \cap B), \quad \Gamma_0^0(-X \cap A \cap -B), \quad \Gamma_0^0(-X \cap -A \cap B).$$

Hence we obtain the desired condition

$$\Gamma_2^0(A \cap B) \wedge \Gamma_1^0(A \cap -B) \wedge \Gamma_1^0(-A \cap B).$$

In other terms, $A \cap -B$ and $B \cap -A$ have to be non-empty and $A \cap B$ has to contain at least two elements.

Exercises

1. Assuming that the set 1 is infinite and that $A_1, \ldots, A_n$ are finite, describe a method of obtaining necessary and sufficient conditions for the existence of a finite set X satisfying the conjunction of conditions (i).

2. Let $\boldsymbol{I}$ be the unit n-dimensional cube, that is, the set of sequences $(x_1, \ldots, x_n)$ such that $0 \leqslant x_i \leqslant 1$ $(i = 1, 2, \ldots, n)$. Let $\boldsymbol{I}_m$ consist of those sequences $(x_1, \ldots, x_n) \in \boldsymbol{I}$ where $1/2 \leqslant x_m \leqslant 1$. Show that the sets $\boldsymbol{I}_1, \ldots, \boldsymbol{I}_n$ are independent. Give a geometrical interpretation for $n = 2$ and $n = 3$.

§8. Applications of the algebra of sets to topology[1])

In order to illustrate applications which the calculus developed in the preceding sections has outside of the general theory of sets, we shall examine the axioms of general topology and apply the algebra of sets to establish several results.

In general topology we study a set 1, called the *space*, whose elements are called *points*. We assume, moreover, that to every set A contained in 1 there corresponds a set $\bar{A}$ also contained in 1 and called the *closure* of A. The space 1 is called *topological* if it satisfies the following axioms (see also p. 116)[2])

$$\overline{A \cup B} = \bar{A} \cup \bar{B}, \tag{1}$$

$$\bar{\bar{A}} = \bar{A}, \tag{2}$$

$$A \subset \bar{A}, \tag{3}$$

$$\bar{\emptyset} = \emptyset. \tag{4}$$

In axioms (1)–(3) the letters A and B denote arbitrary subsets of the space 1.

Axioms (1)–(4) are satisfied if, for example, 1 is the set of points of the plane and if the closure operation $\bar{A}$ consists of adding to the set A all points p such that every disc around p contains elements of A. This interpretation will be referred to as the *natural interpretation* of the axioms (1)–(4).

We shall show how, using only laws of the calculus of sets, it is possible to deduce a variety of properties of the closure operation.

$$\bar{1} = 1. \tag{5}$$

PROOF. For every A we have $\bar{A} \subset 1$, and by axiom (3), $1 \subset \bar{1}$.

[1]) The topological calculus presented in this section originated with Kuratowski [6]; see also Kuratowski [1].

[2]) In § 8 we apply not only axioms I, A, B, C but also axioms (1)–(4). However, we can deduce all theorems given in this section from the complete axiomatic system of set theory given in Chapter II, treating axioms (1)–(4) as assumptions about the operation of closure.

(6) $$\bar{A}-\bar{B} \subset \overline{A-B}.$$

PROOF. From $B \cup (A-B) = A \cup B$ applying axiom (1) we obtain $\bar{B} \cup \overline{A-B} = \bar{A} \cup \bar{B}$. This implies that $\bar{A} \subset \bar{B} \cup \overline{A-B}$ and thus

$$\bar{A}-\bar{B} \subset (\bar{B} \cup \overline{A-B})-\bar{B} = \overline{A-B}-\bar{B} \subset \overline{A-B},$$

which proves (6).

(7) $$A \subset B \to \bar{A} \subset \bar{B}.$$

PROOF. $A \subset B$ is equivalent to the equation $A \cup B = B$. By axiom (1), $\bar{A} \cup \bar{B} = \bar{B}$, thus $\bar{A} \subset \bar{B}$ (cf. 3 (5)).

(8) $$\overline{A \cap B} \subset \bar{A} \cap \bar{B}.$$

PROOF. Since $A \cap B \subset A$ and $A \cap B \subset B$, theorem (7) implies $\overline{A \cap B} \subset \bar{A}$ and $\overline{A \cap B} \subset \bar{B}$, from which it follows that $\overline{A \cap B} \subset \bar{A} \cap \bar{B}$.

(9) *If* $A = \bar{A}$ *and* $B = \bar{B}$, *then* $\overline{A \cap B} = A \cap B$.

PROOF. In fact, $A \cap B \subset \overline{A \cap B}$ by axiom (3). But by (8) and by the hypothesis of (9), $\overline{A \cap B} \subset \bar{A} \cap \bar{B} = A \cap B$. Therefore $A \cap B = \overline{A \cap B}$.

We call a set *closed* if it is equal to its closure. Theorem (9) states that the intersection of two closed sets is closed, and axiom (1) that the union of two closed sets is closed.

We call a set *open* if it is the complement of a closed set. By de Morgan's laws it follows that the union and intersection of two open sets is open.

In the natural interpretation of axioms (1)–(4) *closed sets are those sets which contain all their accumulation points* (cf. p. 32). Open sets have the property: for every point p contained in the open set A there exists a disc with centre p entirely contained in A.

The set[1])

$$\mathrm{Int}(A) = 1-\overline{1-A} = A^{c-c}$$

is called the *interior* of the set A. The interior of any set is clearly an open set. # In the natural interpretation of axioms (1)–(4), the set $\mathrm{Int}(A)$ consists exactly of those points p for which there exists a disc with centre p entirely contained in A. #

[1]) Instead of $\bar{A}$ we sometimes write A^-.

(10) $$\operatorname{Int}(A) \subset A.$$

PROOF. By axiom (3), $A^c \subset A^{c-}$, from which, by taking complements, we obtain

$$1 - A^{c-} \subset 1 - A^c,$$

hence

$$A^{c-c} \subset A^{cc} = A.$$

In particular, the relation $\operatorname{Int}(\operatorname{Int}(A)) \subset \operatorname{Int}(A)$ is a special case of (10). This relation may be strengthened as follows:

(11) $$\operatorname{Int}(\operatorname{Int}(A)) = \operatorname{Int}(A).$$

PROOF. It follows from the definition of $\operatorname{Int}(A)$ that

$$\operatorname{Int}(A) = A^{c-c}, \quad \operatorname{Int}(\operatorname{Int}(A)) = [\operatorname{Int}(A)]^{c-c} = [A^{c-c}]^{c-c}.$$

By the law of double complementation we may eliminate two consecutive occurrences of the operation A^c; we thus obtain

$$\operatorname{Int}(\operatorname{Int}(A)) = A^{c--c},$$

and because $A^{c--} = A^{c-}$ by axiom (2), we obtain

$$\operatorname{Int}(\operatorname{Int}(A)) = A^{c-c} = \operatorname{Int}(A).$$

(12) $$\operatorname{Int}(A \cap B) = \operatorname{Int}(A) \cap \operatorname{Int}(B).$$

PROOF. By de Morgan's laws

$$(A \cap B)^{c-c} = (A^c \cup B^c)^{-c},$$

whence by axiom (1)

$$\operatorname{Int}(A \cap B) = (A \cap B)^{c-c} = (A^{c-} \cup B^{c-})^c,$$

and a final application of de Morgan's laws gives

$$\operatorname{Int}(A \cap B) = A^{c-c} \cap B^{c-c} = \operatorname{Int}(A) \cap \operatorname{Int}(B).$$

As a simple consequence of (12) we have

(13) $$A \subset B \to \operatorname{Int}(A) \subset \operatorname{Int}(B).$$

In fact, the assumption $A \subset B$ gives us $A \cap B = A$,

$$\operatorname{Int}(A) = \operatorname{Int}(A \cap B) = \operatorname{Int}(A) \cap \operatorname{Int}(B) \subset \operatorname{Int}(B).$$

(14) $$\overline{\mathrm{Int}\big(\overline{\mathrm{Int}(A)}\big)} = \mathrm{Int}(A).$$

PROOF. By (10)

$$\mathrm{Int}\big(\overline{\mathrm{Int}(A)}\big) \subset \overline{\mathrm{Int}(A)},$$

whence by (7) and (2)

(14_1) $$\overline{\mathrm{Int}\big(\overline{\mathrm{Int}(A)}\big)} \subset \overline{\mathrm{Int}(A)}.$$

On the other hand, by (11), (3), and (13)

$$\mathrm{Int}(A) = \mathrm{Int}\big(\mathrm{Int}(A)\big) \subset \mathrm{Int}\big(\overline{\mathrm{Int}(A)}\big),$$

and by (7) it follows that

(14_2) $$\overline{\mathrm{Int}(A)} \subset \overline{\mathrm{Int}\big(\overline{\mathrm{Int}(A)}\big)}.$$

Inclusions (14_1) and (14_2) imply (14).

Replacing $\mathrm{Int}(X)$ by X^{c-c} in (14), we obtain

(15) $$A^{c-c-c-c-} = A^{c-c-}.$$

Moreover, substituting A^c for A and applying the law of double complementation, we obtain

(16) $$A^{-c-c-c-} = A^{-c-}.\ ^{1)}$$

Equations (15) and (16) show that if we apply in succession the operations of complementation and closure to the set A, then we obtain only a finite number of sets. Namely, if we start with the operation of complementation, then we obtain the sets

$$A,\ A^c,\ A^{c-},\ A^{c-c},\ A^{c-c-},\ A^{c-c-c},\ A^{c-c-c-},\ A^{c-c-c-c}.$$

The next set in this sequence would be $A^{c-c-c-c-}$, but by (15) this set equals A^{c-c-}. If, on the other hand, we start by applying the operation $^-$, then we obtain the sets

$$A^-,\ A^{-c},\ A^{-c-},\ A^{-c-c},\ A^{-c-c-},\ A^{-c-c-c}.$$

The next set would be $A^{-c-c-c-}$, but by (16) it is equal to A^{-c-}.

Hence by applying the operations of complementation and closure to an arbitrary set A we obtain at most 14 distinct sets.

[1]) Formulas (15), (16) and the result quoted on p. 30 were given by Kuratowski 6]. Hintikka [1] extended this result to the case of "nested topologies".

Formulas (17) and (18) will be used in § 9.

(17) *If* $B = X^{-c-}$, *then* $\mathrm{Int}[\overline{\mathrm{Int}(A-B)} \cap B] = \emptyset$.

PROOF. Clearly, $A - B \subset B^c$, whence by (13) and (7)

$$\mathrm{Int}[\overline{\mathrm{Int}(A-B)} \cap B] \subset \mathrm{Int}[\overline{\mathrm{Int}(B^c)} \cap B].$$

Thus it suffices to show that

$$\mathrm{Int}[\overline{\mathrm{Int}(B^c)} \cap B] = \emptyset.$$

Since $\overline{\mathrm{Int}(B^c)} = B^{cc-c-} = B^{-c-} = X^{-c--c-} = X^{-c-c-}$, it follows by formulas (12), (16), and (10) that

$$\mathrm{Int}[\overline{\mathrm{Int}(B^c)} \cap B] = \mathrm{Int}[\overline{\mathrm{Int}(B^c)}] \cap \mathrm{Int}(B) = X^{-c-c-c-c} \cap B^{c-c}$$
$$= X^{-c-c} \cap B^{c-c} = B^c \cap B^{c-c} = \emptyset.$$

(18) *If* $A = \overline{A}$ *or* $B = \overline{B}$ *then* $\overline{\mathrm{Int}(A)} \cup \overline{\mathrm{Int}(B)} = \overline{\mathrm{Int}(A \cup B)}$.

PROOF. From theorem (13) we conclude that $\mathrm{Int}(A) \subset \mathrm{Int}(A \cup B)$ and $\mathrm{Int}(B) \subset \mathrm{Int}(A \cup B)$, which implies that $\mathrm{Int}(A) \cup \mathrm{Int}(B) \subset \mathrm{Int}(A \cup B)$. Applying the closure to both sides of the inclusion, we obtain by (1) and (7)

(18_1) $$\overline{\mathrm{Int}(A)} \cup \overline{\mathrm{Int}(B)} \subset \overline{\mathrm{Int}(A \cup B)}.$$

For the proof of the opposite inclusion we suppose that, for instance, $\overline{B} = B$. We apply the identity

$$A \cup B \cup [1 - (A \cup B)] = 1,$$

from which by axiom (3) it follows that

$$A \cup B \cup \overline{1 - (A \cup B)} = 1,$$

whence

$$B \cup \overline{1 - (A \cup B)} \supset 1 - A.$$

Applying closure to both sides we obtain (since $\overline{B} = B$):

$$B \cup \overline{1 - (A \cup B)} \supset \overline{1 - A},$$

from which it follows that

$$[1 - \overline{1 - A}] \cup B \cup \overline{1 - (A \cup B)} = 1,$$

whence

$$\mathrm{Int}(A) \cup B \cup \overline{1 - (A \cup B)} = 1.$$

It follows from this equation that

$$\mathrm{Int}(A) \cup \overline{1-(A \cup B)} \supset 1-B,$$

and thus

$$\overline{\mathrm{Int}(A)} \cup \overline{1-(A \cup B)} \supset \overline{1-B}.$$

Adding to both sides of this equation the set $1-\overline{1-B} = \mathrm{Int}(B)$ we obtain

$$\overline{\mathrm{Int}(A)} \cup \mathrm{Int}(B) \cup \overline{1-(A \cup B)} = 1,$$

thus

$$\overline{\mathrm{Int}(A)} \cup \mathrm{Int}(B) \supset 1-\overline{1-(A \cup B)} = \mathrm{Int}(A \cup B).$$

Applying closure to both sides of the inclusion we obtain by (1) and (3)

$$(18_2) \qquad \overline{\mathrm{Int}(A)} \cup \overline{\mathrm{Int}(B)} \supset \overline{\mathrm{Int}(A \cup B)}.$$

Inclusions (18_1) and (18_2) prove theorem (18).

Exercises

1. Prove that if the set A is open, then

$$\overline{A \cap \overline{X}} = \overline{A \cap X} \quad \text{for every set } X.$$

2. Let $\mathrm{Fr}(A) = \overline{A} \cap \overline{1-A}$ (the *boundary* of A).

Prove that:

(a) $\mathrm{Fr}(A \cup B) \cup \mathrm{Fr}(A \cap B) \cup [\mathrm{Fr}(A) \cap \mathrm{Fr}(B)] = \mathrm{Fr}(A) \cup \mathrm{Fr}(B)$ [A.H. Stone],

(b) $\mathrm{Fr}(A) = (A \cap \overline{1-A}) \cup (\overline{A}-A)$,

(c) $A \cup \mathrm{Fr}(A) = \overline{A}$,

(d) $\mathrm{Fr}[\mathrm{Int}(A)] \subset \mathrm{Fr}(A)$,

(e) $\overline{\mathrm{Int}[\mathrm{Fr}(A)]} = \overline{A \cap \mathrm{Int}[\mathrm{Fr}(A)]} = \overline{\mathrm{Int}[\mathrm{Fr}(A)]-A}$.

3. We call the set A a *boundary* set if $\overline{1-A} = 1$. The set A is called *nowhere dense* if $\overline{A}$ is boundary.

Prove that:

(a) the union of a boundary set and a nowhere dense set is a boundary set;

(b) the union of two nowhere dense sets is nowhere dense;

(c) in order that the set $\mathrm{Fr}(A)$ be nowhere dense it is necessary and sufficient that A be the union of an open set and a nowhere dense set.

4. Let 1 be a space satisfying besides axioms (1)–(4) the following axiom (where $\{p\}$ denotes the set consisting of the single element p):

$$\overline{\{p\}} = \{p\}.$$

We say that the point p is an *accumulation point* of the set A if $p \in \overline{A-\{p\}}$ (for the plane this condition is equivalent to the condition that $p = \lim_{n=\infty} p_n$, where p_n

$\in A-\{p\}$). By $A^{\backprime}$ we denote the set of all accumulation points of the set A, called the *derivative* of the set A. Prove the formulas:

$$(A\cup B)^{\backprime} = A^{\backprime}\cup B^{\backprime}, \qquad A^{\backprime}-B^{\backprime} \subset (A-B)^{\backprime}, \qquad A^{\backprime\backprime} \subset A^{\backprime}, \qquad \bar{A} = A\cup A^{\backprime}, \qquad \bar{A}^{\backprime} = A^{\backprime}.$$

5. Let 1 denote the space considered in exercise 4. We call the set A *dense in itself* if $A \subset A^{\backprime}$.

Prove that:

(a) if the space 1 is dense in itself, then every open set is also dense in itself;

(b) if sets A and $1-A$ are boundary sets, then 1 is dense in itself;

(c) the sets $\mathrm{Int}[\mathrm{Fr}(A)]$ and $A\cap\mathrm{Int}[\mathrm{Fr}(A)]$ are dense in themselves.

6. Conditions (1)–(3) are equivalent to the condition

$$A\cup\bar{A}\cup\bar{\bar{B}} = \overline{A\cup B} \qquad \text{[Iseki].}$$

§9. Boolean algebras

We shall conclude this chapter with certain considerations of an axiomatic character. If we examine the theorems of §§ 2–8, we notice that the symbol $\in$ does not occur in the majority of them, though of course it does appear in the definitions and proofs. This suggests developing a separate theory to cover that part of the calculus of sets which does not make reference to the $\in$ relation. In this theory we shall speak only of the equality or inequality between objects and terms resulting from these objects by performing certain operations on them. We shall base this theory on a system of axioms, from which we shall be able to prove all the theorems of the preceding sections in which the symbol $\in$ does not occur. This theory, which is called *Boolean algebra*, has applications in many areas of mathematics.[1])

Let K be an arbitrary set of elements, $\triangle$ and $\wedge$ operations of two arguments always performable on elements of K and having values in K. Finally, let o denote a particular element of K. We say that K is a *Boolean ring* or *Boolean algebra* with respect to these operations and to the element o if for arbitrary $a, b, c \in K$ the following equations hold (axioms of Boolean algebra):

[1]) Boolean algebra originated with George Boole, an English mathematician (1813–1864). His work [1] marks the beginning of mathematical logic. There are many textbooks and monographs on Boolean algebra, e.g., Halmos [1], Sikorski [1].

(1) $$a \triangle b = b \triangle a,$$

(2) $$a \triangle (b \triangle c) = (a \triangle b) \triangle c,$$

(3) $$a \triangle o = a,$$

(4) $$a \triangle a = o,$$

(5) $$a \wedge b = b \wedge a,$$

(6) $$a \wedge (b \wedge c) = (a \wedge b) \wedge c,$$

(7) $$a \wedge o = o,$$

(8) $$a \wedge a = a,$$

(9) $$a \wedge (b \triangle c) = (a \wedge b) \triangle (a \wedge c).$$

We define the sum and the difference of elements of K by the equations

$$a \vee b = a \triangle [b \triangle (a \wedge b)],$$
$$a - b = a \triangle (a \wedge b).$$

We call $a \triangle b$ the *symmetric difference* of a and b, $a \wedge b$ the *product* of a and b, and o the *zero element*.[1]

An example of a Boolean algebra is the family of all subsets of a given fixed set 1 where the operations $\triangle$ and $\wedge$ are the set-theoretical operations of symmetric difference and intersection and where o denotes the empty set. We dealt with this interpretation of axioms (1)–(9) in § 6.[2]

More generally, instead of considering all the subsets of the space 1, we may limit ourselves to the consideration of any family of subsets K of 1 where the symmetric difference and intersection of two sets belonging to K also belong to K. Such a family is a Boolean algebra with respect to the same operations as in the preceding example. Each Boolean algebra of the type just described is called a *field of sets*.

We introduce *Boolean polynomials*. Let $x_1, x_2, \ldots$ be arbitrary letters.

[1]) The fact that we are using the same symbols for operations in Boolean algebras and for logical operations should not lead to misunderstanding.

[2]) Similarly as in § 8, our exposition is based not only on the axioms of set theory but also on the axioms of Boolean algebra and, in part, also on topological axioms. As a matter of fact, we can deduce all theorems from the axioms of set theory given in Chapter II, treating the axioms of Boolean algebra as assumptions about the operations Δ, $\wedge$ and the element o and the axioms of topology—as assumptions about the closure operation. Similar remarks apply to §10.

The symbols

(i) o,

(ii) $x_1, x_2, \ldots$

are polynomials; if f and g are polynomials then the expressions

(iii) $(f)\triangle(g)$,

(iv) $(f)\wedge(g)$

are polynomials. A polynomial is to be understood as a sequence of symbols.

Let us suppose that K is a Boolean algebra and that to every letter x_j there corresponds a certain element $a_j \in K$. We define inductively the value of a polynomial with respect to this correlation. The *value of polynomial* (i) is the zero element of the algebra K, the values of polynomials (ii) are the corresponding elements in K; if the values of f and g are the elements a and b, then the value of polynomial (iii) is $a\triangle b$ and the value of (iv) is $a\wedge b$.

The value of the polynomial f is denoted by $f_K(a_1, a_2, \ldots)$; clearly $f_K(a_1, a_2, \ldots) \in K$.

Let the polynomial f have the form $\ldots(h')\triangle(h'')\ldots$, and the polynomial g the form $\ldots(h'')\triangle(h')\ldots$ where the periods denote sequences of symbols which occur both in f and in g, and where h' and h'' are polynomials. In this case we say that the polynomial g is *immediately transformable into* the polynomial f by means of axiom (1). Similarly we define immediate transformability by means of the remaining axioms (2)–(9). We say that the polynomial g is *transformable into* f if there exists a finite sequence of polynomials $f = f_1, f_2, \ldots, f_k = g$ such that for each i $(1 \leqslant i < k)$ the polynomial f_{i+1} is immediately transformable into the polynomial f_i by means of one of the axioms. In this case we write $f \sim g$.

Clearly, $f \sim f$, $f \sim g \to g \sim f$ and $f \sim g \sim h \to f \sim h$. If $f \sim g$ then $f_K(a_1, a_2, \ldots) = g_K(a_1, a_2, \ldots)$ for every Boolean algebra K and arbitrary elements $a_j \in K$.

Polynomials resulting from the expression $f_1\triangle f_2\triangle \ldots \triangle f_k$ (or from the expression $f_1 \wedge f_2 \wedge \ldots \wedge f_k$) by an arbitrary placement of parentheses are mutually transformable into each other by means of axiom (2) (or axiom (6)). For this reason we shall always omit parentheses when writing such polynomials. Moreover, we shall take no notice of the difference in the order of polynomials to which we apply successively either one of the symbols $\triangle$ or $\wedge$.

THEOREM 1: *Every polynomial is either transformable into o or into some polynomial of the form* $s_1 \triangle s_2 \triangle \ldots \triangle s_h$, *where each of the polynomials* s_j *has the form* $x_{i_1} \wedge x_{i_2} \wedge \ldots \wedge x_{i_t}$ $(i_1 < i_2 < \ldots < i_t, t \geqslant 1)$, *and no two components* s_j, s_k $(1 \leqslant j < k \leqslant h)$ *are identical.*[1])

PROOF. The theorem is clear for polynomials (i) and (ii). Assume that it holds for polynomials f and g. If $f \sim o$ (or $g \sim o$), then $(f) \triangle (g) \sim g$ and $(f) \wedge (g) \sim o$ (or $(f) \triangle (g) \sim f$ and $(f) \wedge (g) \sim o$). At this point we may assume that $f \sim s_1 \triangle s_2 \triangle \ldots \triangle s_h$ and $g \sim t_1 \triangle t_2 \triangle \ldots \triangle t_k$. Thus $(f) \triangle (g) \sim s_1 \triangle s_2 \triangle \ldots \triangle s_h \triangle t_1 \triangle t_2 \triangle \ldots \triangle t_k$. By applying (3) and (4) we eliminate all redundant occurrences of components and thus obtain $(f) \triangle (g)$ in the desired form. The theorem, therefore, holds for formula (iii).

In the case of polynomial (iv) we apply axioms (9) and (5) and obtain

$$\begin{aligned}(f) \wedge (g) &\sim [(s_1 \triangle \ldots \triangle s_h) \wedge t_1] \triangle \ldots \triangle [(s_1 \triangle \ldots \triangle s_h) \wedge t_k] \\ &\sim (s_1 \wedge t_1) \triangle \ldots \triangle (s_p \wedge t_q) \triangle \ldots \triangle (s_h \wedge t_k).\end{aligned}$$

By means of (5) and (8) each of the polynomials $s_p \wedge t_q$ is transformable into the product of individual variables. Omitting as in the previous case redundancies we obtain the desired form. The theorem, therefore, holds for formula (iv). Q.E.D.

THEOREM 2: *Let K be the field of all subsets of the non-empty set* 1. *If f is a polynomial such that* $\neg(f \sim o)$, *then there exist sets* $A_1, A_2, \ldots$ *belonging to K such that* $f_K(A_1, A_2, \ldots) \neq 0$.

PROOF. By Theorem 1 we may limit ourselves to the case where f has the form $s_1 \triangle s_2 \triangle \ldots \triangle s_h$ and where each of the polynomials s_j is a product of letters x_i. Let n be the number of distinct letters x_i occurring in f. We shall prove the theorem by induction on n.

For $n = 1$, $f \sim x_i$, thus we may choose any non-empty set for the set A_i. Assume that the theorem holds for all numbers less than n and

[1]) Theorem 1, as well as Theorem 2, is a scheme: for each polynomial f we obtain a separate theorem.

that the polynomial f contains exactly n distinct variables. If one of the variables x_p occurs in each of the expressions s_j, then $f \sim x_p \wedge g$ where g contains less than n variables. By the induction hypothesis there exist sets $A_1, A_2, \ldots$ such that $g_K(A_1, A_2, \ldots) \neq \emptyset$. Replacing A_p by 1 we leave the value of g unchanged (because g does not contain x_p) and we obtain the set $1 \cap g_K(A_1, A_2, \ldots) \neq \emptyset$ as the value of f.

If none of the letters x_p occurs in each of the s_j, then we substitute in f the symbol o everywhere for some arbitrary x_p. Thus we obtain the polynomial g of fewer variables than f and $\neg(g \sim o)$. Hence in this case the theorem follows from the induction hypothesis.

THEOREM 3: *Every equation $f = g$ which is true for arbitrary sets (and even for arbitrary subsets of a given non-empty set) is derivable from axioms* (1)–(9).

PROOF. If the polynomial $f \triangle g$ has the value o for all $A_1, A_2, \ldots$ contained in a non-empty set 1, then $f \triangle g \sim o$ and thus $f \sim g$. Therefore polynomial g arises from f by transformation by means of axioms (1)–(9) and by the general rule of logic which states that equal elements may be substituted for each other.

Theorem 3 shows that the equations derivable from axioms (1)–(9) are identical with the equations which are true for arbitrary sets. Moreover, this theorem provides a mechanical procedure for deciding when an equation of the form $f = g$ is derivable from axioms (1)–(9). Namely, it suffices to reduce the polynomial $f \triangle g$ by the method given in the proof of Theorem 1 and to determine whether or not it is transformed into o.

We introduce an *order relation* in Boolean algebra by the definition:

$$a \leqslant b \equiv a \wedge b = a.$$

THEOREM 4: $a \leqslant b \equiv a \vee b = b$.

PROOF. If $a \wedge b = a$, then $a \vee b = a \triangle b \triangle (a \wedge b) = a \triangle b \triangle a = b$. Conversely, if $a \vee b = b$, then $a \triangle b \triangle (a \wedge b) = b$. Thus $a \triangle b \triangle b \triangle (a \wedge b) = b \triangle b = o$ and $a \triangle o \triangle (a \wedge b) = o$, whence $a \triangle a \triangle (a \wedge b) = a \triangle o = a$ and $o \triangle (a \wedge b) = a$; that is, $a \wedge b = a$.

We call an element i of the Boolean algebra K a *unit* of K if

(10) $\qquad a \wedge i = a \quad$ for all elements $a \in K$.

It is easy to prove that *a unit, if it exists, is unique.*

In an algebra with a unit we define the *complement* of the element a by the equation

$$-a = i \triangle a.$$

Axioms (1)–(9) are very convenient in most calculations but are seldom used to describe a Boolean algebra. In the next theorem we shall present a different system of axioms, which is usually taken as the basis of a Boolean algebra. We shall limit ourself to the consideration of Boolean algebras with a unit.

THEOREM 5: *If K is a Boolean algebra with a unit, then the following equations hold for all $a, b, c \in K$:*

(i) $a \vee b = b \vee a$, (i′) $a \wedge b = b \wedge a$,

(ii) $a \vee (b \vee c) = (a \vee b) \vee c$, (ii′) $a \wedge (b \wedge c) = (a \wedge b) \wedge c$,

(iii) $a \vee o = a$, (iii′) $a \wedge i = a$,

(iv) $a \vee -a = i$, (iv′) $a \wedge -a = o$,

(v) $a \wedge (b \vee c) = (a \wedge b) \vee (a \wedge c)$, (v′) $a \vee (b \wedge c) = (a \vee b) \wedge (a \vee c)$.

PROOF. Equations (i), (i′) (ii), (ii′), (iii), (v), (v′) are true for arbitrary sets and thus are consequences of axioms (1)–(9). Equation (iii′) is identical with (10). We establish (iv′) and (iv) as follows:

$$\begin{aligned} a \wedge -a &= a \wedge (i \triangle a) && \text{by the definition of } -a, \\ &= a \triangle (a \wedge a) && \text{from (9) and (10)} \\ &= a \triangle a && \text{from (8)} \\ &= o && \text{from (4);} \end{aligned}$$

$$\begin{aligned} a \vee -a &= a \triangle (i \triangle a) \triangle [a \wedge (i \triangle a)] && \text{by the definition of sum,} \\ &= a \triangle (i \triangle a) \triangle o && \text{from (iv')} \\ &= (a \triangle a) \triangle i && \text{from (1), (2), (3)} \\ &= o \triangle i && \text{from (4)} \\ &= i && \text{from (3).} \end{aligned}$$

A partial converse to Theorem 5 holds:

THEOREM 6: *If K is a set, o, $i \in K$, and if $\vee$, $\wedge$, $-$ are operations defined on elements of K satisfying equations* (i)–(v′), *then K is a Boolean* *algebra with respect to the operations* $a \triangle b = (a \wedge -b) \vee (b \wedge -a)$, $a \wedge b$ *and the element* o.

The proof of the theorem is not difficult and is left to the reader.

Equations (i)–(v′) are most often used as axioms for a Boolean algebra. In particular, it is worth noting the symmetry of these equations with respect to the operations $\vee$ and $\wedge$.

We shall conclude this section by giving an interesting example of a Boolean algebra.

Let 1 be an arbitrary topological space with a closure operator (see § 8, p. 27). We call $A \subset 1$ a *regular closed set* if

$$A = \overline{\mathrm{Int}(A)}.$$

By $\boldsymbol{K}$ we denote the family of all regular closed sets contained in 1. Clearly, $\emptyset$ and 1 belong to $\boldsymbol{K}$ since

$$\overline{\mathrm{Int}(\emptyset)} = \overline{\emptyset} = \emptyset \quad \text{and} \quad \overline{\mathrm{Int}(1)} = \overline{1} = 1.$$

If $A \in \boldsymbol{K}$ then $\overline{A} = A$, because

$$\overline{A} = \overline{\overline{\mathrm{Int}(A)}} = \overline{\mathrm{Int}(A)} = A.$$

Thus every set belonging to $\boldsymbol{K}$ is closed (see § 8, p. 28). By theorem 8 (18) it follows that if $A, B \in \boldsymbol{K}$ then

$$A \cup B = \overline{\mathrm{Int}(A)} \cup \overline{\mathrm{Int}(B)} = \overline{\mathrm{Int}(A \cup B)},$$

which proves that $A \cup B \in \boldsymbol{K}$.

For $A \in \boldsymbol{K}$ and $B \in \boldsymbol{K}$, let

$$A \odot B = \overline{\mathrm{Int}(A \cap B)}, \qquad A' = \overline{\mathrm{Int}(-A)}, \qquad A \bigcirc B = (A \odot B') \cup (B \odot A').$$

It follows from this definition and from formula 8 (14) that if $A \in \boldsymbol{K}$ and $B \in \boldsymbol{K}$, then $A \odot B \in \boldsymbol{K}$, $A' \in \boldsymbol{K}$ and $A \bigcirc B \in \boldsymbol{K}$.

THEOREM 7: *$\boldsymbol{K}$ is a Boolean algebra with a unit with respect to the operations* $\bigcirc$ *and* $\odot$.

PROOF. It suffices to show that the operations $\cup$, $\odot$ and $'$ satisfy axioms (i)–(v′) of Theorem 5.

Axioms (i)–(iii) are clearly satisfied. Axiom (i′) follows from the equation

$$A \odot B = \overline{\mathrm{Int}(A \cap B)} = \overline{\mathrm{Int}(B \cap A)} = B \odot A.$$

It is equally easy to show that axiom (iii′) holds:

$$A \odot 1 = \overline{\mathrm{Int}(A \cap 1)} = \overline{\mathrm{Int}(A)} = A.$$

To show that axiom (ii′) holds we apply 8 (12) and (8) and obtain

$$\mathrm{Int}[(A \odot B) \cap C] = \mathrm{Int}(A \odot B) \cap \mathrm{Int}(C),$$

$$A \odot B = \overline{\mathrm{Int}(A \cap B)} = \overline{\mathrm{Int}(A) \cap \mathrm{Int}(B)} \subset \overline{\mathrm{Int}(A)} \cap \overline{\mathrm{Int}(B)} = A \cap B;$$

it follows by (10) of § 8 that

(*) $\qquad \mathrm{Int}[(A \odot B) \cap C] \subset (A \odot B) \cap C \subset A \cap B \cap C \subset B \cap C.$

Thus

$$\mathrm{Int}\{\mathrm{Int}[(A \odot B) \cap C]\} \subset \mathrm{Int}(B \cap C);$$

that is (see 8 (11)),

$$\mathrm{Int}[(A \odot B) \cap C] \subset \mathrm{Int}(B \cap C) \subset \overline{\mathrm{Int}(B \cap C)} = B \odot C.$$

Since (by (*)) $\mathrm{Int}[(A \odot B) \cap C] \subset A$, we have

$$\mathrm{Int}[(A \odot B) \cap C] \subset A \cap (B \odot C),$$

whence we obtain

$$\mathrm{Int}\{\mathrm{Int}[(A \odot B) \cap C]\} \subset \mathrm{Int}[A \cap (B \odot C)],$$

and hence

$$\mathrm{Int}[(A \odot B) \cap C] \subset \mathrm{Int}[A \cap (B \odot C)].$$

Taking closure on both sides of the inclusion, we obtain

$$\overline{\mathrm{Int}[(A \odot B) \cap C]} \subset \overline{\mathrm{Int}[A \cap (B \odot C)]},$$

that is, $(A \odot B) \odot C \subset A \odot (B \odot C)$. The opposite inclusion is obtained in an entirely similar manner. Thus we may consider the equation $(A \odot B) \odot C = A \odot (B \odot C)$ as proved.

We examine axiom (v). We have

$$A \odot (B \cup C) = \overline{\mathrm{Int}[A \cap (B \cup C)]} = \overline{\mathrm{Int}[(A \cap B) \cup (A \cap C)]}.$$

The sets A, B and C are closed, thus (see 8 (9)) $\overline{A \cap B} = A \cap B$ and $\overline{A \cap C} = A \cap C$. By 8 (18) we conclude that

$$\begin{aligned}\overline{\mathrm{Int}[(A \cap B) \cup (A \cap C)]} &= \overline{\mathrm{Int}(A \cap B)} \cup \overline{\mathrm{Int}(A \cap C)} \\ &= (A \odot B) \cup (A \odot C).\end{aligned}$$

Thus

$$A \odot (B \cup C) = (A \odot B) \cup (A \odot C).$$

We check axiom (v′) as follows. From the definitions,

$$A \cup (B \odot C) = \overline{\mathrm{Int}(A)} \cup \overline{\mathrm{Int}(B \cap C)}.$$

By 8 (18) the right-hand side of the equation equals $\overline{\mathrm{Int}[A \cup (B \cap C)]}$, which equals $\overline{\mathrm{Int}[(A \cup B) \cap (A \cup C)]}$, that is, $(A \cup B) \odot (A \cup C)$.

Axiom (iv′) is an easy consequence of theorem 8 (17) and of the fact that every regular closed set has the form X^{-c-}.

Finally we prove that axiom (iv) holds. By 8 (18) we have

$$A \cup A' = \overline{\mathrm{Int}(A)} \cup \overline{\mathrm{Int}(-A)} = \overline{\mathrm{Int}(A \cup -A)},$$

since $\overline{A} = A$. Thus we conclude immediately that $A \cup A' = \overline{\mathrm{Int}(1)} = 1$.

Theorem 7 is thus proved.

Let us take the plane as the space 1. Every circle together with its boundary is clearly a regular closed set. Since every non-empty set of the form Int(A) contains some circle, we conclude that the Boolean algebra of regular closed sets in the plane has the following property:

If $A \in \mathbf{K}$ and $A \neq \emptyset$, then there exists B such that $B \in \mathbf{K}$, $\emptyset \neq B \subset A$ and $B \neq A$. #

Exercises

1. From every equation written in terms of variables, the symbols o and i and the operations $\vee$, $\wedge$ and $-$ we obtain a new equation by interchanging the symbols o and i and the operations $\vee$ and $\wedge$. If the original equation is true in Boolean algebra, then so is the equation obtained from it in this way (*Principle of Duality*).

2. Show that axioms (ii) and (ii′) are derivable from axioms (i), (i′), (iii)–(v), (iii′)–(v′) [Huntington].

3. Show that every Boolean algebra is a ring with respect to the operation Δ as addition and $\wedge$ as multiplication and that in this ring the square of each element is equal to this element (we say that multiplication is an *idempotent operation*). If the Boolean algebra has a unit, then so does the ring. Prove conversely that each ring with a unit and with an idempotent multiplication is a Boolean algebra [Stone].

4. An element of a Boolean algebra is called an *atom* if $a \neq o$ and for each x the relation $x \leqslant a$ implies $x = o$ or $x = a$ (thus the algebra of regular closed subsets of a plane has no atoms). Show that the following algebra K has no atoms: Elements of K are finite unions of real intervals without left end-points but with right end-points and the operations $\vee$, $\wedge$, $-$ are respectively $\cup$, $\cap$, and complementation.

5. Prove that the number of elements of a finite Boolean algebra has always the form 2^n.

Hint: n is the number of atoms of the algebra.

6. The set of sentences of any mathematical theory is a Boolean algebra with a unit if we identify sentences which are equivalent to each other on the basis of the propositional calculus; the operations $\vee$, $\wedge$, and $-$ in this algebra correspond to forming disjunctions, conjunctions and negations of sentences. The zero element is any false sentence and the unit element is any true sentence.

§ 10. Lattices[1])

The concept of a lattice is more general than that of a Boolean algebra. Let L be an arbitrary set of elements, upon which are defined the operations $\vee$ and $\wedge$. We say that L is a *lattice* with respect to the operations $\vee$ and $\wedge$ if the following equations hold (axioms of lattice theory)

(1) $$a \vee a = a, \qquad a \wedge a = a,$$

(2) $$a \vee b = b \vee a, \qquad a \wedge b = b \wedge a,$$

(3) $$a \vee (b \vee c) = (a \vee b) \vee c, \qquad a \wedge (b \wedge c) = (a \wedge b) \wedge c,$$

(4) $$a \wedge (a \vee b) = a, \qquad a \vee (a \wedge b) = a.$$

We call a lattice *distributive* if

(5) $$a \wedge (b \vee c) = (a \wedge b) \vee (a \wedge c), \qquad a \vee (b \wedge c) = (a \vee b) \wedge (a \vee c).$$

We introduce an *order relation* between elements of a lattice just as we did for Boolean algebras:

(6) $$a \leqslant b \equiv a \vee b = b$$

or, equivalently,

(7) $$a \leqslant b \equiv a \wedge b = a.$$

Similarly we define the elements o and i (if they exist in the given lattice) as the elements satisfying conditions

(8) $$a \vee o = a, \qquad a \wedge i = a$$

for all $a \in L$.

[1]) The notion of a lattice originated essentially with Dedekind in connection with his number theoretical studies. For a detailed exposition of lattice theory see Birkhoff [1], Hermes [1] as well as books on universal algebras, e.g. Cohn [1].

It is easy to show that o *is the smallest element in the lattice and that i is the largest*, namely,

$$o \leqslant a \leqslant i \tag{9}$$

for every $a \in L$.

Referring to Theorem 5, we observe that *every Boolean algebra with a unit is a distributive lattice with a zero and a unit.* The converse does not hold, as is shown by the following counter-example which itself is important for numerous applications in topology. The family of all closed subsets of an arbitrary topological space is a lattice (with the natural interpretation of the operations: $a \vee b = a \cup b$, $a \wedge b = a \cap b$). However, this family is not in general a Boolean algebra, since the difference of two closed sets need not be closed (for example, when the space is the space of real numbers).

On the other hand, the following theorem holds.

THEOREM: *If A is a distributive lattice with o and i and if for every $a \in A$ there exists an element $-a \in A$ satisfying the equations*

$$a \vee (-a) = i, \quad a \wedge (-a) = o, \tag{10}$$

then (i) *the element $-a$ is unique, and* (ii) *A is a Boolean algebra with a zero and a unit with respect to the operations $\vee$, $\wedge$, and $-$.*

PROOF. Suppose that the element a' also satisfies conditions (10). Then $a' = a' \wedge i = a' \wedge (a \vee -a) = (a' \wedge a) \vee (a' \wedge -a) = o \vee (a' \wedge -a) = a' \wedge -a$. Similarly, $-a = -a \wedge a'$, therefore $a' = -a$.

For the proof of the second part of the theorem it suffices to show that axioms (i)–(v′) from p. 38 are satisfied. Axioms (i), (i′), (ii) and (ii′) hold in every lattice, (iii) and (iii′) follow from the assumption that o and i are zero and unit in A, (iv) and (iv′) follow from the assumption that condition (10) holds, and finally, (v) and (v′) from the assumption that the lattice is distributive.

The concept of *Brouwerian lattice*[1]) is intermediate between that of lattice and Boolean algebra. We call a lattice with a unit *Brouwerian* if

[1]) The term "Brouwerian lattice" was introduced by McKinsey and Tarski [1] in their algebraic study of intuitionistic (Brouwerian) logic. Rasiowa and Sikorski [1] use for the same purpose a dual notion which they call "Heyting lattices".

for arbitrary elements $a, b \in L$ there exists an element of L called the *pseudo-difference* of a and b and denoted by the symbol $a \overset{*}{-} b$ such that

$$(a \overset{*}{-} b \leqslant c) \equiv (a \leqslant b \vee c).$$

The family of closed subsets of a given space considered above is a Brouwerian lattice, where the pseudo-difference of two closed sets A and B is the closed set $\overline{A-B}$.

We denote by $\overset{*}{-} a$ the *pseudo-complement* of a, namely $\overset{*}{-} a = i \overset{*}{-} a$. Notice that, in contrast to the operation of ordinary complementation, the equation $(\overset{*}{-} a) \wedge a = 0$ does not hold. This corresponds to the fact that the law of the excluded middle does not hold in intuitionistic logic. In the topological interpretation this means that the nowhere dense set $\overline{X-A} \cap A$, namely the boundary of A, is not necessarily empty. On the other hand, the validity of the equation $(\overset{*}{-} a) \vee a = i$ corresponds to the law of contradiction in Brouwerian logic.

Examples and exercises

1. The set of natural numbers is a lattice with respect to the operation of taking the greatest common divisor as the operation $\vee$ and the least common multiple for the operation $\wedge$. The formula $a \leqslant b$ means that b is a divisor of a. The number 1 is the unit of the lattice, and there is no zero element.

2. We consider euclidean n-space $\mathscr{E}^n$ and the family L_n of its linear subsets (points, lines, planes and in general, k-dimensional subspaces where $k \leqslant n$) passing through the origin. The family L_n is a lattice with respect to the operations $\vee$ and $\wedge$ defined by: $A \wedge B$ is the intersection of A and B; $A \vee B$ is the least linear subspace of $\mathscr{E}^n$ containing A and B. For example, if A and B are two planes, then $A \cup B$ is a 3-dimensional space if $A \cap B$ is a straight line, and is a 4-dimensional space if $A \cap B$ is a point. The relation $\leqslant$ is the ordinary inclusion relation. The zero element of the lattice L_n is the one-point set consisting of the origin and the unit is the entire space L_n.

The lattice L_n is not distributive but is modular. Namely, we call a lattice *modular* if

$$(a \leqslant c) \rightarrow [a \vee (b \wedge c) = (a \vee b) \wedge c].$$

It is worth noticing that in the lattice L_n every increasing sequence $a_1 < a_2 < \ldots$ contains at most $n+1$ elements.

3. Prove that the formula

$$(a \leqslant c) \rightarrow [a \vee (b \wedge c) \leqslant (a \vee b) \wedge c]$$

holds in every lattice.

4. Prove that the two equations given in (5) are equivalent.

5. Prove that the following formulas hold in every Brouwerian lattice

$$(x \leqslant y) \rightarrow (x \overset{*}{-} z \leqslant y \overset{*}{-} z) \wedge (z \overset{*}{-} y \leqslant z \overset{*}{-} x),$$

$$(x \leqslant y) \equiv (x \overset{*}{-} y = o),$$

$$z \overset{*}{-} (x \wedge y) = (z \overset{*}{-} x) \vee (z \overset{*}{-} y),$$

$$(x \vee y) \overset{*}{-} z = (x \overset{*}{-} z) \vee (y \overset{*}{-} z),$$

$$\overset{*}{-}(\overset{*}{-} x) \leqslant x,$$

$$\overset{*}{-}\overset{*}{-}\overset{*}{-} x = \overset{*}{-} x \qquad \text{[McKinsey–Tarski]}.$$

6. The family of compact subsets of a topological space is a lattice with respect to the ordinary set theoretical operations $\cup$ and $\cap$. If the space itself is not compact, then the lattice does not have a unit; on the other hand, the lattice always has a zero (see p. 137 for the definition of compact space).

CHAPTER II

AXIOMS OF SET THEORY. RELATIONS. FUNCTIONS

§ 1. Set-theoretical formulas. Quantifiers

We shall begin this chapter by reviewing certain logical notions. In Chapter I, § 1, we dealt with sentences which have constant logical values. Now we shall consider *formulas*, i.e. expressions which may contain (free) variables and whose logical values depend on values given to these variables. Instead of discussing these expressions in full generality we shall describe a class K of formulas which will be used in the rest of this book. We shall call K the *class of pure set-theoretical formulas*.

We fix an arbitrary list of symbols $x, y, z, \ldots, X, Y, Z, \ldots$ which will be called *variables*. Class K is defined by induction.

(A) *Expressions of the form given below belong to* K:

$$x \text{ is a set (abbreviated } Z(x)), \quad x \in y, \quad x = y$$

as well as all *expressions differing from them by a choice of variables.*

(B) *If* Φ and Ψ *belong to the class* K, *then so do the expressions* $\Phi \vee \Psi$, $\Phi \wedge \Psi$, $\Phi \rightarrow \Psi$, $\Phi \equiv \Psi$, *and* $\neg\Phi$.

(C) *If* Φ *belongs to* K *and* v *is any variable, then the expressions* $\bigvee_v \Phi$, $\bigwedge_v \Phi$ *belong to* K.

(D) *Every element of* K *arises by a finite application of rules* (A), (B), (C).

All variables occurring in the "atomic" formulas (A) are the free variables of these formulas; formula $\neg\Phi$ has the same free variables as Φ; free variables of $\Phi \vee \Psi$ are variables which are free in Φ or in Ψ and the same is true for formulas $\Phi \wedge \Psi$, $\Phi \rightarrow \Psi$, and $\Phi \equiv \Psi$. Finally, a variable is free in $\bigwedge_v \Phi$ or in $\bigvee_v \Phi$ if it is free in Φ and different from v.

Symbols $\vee$, $\wedge$, $\rightarrow$, $\equiv$, $\neg$ occurring in (B) are of course the familiar propositional connectives which we discussed in Chapter I. Symbols $\bigwedge$ and $\bigvee$ are called the *general* and the *existential quantifiers*. They are abbreviations of the words "for every" and "there exists", respectively. Thus e.g., $\bigwedge_x [(x = y) \rightarrow (y = x)]$ means that, for every x, if $x = y$, then $y = x$; and $\bigvee_x \neg(x = y)$ means that there is an object x which is different from y.

Notice that according to our definitions the variable v which occurs in the formula $\bigwedge_v \Phi$ and in the formula $\bigvee_v \Phi$ is not free in these formulas; we say that this variable is *bound* by the initial quantifier of the formula. Thus quantifiers are operators which change free variables into bound ones. A similar situation is known in the calculus: the expression x^2+y^2 denotes (the value of) a function of two variables but $\int_0^1 (x^2+y^2)\,dx$ denotes (the value of) a function of one variable y.

Formulas which have no free variables are called *sentences*. We can obtain a sentence from a formula by inserting in the front of the formula enough quantifiers to bind all the free variables of the formula.

We shall use the abbreviations $\bigwedge_{x\in X} \Phi(x)$ and $\bigvee_{x\in X} \Phi(x)$ for the expressions $\bigwedge_x [(x \in X) \rightarrow \Phi(x)]$ and $\bigvee_x [(x \in X) \wedge \Phi(x)]$, respectively. The operators $\bigwedge_{x\in X}$ and $\bigvee_{x\in X}$ are called *quantifiers limited to X*. It should be noticed that the truth value of the formula $\bigwedge_{x\in X} \Phi(x)$ is "true" for $X = \emptyset$ whatever the formula $\Phi(x)$ might be. Similarly the truth value of the formula $\bigvee_{x\in X} \Phi(x)$ for $X = \emptyset$ is always "false".

It is sometimes convenient to consider extensions of the class K and use formulas containing additional primitive notions. We shall introduce later in this chapter one additional primitive notion which although not strictly necessary simplifies many statements of set theory. In such cases we modify the definition of class K by adding to formulas listed in (A) new "atomic" formulas. If, e.g., the new primitive notions are certain binary relations $P, Q, \ldots$ then the new atomic formula will be $P(x, y), Q(x, y), \ldots$ All other definitions remain unchanged. The new class of formulas will be denoted by $K[P, Q, \ldots]$ and called the *class of set-theoretical formulas with respect to* $P, Q, \ldots$[1])

Theoretically speaking we should express all theorems given later in this book as set-theoretical sentences in the sense explained above. In practice however we shall use many self explanatory abbreviations which will render the reading of formulas much easier.

Similarly as in the case of the propositional calculus there are many formulas involving quantifiers which are true independently of the meaning of atomic formulas out of which they are constructed and independently of the values given to the free variables. We call such formulas *theorems of the predicate calculus*. We shall give below several examples of such formulas; a systematic exposition of how such formulas can be obtained may be found in textbooks of mathematical logic.

(1) $$\textit{If } a \in A \textit{ then } \bigwedge_{x \in A} \Phi(x) \to \Phi(a) \textit{ and } \Phi(a) \to \bigvee_{x \in A} \Phi(x).$$

The second formula states that to prove an existential sentence of the form $\bigvee_{x \in A} \Phi(x)$ it suffices to find an object a belonging to the set A and satisfying the condition $\Phi(x)$. Such a proof of an existential sentence is called *effective*.[2])

(2) $$\bigwedge_x [\Phi(x) \wedge \Psi(x)] \equiv \Big[\bigwedge_x \Phi(x) \wedge \bigwedge_x \Psi(x)\Big],$$

(3) $$\bigvee_x [\Phi(x) \vee \Psi(x)] \equiv \Big[\bigvee_x \Phi(x) \vee \bigvee_x \Psi(x)\Big].$$

Thus the universal quantifier is distributive over conjunction and the existential quantifier is distributive over disjunction.

[1]) We shall use Greek capitals $\Phi, \Psi, \ldots$ for set-theoretical formulas; we shall often use more complicated symbols like $\Phi(x, y, \ldots, z)$ for formulas in which the variables $x, y, \ldots, z$ are free.

[2]) For a detailed exposition of principles of logic see e.g., Shoenfield [1].

(4) $$\Big[\bigwedge_x \Phi(x) \vee \bigwedge_x \Psi(x)\Big] \rightarrow \Big[\bigwedge_x [\Phi(x) \vee \Psi(x)]\Big],$$

(5) $$\bigvee_x [\Phi(x) \wedge \Psi(x)] \rightarrow \Big[\bigvee_x \Phi(x) \wedge \bigvee_x \Psi(x)\Big].$$

By means of simple counter-examples we conclude that the converse implications are not, in general, true. The universal quantifier is not distributive over disjunction and the existential quantifier is not distributive over conjunction.

(6) $$\neg\Big[\bigwedge_x \Phi(x)\Big] \equiv \bigvee_x [\neg\Phi(x)],$$

(7) $$\neg\Big[\bigvee_x \Phi(x)\Big] \equiv \bigwedge_x [\neg\Phi(x)].$$

Laws (6) and (7) are called *de Morgan's laws*. It follows from them that:

(6′) $$\neg\bigwedge_x [\neg\Phi(x)] \equiv \bigvee_x \Phi(x),$$

(7′) $$\neg\bigvee_x [\neg\Phi(x)] \equiv \bigwedge_x \Phi(x).$$

Hence the existential quantifier can be defined in terms of the universal quantifier (and conversely, the universal quantifier in terms of the existential).

The first equivalence above shows that the existential sentence $\bigvee_x \Phi(x)$ can be proved by deriving a contradiction from the assumption $\bigwedge_x \neg\Phi(x)$. Such proofs of existential sentences are often used but, in general, are not effective, i.e. they do not give any method of constructing the object satisfying the formula $\Phi(x)$.

If p is a sentence, then clearly we have

(8) $$\bigwedge_x p \equiv p,$$

(9) $$\bigvee_x p \equiv p.$$

Moreover, it is easy to check that

(10) $$\bigwedge_x [p \vee \Phi(x)] \equiv \Big[p \vee \bigwedge_x \Phi(x)\Big],$$

(11) $$\bigvee_x [p \wedge \Phi(x)] \equiv \Big[p \wedge \bigvee_x \Phi(x)\Big].$$

By means of the law of the propositional calculus $[p \to \Phi(x)] \equiv [\neg p \vee \vee \Phi(x)]$ it follows easily from (10) that

(12) $$\Big[p \to \bigwedge_x \Phi(x)\Big] \equiv \bigwedge_x [p \to \Phi(x)].$$

Observe that

(13) $$\Big[\Big(\bigwedge_x \Phi(x)\Big) \to p\Big] \equiv \bigvee_x [\Phi(x) \to p].$$

In fact, it follows from (6) that the left-hand side of (13) is equivalent to the disjunction $\bigvee_x \neg\Phi(x) \vee p$, and then, by (9), to the disjunction $\bigvee_x \neg\Phi(x) \vee \bigvee_x p$. By (3) we obtain $\bigvee_x [\neg\Phi(x) \vee p]$ or $\bigvee_x [\Phi(x) \to p]$.

Laws (1)–(13) are also valid for formulas with several free variables. Instead of p in (7)–(13) we can have any formula which does not contain the variable x.

Note the following laws concerning formulas with two free variables:

(14) $$\bigwedge_x \bigwedge_y \Phi(x, y) \equiv \bigwedge_y \bigwedge_x \Phi(x, y),$$

(15) $$\bigvee_x \bigvee_y \Phi(x, y) \equiv \bigvee_y \bigvee_x \Phi(x, y).$$

Thus it makes no difference in which order we write universal or existential quantifiers when they occur together. We usually write $\bigwedge_{xy}$ instead of $\bigwedge_x \bigwedge_y$ and $\bigvee_{xy}$ instead of $\bigvee_x \bigvee_y$.

(16) $$\Big[\bigwedge_x \Phi(x) \vee \bigwedge_x \Psi(x)\Big] \equiv \bigwedge_x \bigwedge_y [\Phi(x) \vee \Psi(y)] \equiv \bigwedge_{xy} [\Phi(x) \vee \Psi(y)],$$

(17) $$\Big[\bigvee_x \Phi(x) \wedge \bigvee_x \Psi(x)\Big] \equiv \bigvee_x \bigvee_y [\Phi(x) \wedge \Psi(y)] \equiv \bigvee_{xy} [\Phi(x) \wedge \Psi(y)].$$

To prove (16) we substitute in (10) $\bigwedge_y \Psi(y)$ for p and observe that $\bigwedge_y \Psi(y) \vee \Phi(x) \equiv \bigwedge_y [\Psi(y) \vee \Phi(x)]$ by means of the same law (10).

The proof of (17) is similar.

$$\bigvee_x \bigwedge_y \Phi(x, y) \rightarrow \bigwedge_y \bigvee_x \Phi(x, y). \tag{18}$$

Proof. Using (1) twice, we obtain $\bigwedge_y \Phi(x, y) \rightarrow \Phi(x, y)$ and $\Phi(x, y) \rightarrow \bigvee_x \Phi(x, y)$. These implications hold for any x, y, hence

$$\bigwedge_x \bigwedge_y \Big[\bigwedge_y \Phi(x, y) \rightarrow \bigvee_x \Phi(x, y)\Big].$$

By means of (12) and (13) we obtain (18).

As an application of (18) we shall discuss the difference between uniform and ordinary convergence of a sequence of functions. By the definition of limit, the sentence $\bigwedge_x [\lim_{n=\infty} f_n(x) = f(x)]$ is equivalent to the following

$$\bigwedge_{\varepsilon>0} \bigwedge_x \bigvee_k \bigwedge_n |f_{n+k}(x) - f(x)| < \varepsilon.$$

Interchanging the quantifiers $\bigwedge_x$ and $\bigvee_k$ we obtain the definition of uniform convergence making k independent of x. The fact that k is independent of x and depends only on ε is apparent in the above formula with the interchanged quantifiers.

The following diagram gives several other theorems concerning interchanging of quantifiers:

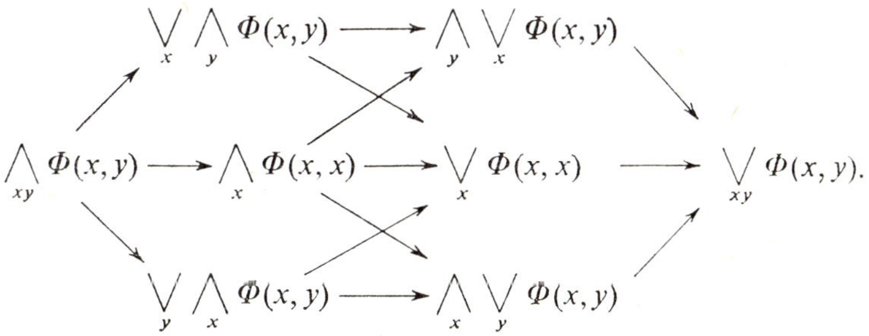

By means of simple counter-examples it is easy to check that none of the above implications can be inverted.

§ 2. Axioms of set theory

We shall introduce axioms upon which we shall base the rest of our exposition of set theory. The new axioms will allow us to form new sets from given sets, and in this sense they do not differ from the axioms in Chapter I. On the other hand, the essential difference between the new axioms and the old ones is that we shall now deal with sets whose elements are also sets.[1])

First of all we retain the axiom of extensionality:

I. (Axiom of extensionality) *If the sets A and B have the same elements, they are identical. In symbols*

$$\bigwedge_x [x \in A \equiv x \in B] \rightarrow (A = B).$$

II. (Axiom of the empty set) *There exists a set $\emptyset$ such that no x is an element of $\emptyset$; symbolically*:

$$\bigvee_P \bigwedge_x (x \notin P).$$

Obviously, there is only one empty set.

II′. (Axiom of pairs)[2]) *For arbitrary a, b, there exists a set which contains only a and b.* In symbols:

$$\bigvee_P \bigwedge_x \{(x \in P) \equiv [(x = a) \vee (x = b)]\}.$$

III. (Axiom of unions) *Let $\boldsymbol{A}$ be a family of sets. There exists a set S such that x is an element of S if and only if x is an element of some set X belonging to $\boldsymbol{A}$.*

Symbolically:

(1) $$x \in S \equiv \bigvee_X [(x \in X) \wedge (X \in \boldsymbol{A})]$$

[1]) To make the terminology clearer we shall use the term a *family of sets* instead of a *set of sets*. Families of sets will be denoted by capital letters printed in bold-faced italics $\boldsymbol{A}, \boldsymbol{B}, \boldsymbol{X}, \boldsymbol{Y}$, etc.

[2]) This axiom can be derived from the remaining ones; therefore we do not give it a separate number.

By axiom I there exists at most one set S satisfying axiom III for a given family of sets $\boldsymbol{A}$. In fact, if besides (1)

$$x \in S_1 \equiv \bigvee_X [(x \in X) \wedge (X \in \boldsymbol{A})],$$

then for any x

$$x \in S_1 \equiv x \in S,$$

and, by axiom I, $S_1 = S$.

Axiom III states that there exists at least one set S satisfying formula (1). We infer that for any $\boldsymbol{A}$ this set is unique. It is called the *union* of the sets belonging to $\boldsymbol{A}$ and is denoted by $\bigcup \boldsymbol{A}$ or $\bigcup_{X \in \boldsymbol{A}} X$.

IV. (AXIOM OF POWER SETS) *For every set A there exists a family of sets $\boldsymbol{P}$ which consists exactly of all the subsets of the set A*:

$$(X \in \boldsymbol{P}) \equiv (X \subset A).$$

It is easy to prove that the set $\boldsymbol{P}$ is uniquely determined by A. This set $\boldsymbol{P}$ is called the *power set* of A and is denoted by $\boldsymbol{P}(A)$.

V. (AXIOM OF INFINITY) *There exists a family of sets $\boldsymbol{A}$ satisfying the conditions*: $\emptyset \in \boldsymbol{A}$; *if $X \in \boldsymbol{A}$, then there exists an element $Y \in \boldsymbol{A}$ such that Y consists exactly of all the elements of X and the set X itself.*

Symbolically:

$$\bigvee_{\boldsymbol{A}} \Big((\emptyset \in \boldsymbol{A}) \wedge \bigwedge_{X \in \boldsymbol{A}} \bigvee_{Y \in \boldsymbol{A}} \bigwedge_x \{(x \in Y) \equiv [(x \in X) \vee (x = X)]\}\Big).$$

Thus the set $\emptyset$ belongs to $\boldsymbol{A}$, moreover, the set N_1 whose unique element is $\emptyset$ also belongs to $\boldsymbol{A}$. Similarly the set N_2 whose elements are $\emptyset$ and N_1 belongs to $\boldsymbol{A}$, etc.

VI. (AXIOM OF CHOICE) *For every family $\boldsymbol{A}$ of disjoint non-empty sets there exists a set B which has exactly one element in common with each set belonging to $\boldsymbol{A}$*:

$$\{\bigwedge_{X, Y \in \boldsymbol{A}} [X \neq \emptyset] \wedge [(X \neq Y) \to (X \cap Y = \emptyset)]\}$$
$$\to \bigvee_B \bigwedge_{E \in \boldsymbol{A}} \bigvee_x \bigwedge_y [(y \in B \cap E) \equiv (y = x)].$$

For easier reading of the formula, observe that the formula $\bigvee_x \bigwedge_y [(y \in B \cap E) \equiv (y = x)]$ states that there exists an element x such that the conditions $y \in B \cap E$ and $y = x$ are equivalent. Thus the element x is the unique element of the intersection $B \cap E$ and the formula above asserts that this intersection contains exactly one element.

Not all mathematicians accept the axiom of choice without reserve, some of them view this axiom with a certain measure of distrust.[1])

Usually, we shall mark theorems which are proved using the axiom of choice with the superscript °. As a rule the axiom of choice will be used only in proofs of theorems which have not yet been proved without the use of this axiom.

For each formula $\Phi(x)$ of the class K we assume the following axiom:[2])

VI'_{Φ}. (AXIOM OF SUBSETS FOR THE FORMULA Φ). *For any set A there exists a set which contains the elements of A satisfying the formula Φ and which contains no other elements.*

Symbolically this axiom can be written in the following form:

There exists a set B such that $\bigwedge_x \{(x \in B) \equiv [(x \in A) \wedge \Phi(x)]\}$

(we suppose that the variable B does not occur in Φ).

[1]) The axiom of choice occupies a rather special place among set theoretical axioms. Although it was subconsciously used very early, it was explicitly formulated as late as 1904 (see Zermelo [1]) and immediately aroused a controversy. Several mathematicians claimed that proofs involving the axiom of choice have a different nature from proofs not involving it, because the axiom of choice is a unique set theoretical principle which states the existence of a set without giving a method of defining ("constructing") it, i.e. is not effective. In the ensuing discussion in which the leading mathematicians at the beginning of the present century took part it became clear that there is no unique intuitive notion of a set. Thus one of the results of the discussion was the growing conviction among mathematicians that it is necessary to build set theory in the axiomatic way. For a detailed account of the discussion mentioned above see Fraenkel, Bar-Hillel and Lévy [1]. Many equivalent formulations of the axiom of choice are known; see Rubin and Rubin [1] and Jech [2]; there is also an extensive bibliography of works dealing with problems of the logical independence of the axiom of choice and of several statements weaker than this axiom but related to it; see Jech [2].

Stressing all the places in which the axiom of choice is used originated with Sierpiński who, since 1918, has published numerous papers dealing with the axiom of choice; see Sierpiński [23].

[2]) This axiom is dependent on the remaining ones, therefore we do not give it a separate number.

If $\Phi(x)$ contains free variables different from x, then they act as parameters upon which B depends.

Clearly, the set B is uniquely determined by Φ. It is denoted by $\{x \in A: \Phi(x)\}$, which is read "the set of x which belong to A and for which $\Phi(x)$".

Finally, for every formula of the class K in which variables z and B do not occur we have the following axiom:

VII$_\Phi$. (AXIOM OF REPLACEMENT FOR THE FORMULA Φ) *If for every x there exists exactly one y such that $\Phi(x, y)$ holds, then for every set A there exists a set B which contains those and only those elements y for which the condition $\Phi(x, y)$ holds for some $x \in A$.*

Symbolically:

$$\{\bigwedge_x \bigvee_z \bigwedge_y [\Phi(x, y) \equiv (y = z)]\} \rightarrow \bigwedge_A \bigvee_B \bigwedge_y [(y \in B) \equiv \bigvee_{x \in A} \Phi(x, y)].$$

The intuitive content of this axiom is as follows. Suppose that the antecedent of the implication holds, namely, for every x there exists exactly one element y satisfying $\Phi(x, y)$. In this case we say that y corresponds to x. The axiom states that for every set A there exists a set B which contains all the elements y corresponding to elements of A and which contains no other elements.

For instance, if $\Phi(X, Y)$ is the formula $Z(X) \wedge (Y = \boldsymbol{P}(X))$, then the element corresponding to the set X is the power set $\boldsymbol{P}(X)$. By the axiom of replacement, for every family of sets $\boldsymbol{A}$ there exists a family of sets $\boldsymbol{B}$ which consists of all the power sets $\boldsymbol{P}(X)$ where $X \in \boldsymbol{A}$.

Observe that we have defined not just one axiom of replacement but actually an infinite number of them. In fact, for every formula belonging to the class K and not involving the variables B and z we have a separate axiom. Similarly, axioms VI$'_\Phi$ form an infinite collection of axioms.

If the uniqueness condition occurring in the antecedent of axiom VII$_\Phi$ is satisfied, then the set B whose existence is asserted in the conclusion of the axiom is unique. The proof follows immediately from axiom I. We call the set B the *image* of A under the formula Φ and we denote it by $\{\Phi\}$ "A.

Axioms I–VI and all of the axioms VII$_\Phi$ where Φ is an arbitrary formula of the class K constitute an infinite system of axioms, which will be denoted by Σ°. Eliminating the axiom of choice from the system Σ° we obtain the axiom system denoted by Σ.

As already mentioned on p. 48, we can introduce new primitive notions $P, Q, \ldots$ into set theory. Then we use axioms I–VI and all axioms of the form VII_Φ, where Φ is an arbitrary formula of the class $K[P, Q, \ldots]$. This system, which includes the system Σ°, will be denoted by $\Sigma^\circ[P, Q, \ldots]$. If we omit the axiom of choice then we shall be dealing with the system which we denote by $\Sigma[P, Q, \ldots]$.

We can fully appraise the meaning which individual axioms have for set theory only after acquainting ourselves with conclusions which follow from them. At this time we shall be content to make a few remarks of a general character.

Axioms III, IV, VI, and VII are axioms of conditional existence, that is, they allow us to conclude the existence of certain sets assuming the existence of others.

Constructions of sets based on axioms III, IV and VII are unique. On the other hand, axiom VI does not assert the existence of a unique set: for a given family A of non-empty disjoint sets there exist, in general, many sets B satisfying the axiom of choice.

Axioms II and V are existential axioms: they postulate the existence of certain sets independently of any assumptions concerning the existence of other sets.

We make a few more remarks of a more general nature. Axioms in mathematical theories can play one of two roles. There are cases where the axioms completely characterize the theory, i.e., they constitute in some sense a definition of the primitive notions of the theory. Such is the case, for instance, in group theory: we define a group as a set and operations satisfying the axioms of group theory. In other cases the axioms formalize only certain chosen properties of the primitive notions of the theory. In this case the purpose of the axioms is not to give a complete description of the primitive notions but rather to give a systematization of the intuitive concept. This second point of view is taken in this book. There arise, of course, problems of a philosophical nature related to establishing the intuitive truth of the axioms. However, we shall not treat these problems here.

Since the intuitive content of the notion of set is not completely characterized by axioms I–VII, it is not surprising that the axioms do not suffice for establishing certain results from intuitive set theory. As an example of an intuitively obvious property of sets which cannot be derived from the axioms I–VII we give the following:

If A is a non-empty family of sets, then there exists a set X such that

$$X \in A \quad \textit{and} \quad X \cap A = \emptyset .$$

Many authors accept the so-called *axiom of regularity*, which states that every family of sets has the property described above.

From the axiom of regularity it follows in particular that $X \in X$ for no X and, more generally, $X_1 \in X_2 \in \ldots \in X_n \in X_1$ for no $X_1, X_2, \ldots, X_n$.

In this book we will not use the axiom of regularity.

In Chapter VIII–X we shall consider certain other axioms independent of the axioms I–VII.[1])

[1]) The first axiomatization of set theory is due to Zermelo [2]; the primitive notions in this system were "set" and $\in$ as in the system Σ°. Zermelo did not have axiom VII and used only axiom VI′ which he called the "*Aussonderungsaxiom*". The formulation of this axiom given by Zermelo was rather ambiguous and gave rise to serious discussions. The formulation which we have adopted was proposed by Skolem [2]. It should be stressed that in this formulation we do not have a single axiom VII and a single axiom VI′ but schemata of axioms, each depending on an arbitrary formula; for this reason it was necessary to formulate correctly the definition of set-theoretical formulas because without this definition the axioms would not be unambiguously determined.

Axiom VII (of replacement) was proposed independently by Mirimanoff [1], Fraenkel [1] and Skolem [1]; Fraenkel's paper was most influential and for this reason the axiom is usually credited to Fraenkel. The axiom of regularity was proposed first by Mirimanoff [1].

Most recent expositions of set theory are based on a system of axioms called the *Zermelo–Fraenkel system* and denoted by ZFC. The system can be obtained from Σ° by adjoining the axiom of regularity as well as the assumption "every x is a set"; in connection with this last axiom the primitive notion of a "set" can be eliminated.

The reason why we did not assume the axiom of regularity is our conviction that it is very unnatural to exclude the possibility that there exist objects which are not sets. It is to be noted that there appear recently papers which explicitly use such objects; see e.g. Barwise [1].

Von Neumann [2] and [3] proposed another way of making precise the unclear axiom VI′ of Zermelo and showed that the axiom of replacement can be similarly reformulated. Instead of using formulas as did Skolem, von Neumann admitted

Exercise

Show that axioms III, IV and VII can be replaced by a single axiom stating the existence of the set $S(\{\Phi\}``P(A))$ for every set A and for every propositional function Φ belonging to K [Hao Wang].

§ 3. Some simple consequences of the axioms

Starting with this section, our exposition of set theory will be based upon the axioms of the system $\Sigma°$.

THEOREM 1: (THE EXISTENCE OF PAIR) *For arbitrary a, b there exists a unique set whose only elements are a and b.*

PROOF. Uniqueness follows from axiom I and existence from axiom II′.

The set, whose uniqueness and existence are stated in Theorem 1, is called the *unordered pair* of elements a, b and is denoted by $\{a, b\}$. If $a = b$ then we simply write $\{a\}$.

THEOREM 2: (THE EXISTENCE OF UNION) *For arbitrary sets A and B there exists a set C such that*

$$(x \in C) \equiv [(x \in A) \vee (x \in B)].$$

In fact, $C = \bigcup_{X \in \{A, B\}} X$.

Theorem 2 asserts that axiom A (p. 6) is a consequence of the axioms Σ.

THEOREM 3: (THE EXISTENCE OF UNORDERED TRIPLES, QUADRUPLES, ETC.) *For arbitrary, $a, b, c, \ldots, m$ there exist sets: $\{a, b, c\}$ whose only elements are a, b and c; $\{a, b, c, d\}$ whose only elements are a, b, c and $d, \ldots$; $\{a, b, \ldots, m\}$ whose only elements are $a, b, \ldots, m$.*

In fact, $\{a, b, c\} = \{a, b\} \cup \{c\}$, $\{a, b, c, d\} = \{a, b, c\} \cup \{d\}$, etc.

The set

$$\langle a, b \rangle = \{\{a\}, \{a, b\}\} \tag{1}$$

a new primitive notion into set theory. This system was later reformulated by Bernays [1] and Gödel [1]. The advantage of the resulting system, called in the literature "GB system", is that it is based on a finite number of axioms. Still another system of axioms was proposed by Morse [1]; see also Kelley [2].

For a critical exposition of the axioms of set theory now in use see Fraenkel, Bar-Hillel and Lévy [1] and from a different point of view Quine [1].

is called an *ordered pair*.[1]) We call a the *first term* of $\langle a, b\rangle$ and b the *second term* of $\langle a, b\rangle$.

THEOREM 4: *In order that $\langle a, b\rangle = \langle c, d\rangle$ it is necessary and sufficient that $a = c$ and $b = d$.*

PROOF. The sufficiency of the condition is obvious. To prove its necessity, suppose that $\langle a, b\rangle = \langle c, d\rangle$. By means of (1), it follows that

$$\{c\} \in \langle a, b\rangle \quad \text{and} \quad \{c, d\} \in \langle a, b\rangle,$$

that is,

(i) $\{c\} = \{a\}$ or (ii) $\{c\} = \{a, b\}$,

and

(iii) $\{c, d\} = \{a\}$ or (iv) $\{c, d\} = \{a, b\}$.

Formula (ii) holds if $a = c = b$. Formulas (iii) and (iv) are then equivalent and it follows that $c = d = a$. Hence we obtain $a = c = b = d$ in which case the theorem holds. Similarly, one can check that the theorem holds for case (iii). It remains to be shown that the theorem holds for cases (i) and (iv). We have then $c = a$ and either $c = b$ or $d = b$. If $c = b$ then (ii) holds and this case has already been considered. If $d = b$ then $a = c$ and $b = d$, which proves the theorem.

COROLLARY: *If $\langle a, b\rangle = \langle b, a\rangle$ then $a = b$.*

By the definition of the set $\{x \in A: \Phi(x)\}$ axiom VI'_Φ implies the following theorem.

THEOREM 5:

$$t \in \{x \in A: \Phi(x)\} \equiv [\Phi(t) \wedge (t \in A)]. \tag{2}$$

In particular, if $\Phi(x) \to (x \in A)$ (in which case we say that the domain of Φ is limited to A), then

$$t \in \{x: \Phi(x)\} \equiv \Phi(t). \tag{3}$$

Equivalence (3) leads easily to the following theorems (with the assumption that the domains of Φ and Ψ are limited to A):

$$\{x: \Phi(x) \vee \Psi(x)\} = \{x: \Phi(x)\} \cup \{x: \Psi(x)\}, \tag{4}$$

$$\{x: \Phi(x) \wedge \Psi(x)\} = \{x: \Phi(x)\} \cap \{x: \Psi(x)\}, \tag{5}$$

$$\{x: \neg\Phi(x)\} = A - \{x: \Phi(x)\}. \tag{6}$$

[1]) The definition of an ordered pair given here is due to Kuratowski [4]; see also Wiener [1], where a closely related definition was also proposed.

As an example we shall prove (4). For this purpose we apply equivalence (3) to the formula $\Phi(x)\vee\Psi(x)$ and we obtain

$$t\in\{x\colon\Phi(x)\vee\Psi(x)\}\equiv[\Phi(t)\vee\Psi(t)]. \tag{i}$$

According to (3) we have

$$\Phi(t)\equiv t\in\{x\colon\Phi(x)\}\quad\text{and}\quad\Psi(t)\equiv t\in\{x\colon\Psi(x)\};$$

thus it follows from (i) that

$$\begin{aligned} t\in\{x\colon\Phi(x)\vee\Psi(x)\} &\equiv t\in\{x\colon\Phi(x)\}\vee t\in\{x\colon\Psi(x)\}\\ &\equiv t\in\{x\colon\Phi(x)\}\cup\{x\colon\Psi(x)\}, \end{aligned}$$

which proves (4).

THEOREM 6: *For every non-empty family of sets A there exists a unique set containing just those elements which are common to all the sets of the family A.*

This set is called the *intersection* of the sets belonging to the family A and is denoted by $\bigcap(A)$ or $\bigcap_{X\in A}X$.

For we have:

$$\bigcap(A)=\{x\in\bigcup(A)\colon\bigwedge_X[(X\in A)\rightarrow(x\in X)]\}.$$

If the family A is composed of sets $X_1, X_2, \ldots, X_n$ (n finite), then $\bigcap(A)=X_1\cap X_2\cap\ldots\cap X_n$. In case $A=\emptyset$, the operation $\bigcap(A)$ is not performable.

We conclude this section with a remark on so-called *antinomies of set theory*. A naïve intuition of set would incline us to accept an axiom (stronger than axiom VI'_{Φ}) stating that for any formula there exists a set B containing those and only those elements which satisfy this formula.

The creator of set theory, Cantor, believed (at least at the beginning of his work) that such an axiom was true.

However, it soon became apparent that the axiom formulated in this way leads to a contradiction (to an *antinomy*).

Let us take as an example the formula

$$\Phi(x) \equiv (x \text{ is a set}) \wedge (x \notin x),$$

which leads to *Russell's antinomy*.[1])

We shall prove

THEOREM 7: *There exists no set Z such that*

$$\bigwedge_x [(x \in Z) \equiv \Phi(x)].$$

PROOF. If such a set existed, then the equivalence

$$(x \in Z) \equiv (x \text{ is a set}) \wedge (x \notin x)$$

would hold. From the assumption that Z is a set we obtain the contradiction $Z \in Z \equiv \neg(Z \in Z)$.

Interdependence of the axioms

We have shown that axiom A (p. 6) follows from Σ. Axiom B (p. 6) follows from Σ as well, because $A - B = \{x \in A: \neg(x \in B)\}$. Axiom C (p. 6) follows directly from axiom II (the axiom of empty set) or from axiom V (the axiom of infinity).

Axiom II′ follows from the other axioms of the system Σ. In fact, let A be a family of sets such that $\emptyset \in A$ and such that there exists at least one non-empty set belonging to A. Such a family exists by the axiom of infinity. The set $\{a, b\}$ is the set $\{\Phi\}$"A where Φ is the formula

$$[(x = \emptyset) \wedge (y = a)] \vee [(x \neq \emptyset) \wedge (y = b)].$$

Axiom VI'_Φ (of subsets) is also a consequence of the other axioms of the system Σ. In fact, let A be a set and $\Phi(x)$ a formula. If $\bigwedge_x [(x \in A) \to \neg\Phi(x)]$, then the empty set satisfies the axiom of subsets. Otherwise, let a be an arbitrary element of A such that $\Phi(a)$. Denote by $\Psi(x, y)$ the formula $[\Phi(x) \wedge (y = x)] \vee [\neg\Phi(x) \wedge (y = a)]$. For every x there exists exactly one y such that $\Psi(x, y)$; namely this element y is x or a, depending on whether $\Phi(x)$ or $\neg\Phi(x)$. The set $\{\Psi\}$"A clearly satisfies the thesis of axiom VI'_Φ.

[1]) Russell's antinomy was first published in an appendix to Frege [2]. For our statement that Cantor believed in the truth of the contradictory axiom: "there is a set X of all objects x satisfying $\Phi(x)$", see Cantor [2].

Exercises

Show that

1. If $X \in A$, then $\bigcap(A) \subset X \subset \bigcup(A)$.

2. $\bigcup(A_1 \cup A_2) = \bigcup(A_1) \cup \bigcup(A_2)$.

3. If $A_1 \cap A_2 \neq \emptyset$, then $\bigcap(A_1 \cap A_2) \supset \bigcap(A_1) \cap \bigcap(A_2)$.

§ 4. Cartesian products. Relations

The *cartesian product* of two sets X and Y is defined to be the set of all ordered pairs $\langle x, y\rangle$ such that $x \in X$ and $y \in Y$.

The existence of this set can be proved as follows. If $x \in X$ and $y \in Y$, then $\{x, y\} \subset X \cup Y$ and $\{x\} \subset X \cup Y$, whence

$$\langle x, y\rangle = \{\{x\}, \{x, y\}\} \in \boldsymbol{P}(\boldsymbol{P}(X \cup Y)).$$

The set

$$\{t \in T: \bigvee_{x\in X} \bigvee_{y\in Y} (t = \langle x, y\rangle)\}, \quad \text{where} \quad T = \boldsymbol{P}(\boldsymbol{P}(X \cup Y)),$$

exists by means of axioms IV, VI′ and Theorem 2. This set contains every ordered pair $\langle x, y\rangle$, where $x \in X$, $y \in Y$, and contains no other elements. Hence this set is the cartesian product of X and Y.

Since there exists at most one set containing exactly the pairs $\langle x, y\rangle$, $x \in X$, $y \in Y$, the cartesian product is uniquely determined by X and Y. This product is denoted by $X \times Y$.

If $X = \emptyset$ or $Y = \emptyset$, then obviously $X \times Y = \emptyset$.

In spite of the arbitrary nature of X and Y, their cartesian product can be treated in geometrical terms: the elements of the set $X \times Y$ are called *points*, the sets X and Y the *coordinate axes*. If $z = \langle x, y\rangle$ then x is called the *abscissa* and y the *ordinate* of z. The fact that the set of points in a plane can be treated as the cartesian product $\mathscr{E} \times \mathscr{E}$ where $\mathscr{E}$ is the set of real numbers justifies the use of this terminology.

Certain properties of cartesian products are similar to the properties of multiplication of numbers. For instance, the distributive laws hold:

$$(X_1 \cup X_2) \times Y = X_1 \times Y \cup X_2 \times Y,$$
$$Y \times (X_1 \cup X_2) = Y \times X_1 \cup Y \times X_2,$$
$$(X_1 - X_2) \times Y = X_1 \times Y - X_2 \times Y,$$
$$Y \times (X_1 - X_2) = Y \times X_1 - Y \times X_2.$$

As an example we shall prove the first of these equations:

$$\begin{aligned}\langle x, y\rangle \in (X_1 \cup X_2)\times Y &\equiv (x \in X_1 \cup X_2) \wedge (y \in Y)\\ &\equiv (x \in X_1 \vee x \in X_2) \wedge (y \in Y)\\ &\equiv (x \in X_1 \wedge y \in Y) \vee (x \in X_2 \wedge y \in Y)\\ &\equiv (\langle x, y\rangle \in X_1 \times Y) \vee (\langle x, y\rangle \in X_2 \times Y)\\ &\equiv \langle x, y\rangle \in (X_1 \times Y \cup X_2 \times Y).\end{aligned}$$

The cartesian product is distributive over intersection:

$$(X_1 \cap X_2)\times Y = (X_1 \times Y) \cap (X_2 \times Y),$$
$$Y\times (X_1 \cap X_2) = (Y\times X_1) \cap (Y\times X_2).$$

The proof is similar to the previous one.

The cartesian product is monotone with respect to the inclusion relation, that is,

(∗) *If* $Y \neq \emptyset$, *then* $(X_1 \subset X_2) \equiv (X_1 \times Y \subset X_2 \times Y) \equiv (Y\times X_1 \subset Y\times X_2)$.

In fact, let $y \in Y$. Suppose that $X_1 \subset X_2$. Since (for $i = 1, 2$)

$$(\langle x, y\rangle \in X_i \times Y) \equiv (x \in X_i) \wedge (y \in Y),$$

we have the following implication

$$(\langle x, y\rangle \in X_1 \times Y) \rightarrow (\langle x, y\rangle \in X_2 \times Y);$$

hence $X_1 \times Y \subset X_2 \times Y$.

Conversely, if $X_1 \times Y \subset X_2 \times Y$ and $y \in Y$, then

$$\begin{aligned}(x \in X_1) \rightarrow (x \in X_1) \wedge (y \in Y) \equiv (\langle x, y\rangle \in X_1 \times Y) &\rightarrow (\langle x, y\rangle \in X_2 \times Y)\\ &\equiv (x \in X_2) \wedge (y \in Y) \rightarrow (x \in X_2);\end{aligned}$$

thus $X_1 \subset X_2$.

The proof of the second part of (∗) is similar.

Using cartesian products, we can perform certain logical transformations. For instance, the formulas (p. 50)

$$\bigwedge_x \bigwedge_y \Phi(x, y) \equiv \bigwedge_{xy} \Phi(x, y) \equiv \bigwedge_z \Phi(z),$$

$$\bigvee_x \bigvee_y \Phi(x, y) \equiv \bigvee_{xy} \Phi(x, y) \equiv \bigvee_z \Phi(z)$$

allow us to replace two consecutive universal or existential quantifiers by one quantifier binding the variable $z = \langle x, y\rangle$ which runs over the cartesian product $X \times Y$.

A subset R of a cartesian product $X \times Y$ is called a (*binary*) *relation*.

Instead of writing $\langle a, b\rangle \in R$, where R denotes a relation, we sometimes write aRb and read: *a is in the relation R to b*, or *the relation R holds between a and b*.

The *left domain* (D_l) (or simply the *domain*) of a relation R is defined to be the set of all x such that $\langle x, y\rangle \in R$ for some y; the *right domain* (D_r) is the set of all y such that $\langle x, y\rangle \in R$ for some x. The right domain of a relation is sometimes called the *range*, or the *counter-domain*, or the *converse domain*. The union $F(R)$ of the left and right domains of R is called the *field* of R.

In geometrical terminology we say that D_l is the *projection* of R on the X-axis and D_r is the projection of R on the Y-axis.

Thus we have

$$(1)\quad D_l = \Big\{x \in X: \bigvee_y \langle x, y\rangle \in R\Big\}, \qquad D_r = \Big\{y \in Y: \bigvee_x \langle x, y\rangle \in R\Big\}.$$

These formulas prove the existence of the sets D_l and D_r.

If the free variables of the formula $\Phi(x, y)$ are limited to the sets X and Y respectively, then the set $R = \{\langle x, y\rangle: \Phi(x, y)\}$ is a relation. Clearly, $\Phi(x, y) \equiv xRy \equiv \langle x, y\rangle \in R$. Hence

THEOREM: *The projection of the set* $\{\langle x, y\rangle: \Phi(x, y)\}$ *on the X-axis is the set* $\Big\{x: \bigvee_y \Phi(x, y)\Big\}$.

The relation $\{\langle x, y\rangle: yRx\}$ is called the *inverse* of R and is denoted by R^i. Obviously, $D_l(R^i) = D_r(R)$ and $D_r(R^i) = D_l(R)$.

The relation $\Big\{\langle x, y\rangle: \bigvee_z (xSz \wedge zRy)\Big\}$ is called the *composition* of R and S and is denoted by $R\circ S$.[1]) Obviously, $D_l(R\circ S) \subset D_l(S)$ and $D_r(R\circ S) \subset D_r(R)$.

[1]) We use this notation instead of the more natural $S\circ R$, because when writing down a transformation, we normally put the symbol of the operation carried out first in the second place (e.g. $\sin(\log x)$).

The operation $\circ$ is associative. In fact,

$$\begin{aligned} x(R\circ S)\circ Ty &\equiv \bigvee_z (xTz \wedge zR\circ Sy) \\ &\equiv \bigvee_z \bigvee_t (xTz \wedge zSt \wedge tRy) \\ &\equiv \bigvee_t \bigvee_z (xTz \wedge zSt \wedge tRy) \\ &\equiv \bigvee_t \Big[\bigvee_z (xTz \wedge zSt) \wedge tRy\Big] \\ &\equiv \bigvee_t (xS\circ Tt \wedge tRy) \\ &\equiv xR\circ(S\circ T)y. \end{aligned}$$

Because of the associativity of $\circ$ we may omit parentheses in expressions of the form $R\circ S\circ \ldots \circ U$.

We shall prove the formula

$$(R\circ S)^i = S^i \circ R^i.$$

In fact,

$$\begin{aligned} x(R\circ S)^i y &\equiv yR\circ Sx \\ &\equiv \bigvee_z (ySz \wedge zRx) \\ &\equiv \bigvee_z (xR^i z \circ zS^i y) \\ &\equiv xS^i \circ R^i y. \end{aligned}$$

Other properties of the operations ^{i} and $\circ$ are given in the exercises.

Examples and exercises

1. Let $X = Y = \mathscr{E}$ (the set of real numbers). The set $\{\langle x, y\rangle: x < y\}$ is that part of the plane which lies above the straight line $x = y$. The set $\{\langle x, y\rangle: y = x^2\}$ is a parabola, its projection on the Y-axis is the set $\{y: \bigvee_x (y = x^2)\}$.

2. Let A be a family of subsets of $X\times Y$. Let $F(Z)$ denote the projection of the set Z (where $Z \subset X\times Y$) on the X-axis and $F(A)$ the family of all projections $F(Z)$, $Z \in A$. Prove that

$$F[\bigcup (A)] = \bigcup [F(A)],$$

i.e. the projection of a union is equal to the union of the projections.

3. Give an example showing that the projection of an intersection may be different from the intersection of the projections.

4. Prove the formulas $(R\cup S)^i = R^i\cup S^i$, $(R\cap S)^i = R^i\cap S^i$ and $(R^i)^i = R$.

5. Prove the formulas

$$(R\cup S)\circ T = (R\circ T)\cup(S\circ T), \quad T\circ(R\cup S) = (T\circ R)\cup(T\circ S),$$

$$(R\cap S)\circ T \subset (R\circ T)\cap(S\circ T), \quad T\circ(R\cap S) \subset (T\circ R)\cap(T\circ S).$$

6. Prove that $(X\times Y)^i = Y\times X$. Compute $(X\times Y)\circ(Z\times T)$.

§ 5. Equivalence relations. Partitions

A relation R is called an *equivalence relation*[1]) if for all $x, y, z \in F(R)$, the following conditions are satisfied:

$$\begin{array}{ll} xRx & (\textit{reflexivity}), \\ xRy \to yRx & (\textit{symmetry}), \\ xRy \wedge yRz \to xRz & (\textit{transitivity}). \end{array}$$

Examples

1. Let x, y be straight lines lying in a plane. Let xRy if and only if x is parallel to y. Then R is an equivalence relation.

2. Let C be a set of Cauchy sequences $\langle a_1, a_2, \ldots, a_n, \ldots\rangle$ of rational numbers. The relation R which holds between two sequences if and only if $\lim(a_n - b_n) = 0$ is an equivalence relation.

3. Let X be the set of real numbers x such that $0 \leqslant x < 1$. The relation R which holds between two numbers $a, b \in X$ if and only if the difference $a-b$ is a rational number is an equivalence relation.[2])

4. Let X be any set, $\boldsymbol{K} = \boldsymbol{P}(X)$ and let $\boldsymbol{I}$ be an ideal in $\boldsymbol{K}$ (see p. 17). The relation $\doteq$ which holds between two sets $X, Y \in \boldsymbol{K}$ if and only if $X \dot{-} Y \in \boldsymbol{I}$ is an equivalence relation.

5. Example 4 can be generalized in the following way. Let $\boldsymbol{K}$ be an arbitrary Boolean algebra and $\boldsymbol{I}$ any subset of $\boldsymbol{K}$ satisfying the conditions:

$$a \leqslant b \in \boldsymbol{I} \to a \in \boldsymbol{I}, \quad (a \in \boldsymbol{I}) \wedge (b \in \boldsymbol{I}) \to (a \vee b \in \boldsymbol{I}).$$

[1]) The notion of an equivalence relation and of equivalence classes was first investigated in full generality by Frege [1].

[2]) Example 3 is due to Vitali [1] who proved that no set of representatives of the relation defined in Example 3 is Lebesgue measurable.

Then $\boldsymbol{I}$ is called an ideal of $\boldsymbol{K}$; the relation $\doteq$ (Example 4) is an equivalence relation.

We shall now give theorems which describe the structure of an arbitrary equivalence relation.

Let C be any set. A family $\boldsymbol{A}$ of subsets of C ($\boldsymbol{A} \subset \boldsymbol{P}(C)$) is called a *partition* of C if $\emptyset \notin \boldsymbol{A}$, $\bigcup(\boldsymbol{A}) = C$ and the sets belonging to $\boldsymbol{A}$ are pairwise disjoint (i.e. for any $X, Y \in \boldsymbol{A}$ either $X = Y$ or $X \cap Y = \emptyset$).

THEOREM 1: *If $\boldsymbol{A}$ is a partition of C, then the relation $R_{\boldsymbol{A}}$ defined by the formula*

$$xR_{\boldsymbol{A}}y \equiv \bigvee_{Y \in \boldsymbol{A}} [(x \in Y) \wedge (y \in Y)]$$

is an equivalence relation whose field is C.

The proof of this theorem is left to the reader.

THEOREM 2: *If $\boldsymbol{A}$ and $\boldsymbol{B}$ are two different partitions of C, then $R_{\boldsymbol{A}} \neq R_{\boldsymbol{B}}$.*

PROOF. Suppose that $R_{\boldsymbol{A}} = R_{\boldsymbol{B}}$; we shall show that $\boldsymbol{A} = \boldsymbol{B}$. Because of the symmetry of the assumptions it suffices to prove that $\boldsymbol{A} \subset \boldsymbol{B}$. So let $Y \in \boldsymbol{A}$ and let $y \in Y$. Since $\bigcup(\boldsymbol{B}) = C$, there exists $Z \in \boldsymbol{B}$ such that $y \in Z$. If $x \in Y$ then $xR_{\boldsymbol{A}}y$ and hence $xR_{\boldsymbol{B}}y$. Because Z is the unique element of $\boldsymbol{B}$ containing y, we have $x \in Z$. Similarly we can show that $x \in Z \to x \in Y$, which proves that $Y = Z$ and hence $Y \in \boldsymbol{B}$.

THEOREM 3: *For any equivalence relation R with field $C \neq \emptyset$ there exists a partition $\boldsymbol{A}$ of the set C such that $R = R_{\boldsymbol{A}}$.*

PROOF. Let

$$\boldsymbol{A} = \left\{Y \subset C: \bigvee_{y \in C} \bigwedge_{u \in C} (uRy \equiv u \in Y)\right\}.$$

Because of the reflexivity of R the elements of the family $\boldsymbol{A}$ are non-empty and $\bigcup(\boldsymbol{A}) = C$. If $Y \in \boldsymbol{A}$ and $Z \in \boldsymbol{A}$, then for some $y, z \in C$ the following formulas hold:

$$\bigvee_{u} (u \in Y \equiv uRy), \qquad \bigwedge_{u} (u \in Z = uRz).$$

From the symmetry and transitivity of R we infer that if Y and Z have a common element, then they are identical. This proves that the family $\boldsymbol{A}$ is a partition of C.

We show now that $R = R_{\boldsymbol{A}}$.

Suppose that uRv. Denoting the set $\{z \in C: zRu\}$ by Y_u, we obtain $Y_u \in A$ and $v \in Y_u$; hence uR_Av and $R \subset R_A$.

Now suppose that uR_Av; then there exist Y in A and an element y such that $\bigwedge_z (zRy \equiv z \in Y)$ and $u \in Y$ and $v \in Y$. Therefore uRy and vRy; by the symmetry and transitivity of R, it follows that uRv. This proves that $R_A \subset R$. Hence $R = R_A$, Q.E.D.

It follows from Theorems 1–3 that every equivalence relation with field $C \neq \emptyset$ defines exactly one partition A of the set C and vice versa.

If $R = R_A$ then sets of the family A are called *equivalence classes* of R. The equivalence class containing an element x is denoted by x/R, the family A itself is denoted by C/R. This family is called the *quotient class of* C *with respect to* R.

Examples

For the relation R of Example 1 each equivalence class consists of all straight lines lying in the same direction (i.e. mutually parallel). For the relation R of Example 2 each equivalence class consists of all sequences of rational numbers convergent to the same real number. Cantor defined real numbers as the equivalence classes with respect to this relation.

A *set of representatives* of an equivalence relation with field C is a subset of C which has exactly one element in common with each equivalence class.

The existence of a set of representatives for any equivalence relation follows from the axiom of choice. Without the axiom of choice we cannot prove the existence of a set of representatives even for very simple relations. Such is the case for the relation of Example 3.

Exercises

1. Let $I = \{x: 0 \leqslant x < 1\}$; for $X \subset I$, let $X(r)$ denote the set of numbers belonging to I and having the form $x+r+n$ where $x \in X$ and n is an integer. Show that if Z is a set of representatives for the relation R of Example 3, then

a) $Z(r) \cap Z(s) = \emptyset$ for all rational numbers r, s $(r \neq s)$;

b) $I = \bigcup_r Z(r)$, where the union is over all rational numbers.

2. Show that the condition $R_A \subset R_B$ is equivalent to the following: every set $Y \in A$ is the union of some family $A' \subset B$.

3. Show that if $\boldsymbol{M}$ is a non-empty family of equivalence relations with common field C, then $\bigcap(\boldsymbol{M})$ is also an equivalence relation with field C.

4. Preserving the notation of Exercise 3 prove that there exists an equivalence relation U with field C such that

a) $R \in \boldsymbol{M} \to R \subset U$;

b) if V is an equivalence relation with field C and $\bigwedge_{R} [R \in \boldsymbol{M} \to R \subset V]$, then $U \subset V$.

5. Assuming in Exercises 3 and 4 that $\boldsymbol{M} = \{R_A, R_B\}$, describe $\bigcup(\boldsymbol{M})$ and U.

§ 6. Functions

A relation $R \subset X \times Y$ is called a *function*[1]) if

(1) $$\bigwedge_{x, y_1, y_2} [xRy_1 \wedge xRy_2 \to (y_1 = y_2)].$$

Functions are denoted by letters $f, g, h, \ldots$. The sets $D_l(f)$ and $D_r(f)$ are called respectively the *domain* and the *range* of the function f. We often write $D(f)$ for $D_l(f)$ and $\mathrm{Rg}(f)$ for $D_r(f)$. The following terminology will be used: if $D_l(f) = X$ and $D_r(f) \subset Y$ then f is called a *mapping* (or a *transformation*) *of* X *into* Y; if, moreover, $D_r(f) = Y$ then f is called a *mapping of* X *onto* Y. When $D_l(f) = X$, we say that the function f *is defined* on X.

The set of all mappings of X into Y is denoted by Y^X. Instead of the formula $f \in Y^X$ we often use the more suggestive formulae $f\colon X \to Y$ or $X \underset{f}{\to} Y$.

If $f \in Y^X$ and $x \in X$, then by the definition of domain there exists at least one element $y \in Y$ such that xfy. On the other hand, it follows from the definition of function that there exists at most one such element. Hence the element y is uniquely determined. It is called the *value* of f at x and is denoted by $f(x)$. Therefore, the formula $y = f(x)$ has the same meaning as xfy.

For f, g belonging to Y^X the following obvious equivalence holds:

$$(f = g) \equiv \bigwedge_{x, y \in X} [f(x) = g(x)].$$

If ordered pairs are identified with points of a plane, the first term with the abscissa and the second term with the ordinate, then it turns out that a function is identical with its graph.

[1]) The definition of a function given in Section 6 is due to Peano [1].

DEFINITION: A function f is said to be a *one-to-one* (or injective) *function* if different elements of the domain have different values under the function f:

(2) $$[f(x_1) = f(x_2)] \to [x_1 = x_2],$$

where x_1 and x_2 are arbitrary elements of the domain.

THEOREM 1: *If $f \in Y^X$ then f^c is a function if and only if f is one-to-one. Moreover, $f^c \in X^{Y_1}$ where Y_1 is the range of f and f^c is also a one-to-one function.*

PROOF. The relation f^c is a function if and only if

$$\bigwedge_{x_1, x_2, y} [yf^c x_1 \wedge yf^c x_2 \to x_1 = x_2],$$

that is, if

$$\bigwedge_{x_1, x_2, y} [y = f(x_1) \wedge y = f(x_2) \to x_1 = x_2].$$

Clearly, this formula is equivalent to (2). The second part of the theorem follows from the formulas for the domain and range of an inverse relation (p. 64).

THEOREM 2: *If $f \in Y^X$ and $g \in Z^Y$ then the relation $g \bigcirc f$ is a function and $g \bigcirc f \in Z^X$ (in other words: if $X \underset{f}{\to} Y \underset{g}{\to} Z$ then $X \underset{g \circ f}{\to} Z$).*

PROOF. The definition of the composition of two relations implies the equivalences

$$\begin{aligned} xg\bigcirc fz &\equiv \bigvee_y [(xfy) \wedge (ygz)] \\ &\equiv \bigvee_y [(f(x) = y) \wedge (g(y) = z)] \\ &\equiv g(f(x)) = z; \end{aligned}$$

it follows that

$$\bigwedge_{x, z_1, z_2} [(xg\bigcirc fz_1) \wedge (xg\bigcirc fz_2) \to z_1 = z_2]$$

and that every element of X belongs to the domain of $g \bigcirc f$.

Since the right domain of this relation is included in the range of g, $g \bigcirc f \in Z^X$.

Theorem 2 implies the following formula

$$g\bigcirc f(x) = g(f(x)) \quad \text{for} \quad x \in X.$$

THEOREM 3: *If $f \in Y^X$, $g \in Z^Y$ and the functions f and g are one-to-one, then their composition is also one-to-one.*

In fact,

$$g(f(x)) = g(f(x')) \to f(x) = f(x') \to x = x'.$$

DEFINITION: A one-to-one function whose domain and range are the same set X is called a *permutation* of the set X.

The simplest permutation of X is the identity permutation I_X, that is, the function defined by the formula $I_X(x) = x$ for all $x \in X$.

THEOREM 4: *If $f \in Y^X$ and f is a one-to-one function, then $f^c \bigcirc f = I_X$ and $f \bigcirc f^c = I_{Y_1}$ where Y_1 is the range of f.*

In fact, the equivalence $f^c(y) = x \equiv f(x) = y$ implies $f^c(f(x)) = x$, thus $f^c \bigcirc f = I_X$. The proof of the second formula is similar.

Let $f \in Y^X$, $g \in Z^X$, $\varphi \in T^Y$ and $\psi \in T^Z$. Hence the range of the function $\varphi \bigcirc f$ is contained in T and the same holds for the function $\psi \bigcirc g$. If $\varphi \bigcirc f = \psi \bigcirc g$ then we say that the following diagram

$$\begin{array}{ccc} X & \xrightarrow{f} & Y \\ {\scriptstyle g}\downarrow & & \downarrow{\scriptstyle \varphi} \\ Z & \xrightarrow[\psi]{} & T \end{array}$$

commutes. This diagram shows that starting with an element $x \in X$ we can obtain the element $\varphi \bigcirc f(x) = \psi \bigcirc g(x)$ in two ways: through an element of the set Y or through an element of the set Z.

DEFINITION: A function g is said to be an *extension* of a function f if $f \subset g$. We also say that f is a *restriction* of g.

THEOREM 5: *In order that $f \subset g$ it is necessary and sufficient that $D_1(f) \subset D_1(g)$ and $f(x) = g(x)$ for all $x \in D_1(f)$.*

PROOF. *Necessity*: Suppose that $f \subset g$. Then $D_1(f) \subset D_1(g)$, for the projection of a subset is a subset of the projection (see p. 65). If $x \in D_1(f)$ and $y = f(x)$ then $\langle x, y\rangle \in f$, hence $\langle x, y\rangle \in g$ and $y = g(x)$.

Sufficiency: Suppose that $D_1(f) \subset D_1(g)$ and $f(x) = g(x)$ for all $x \in D_1(f)$. If $\langle x, y\rangle \in f$ then $y = f(x) = g(x)$, therefore $\langle x, y\rangle \in g$, which shows that $f \subset g$.

The restriction f of g for which $D_1(f) = A$ will be denoted by $g|A$.

The notion of a function should be distinguished from the notion of an operation. By an *operation* we mean a formula $\Phi(x, y)$ with two free variables satisfying the following conditions:

$$\text{(W)} \quad \bigwedge_x \bigvee_y \Phi(x, y), \qquad \bigwedge_{x, y_1, y_2} [\Phi(x, y_1) \wedge \Phi(x, y_2) \rightarrow (y_1 = y_2)].^{1)}$$

These conditions state that for every x there exists exactly one object y such that $\Phi(x, y)$. In general the set of all the pairs $\langle x, y\rangle$ such that $\Phi(x, y)$ does not exist and hence no function f such that $\Phi(x, y) \equiv [y = f(x)]$ exists. For instance, such a function does not exist if $\Phi(x, y)$ is the formula $x = y$. On the other hand, the following theorem holds.

THEOREM 6: *If a formula $\Phi(x, y)$ satisfies conditions* (W) *and A is an arbitrary set, then there exists a function f_A with domain A and such that for arbitrary $x \in A$ and arbitrary y*

$$[y = f_A(x)] \equiv \Phi(x, y).$$

Namely, the required function f_A is the set

$$\{t \in A \times B: \bigvee_{xy} [(t = \langle x, y\rangle) \wedge \Phi(x, y)]\}$$

where B denotes the image of A under the formula Φ (see p. 54).

In particular, if a formula Φ is of the form $\ldots x \ldots = y$ (where on the left-hand side of the equation we have an expression written in terms of the letter x, constants, and operation symbols), then the function f_A will be denoted by $\underset{x \in A}{\boldsymbol{F}} [\ldots x \ldots]$. For example, using this notation,

$$I_X = \underset{x \in X}{\boldsymbol{F}} [x], \quad f = \underset{x \in D(f)}{\boldsymbol{F}} [f(x)].$$

Functions of more than one variable

Let $X \times Y \times Z = X \times (Y \times Z)$, $X \times Y \times Z \times T = X \times (Y \times Z \times T)$ and similarly for any number of sets. If $X = Y = Z$ then instead of $X \times X \times X$ we write X^3 and similarly for $X \times X \times X \times X$. Subsets of the cartesian product of n sets are called *n-ary relations*.

If the domain of a function f is the cartesian product $X \times Y$, then f is said to be a *function of two variables*. Similarly, if the domain of f

[1]) The formula Φ may contain other free variables acting as parameters.

is the cartesian product $X \times Y \times Z$, then we say that f is a function of three variables. Instead of $f(\langle x, y \rangle)$ we write $f(x, y)$.

THEOREM 7: *If a function f is a one-to-one mapping of the set $X \times Y$ onto the set Z, then there exist functions α and β, mapping the set Z onto X and Y respectively, such that $f(\alpha(z), \beta(z)) = z$ for every $z \in Z$.*

PROOF. It suffices to take for α the set of pairs $\langle z, x \rangle$ satisfying the condition $\bigvee_y [f(x, y) = z]$ and for β the set of pairs $\langle z, y \rangle$ satisfying the condition $\bigvee_x [f(x, y) = z]$.

Complex functions and product-functions

Any two functions $f \in Y^X$ and $g \in Z^X$ determine the function $h \in (Y \times Z)^X$, called a *complex function*, denoted by $\langle f, g \rangle$ and defined by the formula

$$h(x) = \langle f(x), g(x) \rangle .$$

More generally, any two functions $f \in Y^X$ and $g \in Z^W$ determine the function $u \in (Y \times Z)^{X \times W}$, called the *product-function*, denoted by $f \times g$ and defined as follows

$$u(x, w) = \langle f(x), g(w) \rangle .$$

To conclude this section we give a formulation of the axiom of choice using the notion of function.

°THEOREM 8: *If A is a non-empty family of sets and $\emptyset \notin A$, then there exists a function $f \in (\bigcup(A))^A$ such that $f(X) \in X$ for every $X \in A$.*

PROOF. Let $h = \underset{X \in A}{F} [\{X\} \times X]$. For $X \in A$ we have thus $h(X) \neq \emptyset$ and, moreover, $h(X) \cap h(Y) = \emptyset$ for $X \neq Y$. Applying the axiom of choice to the family $D_r(h)$, we obtain a set which has exactly one element in common with each set $h(X)$, $X \in A$. As it is easy to show, this set is the required function f.

A function with the properties mentioned in Theorem 8 is called a *choice function* for the family A.

Theorem 8 shows that from the axioms Σ° it is possible to derive the existence of a choice function for an arbitrary non-empty family of sets not containing the empty set. Conversely, it can be shown that the axiom of choice follows from Theorem 8 and the axioms Σ.

Exercise

Let $n \geqslant 3$ and let

$$X = A_0 \cup \ldots \cup A_{n-1},$$

$$B_k = A_{k+1} \cup \ldots \cup A_{k+n-1}, \qquad C_k = A_{k+1} \cup \ldots \cup A_{k+n-2},$$

where the indices are reduced modulo n. Let $f_k \in Y^{B_k}$, $k = 0, \ldots, n-1$, be a system of functions satisfying the condition

$$f_k(x) = f_{k+1}(x) \quad \text{for} \quad x \in C_{k+1}.$$

There exists a function $f \in Y^X$ satisfying the equation $f_k = f|B_k$ for every $k = 0, \ldots, n-1$.

§ 7. Images and inverse images

Let A and B be arbitrary sets and R a relation such that $R \subset A \times B$. For $X \subset A$ let

$$R^1(X) = \Big\{y: \bigvee_{x \in X} (xRy)\Big\}.$$

This set is called the *image* of X under the relation R. Clearly,

$$R^1: \boldsymbol{P}(A) \to \boldsymbol{P}(B).$$

In particular, if f is a function then $f^1(X)$ consists of values of the function f on the set X. We shall write $f^1(X) = \{f(x): x \in X\}$.

The same symbol will be used for operations, e.g. $\{\langle x, y\rangle: x \in X\}$, $\{\bigcup(\boldsymbol{X}): \boldsymbol{X} \in A\}$, etc. As we know, there exists neither a function whose value for any x is the pair $\langle x, y\rangle$ nor a function whose value for any family $\boldsymbol{X}$ is the set $\bigcup(\boldsymbol{X})$. However, every such operation determines a function if we limit its domain to an arbitrary given set (see Theorem 6.6). Thus, strictly speaking, it would be necessary to replace the symbols $\langle x, y\rangle$, $\bigcup(\boldsymbol{X})$, etc., in the formulas $\{\langle x, y\rangle: x \in X\}$, $\{\bigcup(\boldsymbol{X}): \boldsymbol{X} \in A\}$ by symbols for values of functions with domains X and A, respectively.

It follows from the definition of inverse relation (p. 64) that if $Y \subset B$ then the image of Y under the relation R^i is

$$R^{-1}(Y) = \Big\{x: \bigvee_{y \in Y} (yR^i x)\Big\} = \Big\{x: \bigvee_{y \in Y} (xRy)\Big\}.$$

This set is called the *inverse image* of Y under R. If $R = f$ is a function, then

$$f^{-1}(Y) = \{x: \bigvee_{y \in Y} (f(x) = y)\} = \{x: f(x) \in Y\},$$

i.e. the following equivalence holds:

$$x \in f^{-1}(Y) \equiv f(x) \in Y.$$

If Y reduces to the one-element set $\{y\}$, then the set $f^{-1}(Y)$ is called a *coset* of f determined by y. Distinct cosets are always disjoint, the union of all cosets is the domain of f.

We shall now establish several simple properties of images and inverse images.

THEOREM 1: *If $R \subset A \times B$ and X_1, X_2 are subsets of A, then*

$$R^1(X_1) \cup R^1(X_2) = R^1(X_1 \cup X_2), \tag{1}$$

$$X_1 \subset X_2 \to R^1(X_1) \subset R^1(X_2), \tag{2}$$

$$R^1(X_1 \cap X_2) \subset R^1(X_1) \cap R^1(X_2). \tag{3}$$

PROOF. Formula (1) follows from the equivalences

$$\begin{aligned} y \in R^1(X_1 \cup X_2) &\equiv \bigvee_x \{[(x \in X_1) \vee (x \in X_2)] \wedge (xRy)\} \\ &\equiv \bigvee_x [(x \in X_1) \wedge (xRy)] \vee \bigvee_x [(x \in X_2) \wedge (xRy)] \\ &\equiv y \in R^1(X_1) \vee y \in R^1(X_2) \\ &\equiv y \in R^1(X_1) \cup R^1(X_2). \end{aligned}$$

To prove (2) it suffices to notice that if $X_1 \subset X_2$ then $X_2 = X_1 \cup X_2$. Thus by means of (1) it follows that

$$R^1(X_2) = R^1(X_1) \cup R^1(X_2) \supset R^1(X_1).$$

Finally, formula (3) follows from the remark that $X_1 \cap X_2 \subset X_i$ for $i = 1, 2$, whence, by (2), $R^1(X_1 \cap X_2) \subset R^1(X_1)$ and $R^1(X_1 \cap X_2) \subset R^1(X_2)$; and, in turn, we obtain $R^1(X_1 \cap X_2) \subset R^1(X_1) \cap R^1(X_2)$.

THEOREM 2: *If $f \in B^A$ and $Y_1 \subset B$, $Y_2 \subset B$, then*

$$f^{-1}(Y_1 \cup Y_2) = f^{-1}(Y_1) \cup f^{-1}(Y_2), \tag{4}$$

$$f^{-1}(Y_1 \cap Y_2) = f^{-1}(Y_1) \cap f^{-1}(Y_2), \tag{5}$$

$$f^{-1}(Y_1 - Y_2) = f^{-1}(Y_1) - f^{-1}(Y_2). \tag{6}$$

PROOF. (4) is a special case of (1). Formula (5) follows from the equivalence

$$\begin{aligned} x \in f^{-1}(Y_1 \cap Y_2) &\equiv f(x) \in Y_1 \cap Y_2 \\ &\equiv \big(f(x) \in Y_1\big) \wedge \big(f(x) \in Y_2\big) \\ &\equiv \big(x \in f^{-1}(Y_1)\big) \wedge \big(x \in f^{-1}(Y_2)\big) \\ &\equiv \big(x \in f^{-1}(Y_1) \cap f^{-1}(Y_2)\big). \end{aligned}$$

The proof of (6) is similar.

Theorems 1 and 2 show that the operation of forming the image under an arbitrary relation is additive, but it is not multiplicative. On the other hand, the operation of obtaining the inverse image is both additive and multiplicative.

THEOREM 3: *If f: $A \to B$ and if f is a one-to-one function, then for any $X_1, X_2 \subset A$ the following formulas hold*:

$$f^1(X_1 \cap X_2) = f^1(X_1) \cap f^1(X_2), \quad f^1(X_1 - X_2) = f^1(X_1) - f^1(X_2).$$

For the proof, substitute f^1 for f^{-1} in Theorem 2.

THEOREM 4: *If f: $A \to B$, $Y \subset f^1(A)$ and $X \subset A$, then*

$$f^1\big(f^{-1}(Y)\big) = Y, \quad f^{-1}\big(f^1(X)\big) \supset X.$$

The proof of the first formula can be obtained from the equivalence

$$\begin{aligned} y \in f^1\big(f^{-1}(Y)\big) &\equiv \bigvee_x \big[\big(x \in f^{-1}(Y)\big) \wedge \big(y = f(x)\big)\big] \\ &\equiv \bigvee_x \big[\big(f(x) \in Y\big) \wedge \big(y = f(x)\big)\big] \equiv (y \in Y). \end{aligned}$$

The proof of the second formula follows from the implication

$$(x \in X) \to \big(f(x) \in f^1(X)\big) \equiv x \in f^{-1}\big(f_1(X)\big).$$

In the formula just proved the inclusion sign cannot in general be replaced by the equality sign. For instance, if f is a function of the real variable x and $f(x) = x^2$, then for $X = \{x: x \geqslant 0\}$ we have $f^{-1}\big(f^1(X)\big) \neq X$. But for one-to-one functions we obviously have $f^{-1}\big(f^1(X)\big) = X$.

Finally, let us note the following important

THEOREM 5: *If $S \subset A \times B$ and $R \subset B \times C$, then* $(R \circ S)^1(X) = R^1\big(S^1(X)\big)$ *for every set $X \subset A$.*

PROOF.

$$\begin{aligned} y \in (ROS)^1(X) &\equiv \bigvee_{x \in X} (xROSy) \\ &\equiv \bigvee_{x \in X} \bigvee_{z} [(xSz) \wedge (zRy)] \\ &\equiv \bigvee_{z} \bigvee_{x \in X} [(xSz) \wedge (zRy)] \\ &\equiv \bigvee_{z} \big[(z \in S^1(X)) \wedge (zRy)\big] \\ &\equiv y \in R^1\big(S^1(X)\big). \end{aligned}$$

In particular, it follows from Theorem 5 that if $f\colon A \to B$ and $g\colon B \to C$, then $(gOf)^1(X) = g^1\big(f^1(X)\big)$ for every set $X \subset A$.

Exercises

1. Prove that $f^1(X_1) - f^1(X_2) \subset f^1(X_1 - X_2)$ and $f^1\big(X \cap f^{-1}(Y)\big) = f^1(X) \cap Y$.

2. If $g = f|A$, then $g^{-1}(Y) = A \cap f^{-1}(Y)$.

3. A value y of a function f is said to be of *order* n if the set $f^{-1}(\{y\})$ consists of n elements. We say that a *function* f *is of order* $\leqslant n$ if all of its values are of order $\leqslant n$.

Prove that if a function f defined on a set X is of order $\leqslant n$ and $A \subset X$, then the restriction $f|\big(f^{-1}(f^1(A)) - A\big)$ is of order $\leqslant n-1$.

4. We are given a system of $r+1$ disjoint sets $A_0, A_1, \ldots, A_r$ included in X and a function of order $\leqslant n$ defined on X ($n \geqslant r$). Let $B = f^1(A_0) \cap \ldots \cap f^1(A_r)$. Prove that the restriction $f|\big(A_i \cap f^{-1}(B)\big)$ is of order $\leqslant n-r$.

5. Images and inverse images are used in topology, in particular to define the notion of a continuous function.

Let X and Y be two topological spaces and let $f\colon X \to Y$. We say that f is *continuous* if the inverse image of any open set in Y is an open set in X.

Prove that the following conditions are necessary and sufficient for a function f to be continuous:

(a) inverse images of closed sets are closed,

(b) $f^1(\overline{A}) \subset \overline{f^1(A)}$,

(c) $\overline{A} \subset f^{-1}[\overline{f^1(A)}]$,

(d) $\overline{f^{-1}(B)} \subset f^{-1}(\overline{B})$,

(e) $f^1[\overline{f^{-1}(B)}] \subset \overline{B}$,

where $A \subset X$ and $B \subset Y$.

6. Let f be a one-to-one mapping of X onto Y (hence $f^{-1}\colon Y \to X$). We say that f is a *homeomorphism* if f and f^{-1} are continuous.

Show that each of the conditions arising from (b)–(e) by substituting the equality sign for the inclusion sign is a necessary and sufficient condition for a function f to be a homeomorphism.

7. Show by means of an example that the image of an open set need not be open even though the function is continuous.

The same for closed sets.

8. Prove that the composition of two continuous functions is continuous.

9. Let $h = \langle f, g\rangle \in (Y\times Z)^X$ be a complex function. Show that

$$h^1(A) \subset f^1(A)\times g^1(A) \quad \text{where} \quad A\subset X$$

and

$$h^{-1}(B\times C) = f^{-1}(B)\cap g^{-1}(C) \quad \text{where} \quad B\subset Y \text{ and } C\subset Z.$$

10. Let $u = (f\times g) \in (Y\times Z)^{X\times W}$ be a product-function. Show that

$$u^1(A\times C) = f^1(A)\times g^1(C) \quad \text{where} \quad A\subset X \text{ and } C\subset W$$

and

$$u^{-1}(B\times D) = f^{-1}(B)\times g^{-1}(D) \quad \text{where} \quad B\subset Y \text{ and } D\subset Z.$$

§ 8. Functions consistent with a given equivalence relation. Factor Boolean algebras

The construction to be given in this section is one of basic importance in abstract algebra.

Let R be an equivalence relation whose field is X, f a function of two variables belonging to $X^{X\times X}$.

DEFINITION: The function f is *consistent with* R if

$$\bigwedge_{x,x_1,y,y_1} [(xRx_1)\wedge(yRy_1) \rightarrow (f(x,y)Rf(x_1,y_1))].$$

A similar definition can be adopted for a function of arbitrarily many arguments.

It results from the equivalence $xRy \equiv (x\in y/R) \equiv (x/R = y/R)$ that the definition of consistency can be expressed as follows: if $x\in x_1/R$ and $y\in y_1/R$, then $f(x,y)/R = f(x_1,y_1)/R$. In other words, the equivalence class $f(x,y)/R$ depends on the classes x/R and y/R but not on the elements x, y themselves. This implies that there exists a function φ with domain $(X/R)\times(X/R)$ satisfying for any $x, y\in X$ the formula

$$\varphi(x/R, y/R) = f(x,y)/R.$$

Namely, φ is the set of all pairs of the form $\langle\langle k', k''\rangle, k\rangle$, where $k, k', k'' \in X/R$ and

$$\bigvee_{x,y} [(x \in k') \wedge (y \in k'') \wedge (f(x, y) \in k)].$$

We say that the function φ is *induced from f by R.*

The function $k \in (X/R)^X$ defined by the formula $k(x) = x/R$ is called the *canonical mapping* of X onto X/R. The function of two variables k^2 defined by the formula $k^2(x, y) = \langle x/R, y/R\rangle$ is also called the *canonical mapping* of X^2 onto $(X/R)^2$. A similar definition can be given for functions of three and more variables.

THEOREM 1: *If a function $f \in X^{X\times X}$ is consistent with an equivalence relation R and φ is the function induced from f by R, then the diagram*

$$\begin{array}{ccc} X^2 & \xrightarrow{f} & X \\ {\scriptstyle k^2}\downarrow & & \downarrow{\scriptstyle k} \\ (X/R)^2 & \xrightarrow{\varphi} & X/R \end{array}$$

commutes.

PROOF. For any pair $\langle x, y\rangle \in X^2$ the following formulas hold:

$$k \circ f(x, y) = k(f(x, y)) = f(x, y)/R,$$
$$\varphi \circ k^2(x, y) = \varphi(k^2(x, y)) = \varphi(x/R, y/R) = f(x, y)/R.$$

Hence $k \circ f = \varphi \circ k^2$.

Example. Let $X = \boldsymbol{K}$ be a field of sets with unit U, $\boldsymbol{I}$ any ideal in $\boldsymbol{K}$, R the relation $\doteq$ mod $\boldsymbol{I}$ (see p. 17). The set $\boldsymbol{K}/R$ is denoted by $\boldsymbol{K}/\boldsymbol{I}$ and is called a *factor Boolean algebra.*

The functions $f(X, Y) = X \cup Y$, $g(X, Y) = X \cap Y$, and $h(X) = U - X$ are consistent with the relation $\doteq$ (see Exercise I.5.4). The functions induced from f, g, h by the relation $\doteq$ will be denoted by $\vee$, $\wedge$, and $-$, respectively. Hence

$$(X/R) \vee (Y/R) = (X \cup Y)/R, \qquad (X/R) \wedge (Y/R) = (X \cap Y)/R,$$
$$-(X/R) = (U - X)/R.$$

THEOREM 2: *The set $\boldsymbol{K}/\boldsymbol{I}$ is a Boolean algebra with respect to the operations $\vee$, $\wedge$, $-$, with $\emptyset/R$ and U/R as the zero and the unit element, respectively.*

PROOF. It is sufficient to show that the operations $\vee$, $\wedge$, $-$ and the elements $\emptyset/R$ and U/R satisfy axioms (i)–(v'), p. 38. For instance, we check axiom (i). Let $a = X/R$ and $b = Y/R$; then $a \vee b = (X \cup Y)/R$ and $b \vee a = (Y \cup X)/R$ and hence $a \vee b = b \vee a$. The remaining axioms can be checked similarly.

REMARK. The conditions $X \doteq \emptyset \pmod{\boldsymbol{I}}$ and $X \in \boldsymbol{I}$ are equivalent. This proves that $\emptyset/R = \boldsymbol{I}$.

The factor algebras $\boldsymbol{K}/\boldsymbol{I}$ may have properties quite different from those of $\boldsymbol{K}$. Thus the construction leading from $\boldsymbol{K}$ to $\boldsymbol{K}/\boldsymbol{I}$ allows us to build new and interesting examples of rings.

Exercises

1. Generalize the example given above by taking any Boolean algebra as $\boldsymbol{K}$ and any subset of $\boldsymbol{K}$ satisfying the conditions of Example 5.5 as $\boldsymbol{I}$.[1]

2. Let $\boldsymbol{K}$ be the field of all subsets of an infinite set U, and let $\boldsymbol{I}$ be the ideal of all finite subsets of U. Show that the factor ring $\boldsymbol{K}/\boldsymbol{I}$ has no atoms.

§ 9. Order relations

DEFINITION 1: A relation R is said to be an *order relation* if it is reflexive, transitive, and antisymmetric. The last condition means that

$$(xRy) \wedge (yRx) \rightarrow (x = y).$$

A relation which is only reflexive and transitive is said to be a *quasi-order relation*.

A *pre-order relation* is such that is transitive and satisfies the condition $(xRy) \rightarrow (xRx) \wedge (yRy)$ for arbitrary x and y.

Instead of xRy we usually write $x \leqslant_R y$ or $x \leqslant y$. We also say that the field of R is *ordered* (quasi-ordered, pre-ordered) without explicitly mentioning R. It is necessary to remember, however, that an ordering is by no means an intrinsic property of the set. The same set may be ordered by many different relations.

Let X be a set ordered by a relation $\leqslant$. If x, y are elements of X and either $x \leqslant y$ or $y \leqslant x$, then we say that x and y are *comparable*, otherwise *incomparable*. If $Y \subset X$ and any two elements of Y are comparable, then we call Y a *chain* in X; if any two elements of Y are incomparable, then Y is called an *antichain* in X.

[1]) Example 5.5 denotes Example 5 in § 5. Similary, Example II.5.5 denotes Example 5 in § 5, Chapter II, etc.

A relation $\leqslant$ is called *connected* in X if any two elements of X are comparable. Connected order relations are called *linear orderings.*[1]) These relations will be considered in detail in Chapter VI.

Examples

1. Every family of sets is ordered by the inclusion relation. If it is linearly ordered by this relation, then it is called a *monotone family.*

2. Every lattice (in particular, every Boolean algebra) is ordered by the relation $a \leqslant b$.

3. The set of natural numbers is ordered by the relation of divisibility.

4. A family $\boldsymbol{P}$ is said to be a *cover* of a set A if $A = \bigcup(\boldsymbol{P})$. A cover $\boldsymbol{P}_1$ is said to be a *refinement* of a cover $\boldsymbol{P}_2$ if for every $X \in \boldsymbol{P}_1$ there exists $Y \in \boldsymbol{P}_2$ such that $X \subset Y$. The relation R, defined by

$$\boldsymbol{P}_2 R \boldsymbol{P}_1 \equiv (\boldsymbol{P}_1 \text{ is a refinement of } \boldsymbol{P}_2),$$

is a quasi-order relation in the set of all covers of A. It is not, however, an order relation, that is, there may exist two distinct $\boldsymbol{P}_1$ and $\boldsymbol{P}_2$ such that $\boldsymbol{P}_1 R \boldsymbol{P}_2$ and $\boldsymbol{P}_2 R \boldsymbol{P}_1$.

On the other hand, if we limit the field of R to covers which consist of non-empty disjoint sets (such covers are called *partitions*; cf. p. 67), then R is an order relation.

DEFINITION 2: A set A ordered (or quasi-ordered) by the relation $\leqslant$ is said to be *directed* if for every pair $x \in A$ and $y \in A$ there exists $z \in A$ such that $x \leqslant z$ and $y \leqslant z$.

5. Every lattice is a directed set since $x \leqslant x \vee y$ and $y \leqslant x \vee y$. In particular, the family of all subsets of a given set X, as well as the family of all closed subsets of a given topological space, is directed with respect to the inclusion relation (either $\subset$ or $\supset$).

6. The set of all covers of a given set A is directed with respect to the relation R considered in Example 4. For, given two covers $\boldsymbol{P}_1$ and $\boldsymbol{P}_2$, denote by $\boldsymbol{P}_3$ the family of all intersections of the form $X \cap Y$ where

[1]) Linear orderings were considered originally by Cantor [5]. Partial orderings were introduced by Hausdorff [3]. The use of the word "ordering" for what was formerly called partial ordering originated with Bourbaki [1]. The convention is not universally accepted and is often a source of confusion.

$X \in \boldsymbol{P}_1$, $Y \in \boldsymbol{P}_2$. It is easy to check that $\boldsymbol{P}_3$ is a cover and $\boldsymbol{P}_1 R \boldsymbol{P}_3$ as well as $\boldsymbol{P}_2 R \boldsymbol{P}_3$.

DEFINITION 3: An ordered set A is said to be *cofinal* with its subset B if for every $x \in A$ there exists $y \in B$ such that $x \leqslant y$. Analogously we can define *coinitial* sets.

Example. The set of real numbers is cofinal and coinitial with the set of integers.

If an ordered set A contains a greatest element, then A is cofinal with the set composed of this element.

The greatest (least) element should be distinguished from the maximal (minimal) element. Namely, an element x of an ordered set A is said to be *maximal* (*minimal*) if there is no element y in A such that $x < y$ $(y < x)$. In linearly ordered sets the notions of greatest (least) element and of maximal (minimal) element coincide. This is not always the case for arbitrary ordered sets.

DEFINITION 4: Let A be an ordered set, T any set and let $f \in A^T$. An element $u \in A$ is said to be the *least upper bound* of $\{f_t\}$ if $f_t \leqslant u$ and u is the least element having this property:

(i) $\bigwedge_{t \in T} (f_t \leqslant u)$,

(ii) $\bigwedge_{t \in T} (f_t \leqslant v) \rightarrow (u \leqslant v)$.

Replacing $\leqslant$ by $\geqslant$, we obtain the definition of the *greatest lower bound.*

The least upper bound, if it exists, is uniquely determined. For suppose that besides (i) and (ii) we have

(i′) $\bigwedge_{t \in T} (f_t \leqslant u')$,

(ii′) $\bigwedge_{t \in T} (f_t \leqslant v) \rightarrow (u' \leqslant v)$.

Setting $v = u$ in (ii′) and applying (i), we obtain $u' \leqslant u$. Likewise, it follows from (ii) and (i′) that $u \leqslant u'$. Hence $u = u'$, since the relation $\leqslant$ is antisymmetric.

The proof of the uniqueness of the greatest lower bound is similar.

The least upper bound, if it exists, is denoted in the theory of ordered sets by $\bigvee_{t \in T} f_t$, the greatest lower bound by $\bigwedge_{t \in T} f_t$. If T is a finite set $T = \{1, 2, \ldots, n\}$ and $f_1 = a$, $f_2 = b, \ldots, f_n = h$, then the least upper bound of these elements is also denoted by $a \vee b \vee \ldots \vee h$, and the greatest lower bound by $a \wedge b \wedge \ldots \wedge h$.

The greatest lower bound of all elements of A, if it exists, is called the *zero element* and is denoted by 0_A or simply 0. Analogously, the least upper bound of all the elements of A, if it exists, is called the *unit element* and is denoted by 1_A or 1.

Obviously, $a \wedge b \leqslant a$ and $a \wedge b \leqslant b$ if $a \wedge b$ exists; similarly $a \leqslant a \vee b$ and $b \leqslant a \vee b$ provided $a \vee b$ exists (in this case A is a directed set).

If $a \leqslant b$, then $a \vee b$ and $a \wedge b$ exist and they equal b and a, respectively. This implies that if $a \wedge b$ and $a \vee b$ exist for all $a, b \in A$, A is a lattice.

DEFINITION 5: An ordering of a set A is said to be *complete* if for every T and for every $f \in A^T$ there exist $\bigwedge_{t \in T} f_t$ and $\bigvee_{t \in T} f_t$.

Since every lattice is a set ordered by the relation $a \leqslant b$, this definition also explains the meaning of the term "complete lattice."

Examples and exercises

7. $\boldsymbol{P}(X)$ *is a complete lattice* (for an arbitrary set X) *with respect to the operations* $\cap$, $\cup$. The existence of the least upper bound is a consequence of axiom III, § 2, the existence of the greatest lower bound follows from Theorem 3.6.

This example will be used in Chapter IV, § 1.

8. *The family of all closed subsets of an arbitrary topological space is a complete lattice under the operations* $\cap$, $\cup$. In this case $\bigvee_t X_t$ is the closure of the union of the sets X_t, $\bigwedge_t X_t$ is the intersection of X_t.

9. Let A be any set. *The family* $\boldsymbol{P}(A^2)$ (i.e. the family of relations with fields included in A) *is ordered by the inclusion relation.* Prove that the family of all transitive relations is a lattice and describe the meaning of the operations $\wedge$ and $\vee$.

10. The same problem as in 9 for the family of all equivalence relations.

An interesting class of order relations is provided by so-called *trees.*

DEFINITION 6: We say that a set X ordered by the relation $\leqslant$ is a *pseudo-tree* with respect to this relation if X is ordered by $\leqslant$ and moreover for each x in X the set $O_R(x) = \{y: y \leqslant x\}$ of all predecessors of x is a chain.

We shall omit the words "with respect to $\leqslant$" whenever possible.

A pseudo-tree is a *tree* if sets $O_R(x)$ are not only linearly ordered but are well-ordered; this notion will be defined and discussed later.

If T is a pseudo-tree and $S \subset T$ then obviously S is a pseudo-tree with respect to the same relation limited to S; we call S a *sub-pseudo-tree* of T.

Examples

1. Figure 1 gives an example of a pseudo-tree consisting of 15 points. The relation $x \leqslant y$ holds between points x and y if x is connected with y by an arrow and x lies lower than y.

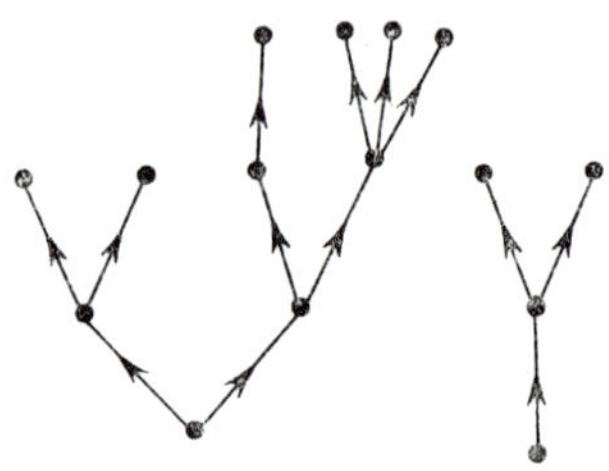

2. Let $f: X \to X$ and let $y \leqslant x$ mean that there is a non-negative integer n such that $y = f^n(x)$ where f^n is the nth iteration of f, i.e. $f^0 = I_X$, $f^1 = f$, $f^2 = f \bigcirc f$, $f^3 = f \bigcirc f \bigcirc f$, etc.

3. Let X be the set consisting of all finite sequences $(a_0, a_1, \ldots, a_n)$ whose terms a_j are either 0 or 1. Let $x \leqslant y$ mean that $x \subset y$. Then X is a (pseudo-)tree; we call it the *full binary tree*.

Let X be a pseudo-tree. If $B \subset X$ is a chain and there is no chain in X which properly contains B, then we say that B is a *branch* of X. We shall prove later that each chain can be extended to a branch (see Ch. VII).

The set of all branches of a pseudo-tree can be provided with a very natural topology which we call the *tree topology*. To define it let T be a pseudo-tree and let T_t be its sub-pseudo-tree consisting of all elements of T which are comparable with t. We define now a closure operation:

For each set A whose elements are branches of T we denote by $\bar{A}$ the set of all branches B of T which have the following property: whenever $B \subset T_t$, then there is a branch $B' \subset T_t$ which belongs to the set A.

It can easily be shown that the set of all branches of T is a topological space under this closure operation.

Exercises

1. Supply the proof of the last statement.

2. The set of all finite sequences whose terms are non-negative integers is a (pseudo-)tree with respect to $\subset$. Show that the set of all branches of this tree provided with the tree topology is homeomorphic with the space of irrational numbers of the interval $(0, 1)$.

3. Prove that each finite (pseudo-) tree is a sub-tree of the full binary tree.

4. Prove that a lattice is a pseudo-tree if and only if it is linearly ordered.

5. Let X be a set with 3 elements. How many mutually non-isomorphic (a) order relations, (b) trees, (c) linear orderings with the field X exist?

6. Let T be a pseudo-tree. We call x an immediate successor of y if $y < x$ but there is no z such that $y < z < x$. Prove that if T is finite and each element y of T has exactly one immediate successor, then T is linearly ordered. Give examples of infinite trees for which this statement is false.

To conclude this section we introduce the important notion of the similarity of ordered sets.

DEFINITION 7: Two sets A and B ordered by the relations R and S respectively are said to be *similar* if there exists a one-to-one function f mapping the set A onto B and satisfying for arbitrary $x, y \in A$ the equivalence

$$xRy \equiv f(x) S f(y).$$

In this case we say that the function establishes the similarity of the sets A and B (under the relations R and S).

For example, defining $f(x) = -x$ for $x \in \mathscr{E}$ we obtain similarity between the set $\mathscr{E}$ ordered by $\leqslant$ and the same set $\mathscr{E}$ ordered by $\geqslant$.

The notion of similarity is a special case of the more general notion of isomorphism which will be treated in the next section.

§10. Relational systems, their isomorphisms and types

Let A be a set, $R_0, R_1, \ldots$ relations respectively of $p_0, p_1, \ldots, p_{k-1}$ arguments in A; in other words, $R_j \subset A^{p_j}$ for $j < k$. The sequence

$$\langle A, R_0, R_1, \ldots, R_{k-1} \rangle$$

is called a *relational system of characteristic* $(p_0, p_1, \ldots, p_{k-1})$ and the set A is called the *universe of the system.*

Relational systems are investigated in many branches of mathematics, especially in algebra. We may, for instance, consider a group as a relational system of characteristic (3) and a ring as a relational system of characteristic (3,3). Boolean algebras (see p. 33) may also be treated as relational systems.

In order to simplify our treatment we shall investigate systems of characteristic (2), that is, systems of the form $\langle A, R \rangle$, where $R \subset A \times A$. However, all proofs can easily be generalized to arbitrary systems.

DEFINITION: Two relational systems $\langle A, R \rangle$ and $\langle B, S \rangle$ are said to be *isomorphic* if there exists a one-to-one function f mapping A onto B such that for all $x, y \in A$

$$xRy \equiv f(s)\, Sf(y).$$

Then we write $\langle A, R \rangle \approx \langle B, S \rangle$ or briefly $R \approx S$ if no confusion can arise about the sets A and B.

The proof of the following theorem is immediate.

THEOREM 1: *The relation* $\approx$ *is reflexive, symmetric, and transitive.*

We shall show that every property of the system $\langle A, R \rangle$ which can be expressed by means of the propositional calculus and quantifiers limited to the universe of the relational system, is also a property of every system isomorphic to $\langle A, R \rangle$. We say that the property in question is *invariant under isomorphism.*

Let Φ be a formula involving free variables x, y. Besides x, y, Φ may involve an arbitrary number of other variables $u_0, u_1, \ldots, u_{k-1}$. Suppose that Φ arises from formulas of the form

(1) $$u_i = u_j,$$

(2) $$\langle u_i, u_j \rangle \in y$$

by means of operations of the propositional calculus and by means of the quantifiers $\bigvee_{u \in x}$ and $\bigwedge_{u \in x}$. Thus the variables x and y are not bounded by quantifiers. For such formulas we have

THEOREM 2:[1]) *If a function f is an isomorphism of the relational systems $\langle A, R\rangle$ and $\langle B, S\rangle$ and if $a_0, a_1, \ldots, a_{k-1} \in A$, then*

$$\Phi(A, R, a_0, \ldots, a_{k-1}) \equiv \Phi\big(B, S, f(a_0), \ldots, f(a_{k-1})\big). \tag{3}$$

PROOF. If Φ is the formula (1), then the equivalence (3) follows from the assumption that f is one-to-one. If Φ is the formula (2), then (3) follows from the assumption that f is an isomorphism.

Suppose now that (3) holds for formulas Φ and Φ'. By the laws of the propositional calculus it follows that (3) also holds for the formulas $\neg\Phi$ and $\Phi \vee \Phi'$. This implies that (3) holds for all the formulas which can be obtained from Φ and Φ' by applying operations of the propositional calculus. Hence, to prove the theorem for all formulas it suffices to show that (3) holds for formulas arising from Φ by applying a quantifier (existential or universal) to Φ. It suffices to consider only one of these quantifiers, for instance the existential quantifier.

Let Ψ be the formula $\bigvee_{u_0} \Phi$ and suppose that $a_1, \ldots, a_{k-1} \in A$. If $\Psi(A, R, a_1, \ldots, a_{k-1})$ then for some a_0 belonging to A we have $\Phi(A, R, a_0, a_1, \ldots, a_{k-1})$. By the induction hypothesis we thus obtain $\Phi\big(B, S, f(a_0), f(a_1), \ldots, f(a_{k-1})\big)$, and it follows that $\Psi\big(B, S, f(a_1), \ldots$ $\ldots, f(a_{k-1})\big)$. Hence we have proved the implication

$$\Psi(A, R, a_1, \ldots, a_{k-1}) \rightarrow \Psi\big(B, S, f(a_1), \ldots, f(a_{k-1})\big).$$

The proof of the converse implication is similar.

Example. The following properties of the system $\langle A, R\rangle$ are by Theorem 2 invariant under isomorphism:

1. *Reflexivity*: $\bigwedge_{x \in A} xRx$.
2. *Irreflexivity*: $\bigwedge_{x \in A} \neg(xRx)$.
3. *Symmetry*: $\bigwedge_{x, y \in A} [xRy \rightarrow yRx]$.
4. *Asymmetry*: $\bigwedge_{x, y \in A} [xRy \rightarrow \neg(yRx)]$.
5. *Antisymmetry*: $\bigwedge_{x, y \in A} [(xRy) \wedge (yRx) \rightarrow (x = y)]$.

[1]) Theorem 2 is a scheme: for each formula we obtain a separate theorem.

6. *Transitivity*: $\bigwedge_{x,y,z\in A} [(xRy) \wedge (yRz) \to (xRz)]$.

7. *Connectedness*: $\bigwedge_{x,y\in A} [(xRy) \vee (x = y) \vee (yRx)]$.

Two isomorphic systems are said to be of the *same type*. The mere use of the word "type" in this expression does not presuppose that there are objects which we call "types." All the theorems of set theory could indeed be expressed without using this notion. However, the introduction of the notion of type simplifies the axiomatic treatment of the theory. Moreover, the use of this notion is justified by the fact that Cantor himself developed set theory using this concept.

In order to deal with the notion of a type in an axiomatic way we introduce a new primitive notion TR. The formula $\alpha \mathrm{TR}\langle A, R\rangle$ is read: *α is the type of the relational system $\langle A, R\rangle$*. We also introduce a new axiom.

AXIOM VIII (OF RELATIONAL SYSTEMS): *For every system $\langle A, R\rangle$ where $R \subset A^2$ there exists exactly one object α such that $\alpha \mathrm{TR}\langle A, R\rangle$. Moreover, for any systems $\langle A, R\rangle$ and $\langle B, S\rangle$ the following formula holds*

$$(\alpha \mathrm{TR}\langle A, R\rangle) \wedge (\beta \mathrm{TR}\langle B, S\rangle) \to [(\alpha = \beta) \equiv \langle A, R\rangle \approx \langle B\ S\rangle].$$

The unique object α such that $\alpha \mathrm{TR}\langle A, R\rangle$ is denoted by $\overline{\langle A, R\rangle}$ or, if there is no confusion, by $\overline{R}$.[1])

The object α is called a *relational type* if and only if there exists some system $\langle A, R\rangle$ such that $\alpha \mathrm{TR}\langle A, R\rangle$.

Exercises

1. Let A have n elements and let r_n be the number of relational types of systems with field A. Show that the number r_n satisfies the inequality

$$2^{n^2}/n! < r_n < 2^{n^2}.$$

2. Prove that $r_2 = 10$ and $r_3 = 104$ [Davis].[2])

[1]) The notion of a type originated with Cantor [4]; however he dealt only with the special case of types of linear orderings and his way of introducing types was objectionable. The general notion appeared first in Russell and Whitehead [1].

Our way of introducing types differs from the way usually taken in expositions based on the system ZFC (see note to p. 57) but is closer to ideas of Cantor. It was shown by Lévy [3] that in the absence of the axiom of regularity one cannot define types in the way used in the system ZFC.

[2]) In connection with Exercise 1 see Davis [1]. He states in his paper that $r_4 = 3044$ and $r_5 = 291\,968$.

CHAPTER III

NATURAL NUMBERS. FINITE AND INFINITE SETS

In this chapter all theorems will be derived from axioms of Σ° (cf. p. 56). As usual, theorems not marked by $^\circ$ do not involve the axiom of choice in their proofs. Throughout this chapter we shall often write 0 instead of $\emptyset$.

§1. Natural numbers[1])

For any set X, let

$$X' = X \cup \{X\}.$$

The set X' will be called the *successor* of X.

THEOREM 1: *There exists exactly one family of sets* N *such that*

(i) $0 \in N$;

(ii) $X \in N \to X' \in N$;

(iii) *if* $\boldsymbol{K}$ *satisfies* (i) *and* (ii), *then* $N \subset \boldsymbol{K}$.

PROOF. It follows from the axiom of infinity that there exists at least one family $\boldsymbol{R}$ satisfying conditions (i) and (ii). Let $\boldsymbol{\Phi}$ be the family of all those subsets of $\boldsymbol{R}$ which satisfy (i) and (ii):

$$\boldsymbol{\Phi} = \{\boldsymbol{S} \subset \boldsymbol{R}\colon 0 \in \boldsymbol{S} \wedge \bigwedge_X (X \in \boldsymbol{S} \to X' \in \boldsymbol{S})\}.$$

It is easy to show that $\bigcap(\boldsymbol{\Phi})$ is the required family.

[1]) The idea of defining natural numbers in set theory goes back to Frege [1], although his way of defining them was very different from that presented in this chapter. Our presentation which is now standard was initiated by von Neumann [1]. A detailed analysis of the problem of what axioms are necessary to justify the laws of arithmetic and other parts of mathematics was made by Bernays in his series of papers [1]–[6].

The elements of N are 0, $\{0\}$, $\{0, \{0\}\}$, etc. These sets can be considered as the counterparts of natural numbers $0, 1, 2, \ldots$, the operation $'$ as the counterpart of $+1$. We shall prove several theorems of arithmetic which hold for the elements of N.[1])

In order to simplify the notation, the elements of N will be denoted by the letters $m, n, p, \ldots$ A set K is said to be *inductive* if it satisfies conditions (i) and (ii).

(1) $$m \in n \to m' \subset n.$$

PROOF. Let $K = \{n: \bigwedge_m (m \in n \to m' \subset n)\}$. To prove (1) it suffices to show that $N \subset K$ or, in other words, to show that the set K is inductive. Condition (i) clearly holds. To prove (ii), let $n \in K$ and $m \in n'$. Hence $m \in n$ or $m = n$. In the first case, $m \subset n$ by the definition of K, in the second case, $m \subset n$ because the sets are equal. Thus $m \subset n'$, and consequently $n' \in K$, which proves the theorem. The proof of (1) is an example of a proof by induction.

(2) $$n \notin n.$$

The proof by induction consists in showing that the set $\{n: n \notin n\}$ is inductive.

(3) $$m' = n' \to m = n.$$

PROOF. It follows from $m' = n'$ that $m \in n'$, thus $m \in n$ or $m = n$ and by (1) $m \subset n$. Similarly we prove that $n \subset m$.

Peano showed that the arithmetic of natural numbers can be based upon the following axioms:

(a) *zero is a natural number*;

(b) *every natural number has a successor*;

[1]) The reader, who feels it unnatural that in our exposition the role of natural numbers is played by sets, can take as natural numbers some objects which are in one-to-one correspondence with the sets belonging to N. Such objects may be, for instance, the types of relational systems $S(n) = \langle n, n \times n \rangle$, where $n \in N$. This method will be used in Chapter V, where we introduce cardinal numbers. Identifying natural numbers with the elements of the family N, we can base our treatment of arithmetic on the axioms Σ° only (mostly even on the axioms Σ without the axiom of replacement); whereas in order to identify them with the types of the systems $S(n)$ we would have to appeal to the axioms Σ° [TR] and VIII.

(c) *zero is not a successor of any natural number*;

(d) *natural numbers having the same successor are equal*;

(e) *a set which contains zero and which contains the successor of every number belonging to this set contains all natural numbers.*

It follows from (i), (ii), (3), (iii) and $n' \neq 0$ that the elements of the set N satisfy Peano's axioms.

(4) *For arbitrary* m, n *exactly one of the following formulas holds*:

$$m \in n, \qquad m = n, \qquad n \in m.$$

PROOF. (1) and (2) imply that every two of the above conditions are mutually contradictory. To prove that for every pair m, n one of these formulas holds, we use induction. Let

$$K(n) = \{m \colon m \in n \vee m = n \vee n \in m\}.$$

Theorem (4) is equivalent to $\bigwedge_n (N \subset K(n))$, thus it suffices to prove that every set $K(n)$ is inductive.

The set $K(0)$ is inductive, for $K(0)$ consists of the set 0 and of those m for which $0 \in m$ and it is obvious that $0 \in m \to 0 \in m'$. Suppose that the set $K(n)$ is inductive, i.e. that $N \subset K(n)$. We shall prove that $K(n')$ is also inductive.

Condition (i): $n' \in N \subset K(0)$ implies that $n' \in 0 \vee n' = 0 \vee 0 \in n'$. Since the first two components of this disjunction are false, $0 \in n'$ and $0 \in K(n')$.

Condition (ii): Suppose that $m \in K(n')$, that is, either $m \in n'$, $m = n'$ or $n' \in m$. In the second and the third case we obviously have $n' \in m'$ and hence $m' \in K(n')$. In the first case either $m = n$ or $m \in n$ holds. If $m = n$ then $m' = n'$ and hence $m' \in K(n')$. If $m \in n$ then $m \in K(n)$ and hence $m' \in K(n)$, for $K(n)$ is inductive by assumption. We thus obtain $m' \in n \vee m' = n \vee n' \in m$. The third component of this disjunction is false, for it implies $n \in n$, which contradicts (2). Hence we have only the two possibilites, $m' \in n$ and $m' = n$, which since $n \subset n'$, prove that $m' \in K(n')$. This completes the proof of theorem (4).

(5) *The set* $Z = \{m \colon n \in m\}$ *is identical with the intersection* P *of all families* $K \subset N$ *such that* $n' \in K$ *and* K *satisfies* (ii).

PROOF. Since $m \subset m'$, the set Z satisfies (ii). This proves that $P \subset Z$, for obviously $n' \in Z$. It remains to be shown that if K satisfies (ii) and $n' \in K$, then $Z \subset K$. For this purpose, let $L = \{n: n \in Z \rightarrow n \in K\}$. It suffices to show that $N \subset L$ or, in other words, that L is an inductive family.

Condition (i) obviously holds, for $0 \notin Z$.

To prove that L satisfies (ii), suppose that $m \in L$. This means that either $m \in K$ or $m \notin Z$. In the first case $m' \in K$, for K satisfies (ii), and we obtain $m' \in L$. The second case splits into three subcases depending upon whether $n \in m$, $n = m$ or $m \in n$. The first subcase contradicts the assumption $m \notin Z$. The second subcase implies $m' = n'$ and hence $m' \in K$ and finally $m' \in L$, for $K \subset L$. In the last subcase we have by (4) either $m' \in n$ or $m' = n$ or $n \in m'$. If either $m' \in n$ or $m' = n$, then by (4) $n \notin m'$; hence $m' \notin Z$ and finally $m' \in L$. The condition $n \in m'$ leads to a contradiction, for it implies that $n \in m \vee n = m$ whereas by assumption we have $m \in n$.

In ordinary arithmetic the set of all numbers greater than n is defined to be the common part of all sets which contain the successor of n and which contain the successor of every number b which they contain. Theorem (5) shows that the membership relation $\in$ in N is the counterpart of the relation "less than" between numbers. We shall often write $m < n$ or $m \leqslant n$ instead of $m \in n$ or $m \in n'$, respectively.

The existence of the set N allows us to define in set theory notions analogous to those found in arithmetic and analysis. For example, a function f whose domain is the set N is called an *infinite sequence* and is sometimes denoted by $(f_0, f_1, \ldots, f_n, \ldots)$. If $n \in N$ then a function with domain n is said to be a *finite sequence of n terms*.

The set of all infinite sequences whose terms belong to A is clearly A^N; the set of all finite sequences of n terms in A is A^n. The set of all finite sequences with terms in A can be defined as

$$\{R \subset N \times A: (R \text{ is a function}) \wedge \bigvee_{n \in N} (D_1(R) = n)\}.$$

This definition implies the existence of the set of all finite sequences with terms in A.

Exercises

1. Show that if $m, n \in N$, then $m \in n \equiv (m \subset n) \wedge (m \neq n)$.

2. Show that if $0 \neq K \subset N$, then $\bigcap (K) \in N \cap K$.

3. Prove that every non-empty family $K \subset N$ contains an element k such that $k \cap K = 0$ (cf. the axiom of regularity, p. 57).

§2. Definitions by induction

Inductive definitions are the most characteristic feature of the arithmetic of natural numbers. The simplest case is the definition of a sequence φ (with terms belonging to a certain set Z) satisfying the following conditions:

(a) $$\varphi(0) = z, \quad \varphi(n') = e(\varphi(n), n),$$

where $z \in Z$ and e is a function mapping $Z \times N$ into Z.

More generally, we consider a mapping f of the cartesian product $Z \times N \times A$ into Z and seek a function $\varphi \in Z^{N \times A}$ satisfying the conditions:

(b) $$\varphi(0, a) = g(a), \quad \varphi(n', a) = f(\varphi(n, a), n, a),$$

where $g \in Z^A$. This is a definition by induction with parameter a ranging over the set A.

Schemes (a) and (b) correspond to induction "from n to $n+1$", i.e. $\varphi(n')$ or $\varphi(n', a)$ depends upon $\varphi(n)$ or $\varphi(n, a)$, respectively. More generally, $\varphi(n')$ may depend upon all values $\varphi(m)$ where $m \leqslant n$ (i.e. $m \in n'$). In the case of induction with parameter, $\varphi(n')$ may depend upon all values $\varphi(m, a)$, where $m \leqslant n'$; or even upon all values $\varphi(m, b)$, where $m \leqslant n'$ and $b \in A$. In this way we obtain the following schemes of definitions by induction:

(c) $$\varphi(0) = z, \quad \varphi(n') = h(\varphi | n', n),$$

(d) $$\varphi(0, a) = g(a), \quad \varphi(n', a) = H(\varphi | (n' \times A), n, a).$$

In the scheme (c), $z \in Z$ and $h \in Z^{C \times N}$, where C is the set of finite sequences whose terms belong to Z; in the scheme (d), $g \in Z^A$ and

$H \in Z^{T\times N\times A}$, where T is the set of functions whose domains are included in $N\times A$ and whose values belong to Z.[1])

Examples of definitions by induction

1. The function $m+n$:

$$m+0 = m, \qquad m+n' = (m+n)'.$$

This definition is obtained from (b) if we set $Z = A = N$, $g(a) = a$, $f(p, n, a) = p'$.

2. The function $\binom{n}{2}$:

$$\binom{0}{2} = 0, \qquad \binom{n'}{2} = \binom{n}{2}+n.$$

This follows from (a) if we set $Z = N$, $z = 0$, $e(p, n) = p+n$.

3. Let $Z = A = X^X$, $g(a) = I_X$ and $f(u, n, a) = u\bigcirc a$ in (b). Then (b) takes on the form

$$\varphi(0, a) = I_X, \qquad \varphi(n', a) = \varphi(n, a)\bigcirc a.$$

The function $\varphi(n, a)$ is denoted by a^n and is called nth *iteration of the function a*. Thus we have:

$$a^0(x) = x, \qquad a^{n'}(x) = a^n\big(a(x)\big) \quad \text{for } x \in X,\ a \in X^X \text{ and } n \in N.$$

4. Let $A = N^N$. Let $g(a) = a_0$ and $f(u, n, a) = u+a_{n'}$ in (b). Then (b) takes on the form

$$\varphi(0, a) = a_0, \qquad \varphi(n', a) = \varphi(n, a)+a_{n'}.$$

The function defined in this way is denoted by $\sum_{i=0}^{n} a_i$. Similarly we define $\prod_{i=0}^{n} a_i$, $\max_{i\leqslant n} a_i$.

It is clear that the scheme (d) is the most general of all the schemes discussed above. By appropriate choice of functions we can obtain

[1]) Scheme (c) could be generalized by assuming that the domain of the function h is not the whole set $C\times N$, but only the set of pairs of the form (c, n) where $c \in Z^n$. However, this generalization is not of great importance.

The theory of inductive definitions was first presented by Dedekind [1].

from (d) any of the schemes (a)–(d). For example, taking the function defined by

$$H(c, n, a) = f\big(c(n, a), n, a\big) \quad \text{for} \quad a \in A,\ n \in N,\ c \in Z^{N \times A}$$

as H in (d), we obtain (b).

We shall now show that, conversely, the scheme (d) can be obtained from (a). Let g and H be functions belonging to Z^A and $Z^{T \times N \times A}$ respectively, and let φ be a function satisfying (d). We shall show that the sequence $\Psi: \Psi_n = \varphi|(n' \times A)$ can be defined by (a).

Obviously, $\Psi_n \in T$ for every n. The first term of the sequence Ψ is equal to $\varphi|(0' \times A)$, i.e. to the set

$$z^* = \big\{\big\langle\langle 0, a\rangle, g(a)\big\rangle : a \in A\big\}.$$

The relation between $\Psi_{n'}$ and Ψ_n is given by the formula

$$\Psi_{n'} = \Psi_n \cup \varphi|(\{n'\} \times A),$$

where the second component is

$$\big\{\big\langle\langle n', a\rangle, \varphi(n', a)\big\rangle : a \in A\big\} = \big\{\big\langle\langle n', a\rangle, H(\Psi_n, n, a)\big\rangle : a \in A\big\}.$$

Thus we see that the sequence Ψ can be defined by (a) if we substitute T for Z, z^* for z and let

$$e(c, n) = c \cup \big\{\big\langle\langle n', a\rangle, H(c, n, a)\big\rangle : a \in A\big\} \quad \text{for} \quad c \in T.$$

Now we shall prove the existence and uniqueness of the function satisfying (a). This theorem shows that we are entitled to use definitions by induction of the type (a). According to the remark made above, this will imply the existence of functions satisfying the formulas (b), (c), and (d). Since the uniqueness of such functions can be proved in the same manner as for (a), we shall use in the sequel definitions by induction of any of the types (a)–(d).

THEOREM 1: *If Z is any set, $z \in Z$ and $e \in Z^{Z \times N}$, then there exists exactly one sequence φ satisfying formulas* (a).

PROOF. *Uniqueness*: Suppose that φ_1 and φ_2 satisfy (a) and let $K = \{n: \varphi_1(n) = \varphi_2(n)\}$. Then (a) implies that K is inductive. Hence $N \subset K$ and $\varphi_1 = \varphi_2$.

Existence: Let $\Phi(z, n, t)$ be the formula $e(z, n) = t$ and let $\Psi(n, z, F)$ be the following formula

$$(F \text{ is a function}) \wedge \big(D_1(F) = n'\big) \wedge \big(F(0) = z\big) \wedge \\ \wedge \bigwedge_{m \in n} \Phi\big(F(m), m, F(m')\big).$$

In other words, F is a function defined on the set of numbers $\leqslant n$ such that $F(0) = z$ and $F(m') = e\big(F(m), m\big)$ for all $m < n$.

We prove by induction that there exists exactly one function F_n such that $\Psi(n, z, F_n)$. The proof of uniqueness of this function is similar to that given in the first part of Theorem 1. The existence of F_n can be proved as follows: for $n = 0$ it suffices to take $\{\langle 0, z\rangle\}$ as F_n; if $n \in N$ and F_n satisfies $\Psi(n, z, F_n)$, then $F_{n'} = F_n \cup \{\langle n', e\big(F_n(n), n\big)\rangle\}$ satisfies the condition $\Psi(n', z, F_{n'})$.

Now, we take as φ the set of pairs $\langle n, s\rangle$ such that $n \in N$, $s \in Z$ and

$$\bigvee_F [\Psi(n, z, F) \wedge \big(s = F(n)\big)].$$

Since F is the unique function satisfying $\Psi(n, z, F)$, it follows that φ is a function. For $n = 0$ we have $\varphi(0) = F_0(0) = z$; if $n \in N$, then $\varphi(n') = F_{n'}(n') = e\big(F_n(n), n\big)$ by the definition of F_n; hence we obtain $\varphi(n') = e\big(\varphi(n), n\big)$. Theorem 1 is thus proved.

We frequently define not one but several functions (with the same range Z) by a simultaneous induction:

$$\varphi(0) = z, \qquad \psi(0) = t,$$
$$\varphi(n') = f\big(\varphi(n), \psi(n), n\big), \quad \psi(n') = g\big(\varphi(n), \psi(n), n\big),$$

where $z, t \in Z$ and $f, g \in Z^{Z \times Z \times N}$.

This kind of definition can be reduced to the previous one. It suffices to notice that the sequence $\vartheta_n = \langle \varphi_n, \psi_n\rangle$ satisfies the formulas

$$\vartheta_0 = \langle z, t\rangle, \qquad \vartheta_{n'} = e(\vartheta_n, n),$$

where we set $e(u, n) = \big\langle f\big(K(u), L(u), n\big), g\big(K(u), L(u), n\big)\big\rangle$, and K, L denote functions such that $K(\langle x, y\rangle) = x$ and $L(\langle x, y\rangle) = y$, respectively. Thus the function ϑ is defined by induction by means of (a). We now define φ and ψ by

$$\varphi(n) = K(\vartheta_n) \quad \text{and} \quad \psi(n) = L(\vartheta_n).$$

The theorem on inductive definitions can be generalized to the case of operations. We shall discuss only one special case. Let Φ be a formula such that

$$\bigwedge_{z} \bigwedge_{n \in N} \bigvee_{t} \Phi(z, n, t),$$

$$\bigwedge_{z} \bigwedge_{n \in N} \bigwedge_{t_1, t_2} [\Phi(z, n, t_1) \wedge \Phi(z, n, t_2) \to t_1 = t_2].$$

THEOREM 2: [1]) *For any set S there exists exactly one sequence* φ *such that*

$$\varphi_0 = S \quad \textit{and} \quad \bigwedge_{n \in N} \Phi(\varphi_n, n, \varphi_{n'}).$$

PROOF. Uniqueness can be proved as in Theorem 1.

To prove the existence of φ, let us consider the following formula $\Psi^*(n, S, F)$.

$$(F \text{ is a function}) \wedge (D_1(F) = n') \wedge (F(0) = S) \wedge \bigwedge_{m \in n} \Phi(F(m), m, F(m')).$$

As in the proof of Theorem 1, it can be shown that there exists exactly one function F_n such that $\Psi^*(n, S, F_n)$. To proceed further we must make certain that there exists a set containing all the elements of the form $F_n(n)$ where $n \in N$. (In the case considered in Theorem 1 this set is Z, for the domain of the last variable of the formula Φ which we used in the proof of Theorem 1 was limited to the set Z.) In the case under consideration, the existence of the required set Z follows from the axiom of replacement.

In fact, the uniqueness of F_n implies that the formula

$$\bigvee_{F} [\Psi^*(n, S, F) \wedge (y = F(n))]$$

satisfies the assumption of axiom VII. Hence by means of axiom VII the image of N obtained by this formula exists. This image is the required set Z containing all the elements $F_n(n)$.

The remainder of the proof is analogous to that of Theorem 1.

Example. Let $\Phi(S, t)$ be the formula $t = \boldsymbol{P}(S)$. Thus for any set S there exists a sequence φ such that $\varphi_0 = S$ and $\varphi_{n'} = \boldsymbol{P}(\varphi_n)$ for every natural number n.

[1]) Theorem 2 is a scheme: for each formula we have a separate theorem.

§3. The mapping J of the set $N\times N$ onto N and related mappings

Using definitions by induction, we shall now define several mappings important in the sequel.

1. *The mapping J of the set $N\times N$ onto N.* Let for $x, y \in N$

$$J(x, y) = \binom{x+y+1}{2}+x.$$

THEOREM 1: *J is a one-to-one mapping of $N\times N$ onto N.*

PROOF. Suppose that $J(x, y) = J(a, b)$. We shall first prove that $x = a$. In fact, if we suppose $x > a$, then $x = a+r$, $r > 0$. Thus we would obtain

$$\binom{a+r+y+1}{2}+r = \binom{a+b+1}{2}. \tag{1}$$

This implies $b > r+y$, for $\binom{x}{2}$ is an increasing function. Hence $b = r+ +y+s$ where $s > 0$. Substituting this value for b in (1) and letting $c = a+r+y+1$, we obtain $\binom{c}{2}+r = \binom{c+s}{2}$. But this is not true for $r < c$, since $\binom{c}{2}+r < \binom{c}{2}+c = \binom{c+1}{2} \leqslant \binom{c+s}{2}$. In the same way it can be shown that $x < a$ does not hold. Hence $x = a$ and we obtain $\binom{a+y+1}{2} = \binom{a+b+1}{2}$. If $y < b$ then we would have $b = y+t$, $t > 0$; and we would obtain $\binom{a+b+1}{2} \geqslant \binom{a+y+2}{2} > \binom{a+y+1}{2}$. Likewise we can derive a contradiction from the assumption that $y > b$. Therefore the function J is one-to-one.

Now we shall prove that the range Z of J is identical with N. It follows from $J(0, 0) = 0$ and $J(0, 1) = 1$ that $0, 1 \in Z$. Suppose that $n \in Z$, i.e. that $n = J(x, y)$ for some x and y. If $y > 0$ then

$$n+1 = J(x, y)+1 = \binom{x+y+1}{2}+x+1 = J(x+1, y-1) \in Z.$$

If $y = 0$ then

$$n = \binom{x}{2}+x = \binom{x+1}{2}, \quad \text{thus} \quad n+1 = \binom{x+1}{2}+1.$$

Assuming that $x > 0$, we can write $\binom{x+1}{2}+1$ in the form $\binom{1+(x-1)+1}{2}$

$+1 = J(1, x-1)$; hence $n+1 \in Z$. Finally, if $x = y = 0$ then $n = 0$ and $n+1 = 1$. Hence $n+1 \in Z$. Theorem 1 is thus proved.

THEOREM 2: *There exist functions K, L mapping N onto N such that $J(K(x), L(x)) = x$. Moreover, these functions satisfy the inequalities*

$$K(x) \leqslant x, \quad L(x) \leqslant x. \tag{2}$$

The existence of the functions K and L follows from Theorem II.6.7; the inequalities follow from $x \leqslant J(x, y)$ and $y \leqslant J(x, y)$.

REMARK: The intuitive meaning of the functions J, K, L can be illustrated by arranging the pairs $\langle x, y\rangle$ of natural numbers into the following infinite array:

$$\begin{matrix} \langle 0,0\rangle & \langle 0,1\rangle & \langle 0,2\rangle & \ldots \\ \langle 1,0\rangle & \langle 1,1\rangle & \langle 1,2\rangle & \ldots \\ \langle 2,0\rangle & \langle 2,1\rangle & \langle 2,2\rangle & \ldots \\ \ldots & \ldots & \ldots & \ldots \end{matrix} \tag{3}$$

and then ordering them in the sequence

$$\langle 0,0\rangle, \langle 0,1\rangle, \langle 1,0\rangle, \langle 0,2\rangle, \langle 1,1\rangle\ \langle 2,0\rangle, \ldots \tag{4}$$

The pair $\langle x, y\rangle$ occurs in the $J(x, y)$th position in this sequence. The nth term of this sequence occurs in the $(K(n)+1)$th row and the $(L(n)+1)$th column of (3).

2. *The mapping of the set N^{n+1} onto N.* We shall define by induction a sequence of one-to-one functions such that the kth term of this sequence (denoted by τ_k) is a one-to-one mapping of the set N^{k+1} onto N. Identifying every one-term sequence with its only term, we let

$$\tau_0(x) = x \quad \text{for} \quad x \in N,$$

$$\tau_{k+1}(e) = J(\tau_k(e|k'), e_{k+1}) \quad \text{for} \quad e \in N^{k+2}.$$

THEOREM 3: *The function τ_k maps N^{k+1} onto N and is one-to-one.*

PROOF. For $k = 0$ the theorem is obvious. Suppose now that it holds for the number k. If $e \in N^{k+2}$ then $e|k' \in N^{k+1}$, whence $\tau_k(e|k') \in N$ and, by definition, $\tau_{k+1}(e) \in N$. The function τ_{k+1} thus maps N^{k+1} into N.

The fact that τ_{k+1} is one-to-one follows from the implications:

$$\tau_{k+1}(e) = \tau_{k+1}(e^*) \to \big(\tau_k(e|k') = \tau_k(e^*|k')\big) \wedge (e_{k+1} = e^*_{k+1})$$

$$\to (e|k' = e^*|k') \wedge (e_{k+1} = e^*_{k+1}) \to (e = e^*).$$

It remains to be shown that for every number n there exists a sequence $e \in N^{k+2}$ such that $\tau_k(e) = n$. By the inductive assumption there exists a sequence $f \in N^{k+1}$ such that $\tau_k(f) = K(n)$. We let e to be the sequence whose $k+1$ initial terms coincide with those of f and whose last term is $L(n)$. For this sequence e the following formula holds by definition:

$$\tau_{k+1}(e) = J\big(\tau_k(f), L(n)\big) = J\big(K(n), L(n)\big) = n.$$

3. *The mapping of the set of all finite sequences of natural numbers onto the set N.* Let for $e \in N^{k+1}$

$$\sigma_0(e) = J\big(k, \tau_k(e)\big).$$

This function is a one-to-one mapping of the set of all non-empty finite sequences of natural numbers onto N. We have the following:

THEOREM 4: *There exists a one-to-one mapping σ of the set $\mathcal{S}$ of all finite sequences of natural numbers onto N which satisfies the condition $\sigma(0) = 0$.*

To prove this it suffices to let $\sigma(e) = 1+\sigma_0(e)$ for non-empty sequences e and $\sigma(0) = 0$.

4. *The mapping J' of the set $N^{n+1} \times N^{n+1}$ onto the set N^{n+1} and of the set $N^N \times N^N$ onto the set N^N.* Let n be a natural number. For e, $f \in N^{n+1}$, let

$$J'(e,f) = \underset{k\leqslant n}{F}[J(e_k, f_k)],$$

$$K'(e) = \underset{k\leqslant n}{F}[K(e_k)], \quad L'(e) = \underset{k\leqslant n}{F}[L(e_k)].$$

Hence if e and f are sequences with $n+1$ terms e_k and f_k $(k \leqslant n)$, respectively, then $J'(e,f)$ is the sequence of $n+1$ terms $J(e_k, f_k)$ $(k \leqslant n)$; and $K'(e)$ and $L'(e)$ are sequences with the terms $K(e_k)$ and $L(e_k)$ $(k \leqslant n)$, respectively.

Similar definitions will be applied to infinite sequences of natural numbers. Let φ, $\psi \in N^N$ and let

$$J^*(\varphi, \psi) = \underset{k \in N}{F}[J(\varphi_k, \psi_k)],$$

$$K^*(\varphi) = \underset{k \in N}{F}[K(\varphi_k)], \quad L^*(\varphi) = \underset{k \in N}{F}[L(\varphi_k)].$$

$J^*(\varphi, \psi)$ is thus an infinite sequence whose kth term is $J(\varphi_k, \psi_k)$, $K^*(\varphi)$ and $L^*(\varphi)$ are infinite sequences whose kth terms are $K(\varphi_k)$ and $L(\varphi_k)$, respectively.

THEOREM 5: *The function J' is a one-to-one mapping of the set $N^{n+1} \times \times N^{n+1}$ onto N^{n+1}; the function J^* is a one-to-one mapping of $N^N \times N^N$ onto N^N.*

THEOREM 6: *For any $e \in N^{n+1}$ and for any $\varphi \in N^N$ the following formulas hold*:

$$J'\big(K'(e), L'(e)\big) = e, \quad J^*\big(K^*(\varphi), L^*(\varphi)\big) = \varphi.$$

The proof of these theorems is left to the reader.

5. *The mapping of the set N^N onto the set $(N^N)^N$.* For $k \in N$ and for $\varphi \in N^N$ let

$$\varphi^{(k)} = \underset{n \in N}{F}[\varphi_{J(k,n)}];$$

thus the sequence $\varphi^{(k)}$ has as terms $\varphi_{J(k,0)}, \varphi_{J(k,1)}, \varphi_{J(k,2)}, \ldots$.

THEOREM 7: *The function M: $\varphi \rightarrow \underset{k \in N}{F}[\varphi^{(k)}]$ (which associates with every sequence φ the sequence $\varphi^{(0)}, \varphi^{(1)}, \varphi^{(2)}, \ldots$) is a one-to-one mapping of the set N^N onto the set $(N^N)^N$.*

PROOF. For every $\varphi \in N^N$ we clearly have $M(\varphi) \in \big(N^N\big)^N$. The function M is one-to-one. In fact, it follows from $M(\varphi) = M(\psi)$ that $\varphi^{(k)} = \psi^{(k)}$ for every natural number k. Thus $\varphi_{J(k,m)} = \psi_{J(k,m)}$ for any k, m, and by letting $k = K(n)$ and $m = L(n)$ we obtain $\varphi_n = \psi_n$ for all n. Finally, every element of the set $(N^N)^N$, that is, every infinite sequence t whose terms t_k are elements of N^N for all natural k, can be represented as $M(\varphi)$ for some φ. In fact, if φ is the sequence $\varphi_n = t_{K(n)}\big(L(n)\big)$, then $\varphi^{(k)}$ is the sequence whose nth term is $t_{K(J(k,n))}\big(L\big(J(k,n)\big)\big) = t_k(n)$, i.e. $\varphi^{(k)} = t_k$ for arbitrary k. Hence $M(\varphi) = t$. Q.E.D.

The theory presented in this and the two preceding sections shows that notions of arithmetic are reducible to set-theoretical ones. Thus e.g., the sentence $2+2 = 4$ is expressible by means of the set-theoretical formulas defined in Section II.1. Thus we can freely use in set theory the notions of arithmetic.

In later sections we shall occasionally use various other mathematical notions such as real numbers, real functions, etc., assuming that they have been defined in set theory by means of the notions of arithmetic. The details of these definitions need not be given here since they are sufficiently well known.

§ 4. Finite and infinite sets

The notions introduced in § 1 and § 2 of this chapter allow us to derive basic properties of finite and infinite sets from the axioms of set theory.

DEFINITION: We say that a *set X has n elements* and we write $|X| = n$ (where $n \in N$) if there exists a one-to-one sequence with domain n and range X. Such a sequence is called a *one-to-one sequence with n terms.*

A set X is *finite* if $|X| = n$ for some $n \in N$, otherwise we say that the set X is *infinite.*

A set X has 0 elements if and only if X is empty, for the only one-to-one sequence with 0 terms is the empty sequence.

For every $p \in N$ the set p has p elements; in fact, the function I_p defined by $I_p(x) = x$ for every $x \in p$ is a one-to-one sequence of p terms whose range is p.

THEOREM 1: *If f is a one-to-one mapping of the set X onto the set Y, then $|X| = n$ if and only if $|Y| = n$.*

PROOF. If e is a one-to-one sequence of n terms with range X, then $f \circ e$ is also a one-to-one sequence (see p. 71) of n terms whose range is Y.

LEMMA: *If f is a one-to-one function, $D(f) = X \cup \{a\}$, $\mathrm{Rg}(f) = Y \cup \cup \{b\}$ and $a \notin X$, $b \notin Y$, then there exists a one-to-one function g such that $X \to Y$ and such that Y is the range of g.*

PROOF. Let $f(a) = a_1$, $f^c(b) = b_1$. If $a_1 = b$ then $b_1 = a$ and the function f maps X onto Y; thus it suffices to take $g = f|X$.

If $a_1 \neq b$ then $a_1 \in Y$ and similarly $b_1 \in X$. In this case, g is defined as follows:

$$g(x) = f(x) \quad \text{if} \quad x \neq b_1, \quad g(b_1) = a_1.$$

Checking that this function satisfies the lemma is left to the reader.

THEOREM 2: *Let $n \in N$. The following conditions are equivalent*:

(i) $|X| = n'$.

(ii) *There exist a set $X_1 \subset X$ and an element $a_1 \notin X_1$ such that $|X_1| = n$ and $X = X_1 \cup \{a_1\}$.*

(iii) *$X \neq \emptyset$ and for every X_2 and a_2, if $a_2 \notin X_2$ and $X = X_2 \cup \{a_2\}$ then $|X_2| = n$.*

PROOF. (i) → (ii). Let e be a one-to-one sequence of n' terms with range X. Condition (ii) is satisfied when $a = e_n$, $X_1 = X - \{a\}$.

(ii) → (iii). The condition $X \neq \emptyset$ is an immediate consequence of (ii). Letting X_1 and a_1 denote respectively the set and element satysfying (ii), we infer that $X_2 \cup \{a_2\} = X_1 \cup \{a_1\}$; and thus there exists a one-to-one function mapping $X_1 \cup \{a_1\}$ onto $X_2 \cup \{a_2\}$. By means of the lemma there exists a one-to-one function g mapping X_1 onto X_2 and hence, by Theorem 1, $|X_2| = n$.

(iii) → (i). Let a be any element of X and let $X_1 = X - \{a\}$. By (iii) we have $|X_1| = n$ and thus X_1 is the range of a one-to-one sequence e of n terms. The sequence e' of n' terms defined by $e'_p = e_p$ for $p \in n$, $e'_n = a$, is one-to-one and its range is X. Hence $|X| = n'$.

THEOREM 3: *If $|X| = m$ and $|Y| = n$, then $m \leqslant n$ if and only if there exists a set $Y_1 \subset Y$ such that the set X is a one-to-one image of Y_1.*

PROOF. If $m \leqslant n$ then $m \subset n$ (see pp. 90 and 92). Suppose that Y is the range of a one-to-one sequence e of n terms and X is the range of a one-to-one sequence f of m terms. The function $e \circ f^c$ is thus one-to-one and maps X onto a subset of Y (for $f^{c1}(X) = m \subset n$).

Suppose now that there exist a set $Y_1 \subset Y$ and a one-to-one function f mapping Y_1 onto X. To prove that $m \leqslant n$ we shall use induction on n.

If $n = 0$ then $Y = \emptyset$; hence $Y_1 = X = \emptyset$ and, moreover, $m = 0$ and $m \leqslant n$. Suppose that the theorem holds for some $n \in N$ and let $|Y| = n'$. Hence $Y = Y_2 \cup \{a\}$, where $|Y_2| = n$ and $a \notin Y_2$. Since $X = f^1(Y_1)$, we have $f^1(X) = f^1(Y_2 \cap Y_1) \cup f^1(\{a\} \cap Y_1)$. The set $\{a\} \cap Y_1$ is either empty or equal to $\{a\}$. In the first case we have $|Y_2 \cap Y| = m$. The formulas $Y_2 \cap Y_1 \subset Y_1$ and $|Y_1| = m$ show that the inductive hypothesis holds; hence $m \leqslant n$ and thus $m < n'$. In the second case Theorem 2 implies $m = p'$ where p is a number such that $|Y_2 \cap Y_1| = p$. By the inductive hypothesis, $p \leqslant n$, hence $m = p' \leqslant n'$. Thus the theorem is proved.

THEOREM 4: *If* $|X| = m$, $|Y| = n$ *and* $X \cap Y = \emptyset$, *then* $|X \cup Y| = m+n$.

PROOF. The proof is by induction on n. For $n = 0$ we have $Y = \emptyset$, and the theorem holds. Suppose that the theorem holds for the number n and let $|Y| = n'$. Thus $Y = Y_1 \cup \{a\}$ where $a \notin Y_1$ and hence $X \cup Y = (X \cup Y_1) \cup \{a\}$, where $a \notin Y$. It follows from the inductive hypothesis that $|X \cup Y_1| = m+n$ and, by Theorem 2, $|X \cup Y_1 \cup \{a\}| = (m+n)'$. This proves the theorem because $(m+n)' = m+n'$ by the definition of addition (see p. 94).

COROLLARY 5: *For an arbitrary U the finite sets contained in U form an ideal.*

In fact, a subset of a finite set is finite by Theorem 3, and the union of finite sets is finite by Theorems 3 and 4.

THEOREM 6: *If* $|X| = m$ *and* $|Y| = n$, *then the set X is a one-to-one image of Y if and only if* $m = n$.

PROOF. If $m = n$, then X is obviously a one-to-one image of Y, because there exist one-to-one sequences mapping the set n onto X and Y. Now suppose that X is a one-to-one image of Y. Then m is a one-to-one image of n. We shall prove by induction that $m = n$.

For $n = 0$ the theorem is obvious. Suppose that it is true for some n and let m be a one-to-one image of n'. Hence $m \neq 0$ and we may assume $m = p'$. Because $p' = p \cup \{p\}$ and $n' = n \cup \{n\}$, p is, according to the lemma, a one-to-one image of n. Hence by the inductive hypothesis $p = n$ and finally $m = p' = n'$, which proves the theorem.

COROLLARY 7: (The so-called *drawer principle of Dirichlet*) *If* $|X| = m$, $|Y| = n$, $m > n$ *and f is a function such that* $f^1(X) = Y$, *then the function f is not one-to-one.*

Dirichlet formulated this theorem as follows:

If m objects are put into n drawers and m > n, then at least one of the drawers contains at least two objects.

Obviously, our function f is the function assigning to each object the drawer in which it is contained.

We shall now apply the theorems just proved to draw certain conclusions about infinite sets.

THEOREM 8: *If a set X is infinite and $X \subset Y$, then the set Y is infinite.*

This is an immediate consequence of Theorem 3.

THEOREM 9: *If X is infinite and Y finite, then the difference $X - Y$ is infinite.*

This follows from Theorem 4.

THEOREM 10: *The set N is infinite.*

PROOF. By way of a contradiction suppose that $|N| = n$ where $n \in N$. Since $n' \subset N$, we infer from Theorem 3 that the set n' has m elements, where m is a certain element of N such that $m \leqslant n$. Because $|p| = p$ for every $p \in N$, we obtain $n' \leqslant n$, which contradicts formula (4) (p. 91).

COROLLARY 11: *The range of a one-to-one infinite sequence is infinite.*

In fact, such a set is a one-to-one image of N. If it were finite then, by Theorem 1, N would also be finite.

From this theorem and from Theorem 8 we obtain

THEOREM 12: *If a set X contains a subset which is the set of terms of an infinite one-to-one sequence, then X is infinite.*

A set satisfying the hypothesis of Theorem 12 is called a *Dedekind infinite set.*[1]) Theorem 12 can thus be expressed as follows:

If a set is Dedekind infinite, then it is infinite.

The converse theorem is also true, but its proof requires the axiom of choice.

°THEOREM 13: *If a set X is infinite, then it is Dedekind infinite.*

PROOF. Let f be a choice function for the family $\boldsymbol{P}(X) - \{\emptyset\}$. We extend f by letting $f(\emptyset) = x$, where x is an arbitrary fixed element of

[1]) The definition of Dedekind infinite sets is due to Dedekind [1].

X. Thus f assigns an element of X to every subset of X and in particular $f(Y) \in Y$ if $Y \neq \emptyset$.

Now let us define by induction two sequences $A \in (\boldsymbol{P}(X))^N$ and $a \in X^N$:

$$a_0 = x, \qquad A_0 = \{x\},$$
$$a_{n'} = f(X - A_n), \qquad A_{n'} = A_n \cup \{a_{n'}\}.$$

We can prove by induction that the set A_n is finite and $A_m \subset A_n$ for $m \leqslant n$. Hence $X - A_n \neq \emptyset$ and thus $a_{n'} \in X - A_n$ and $a_n \in A_n$ for every $n \in N$. The sequence a is one-to-one. In fact, if $k < j$ then $a_k \in A_k$, but $a_j \notin A_k$, for, letting $j = n'$, we have $A_k \subset A_n$ and $a_j = a_{n'} \in X - A_n \subset X - A_k$. Hence $a_j \neq a_k$. The set X is therefore Dedekind infinite, because it contains the subset $\{a_n : n \in N\}$ which is the set of terms of an infinite one-to-one sequence.

Exercises

1. If $f \in X^Y$, Y is an infinite set and X is finite, then at least one of the cosets of f is infinite (*Dirichlet's principle for infinite sets*).

2. If $|X| = m$ and $|Y| = n$, then $|X \times Y| = m \cdot n$ and $|X^Y| = m^n$.

3. If $|X_i| = m$ for $i = 1, 2, 3$, $|X_j \cap X_k| = n_{jk}$ for $j, k = 1, 2, 3$, $|X_1 \cap X_2 \cap X_3| = n_{123}$, then

$$|X_1 \cup X_2 \cup X_3| = n_1 + n_2 + n_3 - n_{12} - n_{23} - n_{31} + n_{123}.$$

Generalize this formula to the case of an arbitrary (finite) number of sets.

4. Prove that a set $X \subset N$ is infinite if and only if $\bigwedge_m \bigvee_{x \in X} (m < x)$.

CHAPTER IV

GENERALIZED UNION, INTERSECTION AND CARTESIAN PRODUCT

In the present chapter our treatment will be based upon the axioms Σ° as in the preceding chapters. Theorems which are not marked by the symbol $^\circ$ are theorems of the system Σ.

The purpose of this chapter is to generalize the operations of union, intersection and cartesian product for an arbitrary number of sets.

§ 1. Set-valued functions. Generalized union and intersection

Let F be a function from a non-empty set T into the family of all subsets of a given fixed set $\mathscr{X}$. Thus $F(t) \in \big(P(X)\big)$. Instead of $F(t)$ we shall write F_t.

Let W be the range of F, that is, the family of sets F_t where $t \in T$. The union of the sets belonging to the family W is denoted by $\bigcup_t F_t$ or $\bigcup(W)$ (see p. 53), the intersection is denoted by $\bigcap_t F_t$ or $\bigcap(W)$ (see p. 60).

It is easy to show that

(1) $$x \in \bigcup_t F_t \equiv \bigvee_t (x \in F_t), \qquad x \in \bigcap_t F_t \equiv \bigwedge_t (x \in F_t).$$

If the set T consists of the single element a, then

$$\bigcup_{t \in T} F_t = F_a = \bigcap_{t \in T} F_t;$$

on the other hand, if T consists of two elements a and b, then

$$\bigcup_{t \in T} F_t = F_a \cup F_b \quad \text{and} \quad \bigcap_{t \in T} F_t = F_a \cap F_b.$$

Thus these notions are indeed generalizations of the notions of union and intersection of sets to the case of an arbitrary family of sets.

It follows from (1) that the following equations hold for arbitrary formula $\Phi(x, y)$ of two variables with a limited domain:

$$\bigcup_y \{x: \Phi(x, y)\} = \{x: \bigvee_y \Phi(x, y)\},$$
(2)
$$\bigcap_y \{x: \Phi(x, y)\} = \{x: \bigwedge_y \Phi(x, y)\}.$$

In fact, let $F_y = \{x: \Phi(x, y)\}$; we obtain

$$\Phi(z, y) \equiv z \in \{x: \Phi(x, y)\} \equiv z \in F_y,$$

and thus

$$z \in \{x: \bigvee_y \Phi(x, y)\} \equiv \bigvee_y \Phi(z, y)$$
$$\equiv \bigvee_y (z \in F_y) \equiv z \in \bigcup_y F_y \equiv z \in \bigcup_y \{x: \Phi(x, y)\}.$$

The proof of the second equation in (2) is similar.

By (1), the formulas concerning quantifiers given in II.1 lead to the following formulas for the generalized operations:

(3) $$\bigcap_t F_t \subset F_t \subset \bigcup_t F_t,$$

(4) $$\bigcap_t (F_t \cap G_t) = \bigcap_t F_t \cap \bigcap_t G_t,$$

(5) $$\bigcup_t (F_t \cup G_t) = \bigcup_t F_t \cup \bigcup_t G_t,$$

(6) $$\bigcap_t F_t \cup \bigcap_t G_t = \bigcup_{ts} (F_t \cup G_s) \subset \bigcap_t (F_t \cup G_t),$$

(7) $$\bigcup_t (F_t \cap G_t) \subset \bigcup_{ts} (F_t \cap G_s) = \bigcup_t F_t \cap \bigcup_t G_t,$$

(8) $$-(\bigcap_t F_t) = \bigcup_t (-F_t),$$

(9) $$-(\bigcup_t F_t) = \bigcap_t (-F_t),$$

(10) $$\bigcap_t (A \cup F_t) = A \cup \bigcap_t F_t,$$

(11) $$\bigcup_t (A \cap F_t) = A \cap \bigcup_t F_t,$$

(12) $$[\bigwedge_t (A \subset F_t)] \rightarrow [A \subset (\bigcap_t F_t)],$$

(13) $$[\bigwedge_t (F_t \subset A)] \rightarrow [(\bigcup_t F_t) \subset A].$$

In formulas (8) and (9) the symbol "$-$" denotes complementation with respect to the set $\mathscr{X}$.

The proofs of the formulas above follow directly from the respective formulas in II.1. As an example we prove de Morgan's law (8):

$$\begin{aligned} x \in -(\bigcap_t F_t) &\equiv \neg (x \in \bigcap_t F_t) \\ &\equiv \neg [\bigwedge_t (x \in F_t)] \\ &\equiv \bigvee_t (x \in -F_t) \\ &\equiv x \in \bigcup_t (-F_t), \end{aligned}$$

where we apply successively formulas I.2(2), II.1(6), and (1) above.

The diagram on page 51 also leads to formulas for the generalized operations. It suffices to replace the implication sign $\rightarrow$ by the inclusion sign $\subset$, and Φ by a function F of two arguments having sets as values. In particular, the following important formula holds:

(14) $$\bigcup_t \bigcap_s F_{ts} \subset \bigcap_s \bigcup_t F_{ts}.$$

This inclusion cannot in general be reversed (see p. 51).

THEOREM 1: *The union* $\bigcup_t F_t$ *is the unique set* S *satisfying the conditions*:

(15) $$\bigwedge_t (F_t \subset S),$$

(16) $$\bigwedge_X \{[\bigwedge_t (F_t \subset X)] \rightarrow (S \subset X)\}.$$

The intersection $\bigcap_t F_t$ *is the unique set* P *satisfying the conditions*:

(15′) $$\bigwedge_t (P \subset F_t),$$

(16′) $$\bigwedge_X \{[\bigwedge_t (X \subset F_t)] \rightarrow (X \subset P)\}.$$

In other words, the union $\bigcup_t F_t$ is the *smallest* set containing all the sets F_t and the intersection $\bigcap_t F_t$ is the *largest* set included in each of the sets F_t.

PROOF. It follows from (3) and (13) that the union $\bigcap_t F_t$ satisfies conditions (15) and (16). Conversely, assuming that the set S satisfies these conditions, we infer from (15) and (13) that $\bigcup_t F_t \subset S$. Setting $X = \bigcup_t F_t$ in (16) and applying (3), we obtain $S \subset \bigcup_t F_t$. Hence $S = \bigcup_t F_t$.

The proof for intersection is similar.

THEOREM 2: (GENERALIZED ASSOCIATIVE LAWS) *If* $T = \bigcup_{u\in U} H_u$ *where* H *is a set-valued function with domain* U (i.e. $H \in (\mathbf{P}(T)^U)$), *then*

$$\bigcup_{t\in T} F_t = \bigcup_{u\in U} \bigcup_{t\in H_u} F_t, \tag{17}$$

$$\bigcap_{t\in T} F_t = \bigcap_{u\in U} \bigcap_{t\in H_u} F_t. \tag{18}$$

PROOF. Letting

$$S = \bigcup_{t\in T} F_t \quad \text{and} \quad S_u = \bigcup_{t\in H_u} F_t,$$

we reduce equation (17) to the form

$$S = \bigcup_{u\in U} S_u. \tag{19}$$

By assumption we have $S \supset F_t$ for every $t \in T$, in particular for every $t \in H_u$. Thus $S \supset S_u$ by Theorem 1. On the order hand, suppose that $X \supset S_u$ for arbitrary $u \in U$. If $t \in T$, then there exists $u \in U$ such that $t \in H_u$, whence it follows that $S_u \supset F_t$, and thus $X \supset F_t$. Since t is arbitrary, we conclude that $X \supset S$. Applying Theorem 1 we obtain (19).

The proof of (18) is similar.

THEOREM 3: (GENERALIZED COMMUTATIVE LAWS) *If* φ *is a permutation of the elements of a set* T, *then*

$$\bigcup_{t\in T} F_t = \bigcup_{t\in T} F_{\varphi(t)}, \quad \bigcap_{t\in T} F_t = \bigcap_{t\in T} F_{\varphi(t)}.$$

PROOF. Let $S = \bigcup_t F_{\varphi(t)}$. If $t \in T$, then $t = \varphi(\varphi^i(t))$, and because $S \supset F_{\varphi(u)}$ for arbitrary $u \in T$, in particular for $u = \varphi^i(t)$, we have $S \supset F_t$. Conversely, if X is a set such that $X \supset F_t$ for $t \in T$, then $X \supset F_{\varphi(t)}$, because $\varphi(t) \in T$. Thus $X \supset S$, which shows that S is the smallest set containing all the sets F_t (i.e. $S_t = \bigcup_t F_t$).

The proof of the second formula is similar.

THEOREM 4: (GENERALIZED DISTRIBUTIVE LAWS) [1]) *If*

$$M = \bigcup_{u \in U} T_u \quad \textit{and} \quad \boldsymbol{K} = \{Y \in \boldsymbol{P}(M): \bigwedge_{u \in U} (Y \cap T_u \neq \emptyset)\},$$

then

$$\bigcap_{u \in U} \bigcup_{t \in T_u} F_t = \bigcup_{Y \in \boldsymbol{K}} \bigcap_{t \in Y} F_t, \tag{20}$$

$$\bigcup_{u \in U} \bigcap_{t \in T_u} F_t = \bigcap_{Y \in \boldsymbol{K}} \bigcup_{t \in Y} F_t. \tag{21}$$

PROOF. Suppose that $Y \in \boldsymbol{K}$ and $u \in U$. By the definition of the family $\boldsymbol{K}$ we have $Y \cap T_u \neq \emptyset$; thus there exists $t_0 \in Y \cap T_u$. This implies by (3) that

$$\bigcap_{t \in Y} F_t \subset F_{t_0} \subset \bigcup_{t \in T_u} F_t.$$

Since this inclusion holds for any $u \in U$ (where Y is constant), we infer from Theorem 1 that

$$\bigcap_{t \in Y} F_t \subset \bigcap_{u \in U} \bigcup_{t \in T_u} F_t.$$

Since Y is arbitrary, we obtain by (3) the following inclusion:

$$\bigcup_{Y \in \boldsymbol{K}} \bigcap_{t \in Y} F_t \subset \bigcap_{u \in U} \bigcup_{t \in T_u} F_t. \tag{22}$$

To prove the opposite inclusion, suppose that

$$a \in \bigcap_{u \in U} \bigcup_{t \in T_u} F_t. \tag{23}$$

Let

$$Y = \{t \in M: a \in F_t\}. \tag{24}$$

[1]) The formulation of the general distributive law given in Theorem 4 is due to Tarski [5].

If $u \in U$ then by (23) $a \in \bigcup_{t \in T_u} F_t$. Thus there exists $t \in T_u$ such that $a \in F_t$; hence $t \in Y$, which proves that $Y \cap T_u \neq \emptyset$. By the definition of $\boldsymbol{K}$ we have $Y \in \boldsymbol{K}$. It now follows from (24) that $\bigwedge_{t \in Y} (a \in F_t)$; that is, $a \in \bigcap_{t \in Y} F_t$. This shows that

$$a \in \bigcup_{Y \in \boldsymbol{K}} \bigcap_{t \in Y} F_t. \tag{25}$$

This together with (22) gives (20). To prove (21), replace F_t in (20) by $S - F_t$, where $S = \bigcup_{t \in M} F_t$. Then we obtain

$$\bigcap_{u \in U} \bigcup_{t \in T_u} (S - F_t) = \bigcup_{Y \in \boldsymbol{K}} \bigcap_{t \in Y} (S - F_t),$$

whence, by de Morgan's laws (8) and (9) and by $-(-F_t) = F_t$, we obtain (21).

A more familiar form of the generalized distributive laws is

$$°(20') \qquad \bigcap_{u \in U} \bigcup_{t \in T_u} F_t = \bigcup_{f \in \boldsymbol{L}} \bigcap_{u \in U} F_{f(u)},$$

$$°(21') \qquad \bigcup_{u \in U} \bigcap_{t \in T_u} F_t = \bigcap_{f \in \boldsymbol{L}} \bigcup_{u \in U} F_{f(u)}.$$

where $\boldsymbol{L}$ is the set of all choice functions f for the family $\mathrm{Rg}(T) = \{T_u\colon u \in U\}$.

We shall prove only the first formula. Consider the families

$$\boldsymbol{A} = \Big\{\bigcap_{u \in U} F_{f(u)}\colon f \in \boldsymbol{L}\Big\} \quad \text{and} \quad \boldsymbol{B} = \Big\{\bigcap_{Y \in \boldsymbol{K}} F_t\colon Y \in \boldsymbol{K}\Big\}.$$

In view of (20) the formula to be proved takes on the form $\bigcup(\boldsymbol{A}) = \bigcup(\boldsymbol{B})$ and so it is sufficient to show that $\boldsymbol{A} = \boldsymbol{B}$. Now if $f \in \boldsymbol{L}$ and if we put $Y = \mathrm{Rg}(f)$, then we obtain

$$\bigcap_{u \in U} F_{f(u)} = \bigcap_{t \in Y} F_t. \tag{$*$}$$

Since $Y \in \boldsymbol{K}$, we obtain therefore $\boldsymbol{A} \subset \boldsymbol{B}$. Conversely, if $Y \in \boldsymbol{K}$ and we denote by f_0 a choice function for the family $\{T_u \cap Y\colon u \in U\}$, then $f_0 \in \boldsymbol{L}$ and, since ($*$) is true, we infer $\boldsymbol{B} \subset \boldsymbol{A}$.

It should be stressed that equations (20′) and (21′) require the axiom of choice in the absence of which we could not claim that the function

f_0 exists. Equations (20) and (21) on the contrary are provable without the choice axiom.

We shall now generalize formulas II.8 (1)–(4) concerning images and inverse images of finite unions and intersections, to the case of arbitrary unions and intersections.

THEOREM 5: *Let* $F \in (\boldsymbol{P}(X))^T$ *and let* $f \in Y^{\mathfrak{X}}$. *Then*

$$f^1(\bigcup_t F_t) = \bigcup_t f^1(F_t), \tag{26}$$

$$f^1(\bigcap_t F_t) \subset \bigcap_t f^1(F_t). \tag{27}$$

If the function f is one-to-one, then the inclusion sign in (27) *can be replaced by the identity sign.*

PROOF. It follows from the definition of image that

$$\begin{aligned} y \in f^1(\bigcup_t F_t) &\equiv \bigvee_x [(x \in \bigcup_t F_t) \wedge (y = f(x))] \\ &\equiv \bigvee_x [\bigvee_t ((x \in F_t) \wedge (y = f(x)))] \\ &\equiv \bigvee_t [\bigvee_x ((x \in F_t) \wedge (y = f(x)))] \\ &\equiv \bigvee_t (y \in f^1(F_t)) \equiv y \in \bigcup_t f^1(F_t), \end{aligned}$$

which proves (26). Similarly, by means of II.1 (18) we obtain the following equivalences:

$$\begin{aligned} y \in f^1(\bigcap_t F_t) &\equiv \bigvee_x [(x \in \bigcap_t F_t) \wedge (y = f(x))] \\ &\equiv \bigvee_x \bigwedge_t [(x \in F_t) \wedge (y = f(x))] \\ &\rightarrow \bigwedge_t \bigvee_x [(x \in F_t) \wedge (y = f(x))] \\ &\equiv \bigwedge_t (y \in f^1(F_t)) \equiv y \in \bigcap_t f^1(F_t), \end{aligned}$$

whence it follows that (27) holds.

If the function f is one-to-one, then using (27) for the inverse function f^{-1} and for the sets $f^1(F_t)$ we obtain

$$f^{-1}(\bigcap_t f^1(F_t)) \subset \bigcap_t f^{-1}(f^1(F_t)) = \bigcap_t F_t,$$

and by II.7 (2) it follows that

$$\bigcap_t f^1(F_t) \subset f^1\bigl(\bigcap_t (F_t)\bigr).$$

Since (27) also holds, Theorem 5 is proved.

THEOREM 6: *If* $G \in (\boldsymbol{P}(Y))^T$ *and* $f \in Y^{\mathfrak{X}}$ *then*

(28) $$f^{-1}\bigl(\bigcup_t G_t\bigr) = \bigcup_t f^{-1}(G_t),$$

(29) $$f^{-1}\bigl(\bigcap_t G_t\bigr) = \bigcap_t f^{-1}(G_t).$$

The proof can be obtained from the following equivalences which are consequences of the definition of the inverse image (see p. 75).

$$y \in f^{-1}\bigl(\bigcup_t G_t\bigr) \equiv f(y) \in \bigcup_t G_t \equiv \bigvee_t [f(y) \in G_t]$$

$$\equiv \bigvee_t [y \in f^{-1}(G_t)] \equiv y \in \bigcup_t f^{-1}(G_t);$$

$$y \in f^{-1}\bigl(\bigcap_t G_t\bigr) \equiv f(y) \in \bigcap_t G_t \equiv \bigwedge_t [f(y) \in G_t]$$

$$\equiv \bigwedge_t [y \in f^{-1}(G_t)] \equiv y \in \bigcap_t f^{-1}(G_t).$$

Formulas (26) and (28) assert the additivity of the operation of forming images and inverse images. Formula (29) asserts that the operation of forming inverse images is multiplicative. The operation of forming images is multiplicative, however, only for one-to-one functions.

Examples

Let the set 1 be a topological space (see I.8).

1. *If F is a function whose values are closed sets* (see p. 28), *then the intersection* $P = \bigcap_t F_t$ *is also a closed set.*

PROOF. Since $P \subset F_t$, we have $\overline{P} \subset \overline{F_t}$ for every t; thus $\overline{P} \subset F_t$, because $\overline{F_t} = F_t$. This implies that $\overline{P} \subset \bigcap_t F_t = P$, hence $\overline{P} = P$, for $P \subset \overline{P}$ by axiom I.8 (3).

2. *If G is a function whose values are open sets, then the union* $S = \bigcup_t G_t$ *is an open set.*

PROOF. The sets $1-G_t$ are closed, thus the intersection $\bigcap_t (1-G_t)$ is also closed. By de Morgan's law (9) the set $1-S$ is closed; hence the set S is open.

3. *If D is a function whose values are regular closed sets*[1]) (cf. p. 39), *then the set $S_0 = \overline{\bigcup_t D_t}$ is a regular closed set containing all the sets D_t. Moreover, every regular closed set containing all the sets D_t also contains the set S_0.*

PROOF. Clearly, $S_0 \supset D_t$, so that $\mathrm{Int}(S_0) \supset \mathrm{Int}(D_t)$; thus

$$\text{(i)} \qquad \overline{\mathrm{Int}(S_0)} \supset \overline{\mathrm{Int}(D_t)} = D_t.$$

Since t is arbitrary, we infer by Theorem 1 that

$$\overline{\mathrm{Int}(S_0)} \supset \bigcup_t D_t \quad \text{and} \quad \overline{\mathrm{Int}(S_0)} \supset \overline{\bigcup_t D_t} = S_0.$$

On the other hand, $\mathrm{Int}(S_0) \subset S_0$. Thus

$$\overline{\mathrm{Int}(S_0)} \subset \bar{S}_0 = S_0,$$

which proves that $S_0 = \overline{\mathrm{Int}(S_0)}$. Hence the set S_0 is regular closed. It follows from (i) that S_0 contains each set D_t.

If Z is a regular closed set and $Z \supset D_t$ for every t, then

$$Z \supset \bigcup_t D_t, \quad \text{thus} \quad Z = \bar{Z} \supset \overline{\bigcup_t D_t} = S_0.$$

4. *If D is a function whose values are regular closed sets, then the set $P_0 = \overline{\mathrm{Int}\left(\bigcap_t D_t\right)}$ is a regular closed set included in each set D_t. Moreover, every regular closed set included in each set D_t is also included in P_0.*

PROOF. Let $X = \bigcap_t D_t$. We thus have

$$P_0 = \overline{\mathrm{Int}(X)} = X^{c-c-} \quad \text{hence} \quad \overline{\mathrm{Int}(P_0)} = X^{c-c-c-c-}.$$

Applying formula I. 8 (15), we have

$$\overline{\mathrm{Int}(P_0)} = X^{c-c-} = P_0.$$

[1]) The notion of a regular closed set is due to Lebesgue [1]. Theorems 3, 4, 5 were proved by Kuratowski [6]. Regular closed sets and their complements called the *regular open sets* have found numerous applications in particular in proofs of independence of set theoretic hypotheses from the axioms. See e.g. Rosser [1] and Takeuti and Zaring [1]. The regular sets are called also *domains* (closed or open).

Hence the set P_0 is regular closed.

Since $X \subset D_t$, we have $\mathrm{Int}(X) \subset \mathrm{Int}(D_t)$ and $\overline{\mathrm{Int}(X)} \subset \overline{\mathrm{Int}(D_t)}$, that is, $P_0 \subset \overline{\mathrm{Int}(D_t)} = D_t$ for every t.

Finally, if Z is a regular closed set and $Z \subset D_t$ for every t, then $Z \subset \bigcap_t D_t$. Hence $\mathrm{Int}(Z) \subset \mathrm{Int}(X)$ and $Z = \overline{\mathrm{Int}(Z)} \subset \overline{\mathrm{Int}(X)} = P_0$.

5. As a result of the theorems proved in Examples 1 and 2, it is possible to define a topological space by taking as primitive notion either that of an open set or that of a closed set instead of closure.

Namely, we may conceive of a *topological space* as a set with a distinguished family of subsets $\boldsymbol{F}$. Subsets belonging to the family $\boldsymbol{F}$ are called *closed sets*. We suppose that $\boldsymbol{F}$ satisfies two conditions:

(i) *If* $\boldsymbol{W} \subset \boldsymbol{F}$, *then* $\bigcap(\boldsymbol{W}) \in \boldsymbol{F}$ (*that is, the intersection of an arbitrary family of closed sets is closed*).

(ii) *If a family* $\boldsymbol{W}$ *is finite and* $\boldsymbol{W} \subset \boldsymbol{F}$, *then* $\bigcup(\boldsymbol{W}) \in \boldsymbol{F}$ (*that is, the union of a finite number of closed set is closed*).

We obviously assume that $\bigcap \emptyset$ is the whole space.

If we take the notion of an open set as primitive, then denoting the family of open sets by $\boldsymbol{G}$ we assume axioms dual to (i) and (ii):

(i′) *If* $\boldsymbol{W} \subset \boldsymbol{G}$ *then* $\bigcup(\boldsymbol{W}) \in \boldsymbol{G}$.

(ii′) *If a family* $\boldsymbol{W}$ *is finite and* $\boldsymbol{W} \subset \boldsymbol{G}$, *then* $\bigcap(\boldsymbol{W}) \in \boldsymbol{G}$.

The system of axioms (i)–(ii) is equivalent to the system I. 8 (1)–(4). The axioms (1)–(4) are satisfied if we define $\bar{A}$ by the formula $\bar{A} = \bigcap(\boldsymbol{W}_A)$, where $\boldsymbol{W}_A$ is the family of all closed sets containing the set A. Then we have $(A = \bar{A}) \equiv (A \in \boldsymbol{F})$.

A similar remark can be made for the system (i′)–(ii′).

6. A family $\boldsymbol{R} \subset \boldsymbol{F}$ is said to be a *closed base* for the topological space if for every $A \in \boldsymbol{F}$ there exists $\boldsymbol{W} \subset \boldsymbol{R}$ such that $A = \bigcap(\boldsymbol{W})$. A family $\boldsymbol{R} \subset \boldsymbol{F}$ is a *closed subbase* if the family of all finite unions of the sets belonging to $\boldsymbol{R}$ is a closed base.

7. The notion of *open base* and *subbase* can be defined dually replacing $\boldsymbol{F}$ by $\boldsymbol{G}$ (= the family of open sets), intersection by union and union by intersection.

Exercises

1. Let $F \in (P(\mathscr{X}))^T$, $f \in \mathscr{Y}^{\mathscr{X}}$ and $\mathscr{X} = \bigcup_t F_t$. Let $f_t = f|F_t$. Prove that

$$f^{-1}(Y) = \bigcup_t f_t^{-1}(Y) \quad \text{for every } Y \subset \mathscr{Y}.$$

2. Prove that

$$(\bigcup_t F_t) \times (\bigcup_u G_u) = \bigcup_{t,u} (F_t \times G_u),$$

$$(\bigcap_t F_t) \times (\bigcap_u G_u) = \bigcap_{t,u} (F_t \times G_u).$$

3. Let T be any set and let $\boldsymbol{K} \subset \boldsymbol{P}(T)$. Let the operation $D_{\boldsymbol{K}}$ on $F \in (\boldsymbol{P}(X))^T$ be defined by the formula

$$x \in D_{\boldsymbol{K}}(F) \equiv \{t: x \in F_t\} \in \boldsymbol{K}.$$

Find $\boldsymbol{K}$ for which the operation $D_{\boldsymbol{K}}$ coincides with the operations of union and intersection discussed above.

4. Show that if $\boldsymbol{I}$ is an ideal in $\boldsymbol{P}(T)$ and $\boldsymbol{K} = \boldsymbol{P}(T) - \boldsymbol{I}$, then the operation $D_{\boldsymbol{K}}$ is distributive over finite unions; that is,

$$\bigwedge_t (F_t = G_t \cup H_t) \to (D_{\boldsymbol{K}}(F) = D_{\boldsymbol{K}}(G) \cup D_{\boldsymbol{K}}(H)).$$

5. Prove that the family of all intervals $r < x < s$, where r and s are rational numbers, is a base for the space $\mathscr{E}$ of real numbers.

Prove that the sets $\{x: r < x\}$ and $\{x: x < r\}$, where r is rational, form an open subbase for this space.

6. Let X be any set and $\boldsymbol{R}$ be any family of its subsets. Prove that the set X can be considered as a topological space with the family $\boldsymbol{R}$ as an open subbase (resp. closed subbase).

7. If X is a topological space and R is an equivalence relation with field X, then X/R becomes a topological space when we assume that a set $\boldsymbol{U} \subset X/R$ is open if and only if the union $S(U) = \bigcup_{Z \in U} \boldsymbol{Z}$ is an open set in X.

8. Prove that the canonical mapping $X \to X/R$ is continuous if X/R has the quotient topology defined in Exercise 7.

§ 2. Operations on infinite sequences of sets

We shall now consider a special case of the previous operations; namely, where the domain T of the function F coincides with N, that is, where F is an infinite sequence of sets. In analogy with infinite series and products of real numbers, we write

$$\bigcup_n F_n \text{ or } \bigcup_{n=0}^{\infty} F_n \text{ or } F_0 \cup F_1 \cup \ldots \quad \text{instead of} \quad \bigcup_{n \in N} F_n;$$

$$\bigcap_n F_n \text{ or } \bigcap_{n=0}^{\infty} F_n \text{ or } F_0 \cap F_1 \cap \ldots \quad \text{instead of} \quad \bigcap_{n \in N} F_n.$$

The following formulas follow immediately from formulas 1 (2)

(1)
$$\bigcup_{n=0}^{\infty} \{x\colon \Phi(n, x)\} = \{x\colon \bigvee_{n=0}^{\infty} \Phi(n, x)\},$$
$$\bigcap_{n=0}^{\infty} \{x\colon \Phi(n, x)\} = \{x\colon \bigwedge_{n=0}^{\infty} \Phi(n, x)\},$$

where $\Phi(n, x)$ is a formula with two free variables, n is limited to N and x to a given set X.

Besides infinite union and intersection we consider the operations

$\operatorname*{Lim\,sup}_{n=\infty} F_n$ (*limit superior* of the sequence $F_0, F_1, \ldots$),

$\operatorname*{Lim\,inf}_{n=\infty} F_n$ (*limit inferior* of the sequence $F_0, F_1, \ldots$),

defined as follows:

$$\operatorname*{Lim\,sup}_{n=\infty} F_n = \bigcap_{n=0}^{\infty} \bigcup_{k=0}^{\infty} F_{n+k}, \qquad \operatorname*{Lim\,inf}_{n=\infty} F_n = \bigcup_{n=0}^{\infty} \bigcap_{k=0}^{\infty} F_{n+k}.$$

It is easy to check that $\operatorname{Lim\,sup} F_n$ is the set of those elements x which belong to F_n for infinitely many n. Analogously, x belongs to $\operatorname{Lim\,inf} F_n$ if and only if it belongs to F_n for almost all n; that is, if it belongs to all but a finite number of the F_n.

It is easily seen that

(2)
$$\operatorname*{Lim\,inf}_{n=\infty} F_n \subset \operatorname*{Lim\,sup}_{n=\infty} F_n.$$

(see formula II.1 (18)).

If the inclusion sign in (2) can be replaced by the equality sign, that is, if the superior and inferior limits are equal, then their common value is denoted by

$$\operatorname*{Lim}_{n=\infty} F_n,$$

and is called the *limit* of the sequence $F_0, F_1, \ldots$. In this case we also say that the sequence is *convergent*.

This terminology is similar to that used in the theory of real numbers. In order to emphasize this analogy, let us consider the notion of the characteristic function of a given set.

Let the set 1 be given and $X \subset 1$. The function with domain 1

$$f_X(x) = \begin{cases} 1 & \text{if} \quad x \in X, \\ 0 & \text{if} \quad x \in 1 - X \end{cases} \tag{3}$$

is said to be the *characteristic function* of the set X.[1])

It is easy to show that the sequence $F_0, F_1, \ldots$ of subsets of 1 is convergent if and only if the sequence of the characteristic functions of these sets converges to the characteristic function of $\operatorname*{Lim}_{n=\infty} F_n$.

It is also easy to show that the following conditions are equivalent:

$$\operatorname*{Lim}_{n=\infty}(F_n \dot{-} A) = \emptyset, \tag{4}$$

$$\operatorname*{Lim}_{n=\infty} F_n = A, \tag{4'}$$

where the sign $\dot{-}$ denotes the symmetric difference of two sets. The same equivalence holds for real numbers if we replace $F_n \dot{-} A$ by $|F_n - A|$.

PROOF. Condition (4) is equivalent to the following: every element x belongs to $F_n \dot{-} A$ for at most finitely many n. In other words, for every x there exists n_0 such that $n > n_0$ implies

$$x \in F_n \equiv x \in A. \tag{5}$$

Suppose that $x \in \operatorname{Lim\,sup} F_n$, i.e. that x belongs to F_n for infinitely many n. It follows from (5) that $x \in A$ and that $x \in F_n$ for all $n > n_0$; that is, $x \in \operatorname{Lim\,inf} F_n$. Thus we have proved that (4) implies

$$\operatorname*{Lim\,sup}_{n=\infty} F_n \subset A \subset \operatorname*{Lim\,inf}_{n=\infty} F_n, \tag{6}$$

from which (4′) follows by (2).

Conversely, suppose that (6) holds and $x \in A$. Thus $x \in \operatorname{Lim\,inf} F_n$ and $x \in F_n$ for all n greater than some n_0. If, on the other hand, $x \notin A$,

[1]) The characteristic functions of a set were introduced by de la Vallée Poussin [1]. They proved very convenient in the theory of real functions. Analogies between limits of real sequences and sequences of sets expressed e.g. in formulas (4) and (4′) were stressed by Marczewski [3].

then $x \notin \operatorname{Limsup} F_n$ and hence $x \notin F_n$ from an n_0 on. Hence condition (6) implies that (5) holds for every x if $n > n_0$.

Exercises

1. Prove that the characteristic function defined by (3) satisfies the following conditions:

(a) $f_{\emptyset}(x) = 0$,

(b) $f_1(x) = 1$,

(c) $f_{-X}(x) = 1 - f_X(x)$,

(d) $f_{A \cap B}(x) = f_A(x) \cdot f_B(x)$,

(e) $f_{A-B}(x) = f_A(x) - f_{A \cap B}(x)$.

2. Prove that if $F_0 \subset F_1 \subset F_2 \subset \ldots$, then $\bigcup_{n=0}^{\infty} F_n = \operatorname{Lim}_{n=\infty} F_n$.

3. Prove that if $F_0 \supset F_1 \supset \ldots$, then $\bigcap_{n=0}^{\infty} F_n = \operatorname{Lim}_{n=\infty} F_n$.

4. Prove that if $F_0 = 1$, then

$$1 = (F_0 - F_1) \cup (F_1 - F_2) \cup (F_2 - F_3) \cup \ldots \cup \bigcap_{n=0}^{\infty} F_n.$$

If, moreover, $F_0 \supset F_1 \supset F_2 \supset \ldots$, then

$$(F_1 - F_2) \cup (F_3 - F_4) \cup \ldots \cup \bigcap_{n=0}^{\infty} F_n = 1 - [(F_0 - F_1) \cup (F_2 - F_3) \cup \ldots].$$

5. Prove that if $k_1 < k_2 < \ldots$, then

$$\operatorname{Liminf}_{n=\infty} F_n \subset \operatorname{Liminf}_{n=\infty} F_{k_n}, \quad \operatorname{Limsup}_{n=\infty} F_{k_n} \subset \operatorname{Limsup}_{n=\infty} F_n.$$

6. Prove that if $\bigcap_{n=1}^{\infty} A_n \cap \bigcap_{n=1}^{\infty} B_n = \emptyset$, then

$$\bigcap_{n=1}^{\infty} A_n \subset \bigcup_{n=1}^{\infty} [A_n \cap (B_{n-1} - B_n)]$$

where $B_0 = 1$.

7. $\bigcup_{n=1}^{\infty} A_n = \bigcup_{n=1}^{\infty} B_n$, where $B_1 = A_1$ and $B_n = A_n - (A_1 \cup \ldots \cup A_{n-1})$ for $n > 1$.

8. Let $\bigcup_n \bigcap_m B_{n,m} = A = \bigcap_m \bigcup_n C_{n,m}$ and let $\bigcup_{n,m} C_{n,m+1} \subset \bigcup_{n,m} C_{n,m}$. Prove that

$$A = \operatorname{Lim}_{n=\infty} A_n, \quad \text{where} \quad A_n = \bigcup_{k=1}^{\infty} (B_{k,1} \cap \ldots \cap B_{k,n}) \cap (C_{1,k} \cup \ldots \cup C_{n,k}).$$

9. Prove that

(a) $\operatorname{Liminf}(-A_n) = -\operatorname{Limsup} A_n$.

(b) $\operatorname{Lim}(-A_n) = -(\operatorname{Lim} A_n)$,
(c) $\operatorname{Liminf}(A_n \cap B_n) = \operatorname{Liminf} A_n \cap \operatorname{Liminf} B_n$,
(d) $\operatorname{Limsup}(A_n \cup B_n) = \operatorname{Limsup} A_n \cup \operatorname{Limsup} B_n$,
(e) $\bigcap_{n=1}^{\infty} A_n \subset \operatorname{Liminf} A_n \subset \operatorname{Limsup} A_n \subset \bigcup_{n=1}^{\infty} A_n$,
(f) $\operatorname{Liminf} A_n \cup \operatorname{Liminf} B_n \subset \operatorname{Liminf}(A_n \cup B_n)$,
(g) $\operatorname{Limsup}(A_n \cap B_n) \subset \operatorname{Limsup} A_n \cap \operatorname{Limsup} B_n$,
(h) $A \dot{-} \operatorname{Liminf} A_n \subset \operatorname{Limsup}(A \dot{-} A_n)$,
$A \dot{-} \operatorname{Limsup} A_n \subset \operatorname{Limsup}(A \dot{-} A_n)$;

show that the opposite inclusions do not hold in general.

10. A function f from sets into sets is said to be *continuous* if for every convergent sequence $F_1, F_2, \ldots$ the following identity holds:

$$f(\operatorname*{Lim}_{n=\infty} F_n) = \operatorname*{Lim}_{n=\infty} f(F_n).$$

Show that the functions $X \cup Y$, $X \cap Y$, $-X$ and generally $\bigcup_n F_n$ and $\bigcap_n F_n$ are continuous with respect to each variable.

11. Prove the following condition for a sequence F_n to be convergent: for every sequence of pairs $\langle m_i, n_i \rangle$ such that $\lim_{i=\infty} m_i = \lim_{i=\infty} n_i = \infty$, we have

$$\bigcap_i (F_{m_i} \dot{-} F_{n_i}) = \emptyset.$$

[Marczewski]

12. If $\boldsymbol{K}$ is a family of subsets of N such that the complement of every set in $\boldsymbol{K}$ is finite, $D_{\boldsymbol{K}}(F) = \operatorname{Liminf} F_n$; if $\boldsymbol{K}$ is a family of infinite subsets of N then $D_{\boldsymbol{K}}(F) = \operatorname{Limsup} F_n$ (see § 1, Exercise 3). Using this result, generalize the operations Limsup, Liminf for the case where the argument is a function defined on an arbitrary set T (not necessarily on the set N).

§ 3. Families of sets closed under given operations

Let X be a fixed set and f a function of an arbitrary number of variables, where each variable ranges over the subsets of X. For simplicity let us suppose that f is a function of two variables; that is, the domain of f is the cartesian product $\boldsymbol{P}(X) \times \boldsymbol{P}(X)$.

A family $\boldsymbol{R} \subset \boldsymbol{P}(X)$ is said to be *closed under a given function* f if

$$\bigwedge_{Y_1, Y_2} [(Y_1 \in \boldsymbol{R}) \wedge (Y_2 \in \boldsymbol{R}) \rightarrow (f(Y_1, Y_2) \in \boldsymbol{R})].$$

THEOREM 1: *For each family* $\boldsymbol{R} \subset \boldsymbol{P}(X)$ *there exists a family* $\boldsymbol{R}_1$ *such that*: 1. $\boldsymbol{R} \subset \boldsymbol{R}_1 \subset \boldsymbol{P}(X)$; 2. *the family* $\boldsymbol{R}_1$ *is closed under the operation* f;

3. *the family* $\boldsymbol{R}_1$ *is the least family satisfying conditions* 1 *and* 2, *that is, if* $\boldsymbol{R}'$ *satisfies the following two conditions*

$$(1)\quad \boldsymbol{R} \subset \boldsymbol{R}' \subset \boldsymbol{P}(X) \text{ and } \bigwedge_{Y_1, Y_2} [(Y_1 \in \boldsymbol{R}') \wedge (Y_2 \in \boldsymbol{R}') \to (f(Y_1, Y_2) \in \boldsymbol{R}')],$$

then $\boldsymbol{R}_1 \subset \boldsymbol{R}'$.

PROOF. Let $\boldsymbol{K}$ be the set of all families $\boldsymbol{R}'$ satisfying (1). $\boldsymbol{K}$ is a non-empty set, for $\boldsymbol{P}(X) \in \boldsymbol{K}$. The required family is the intersection $\bigcap_{\boldsymbol{R}' \in \boldsymbol{K}} \boldsymbol{R}'$.

The family $\boldsymbol{R}_1$ satisfying conditions 1–3 is uniquely determined. In fact, if $\boldsymbol{R}_2$ also satisfies the same conditions, then $\boldsymbol{R}_1 \subset \boldsymbol{R}_2$, since $\boldsymbol{R}_1$ is the least such family. Similarly we obtain $\boldsymbol{R}_2 \subset \boldsymbol{R}_1$. Hence $\boldsymbol{R}_1 = \boldsymbol{R}_2$. We denote this family by $\boldsymbol{R}^*$.

THEOREM 2: *For arbitrary families* $\boldsymbol{R}, \boldsymbol{R}_1$ *and* $\boldsymbol{R}_2$ *the following conditions hold*

(i) $\boldsymbol{R} \subset \boldsymbol{R}^*$,

(ii) $\boldsymbol{R}_1 \subset \boldsymbol{R}_2 \to \boldsymbol{R}_1^* \subset \boldsymbol{R}_2^*$,

(iii) $\boldsymbol{R}^{**} = \boldsymbol{R}^*$.

PROOF. Formula (i) follows from Theorem 1 (condition 1). Formula (ii) follows from the fact that $\boldsymbol{R}_2^*$ is a family closed under f and containing $\boldsymbol{R}_1$, thus by minimality, $\boldsymbol{R}_2^* \supset \boldsymbol{R}_1^*$. Finally, condition (iii) can be proved as follows: (i) implies $\boldsymbol{R}^* \subset \boldsymbol{R}^{**}$; since $\boldsymbol{R}^* \supset \boldsymbol{R}^*$ and $\boldsymbol{R}^*$ is closed under f, we obtain $\boldsymbol{R}^{**} \subset \boldsymbol{R}^*$ by minimality.

Theorems similar to 1 and 2 also hold for the case where there is given not one function f but an arbitrary family of such functions and $\boldsymbol{R}^*$ denotes the least family containing $\boldsymbol{R}$ and closed under all these functions. Moreover, the domains of these functions may be sequences of subsets of X. We shall not, however, formulate all of these generalizations.

Example 1. Let f denote the union of sets, i.e. $f(Y_1, Y_2) = Y_1 \cup Y_2$. The least family of sets containing $\boldsymbol{R}$ and closed under f is denoted by $\boldsymbol{R}_s$. This family consists of finite unions of the form $\bigcup_{i<n} Y_i$ where $n \in N$, $n \neq 0$ and $Y = (Y_0, Y_1, \ldots, Y_{n-1})$ is a sequence of sets belonging to $\boldsymbol{R}$; in other words, $Y \in \boldsymbol{R}^n$.

Similarly, if g is the function defined by $g(Y_1, Y_2) = Y_1 \cap Y_2$, then the least family containing $\boldsymbol{R}$ and closed under g is denoted by $\boldsymbol{R}_d$. This family consists of all intersections of the form $\bigcap_{i<n} Y_i$ where $n \in N$, $n \neq 0$, $Y \in \boldsymbol{R}^n$.

Example 2. By a *lattice of sets* we mean any family of sets closed under the operations $A \cup B$ and $A \cap B$. If the lattice contains the space and is also closed under the operation $A - B$ then it is called an *algebra of sets*. The least lattice of sets containing a given family $\boldsymbol{R}$ of sets is said to be *generated by* $\boldsymbol{R}$.

THEOREM 3: *The lattice of sets generated by* $\boldsymbol{R}$ *is identical with the family* $\boldsymbol{R}_{sd}$. *Moreover*, $\boldsymbol{R}_{sd} = \boldsymbol{R}_{ds}$.

PROOF. First we prove the second part of the theorem. Let $Z \in \boldsymbol{R}_{sd}$, that is, $Z = \bigcap_{i<n} Y_i$, where $n \in N$, $n \neq 0$ and $Y_i \in \boldsymbol{R}_s$ for $i < n$. We show by induction that $Z \in \boldsymbol{R}_{ds}$. For $n = 1$ we have $Z = Y_0 \in \boldsymbol{R}_s$ and $R_s \subset \boldsymbol{R}_{ds}$ because $\boldsymbol{R} \subset \boldsymbol{R}_d$. Suppose that the theorem holds for $n = k$ and let Z be the intersection of $k+1$ components Y_i belonging to $\boldsymbol{R}_s$. In particular, let $Y_k = \bigcup_{j<m} T_j$ where $m \in N$, $m \neq 0$ and $T_j \in \boldsymbol{R}$ for $j < m$. Let $Z' = \bigcap_{i<k} Y_i$. By the induction hypothesis, $Z' \in \boldsymbol{R}_{ds}$, thus $Z' = \bigcup_{h<p} V_h$ where $p \in N$, $p \neq 0$ and $V_h \in \boldsymbol{R}_d$ for $h < p$. Since $Z = Z' \cap Y_k$, we have

$$Z = Z' \cap \bigcup_{j<m} T_j = \bigcup_{j<m} (Z' \cap T_j) = \bigcup_{j<m} \bigcup_{h<p} (V_h \cap T_j).$$

It follows now from $V_h \cap T_j \in \boldsymbol{R}_d$ that $Z \in \boldsymbol{R}_{ds}$. We have thus proved that $\boldsymbol{R}_{sd} \subset \boldsymbol{R}_{ds}$. In a similar way we prove the opposite inclusion.

Now let us show that $\boldsymbol{R}_{sd}$ is the lattice of sets generated by $\boldsymbol{R}$. It is clear that the family $\boldsymbol{R}_{sd}$ is included in this lattice, for the operations of union and intersection do not lead out of the lattice. On the other hand, $\boldsymbol{R}_{sd} = (\boldsymbol{R}_{sd})_d = \boldsymbol{R}_{ds} = (\boldsymbol{R}_{ds})_s$; thus the family $\boldsymbol{R}_{sd}$ is closed under both union and intersection. Hence it contains the lattice generated by $\boldsymbol{R}$.

Exercises

1. Let $\boldsymbol{R}_r$ be the least family of subsets of X closed under the operation $Y_1 - Y_2$. Prove that

(a) $\boldsymbol{R}_d \subset \boldsymbol{R}_r$;

(b) if $X \in \boldsymbol{R}$ then $\boldsymbol{R}_s \subset \boldsymbol{R}_r$.

Show that the assumption $X \in \boldsymbol{R}$ in theorem (b) is essential.

2. Prove that the least field of sets containing $\boldsymbol{R}$ is $(\boldsymbol{R} \cup c\boldsymbol{R})_{sd}$ where $c\boldsymbol{R} = \{X - Y: Y \in \boldsymbol{R}\}$.

3. Let $\boldsymbol{R}_\Sigma$, and $\boldsymbol{R}_\Delta$ denote the least families containing $\boldsymbol{R}$ such that for every non-empty family $\boldsymbol{S} \subset \boldsymbol{R}$ we have respectively $\bigcup_{Y \in \boldsymbol{S}} Y \in \boldsymbol{R}_\Sigma$, and $\bigcap_{Y \in \boldsymbol{S}} Y \in \boldsymbol{R}_\Delta$. Prove that $\boldsymbol{R}_{\Sigma\Delta} = \boldsymbol{R}_{\Delta\Sigma}$.

Hint. Use Theorem 1.4.

§ 4. σ-additive and δ-multiplicative families of sets

A family $\boldsymbol{R}$ of sets is said to be *σ-additive* (resp. *δ-multiplicative*) if for every sequence $H \in \boldsymbol{R}^N$ the formula $\bigcup_n H_n \in \boldsymbol{R}$ (resp. $\bigcap_n H_n \in \boldsymbol{R}$) holds.

A σ-additive lattice of sets is called briefly *σ-lattice*. Symmetrically, a *δ-lattice* of sets is a δ-multiplicative lattice of sets. We shall assume that the σ-lattices under consideration (as well as the δ-lattices) contain $\emptyset$ and X as members.

We refer to a σ-additive algebra of sets as a σ-algebra. Thus a *σ-algebra* is a family of sets closed under the operations $A - B$, $A \cap B$ and $\bigcup_n A_n$ (and by the de Morgan rule—under the operation $\bigcap_n A_n$; of course, the term δ-algebra can be used instead of σ-algebra).

The next theorems follow from Theorems 3.1 and 3.2 generalized to the case of functions whose domains are sequences of sets.

THEOREM 1: *For every family* $\boldsymbol{R}$ *there exists a unique family* Bor($\boldsymbol{R}$) *which is σ-additive, δ-multiplicative, contains* $\boldsymbol{R}$ *and is the least family with these properties.*

THEOREM 2: *For every family* $\boldsymbol{R}$ *the following formulas hold*:

(1) $$\boldsymbol{R} \subset \mathrm{Bor}(\boldsymbol{R});$$

(2) $$\boldsymbol{R}_1 \subset \boldsymbol{R}_2 \to \mathrm{Bor}(\boldsymbol{R}_1) \subset \mathrm{Bor}(\boldsymbol{R}_2),$$

(3) $$\mathrm{Bor}\big(\mathrm{Bor}(\boldsymbol{R})\big) = \mathrm{Bor}(\boldsymbol{R}).$$

Performing the operation $\bigcup_n$ resp. $\bigcap_n$ on sequences whose terms belong to Bor($\boldsymbol{R}$), we obtain sets belonging to Bor($\boldsymbol{R}$). This remark allows us to obtain a classification of sets belonging to Bor($\boldsymbol{R}$). Namely, for any family $\boldsymbol{R}$ let $\boldsymbol{R}_\sigma$ denote the family of sets of the form $\bigcup_n H_n$ where $H \in \boldsymbol{R}^N$ and $\boldsymbol{R}_\delta$ the family of sets of the form $\bigcap_n H_n$, where $H \in \boldsymbol{R}^N$. It is clear that $\boldsymbol{R} \subset \boldsymbol{R}' \to (\boldsymbol{R}_\sigma \subset \boldsymbol{R}'_\sigma) \wedge (\boldsymbol{R}_\delta \subset \boldsymbol{R}'_\delta)$.

We define a σ-additive family as a family $\boldsymbol{R}$ such that $\boldsymbol{R} = \boldsymbol{R}_\sigma$ and a σ-multiplicative family as a family $\boldsymbol{R}$ such that $\boldsymbol{R} = \boldsymbol{R}_\delta$. Since Bor ($\boldsymbol{R}$) is both σ-additive and σ-multiplicative, we obtain $(\mathrm{Bor}(\boldsymbol{R}))_\sigma = \mathrm{Bor}(\boldsymbol{R}) = (\mathrm{Bor}(\boldsymbol{R}))_\delta$. From $\boldsymbol{R} \subset \mathrm{Bor}(\boldsymbol{R})$ we have:

THEOREM 3: *The family* Bor($\boldsymbol{R}$) *contains each of the following families*

$$\boldsymbol{R}_\sigma, \boldsymbol{R}_{\sigma\delta}, \boldsymbol{R}_{\sigma\delta\sigma}, \ldots,$$

$$\boldsymbol{R}_\delta, \boldsymbol{R}_{\delta\sigma}, \boldsymbol{R}_{\delta\sigma\delta}, \ldots$$

In general, no two of these families are equal; moreover, they do not exhaust the whole family Bor($\boldsymbol{R}$).

We shall describe a method which often allows us to decide whether or not an individual set defined by a formula belongs to Bor($\boldsymbol{R}$).

Let $\Phi(i, j, \ldots, k, x)$ be a formula in which the range of $i, j, \ldots, k$ is limited to N. Let

$$Z_{i,j,\ldots,k} = \{x \colon \Phi(i, j, \ldots, k, x)\},$$

$$W = \{x \colon \Omega'_i \Omega''_j \ldots \Omega^{(h)}_k \Phi(i, j, \ldots, k, x)\},$$

where each of the symbols $\Omega', \Omega'', \ldots, \Omega^{(h)}$ is either the universal or the existential quantifier. We have

THEOREM 4: *If for any* $i, j, \ldots, k \in N$ *the set* $Z_{i,j,\ldots,k}$ *belongs to* Bor($\boldsymbol{R}$), *then* $W \in \mathrm{Bor}(\boldsymbol{R})$.[1])

PROOF. The proof is by induction on the number of quantifiers. If $h = 0$, then $W = Z_{i,j,\ldots,k}$; thus $W \in \mathrm{Bor}(\boldsymbol{R})$ by assumption. If the theorem holds for $h-1$ quantifiers, then each of the sets

$$W_i = \{x \colon \Omega''_j \ldots \Omega^{(h)}_k \Phi(i, j, \ldots, k, x)\}$$

[1]) Theorem 4 gives a simple and easily applicable method of proving that explicitly definable sets are Borel. The method was invented by Kuratowski and Tarski [1]; see also Kuratowski [7]. The method is also applicable to the projective sets; see Chapter XIII.

belongs to Bor($\boldsymbol{R}$). Since $W = \bigcup_{i \in N} W_i$ when Ω' is the existential quantifier, and $W = \bigcap_{i \in N} W_i$ when Ω' is the universal quantifier, we have in both cases $W \in \mathrm{Bor}(\boldsymbol{R})$. Q.E.D.

We obtain important examples by taking as $\boldsymbol{R}$ the family $\boldsymbol{F}$ of closed sets in an arbitrary topological space X. In this case, Bor($\boldsymbol{R}$) is called the *family of Borel sets* of the space X.[1])

As an example we consider a sequence f_n of continuous functions and show that points x for which the sequence $f_n(x)$ converges is an $\boldsymbol{F}_{\sigma\delta}$-set.

The Cauchy condition for the convergence of a sequence of real numbers $a_1, a_2, \ldots, a_n, \ldots$ can be written in the following form

$$\bigwedge_k \bigvee_m \bigwedge_i [|a_{m+i} - a_m| \leqslant 1/k].$$

This implies that the set Z of points at which the values of the sequence of continuous functions $f_1, f_2, \ldots, f_n, \ldots$ converge is

$$Z = \{x\colon \bigwedge_k \bigvee_m \bigwedge_i [|f_{m+i}(x) - f_m(x)| \leqslant 1/k]\}. \tag{i}$$

Letting

$$Z_{k,m,i} = \{x\colon |f_{m+i}(x) - f_m(x)| \leqslant 1/k\},$$

we infer from (i) that

$$Z = \bigcap_{k=1}^{\infty} \bigcup_{m=1}^{\infty} \bigcap_{i=1}^{\infty} Z_{k,m,i}.$$

Since the set $Z_{k,m,i}$ (for fixed indices) is closed (because the functions considered are continuous), the set Z is $\boldsymbol{F}_{\sigma\delta}$.

Exercises

1. Prove that

(a) the intersection of two $\boldsymbol{F}_\sigma$-sets is an $\boldsymbol{F}_\sigma$-set,

(b) the union of an infinite sequence of $\boldsymbol{F}_\sigma$-sets is an $\boldsymbol{F}_\sigma$-set.

Prove the analogous properties for $\boldsymbol{F}_{\sigma\delta}$-sets and for Borel sets.

2. Prove that every set open in $\mathscr{E}^n$ is an $\boldsymbol{F}_\sigma$-set.

3. Prove that the Borel sets in $\mathscr{E}^n$ constitute the least σ-additive and δ-multiplicative family containing the family of all open sets.

[1]) After the name of the French mathematician E. Borel who first investigated them.

4. Prove the formulas which are obtained from (1)–(3) replacing Bor($\boldsymbol{R}$) by $\boldsymbol{R}_\sigma$ or by $\boldsymbol{R}_\delta$.

5. Give examples of (finite) families $\boldsymbol{R}$ such that $\boldsymbol{R} = \boldsymbol{R}_\sigma = \boldsymbol{R}_{\sigma\delta}$ and examples of families $\boldsymbol{R}$ such that $\boldsymbol{R} \neq \boldsymbol{R}_\sigma \neq \boldsymbol{R}_{\sigma\delta} = \boldsymbol{R}_{\sigma\delta\sigma}$.[1])

6. An ideal $\boldsymbol{I}$ is called a *σ-ideal* if it is closed under countable unions. Show that

$$\text{if } A_n \doteq B_n \bmod \boldsymbol{I}, \text{ then } \Big(\bigcup_n A_n\Big) \doteq \Big(\bigcup_n B_n\Big) \bmod \boldsymbol{I}.$$

Hint. Show that

$$\bigcup_n A_n - \bigcup_n B_n \subset \Big[\bigcup_n (A_n - B_n)\Big] \in \boldsymbol{I}.$$

7. Let $\boldsymbol{I}$ be a σ-ideal, $\boldsymbol{A}$ a σ-lattice and

$$L = \{E\colon \bigvee_{G\in\boldsymbol{A}} E \doteq G \bmod \boldsymbol{I}\}.$$

Show that L is a σ-lattice.

§ 5. Reduction and separation properties

We say that the family $\boldsymbol{R}$ has the *finite reduction property* if for each pair A_1, A_2 of members of $\boldsymbol{R}$ there exists a pair B_1, B_2 of members of $\boldsymbol{R}$ such that

(1) $\quad B_1 \subset A_1, \quad B_2 \subset A_2, \quad B_1 \cup B_2 = A_1 \cup A_2 \quad$ and $\quad B_1 \cap B_2 = \emptyset.$

The *countable* reduction property means that for each sequence $A_1, A_2, \ldots$ of members of $\boldsymbol{R}$ there is a sequence $B_1, B_2, \ldots$ of disjoint members of $\boldsymbol{R}$ such that

(2) $\quad B_n \subset A_n$ for each $n \in N$, $\quad$ and $\quad \bigcup_n B_n = \bigcup_n A_n.$

THEOREM 1: *If $\boldsymbol{R}$ is a field, then $\boldsymbol{R}_\sigma$ has the countable reduction property.*

PROOF. Let $A_n \in \boldsymbol{R}_\sigma$ for $n \in N$. Thus

$$A_n = A_{n0} \cup A_{n1} \cup \ldots \cup A_{nm} \cup \ldots, \qquad \text{where} \qquad A_{nm} \in \boldsymbol{R}.$$

Arrange the double sequence $\{n, m\}$ in a single sequence and let $j = j(n, m)$ be the integer corresponding to the pair $\langle n, m\rangle$ (see III § 3).

[1]) Exercise 5 is connected with a problem proposed by Kolmogoroff; see Sierpiński [3], p. 171.

Let

$$C_{nm} = A_{nm} - \bigcup A_{kl},$$

where the union $\bigcup$ ranges over all pairs $\langle k, l \rangle$ such that

$$j(k, l) < j(n, m).$$

Clearly, the sets C_{nm} are disjoint and thus the sets

$$B_n = C_{n0} \cup C_{n1} \cup C_{n2} \cup \dots$$

are the required sets.

REMARK: If the additional condition

$$(3) \qquad \bigcup_n A_n = X,$$

is fulfilled, then we have $(X - B_n) \in \boldsymbol{R}_\sigma$.

COROLLARY 1: *Each σ-algebra has the countable reduction property.*

COROLLARY 2: (SEPARATION THEOREM)[1]) *If* $\boldsymbol{R}$ *is a field and if* $F_0, F_1, \dots$ *is a sequence of members of the complement of* $\boldsymbol{R}_\sigma$ *such that* $\bigcap_n F_n = \emptyset$, *then there exists a sequence of sets* $E_0, E_1, \dots$ *which, simultaneously with their complements, are members of* $\boldsymbol{R}_\sigma$ *and*

$$F_n \subset E_n \text{ for each } n \in N, \text{ and } \bigcap_n E_n = \emptyset.$$

It suffices to put $F_n = X - A_n$ and $E_n = X - B_n$.

Exercises

Call a (metric separable) space 0-*dimensional* if it contains a countable base composed of closed-open sets.

Show that in a 0-dimensional space

(i) every open set is a countable union of closed-open sets;

(ii) if A and B are closed and disjoint, there exist two disjoint closed-open sets G and H such that $A \subset G$ and $B \subset H$;

(iii) apply Corollary 2 to show that given a finite system $A_0, \dots, A_k$ of disjoint closed subsets of a 0-dimensional space, there exists a system $F_0, \dots, F_k$ of disjoint closed-open sets such that

$$X = F_0 \cup \dots \cup F_k \quad \text{and} \quad A_i \subset F_i \text{ for } i = 0, 1, \dots, k.$$

[1]) The separation property was defined and studied by Lusin [4] in connection with the theory of analytic and projective sets. The reduction property was introduced by Kuratowski [10].

§ 6. Generalized cartesian products

As in § 1, let F be a function whose values are subsets of the set X and whose domain is a set $T \neq \emptyset$.

DEFINITION: The *cartesian product* $\prod_{t \in T} F_t$ is the set of all functions f whose domain is T and which satisfy the condition $f(t) \in F_t$ for every $t \in T$. That is,

$$\prod_{t \in T} F_t = \{f \in \Phi : \bigwedge_{t \in T} [f(t) \in F_t]\}, \quad \text{where} \quad \Phi = X^T.$$

If $T = N$ we write $\prod_{n=0}^{\infty} F_n$ instead of $\prod_{n \in N} F_n$. The elements of this cartesian product are sequences φ such that $\varphi_n \in F_n$ for $n \in N$.

If all the sets F_t are identical, $F_t = Y$, then we have $\prod_{t \in T} F_t = Y^T$. In this case the symbol $\prod_{t \in T} F_t$ denotes the set of functions with domain T and range $\boldsymbol{P}(Y)$.

The set Y^T is called the *cartesian power* of the set Y.

For $Y \subset \prod_{t \in T} F_t$ let Y^t denote the projection of Y on F_t. Thus Y^t is the set $\{f(t) : f \in Y\}$. Clearly, $Y_1 \subset Y_2 \rightarrow Y_1^t \subset Y_2^t$ for every $t \in T$.

REMARK: Let $T = \{1, 2\}$. The cartesian product $\prod_{t \in T} F_t$ and the product $F_1 \times F_2$ are not identical. In fact, the first product has as elements two-term sequences, the second product, ordered pairs. These two notions are distinct. In practice, however, the distinction between these two kinds of products is inessential, for we can always associate with every pair $\langle x, y \rangle$ of $F_1 \times F_2$ the sequence $\{\langle 1, x \rangle, \langle 2, y \rangle\}$ belonging to $\prod_{t \in T} F_t$ in a one-to-one manner.

If $F_{t_0} = \emptyset$ *for some* t_0, *then* $\prod_{t \in T} F_t = \emptyset$.

In fact, if $f \in \prod_{t \in T} F_t$, then $f(t_0) \in F_{t_0}$; thus $F_{t_0} \neq \emptyset$.

THEOREM 1: *If a set T has a finite number of elements and $F_t \neq \emptyset$ for every $t \in T$, then $\prod_{t \in T} F_t \neq \emptyset$.*

The proof is by induction on the number of elements of T. If T

consists of one element, then the theorem clearly holds. Suppose that it holds for the case where T consists of n elements. Let $T_1 = T \cup \{a\}$ where $a \notin T$. Suppose further that $F_t \neq \emptyset$ for $t \in T_1$. We shall show that $\prod_{t \in T_1} F_t \neq \emptyset$. In fact, the induction hypothesis gives $\prod_{t \in T} F_t \neq \emptyset$; thus let $f \in \prod_{t \in T} F_t$. Let $t_0 \in F_a$. The set $f_1 = f \cup \{\langle a, t_0 \rangle\}$ is a function belonging to $\prod_{t \in T_1} F_t$. Consequently, the set $\prod_{t \in T_1} F_t$ is non-empty.

Theorem 1 also holds for arbitrary T if all factors F_t are equal.

THEOREM 2: *If $Y \neq \emptyset$, then $Y^T \neq \emptyset$.*

In the general case the proof that the cartesian product is non-empty requires the axiom of choice.

°THEOREM 3: *If $F_t \neq \emptyset$ for $t \in T$, then $\prod_{t \in T} F_t \neq \emptyset$.*[1])

PROOF. A choice function for the family $\{F_t \colon t \in T\}$ is an element of $\prod_{t \in T} F_t$.

In applications of cartesian products (e.g. in algebra, in topology) we deal mostly with cases where certain operations are defined on the sets F_t or where the sets F_t are topological spaces. We discuss first the case where only one operation is defined on each set F_t. For convenience, we assume that this operation is binary. In other words, we suppose that besides the function $F \in (\boldsymbol{P}(X))^T$ we are given a function G such that $G_t \in (F_t)^{F_t \times F_t}$ for every $t \in T$.

The function G induces a binary operation φ on the elements of the cartesian product $\prod_{t \in T} F_t$. Namely, we let for $f, g \in \prod_{t \in T} F_t$

$$\varphi(f, g) = \underset{t \in T}{\boldsymbol{F}}\left[G_t\big(f(t), g(t)\big)\right].$$

Thus $\varphi(f, g)$ is that element h of the cartesian product for which $h(t) = G_t\big(f(t), g(t)\big)$ for every t. The operation φ is called the *cartesian product of the operations G_t.*

[1]) B. Russell took Theorem 3 as an axiom instead of the axiom of choice. He called this axiom the *multiplicative axiom*. See B. Russell and A. N. Whitehead, [1], p. 536.

In a similar way we define the cartesian product of *relations*. Let R be a function such that (for every $t \in T$) R_t is a relation with its field included in F_t. The *cartesian product of these relations* is the relation ϱ, whose field is included in $\prod_{t \in T} F_t$, such that

$$\langle f, g \rangle \in \varrho \equiv \bigwedge_t [\langle f(t), g(t) \rangle \in R_t].$$

It should be pointed out that ϱ is not the cartesian product $\prod_{t \in T} R_t$, because ϱ is a binary relation; that is, ϱ is a set of ordered pairs, whereas $\prod_{t \in T} R_t$ is a set of functions. However, we can associate in a natural way the cartesian product $\prod_{t \in T} R_t$ with the relation which holds between functions f and g if and only if the function h defined by $h(t) = \langle f(t), g(t) \rangle$ (that is, the function $\underset{t \in T}{F}[\langle f(t), g(t) \rangle]$) belongs to $\prod_{t \in T} R_t$. This relation coincides with the relation ϱ defined above.

Clearly, the definitions can be applied without modifications to the case where not just one but several operations (or relations) are defined on each F_t.

Example. Suppose that F_t is a Boolean algebra with respect to the operations $\vee_t$, $\wedge_t$, $-_t$ and the elements 0_t, 1_t. Let $\vee$, $\wedge$, $-$ denote the cartesian products of the operations $\vee_t$ $\wedge_t$, $-_t$, respectively, and let ξ, ι denote the functions such that $\xi(t) = 0_t$ and $\iota(t) = 1_t$ for all $t \in T$.

THEOREM 4: *The cartesian product $\prod_{t \in T} F_t$ is a Boolean algebra under the operations $\vee$, $\wedge$, $-$, ξ, ι.*

PROOF. To prove the theorem it suffices to check the axioms given in I, §9. For instance, we check axiom (iv). Let $f \in \prod_{t \in T} F_t$; it follows from the definition that $f \vee -f$ is the function g such that $g(t) = f(t) \vee \vee -_t f(t)$ for every t. Since axiom (iv) holds in F_t, we have $g(t) = 1_t$, thus $g = \iota$.

The Boolean algebra $\prod_{t \in T} F_t$ is called the *direct product* of the Boolean algebras F_t.

In a similar way we can define the direct product of groups, rings, and other algebraic systems.

The notion of a *complex function* determined by two functions $f \in Y^X$, $g \in Z^X$ (see Ch. II) can be generalized to any set of functions $f_t \in Y_t^X$ where $t \in T$. Namely

$$h \in (\prod_{t \in T} Y_t)^X$$

is defined as follows:

$$h(x) = \{f_t(x)\}_{t \in T}.$$

Thus

$$[w = h(x)] \equiv \bigwedge_t [w(t) = f_t(x)].$$

Similarly, the notion of a *product-function* determined by the set of functions $f_t \in Y_t^{X_t}$, where $t \in T$, is defined as follows

$$u\{x_t\} = \{f_t(x)\}_{t \in T}.$$

Thus $u\colon \prod_{t \in T} X_t \to \prod_{t \in T} Y_t$ and, for $z \in \prod_{t \in T} X_t$, we have

$$[v = u(z)] \equiv \bigwedge_t [v(t) = f_t(z(t))].$$

Exercises

1. Prove that a Boolean algebra of 2^n elements is the direct product of n Boolean algebras of two elements.

Hint: Use induction on n.

2. Prove that if the relations R_t are (a) reflexive, (b) symmetric, (c) transitive, then their cartesian product has the same properties. Give an example of a property which does not hold for cartesian products although it holds for the individual factors. In other words, find a function R and a property which holds for all relations R_t but does not hold for their cartesian product.

In the following two exercises we assume that H is a function such that $H_t \in F_t^{F_t \times F_t}$, R_t is an equivalence relation in F_t, φ is the cartesian product of the operations H_t and ϱ is the cartesian product of the relations R_t.

3. Show that if each function H_t is consistent with R_t, then φ is consistent with ϱ.

4. Let X be the function such that, for every t, X_t is the quotient function induced from H_t by means of R_t and let χ be the quotient function induced from φ by ϱ. Show that the following diagram is commutative:

$$\begin{array}{ccc} H & \longrightarrow & \varphi \\ \downarrow & & \downarrow \\ X & \longrightarrow & \chi \end{array}.$$

5. Let $f_t \in Y_t^X$ for $t \in T$, and let h be the *complex function* determined by the functions f_t. Show that (comp. Exercise II.7.9)

$$h^1(A) \subset \prod_{t\in T} f_t^1(A) \quad \text{where} \quad A \subset X,$$

$$h^{-1}\left(\prod_{t\in T} B_t\right) = \bigcap_{t\in T} f_t^{-1}(B_t) \quad \text{where} \quad B_t \subset Y_t;$$

in particular, if $B_t = Y_t$ for all t except t_0, we have

$$h^{-1}\left(\prod_{t\in T} B_t\right) = f_{t_0}^{-1}(B_{t_0}).$$

6. Let $f_t \in Y_t^{X_t}$ and let u be the *product-function* determined by these functions for $t \in T$. Show that (comp. Exercise II.7.10)

$$u^1\left(\prod_{t\in T} A_t\right) = \prod_{t\in T} f_t^1(A_t) \quad \text{where} \quad A_t \subset Y_t,$$

$$u^{-1}\left(\prod_{t\in T} B_t\right) = \prod_{t\in T} f_t^{-1}(B_t) \quad \text{where} \quad B_t \subset Y_t;$$

in particular, if $\varphi \in \prod_{t\in T} Y_t$, we have

$$u^{-1}(\varphi) = \prod_{t\in T} f_t^{-1}(\varphi(t)).$$

7. Let $f_t\colon X_t \to Y$ for $t \in T$. Put

$$Z = \left\{z\colon \bigwedge_{tt'} f_t(z^t) = f_{t'}(z^{t'})\right\}$$

and for fixed t_0 denote by π the projection of the space $\prod_{t\in T} X_t$ on the X_{t_0} axis. Show that

$$f_{t_0}^1[\pi(Z)] = \bigcap_{t\in T} f_t^1(X_t)$$

(see Kuratowski [1], p. 153).

§ 7. Cartesian products of topological spaces

For every $t \in T$, let F_t be a topological space and let $C_t X$ denote the closure of the set $X \subset F_t$ in the space F_t. Hence C is a function such that

$$C_t \in \boldsymbol{P}(F_t)^{\boldsymbol{P}(F_t)} \quad \text{for every } t.$$

Clearly, there are many different ways of defining the closure operation on the space $\prod_{t\in T} F_t$. In fact, an arbitrary set can be made into a topological space in many different ways. Here we shall discuss one of the special topologies on the space $\prod_{t\in T} F_t$, introduced by Tychonoff [1].

Let S be a finite subset of T and let G_s be an open set in the space F_s for every $s \in S$. We define the neighborhood determined by S and by the sets G_s to be the following subset of the cartesian product $\Pi = \prod_{t \in T} F_t$:

$$\Gamma = \{f \in \Pi : \bigwedge_{s \in S} f(s) \in G_s\}.$$

We shall prove that the intersection of two neighborhoods is a neighborhood. In fact, if Γ is the neighborhood determined by a finite set S and by open sets G_s ($s \in S$) and Γ' is the neighborhood determined by a finite set S' and by open sets G'_s ($s \in S'$), then

$$\begin{aligned}\Gamma \cap \Gamma' &= \{f \in \Pi : \bigwedge_{s \in S} \bigwedge_{s' \in S'} (f(s) \in G_s) \wedge (f(s') \in G'_{s'})\} \\ &= \{f \in \Pi : \bigwedge_{s \in S - S'} (f(s) \in G_s) \wedge \bigwedge_{s' \in S' - S} (f(s') \in G'_{s'}) \wedge \\ &\qquad \wedge \bigwedge_{s \in S \cap S'} (f(s) \in G_s \cap G'_{s'})\}.\end{aligned}$$

Thus $\Gamma \cap \Gamma'$ is the neighborhood determined by the finite set $S \cup S'$ and by the open sets G''_t, where the sets G''_t are defined as follows: $G''_t = G_t$ for $t \in S - S'$, $G''_t = G'_t$ for $t \in S' - S$, and $G''_t = G_t \cap G'_t$ for $t \in S \cap S'$.

We define the closure of a set $X \subset \Pi$ to be the set CX of all $f \in \Pi$ such that each neighborhood Γ containing f also contains at least one element of X:

$$(*) \qquad \bigwedge_{\Gamma} [(\Gamma \text{ is a neighborhood}) \wedge (f \in \Gamma) \rightarrow (\Gamma \cap X \neq \emptyset)].$$

THEOREM 1: *The cartesian product $\prod_{t \in T} F_t$ is a topological space with respect to the closure defined by $\overline{X} = CX$ for $X \subset \prod_{t \in T} F_t$.*

PROOF. It is necessary to check that axioms I.8 (1)–(4) hold.

Axioms (3) and (4) clearly hold.

Axiom (1). Let $f \in \overline{A}$; thus every neighborhood Γ containing f also contains at least one element of A. Hence $\Gamma \cap (A \cup B) \neq \emptyset$, which implies $f \in \overline{A \cup B}$, and thus $\overline{A} \subset \overline{A \cup B}$. Similarly, $\overline{B} \subset \overline{A \cup B}$. Hence $\overline{A} \cup \overline{B} \subset \overline{A \cup B}$.

Now suppose that $f \in \overline{A \cup B}$ and $f \notin \overline{A}$. Thus for every neighborhood Γ containing f we have $\Gamma \cap (A \cup B) \neq \emptyset$ and for some Γ_0 containing f we have $\Gamma_0 \cap A = \emptyset$. If Γ is an arbitrary neighborhood containing f, then $\Gamma \cap \Gamma_0$ is also a neighborhood containing f. Therefore $\Gamma \cap \Gamma_0 \cap \cap (A \cup B) \neq \emptyset$, whence $\Gamma \cap \Gamma_0 \cap B \neq \emptyset$ and hence $\Gamma \cap B \neq \emptyset$. This shows that $f \in \overline{B}$.

Axiom (2). It suffices to show that $\overline{\overline{X}} \subset \overline{X}$. Let $f \in \overline{\overline{X}}$ and let Γ be any neighborhood containing f. Thus $\Gamma \cap \overline{X} \neq \emptyset$; let $g \in \Gamma \cap \overline{X}$. Hence Γ is a neighborhood containing g. Since $g \in \overline{X}$, we have $\Gamma \cap X \neq \emptyset$. This shows that the condition (*) holds; hence $f \in \overline{X}$.

Examples of cartesian products of topological spaces

1. *The Cantor set.* This is the set $C = \{0, 1\}^N$ or, in other words, the cartesian power of a two-element set. If we define a topology on the set $\{0, 1\}$ by letting $\overline{X} = X$ (the discrete topology), then C becomes a topological space with the Tychonoff topology.

By assigning the real number $\sum_{n=0}^{\infty} 2f(n)/3^{n+1}$ to the element $f \in C$ we obtain a one-to-one mapping φ of the set C onto the set of those real numbers of the closed interval $[0, 1]$ whose triadic expansion contains only the digits 0 and 2.

2. The *generalized Cantor set* C_T is the cartesian power $\{0, 1\}^T$. The Tychonoff topology may be defined on this set similarly as on the set C.

The generalized Cantor set may also be defined as the set of all characteristic functions of subsets of T. In practice, we may identify the set C_T with $\boldsymbol{P}(T)$. Therefore sometimes we shall treat the family $\boldsymbol{P}(T)$ as a topological space. Similarly, the elements of the set $C_{T \times T} = \{0, 1\}^{T \times T}$ can be identified with the set of all relations on T. In fact, each element of the set $C_{T \times T}$ is a characteristic function of a set of ordered pairs of elements of T.

THEOREM 2: *The family* $K_t = \{X \subset T\colon t \in X\}$ *is both open and closed in* C_T; *the family* $\{R \subset T \times T\colon tRs\}$ *is both closed and open in* $C_{T \times T}$.

PROOF: Let Γ denote the neighborhood in C_T, determined by the set $S = \{t\}$ and by the open set $G_t = \{1\}$. Then $f \in \Gamma \equiv f(t) = 1$;

thus Γ consists of the characteristic functions of the sets belonging to the family K_t. Hence this family is an open set. Likewise, the neighborhood determined by the set S and the open set $G_t' = \{0\}$ consists of the characteristic functions of the sets belonging to the family $\boldsymbol{P}(T)-K_t$. This shows that the family K_t is closed in C_T.

The second part of the theorem follows from the first.

3. *The Baire space*. This is the cartesian power N^N or, in other words, the set of infinite sequences of natural numbers. The topology in N^N is defined as the Tychonoff topology, where we define the closure operation in N to be $\overline{X} = X$.

If $a = (a_0, a_1, \ldots, a_{n-1})$ is a sequence of n terms ($a \in N^n$), then the set $N_a = N_{a_0, \ldots, a_{n-1}} = \{e\colon e|n = a\}$ is both open and closed in N^N. To see this, we notice that N_a is the set of sequences satisfying the conditions $\varphi_j = a_j$ where $j < n$; it thus coincides with the neighborhood Γ in N determined by the set $S = \{0, 1, \ldots, n-1\}$ and the open sets $G_j' = \{a_j\}$ for $j < n$. The complement of Γ is open in N^N, because it coincides with the union of neighborhoods determined by the sets $\{j\}$ and by the open sets $G_j' = N - \{a_j\}$, $j < n$.

Assigning the number

$$x = \frac{1\ |}{|\ \varphi_0+1} + \frac{1\ |}{|\ \varphi_1+1} + \frac{1\ |}{|\ \varphi_2+1} + \ldots$$

to the element $\varphi \in N^N$, we obtain a one-to-one mapping of the space N^N onto the set of irrational numbers in the open interval $(0, 1)$. In practice, we may identify the Baire space with the set of irrational numbers in the open interval $(0, 1)$.

4. The *cartesian product of a finite number of spaces*. The construction described in this section is used both in the case in which the set T in the formula $X = \prod_{t \in T} F_t$ is finite and also in the case in which T is infinite.

If T is a finite set, for instance $T = \{0, 1, \ldots, n-1\}$, then the cartesian products $\prod_{t<n} V_t$, where for every t the set V_t is open in F_t, form an open base of $\prod_{t<n} F_t$.

5. The *n-dimensional Euclidean space* is the product $\mathscr{E}^n$ where $\mathscr{E}$ is the space of real numbers.

6. The *Hilbert cube* is the product I^N where I denotes the closed interval $0 \leqslant x \leqslant 1$.

In later chapters we shall make use of the following theorem.

THEOREM 3: *If* $X = \prod_{t \in T} F_t$ *is a cartesian product of topological spaces (with Tychonoff topology),* $Z_t \subset F_t$ *and* Z_t *is a closed set in* F_t *for every* $t \in T$, *then the set* $\prod_{t \in T} Z_t$ *is also closed.*

PROOF. Let $P = \prod_{t \in T} Z_t$ and let $f \notin P$. For some $s \in T$ we thus have $f(s) \notin Z_s$. The element of the subbase of X, determined by the one-element set $\{s\}$ and by the open set $F_s - Z_s$, contains f and is disjoint from P.

Exercises

1. The set of reflexive relations whose fields are included in T is closed in $C_{T \times T}$. Prove that the same holds for the sets of symmetric relations, the set of transitive relations and the set of equivalence relations.

2. Prove that each neighborhood in the Baire space contains some neighborhood of the form N_a, where a is a finite sequence.

3. Show that the set $\{X \subset N: X \text{ is a finite set}\}$ is a Borel set in the space C_N.

4. Let I be any ideal in $P(T)$. Show that taking as neighborhoods the sets of the form $\{g: \bigwedge_{s \in S} (g(s) \in G_s)\}$, where $S \in I$ and G_s are open sets in F_s for $s \in S$ and defining the closure operation by (*), we obtain the function satisfying the axioms of topology I.8 (1)–(4).

§ 8. The Tychonoff theorem

A family R of subsets of the set X is said *to have the finite intersection property* if every finite subfamily of R has a non-empty intersection.

A topological space X (that is, a set with an operation defined on subsets of X and satisfying axioms I.8 (1)–(4)) is said to be *compact* if every family of closed subsets of X which has the finite intersection property has a non-empty intersection. This definition implies that a topological space is compact if and only if every family of open sets whose union is X contains a finite subfamily whose union is also X.

Although the following theorem belongs to topology rather than to general set theory, we give it here because it has numerous applications in mathematical theories in general and in set theory in particular. Moreover, the means used in proving this theorem involve little more than set-theoretical techniques.

°THEOREM 1: (Tychonoff) *If for every* $t \in T$ *the space* F_t *is compact, then the space* $P = \prod_{t \in T} F_t$ *is also compact (relative to the Tychonoff topology).*

In the proof of this theorem we make use of a lemma to be proved in Chapter VII.[1])

°LEMMA: *If* $\boldsymbol{R}_0$ *is a family of subsets of* X *with the finite intersection property, then there exists the maximal family* $\boldsymbol{R} \subset \boldsymbol{P}(X)$ *containing* $\boldsymbol{R}_0$ *which also has the finite intersection property. That is to say, every family of subsets of* X *containing* $\boldsymbol{R}$ *and different from* $\boldsymbol{R}$ *contains a finite subfamily with empty intersection.*

We shall make use of the following two properties of maximal families $\boldsymbol{R}$ with the finite intersection property.

(i) *If* $A \in \boldsymbol{R}$ *and* $B \in \boldsymbol{R}$, *then* $A \cap B \in \boldsymbol{R}$.

Suppose the contrary. Then the family arising from $\boldsymbol{R}$ by adding to it $A \cap B$ does not have the finite intersection property. Thus it contains a finite subfamily with empty intersection. Clearly the set $A \cap B$ belongs to this subfamily. Hence we conclude that there exists a finite subfamily $\boldsymbol{R}' \subset \boldsymbol{R}$ such that $A \cap B \cap \bigcap_{Y \in \boldsymbol{R}'} Y = \emptyset$, which contradicts the assumption that the family $\boldsymbol{R}$ has the finite intersection property.

(ii) *If* $A \subset X$ *and* $A \cap Y \neq \emptyset$ *for every* $Y \in \boldsymbol{R}$, *then* $A \in \boldsymbol{R}$.

In fact, if $A \notin \boldsymbol{R}$, then the family $\boldsymbol{R} \cup \{A\}$ does not have the finite intersection property. Thus there exists a finite subfamily $\boldsymbol{R}' \subset \boldsymbol{R}$ such that $A \cap \bigcap_{Y \in \boldsymbol{R}'} Y = \emptyset$. The intersection $\bigcap_{Y \in \boldsymbol{R}'} Y$ belongs to $\boldsymbol{R}$ by (i). This contradicts the assumption that $A \cap Y \neq \emptyset$ for every $Y \in \boldsymbol{R}$.

Now to prove the Tychonoff theorem, let $\boldsymbol{R}_0$ be a family of closed

[1]) The original proof of Tychonoff theorem was published in Tychonoff [1]. The proof given here is taken from Bourbaki [2]. It was shown by Kelley [1] that Tychonoff theorem is equivalent to the axiom of choice on the basis of axioms $\Sigma[TR]$.

subsets of P with the finite intersection property. Let $\boldsymbol{R}$ denote any maximal family of subsets of P with the finite intersection property, containing $\boldsymbol{R}_0$. For the proof of the theorem it now suffices to show that $\bigcap_{Y\in\boldsymbol{R}} \overline{Y} \neq \emptyset$.

For arbitrary $Z \subset P$ let Z^t denote the projection of Z into F_t and let $\boldsymbol{R}^t = \{\overline{Y_j^t}: Y \in \boldsymbol{R}\}$. The family $\boldsymbol{R}^t$ consists of closed subsets of F_t. If the sets $\overline{Y_t^j}$, $j < n$, belong to $\boldsymbol{R}_i^t$ and $Y_j \in \boldsymbol{R}$, then $\bigcap_{j<n} Y_j \neq \emptyset$ because the family $\boldsymbol{R}$ has the finite intersection property. This implies that $\bigcap_{j<n} Y_j^t \neq \emptyset$ (see p. 129), whence $\bigcap_{j<n} \overline{Y_j^t} \neq \emptyset$. Thus the family $\boldsymbol{R}$ has the finite intersection property. From the assumption that every F_t is compact we infer that $\bigcap_{Y\in\boldsymbol{R}} \overline{Y_i^t} \neq \emptyset$. It now follows that there is an $f \in P$ such that for every $t \in T$, $f(t) \in \bigcap_{Y\in\boldsymbol{R}} \overline{Y^t}$. We shall prove that $f \in \bigcap_{Y\in\boldsymbol{R}} \overline{Y}$.

For this purpose suppose that $Y \in \boldsymbol{R}$ and that Γ is a neighborhood containing f. We have to show that $\Gamma \cap Y \neq \emptyset$.

Let Γ be the neighborhood determined by a finite set $S \subset T$ and open sets $G_s \subset F_s$ $(s \in S)$, where clearly $f(s) \in G_s$ for $s \in S$. Letting $\Gamma_s = \{g: g(s) \in G_s\}$, we have $\Gamma = \bigcap_{s\in S} \Gamma_s$.

If Z is an arbitrary set belonging to $\boldsymbol{R}$, then $f(s) \in \overline{Z^s}$; thus $G_s \cap Z_s^s \neq \emptyset$. This means that there exists $z_s \in G_s$ such that for some function $g \in Z$ we have $g(s) = z_s$; hence $g \in \Gamma_s$. Thus for any $Z \in \boldsymbol{R}$ we have $\Gamma_s \cap Z \neq \emptyset$.

By (ii) it follows that $\Gamma_s \in \boldsymbol{R}$ and by (i) $\bigcap_{s\in S} \Gamma_s \in \boldsymbol{R}$; that is, $\Gamma \in \boldsymbol{R}$, which implies $\Gamma \cap Y \neq \emptyset$. Thus every neighborhood containing f has elements in common with Y, consequently $f \in \overline{Y}$.

Examples

1. The sets C, C_T, $C_{T\times T}$ are compact.
2. Let $\Phi(R, x_1, \ldots, x_n)$ be a formula constructed from the formulas

$$(*) \qquad x_i = x_j, \qquad \langle x_i, x_j \rangle \in R$$

by applying only the operations of the propositional calculus. Such formulas are called *open*. For $a_1, \ldots, a_n \in T$ and for an arbitrary open formula Φ let

$$Z_\Phi = Z_\Phi(a_1, \ldots, a_n) = \{R \in T \times T\colon \Phi(R, a_1, \ldots, a_n)\}.$$

The set Z_Φ is both open and closed in $C_{T\times T}$. This fact follows from Theorem 6.2 when Φ is one of the formulas (*). For other open formulas, this property follows from the relationships between logical and set-theoretical operations as well as from the remark that the finite union, intersection, and complement of sets which are both open and closed are again both open and closed.

Now, let Φ_j be an open formula with the free variables $x_{j1}, x_{j2}, \ldots, x_{jn_j}$ and let $a_{j1}, a_{j2}, \ldots, a_{jn_j}$ be elements of T $(j \in N)$. Since the set $C_{T\times T}$ is compact, we have

THEOREM 2.[1]) *If for every* $k \in N$ *the intersection* $\bigcap_{j<k} Z_{\Phi_j}(a_{j1}, \ldots, a_{jn_j})$ *is non-empty, then the intersection* $\bigcap_{j\in N} Z_{\Phi_j}(a_{j1}, \ldots, a_{jn_j})$ *is also non-empty.*

This theorem asserts that if there exist relations R_k which satisfy the conditions Φ_j for $j < k$ $(k = 1, 2, \ldots)$, then there also exists a "universal" relation R which satisfies all of these conditions.

Exercise

1. Show that the Baire space N^N is not compact.

§ 9. Reduced direct products

By combining the operation of cartesian product with forming of equivalence classes we obtain new operations which have found interesting applications in mathematical logic.

Let T be any set and F a function defined on T whose values are non-empty sets. Let f_t be a function whose domain is $F_t \times F_t$ and whose range is included in F_t. Finally, let R_t be a binary relation with field included in F_t. All arguments below can be generalized for the case in which the number of functions or relations is greater than 1.

[1]) Theorem 2 gives an interesting method of proving existential statements in a non-effective way. See in this connection Łoś and Ryll-Nardzewski [1].

Let $\boldsymbol{I}$ be an ideal in $\boldsymbol{P}(T)$. Define the relation $\sim_I$ in $P = \prod_{t \in T} F_t$ by the formula

$$f \sim_I g \equiv \{t\colon f(t) \neq g(t)\} \in \boldsymbol{I}.$$

THEOREM 1: *The relation* $\sim_I$ *is an equivalence relation in* P.

The reflexivity of $\sim_I$ follows from $\emptyset \in \boldsymbol{I}$, symmetry is obvious, and transitivity follows from the remark that for any $f, g, h \in P$

$$\{t\colon f(t) \neq g(t)\} \subset \{t\colon f(t) \neq h(t)\} \cup \{t\colon h(t) \neq g(t)\}.$$

THEOREM 2: *The cartesian product* φ *of the functions* f_t *is consistent with* $\sim_I$.

PROOF. We have to show that if $e', e'', d', d'' \in P$, then

$$(e' \sim_I e'') \wedge (d' \sim_I d'') \to [\varphi(e', d') \sim_I \varphi(e'', d'')].$$

Let $h' = \varphi(e', d')$, $h'' = \varphi(e'', d'')$ and $A = \{t\colon h'(t) \neq h''(t)\}$. It follows from the definition of φ that $h'(t) = f_t(e'(t), d'(t))$ and similarly $h''(t) = f_t(e''(t), d''(t))$. Hence $t \in A \to [e'(t) \neq e''(t)] \vee \vee [d'(t) \neq d''(t)]$, whence it follows that

$$A \subset \{t\colon e'(t) \neq e''(t)\} \cup \{t\colon d'(t) \neq d''(t)\} \in \boldsymbol{I}.$$

From Theorem 1 and from the definition of the quotient class (Ch. II) it follows that there exists the quotient class $P/\boldsymbol{I}$ of P with respect to the relation $\sim_I$. We call $P/\boldsymbol{I}$ the *direct product of the sets* F_t *reduced* mod $\boldsymbol{I}$ or simply the *reduced product*.[1]) Using Theorem II.8.1 and Theorem 2 proved above, we infer that there exists the function $\varphi/\boldsymbol{I}$ induced from φ by $\sim_I$ and that $\varphi/\boldsymbol{I}$ maps the Cartesian product $(P/\boldsymbol{I}) \times \times (P/\boldsymbol{I})$ into $P/\boldsymbol{I}$. We call $\varphi/\boldsymbol{I}$ the *reduced product of the functions* f_t.

Finally we define in $P/\boldsymbol{I}$ a binary relation $\varrho/\boldsymbol{I}$. It holds between the equivalence classes $e/\boldsymbol{I}$ and $d/\boldsymbol{I}$ of two functions $e, d \in P$ if and only if $\{t\colon \langle e(t), d(t)\rangle \notin R_t\} \in \boldsymbol{I}$. We call $\varrho/\boldsymbol{I}$ the *reduced product of the relations* R_t.

Let $\Phi(x, y, z)$ be an arbitrary formula. The main problem of the theory of reduced products can be stated as follows: When the sets $\{t\colon \Phi(F_t, f_t, R_t)\}$ are known, under what conditions do the set $P/\boldsymbol{I}$, the function $\varphi/\boldsymbol{I}$ and the relation $\varrho/\boldsymbol{I}$ satisfy the formula Φ?

In order to solve this problem, we consider more general formulas

[1]) The notion of a reduced product is due to Łoś [1]. For more references concerning reduced products see Bell and Slomson [1].

$\Phi(x, y, z, u_1, \ldots, u_k)$ with an arbitrary number of free variables. Suppose that $e_1, \ldots, e_k \in P$ and let

$$A_\Phi = \{t\colon \Phi(F_t, f_t, R_t, e_1(t), \ldots, e_k(t))\}$$

(clearly, the set A_Φ depends not only on Φ but also on the elements $e_1, \ldots, e_k$; we do not write $e_1, \ldots, e_k$ in the symbol A_Φ in order to simplify the notation).

With this notation, the following theorems (i)–(iv) hold:

(i) $$A_{\Phi \vee \Psi} \notin \boldsymbol{I} \equiv A_\Phi \notin \boldsymbol{I} \vee A_\Psi \notin \boldsymbol{I}.$$

In fact, $A_{\Phi \vee \Psi} = A_\Phi \cup A_\Psi$; thus (see formula (11), p. 17) $A_{\Phi \vee \Psi} \in \boldsymbol{I} \equiv (A_\Phi \in \boldsymbol{I}) \wedge (A_\Psi \in \boldsymbol{I})$, and (i) follows by de Morgan's laws.

°(ii) *If Θ is the formula $\bigvee_{u_k} \Phi$, then*

$$A_\Theta \notin \boldsymbol{I} \equiv \bigvee_{e_k \in P} [A_\Phi \notin \boldsymbol{I}].$$

In fact, suppose that

$$A_\Theta = \{t\colon \Theta(F_t, f_t, R_t, e_1(t), \ldots, e_{k-1}(t))\} \notin \boldsymbol{I}.$$

For $t \in A_\Theta$ there exists $x \in F_t$ such that

$$\Phi(F_t, f_t, R_t, e_1(t), \ldots, e_{k-1}(t), x).$$

Let X_t denote the set of all these elements x, and let $X_t = F_t$ for $t \notin A_\Theta$. Let e_k be a choice function for the family consisting of all the sets X_t. We have $e_k \in P$ and

(1) $$\Phi(F_t, f_t, R_r, e_1(t), \ldots, e_k(t))$$

for all $t \in A_\Theta$. This implies $A_\Theta \subset A_\Phi$ and thus $A_\Phi \notin \boldsymbol{I}$.

Conversely, if $A_\Phi \notin \boldsymbol{I}$ then formula (1) holds for all $t \in A_\Phi$. Thus for these t

$$\Theta(F_t, f_t, R_t, e_1(t), \ldots, e_{k-1}(t)).$$

This implies that $t \in A_\Theta$. Hence $A_\Phi \subset A_\Theta$ and $A_\Theta \notin \boldsymbol{I}$.

An ideal $\boldsymbol{I}$ is said to be *prime* if for any $X \subset T$ exactly one of the conditions $X \in \boldsymbol{I}$, $T - X \in \boldsymbol{I}$ holds.

The set $\{X \subset T\colon x \notin X\}$ is an example of a prime ideal. In Chapter VII we shall prove that every ideal can be extended to a prime ideal.

(iii) *If $\boldsymbol{I}$ is a prime ideal and Ψ is the formula $\neg\Phi$, then*

$$A_\Psi \notin \boldsymbol{I} \equiv \neg(A_\Phi \notin \boldsymbol{I}).$$

In fact, $A_\Psi = T - A_\Phi$.

°(iv) *If $\boldsymbol{I}$ is a prime ideal, then*

$$A_{\Phi\wedge\Psi} \notin \boldsymbol{I} \equiv (A_\Phi \notin \boldsymbol{I}) \wedge (A_\Psi \notin \boldsymbol{I});$$

moreover, if Ξ is the formula $\bigwedge_{u_k} \Phi$, then

$$A_\Xi \notin \boldsymbol{I} \equiv \bigwedge_{c_k \in P} [A_\Phi \notin \boldsymbol{I}].$$

Theorem (iv) follows from (i)–(iii).

A formula $\Phi(x, y, z, u_1, \ldots, u_k)$ is said to be *elementary* if it can be constructed from the formulas

(a) $u_i = u_j$,

(b) $y(u_i, u_j) = u_k$,[1])

(c) $\langle u_i, u_j\rangle \in z$

by the propositional operations and the quantifiers $\bigvee_{u \in x}$, $\bigwedge_{u \in x}$.

The following theorem solves the problem stated above.

°THEOREM 3: *If $\boldsymbol{I}$ is a prime ideal, $\Phi(x, y, z, u_1, \ldots, u_k)$ is an elementary formula and $e_1, \ldots, e_k$ are arbitrary elements of P, then*

$$(2) \quad \Phi(P/\boldsymbol{I}, \varphi/\boldsymbol{I}, \varrho/\boldsymbol{I}, e_1/\boldsymbol{I}, \ldots, e_k/\boldsymbol{I}) \equiv \{t\colon \Phi(F_t, f_t, R_t, e_1(t), \ldots, e_k(t))\} \notin \boldsymbol{I}.$$

PROOF. If Φ is one of the formulas (a), (b), (c), then (2) holds. In fact, the left-hand side of (2) is equivalent to $e_i/\boldsymbol{I} = e_j/\boldsymbol{I}$ in case (a), to $\varphi/\boldsymbol{I}(e_i/\boldsymbol{I}, e_j/\boldsymbol{I}) = e_h/\boldsymbol{I}$ in case (b), and to $\langle e_i/\boldsymbol{I}, e_j/\boldsymbol{I}\rangle \in \varrho/\boldsymbol{I}$ in case (c). The right-hand side of (2) is then equivalent:

in case (a) to $\{t\colon e_i(t) = e_j(t)\} \notin \boldsymbol{I}$,
in case (b) to $\{t\colon f_t(e_i(t), e_j(t)) = e_h(t)\} \notin \boldsymbol{I}$,
in case (c) to $\{t\colon \langle e_i(t), e_j(t)\rangle \in R_t\} \notin \boldsymbol{I}$.

[1]) We read this formula as: the value of y for the arguments u_i, u_j is u_k. We could also write it as $\langle\langle u_i, u_j\rangle, u_k\rangle \in y$.

From the definitions of the set P/I, the function φ/I, and the relation ϱ/I as well as from the definition of prime ideals it follows that the left-hand and right-hand sides of (2) are equivalent.

In turn, it follows from Theorems (i)–(iv) that if (2) holds for formulas Φ and Ψ, then it also holds for the formulas arising from Φ and Ψ by the propositional operations and the quantifiers $\bigvee_{u \in x}$ and $\bigwedge_{u \in x}$.

In this way Theorem 3 is proved.[1])

°COROLLARY 4: *If $\Phi(x, y, z)$ is an elementary formula, then*

$$\Phi(P/I, \varphi/I, \varrho/I) \equiv \{t\colon \Phi(F_t, f_t, R_t)\} \notin I.$$

This corollary follows from Theorem 3 if we assume that Φ does not contain the variables $u_1, \ldots, u_k$.

°COROLLARY 5: *If $\Phi(x, y, z)$ is an elementary formula and for every t the formula $\Phi(F_t, f_t, R_t)$ holds, then $\Phi(P/I, \varphi/I, \varrho/I)$.*

This corollary follows directly from the previous corollary and from the remark that if I is a prime ideal then $T \notin I$.

Examples

In the following we suppose that I is a prime ideal in $P(T)$.

1. If the relations R_t are reflexive, transitive and satisfy the conditions:

$$\bigwedge_{x, y \in F_t} \{[\langle x, y\rangle \in R_t] \wedge [\langle y, x\rangle \in R_t] \to x = y\},$$

$$\bigwedge_{x, y \in F_t} [(\langle x, y\rangle \in R_t) \vee (\langle y, x\rangle \in R_t)],$$

then the relation ϱ/I satisfies the same conditions.

2. If F_t is a field with respect to the operations of addition f_t and multiplication g_t, then the set P/I is a field with respect to the operations φ/I and ψ/I, where φ and ψ are the cartesian products of the operations f_t and g_t, respectively.

Similarly, if each of the sets F_t is an ordered field with respect to the operations f_t and g_t and with respect to the order relation R_t,

[1]) Theorem 3 which is due to Łoś [1] has found numerous applications which may be found e.g. in Bell and Slomson [1] or Chang and Keisler [1]. See also Chapter X, Section 7.

then P/I is an ordered field with respect to the operations φ/I and ψ/I and the order relation ϱ/I.

These properties follow by Corollary 5 from the remark that the formulas "X is a field under the operations D and M" and "X is a field under these operations ordered by the relation R" are equivalent to elementary formulas.

These examples show that if we are given a system of axioms and a family of models of this system then by forming the reduced product of this family we can obtain a new model of the same axioms. Other applications of reduced products will be given in Chapter X, Section 7.

Exercises

1. By applying the lemma on p. 138, prove the existence of a prime ideal which contains a given ideal $\neq P(T)$.

2. Prove that if each of the sets F_t is a Boolean algebra with respect to the operations $\vee_t$, $\wedge_t$, $-_t$ and the elements 0_t, 1_t and if for every t the condition

$$(*) \qquad \bigwedge_{x} \{(x = 0_t) \vee \bigvee_{x,z} [(x = y \vee_t z) \wedge (y \neq 0_t) \wedge (z \neq 0_t)]\}$$

holds, then the set P/I is a Boolean algebra with respect to the operations $\vee/I$, $\wedge/I$, $-/I$ and the elements ξ/I, ι/I, where $\vee$, $\wedge$, $-$ are the cartesian products of the operations $\vee_t$, $\wedge_t$, $-_t$, and ξ, ι are functions such that $\xi(t) = 0_t$ and $\iota(t) = 1_t$ for every t. Moreover, P/I satisfies condition $(*)$. Show that the ordinary cartesian product of Boolean algebras satisfying $(*)$ may fail to satisfy this condition.

§ 10. Infinite operations in lattices and in Boolean algebras

The theorems in the previous sections of this chapter can be considered as theorems about the lattice $P(X)$ (see Ch. I, §10). As we know, this lattice is a Boolean algebra and a complete lattice. In a natural way the question arises as to whether the theorems in § 1 can be generalized to the case of arbitrary lattices, or complete lattices, or Boolean algebras.

Suppose first that K is any ordered set and $f \in K^T$. The following theorems, analogous to Theorems 1.1–1.3. hold:

THEOREM 1: *The least upper bound $g = \bigvee_{t \in T} f_t$, if it exists, is the unique element of K satisfying the conditions*

$$\bigwedge_{t \in T} (f_t \leqslant g) \quad \textit{and} \quad \bigwedge_{t \in T} (f_t \leqslant a) \rightarrow (g \leqslant a).$$

A similar theorem holds for the greatest lower bound $d = \bigwedge_{t \in T} f_t$.

If the least upper bound $\bigvee_{t \in T} f_t$ exists, it is called the *supremum* of the elements f_t, $t \in T$. If the greatest lower bound $\bigwedge_{t \in T} f_t$ exists, it is called the *infimum* of the elements f_t, $t \in T$.

THEOREM 2: *If* $T = \bigcup_{u \in U} H_u$ *and for every* $u \in U$ *the suprema* $\bigvee_{t \in H_u} f_t = g_u$ *exist and if the supremum* $\bigvee_{t \in T} f_t = g$ *also exists, then the supremum* $\bigvee_{u \in U} g_u$ *exists and is equal to* g *(similarly for the infimum).*

THEOREM 3: *If* φ *is a permutation of the set* T *and the supremum* $\bigvee_{t \in T} f_t = g$ *exists, then the supremum* $\bigvee_{t \in T} f_{\varphi(t)}$ *also exists and is equal to* g *(similarly for the infimum).*

Theorem 1 is a restatement of the definition of the least upper bound. Theorems 2 and 3 can be proved similarly to Theorems 1.2 and 1.3.

The following theorem holds for all ordered sets and is analogous to 1(3).

THEOREM 4: *If the supremum* $\bigvee_{t \in T} f_t = g$ *and the infimum* $\bigwedge_{t \in T} f_t = d$ *exist, then for all* $t \in T$ *we have* $d \leqslant f_t \leqslant g$.

On the other hand, formulas 1(4)–1(11) do not have counterparts for arbitrary ordered sets.

THEOREM 5: *If the set* K *is a lattice,* $p, q \in K^T$, *and if the suprema* $\bigvee_{t \in T} p_t = g_1$ *and* $\bigvee_{t \in T} q_t = g_2$ *exist, then the supremum* $\bigvee_{t \in T} (p_t \vee q_t)$ *exists and is equal to* $g_1 \vee g_2$ *(similarly for the infimum).*

PROOF. $g_1 \vee g_2 \geqslant p_t \vee q_t$ for all $t \in T$. If $\bigwedge_{t \in T} (x \geqslant p_t \vee q_t)$, then also $\bigwedge_{t \in T} (x \geqslant p_t)$, whence $x \geqslant g_1$; and similarly $x \geqslant g_2$; hence $x \geqslant g_1 \vee g_2$.

The assumption that K is a lattice has been made in this theorem in order to ensure the existence of $p_t \vee q_t$ and $g_1 \vee g_2$.

THEOREM 6: *If* K *is a lattice and the suprema* $\bigvee_{t \in T} f_t$ *and* $\bigvee_{t \in T} (a \wedge f_t)$

exist, then

$$\bigvee_{t\in T}(a\wedge f_t) \leqslant a\wedge \bigvee_{t\in T} f_t$$

(similarly for the infimum).

PROOF. For every $t\in T$ we have $a\wedge f_t \leqslant a$ and $a\wedge f_t \leqslant f_t \leqslant \bigvee_{t\in T} f_t$; thus $a\wedge f_t \leqslant a\wedge \bigvee_{t\in T} f_t$, whence the required formula follows.

The inequality sign in Theorem 6 cannot in general be replaced by the equality sign even in the case of complete lattices. However, the following theorem holds.

THEOREM 7: *If K is a Boolean algebra and the supremum $\bigvee_{t\in T} f_t$ exists, then for any $a\in K$ the supremum $\bigvee_{t\in T}(a\wedge f_t)$ exists and is identical to $a\wedge\bigvee_{t\in T} f_t$ (similarly for the infimum).*

PROOF. Since $a\wedge f_t \leqslant a\wedge \bigvee_{t\in T} f_t$ for every $t\in T$, it suffices to show that if $\bigwedge_{t\in T}(a\wedge f_t \leqslant x)$, then $a\wedge\bigvee_{t\in T} f_t \leqslant x$. From the assumptions it follows that $-a\vee(a\wedge f_t) \leqslant -a\vee x$; hence $f_t \leqslant -a\vee x$ for arbitrary $t\in T$. We obtain $\bigvee_{t\in T} f_t \leqslant -a\vee x$, thus $a\wedge\bigvee_{t\in T} f_t \leqslant a\wedge(-a\vee x) \leqslant x$.

For Boolean algebras the following theorem holds (de Morgan's law).

THEOREM 8. *If the supremum $\bigvee_{t\in T} f_t = g$ exists, then so does the infimum $\bigwedge_{t\in T}(-f_t)$ and is equal to $-g$ (similarly with supremum and infimum interchanged).*

PROOF. Since $f_t \leqslant g$, we have $-g \leqslant -f_t$ for every $t\in T$. If $x \leqslant -f_t$ for every $t\in T$, then $f_t \leqslant -x$; hence $g \leqslant -x$ and $x \leqslant -g$, whence we obtain $\bigwedge_{t\in T}(-f_t) = -g$.

As the theorems above show, all basic theorems in § 1 can be generalized to the case of complete Boolean algebras. For non-complete Boolean algebras the theorems hold if we make the additional assumption that all the necessary suprema and infima exist. It is interesting to notice that the distributive law stated in Theorem 7 does not involve the complement sign, yet it can be proved only for Boolean algebras. The general distributive law given in Theorem 1.4 is even more peculiar. We shall prove that Boolean algebras of the form $\boldsymbol{P}(X)$ are, in fact, the only Boolean algebras for which this theorem holds. First let us assume two definitions.

DEFINITION 1: A Boolean algebra $\boldsymbol{K}$ is said to be *distributive* if it is complete and, moreover, if for every set M and for every function $f\colon M \to \boldsymbol{K}$ and for every partition of M into non-empty sets $M = \bigcup_{u \in U} T_u$ the following identity holds:

$$\bigwedge_{u \in U} \bigvee_{t \in T_u} f_t = \bigvee_{Y \in K} \bigwedge_{t \in Y} f_t, \tag{1}$$

where

$$K = \{Y \in \boldsymbol{P}(M) \colon \bigwedge_{u \in U} (Y \cap T_u \neq \emptyset)\}. \tag{2}$$

DEFINITION 2: An element a is said to be an *atom* of a Boolean algebra K if $a \in K$, $a \neq o$ and $x < a \to x = o$. A Boolean algebra K is said to be *atomic* if for every element $x \neq o$ there is at least one atom a such that $a \leqslant x$.

THEOREM 9:[1]) *Every complete and atomic Boolean algebra K is isomorphic to the field $\mathbf{P}(A)$ where A is the set of atoms of K. Namely, there exists a one-to-one mapping Φ of K onto $\mathbf{P}(A)$ such that*

$$\Phi\left(\bigvee_{t \in T} f_t\right) = \bigcup_{t \in T} \Phi(f_t), \tag{3}$$

$$\Phi\left(\bigwedge_{t \in T} f_t\right) = \bigcap_{t \in T} \Phi(f_t) \tag{4}$$

for any set T and for any function $f \in K^T$.

PROOF. Let for $x \in K$

$$\Phi(x) = \{a \in A \colon a \leqslant x\}.$$

This formula defines a function whose domain is K and whose values are subsets of A. Clearly, $x \leqslant y \to \Phi(x) \subset \Phi(y)$.

[1]) See Tarski [5].

The function Φ is one-to-one. For suppose that x and y are elements of the algebra K such that $x \triangle y \neq o$. We can assume that $x-y \neq o$. By the assumption that K is atomic, there exists an atom a such that $a \leqslant x-y$. It follows from the formulas $a \wedge x \leqslant a$ and $a \wedge y \leqslant a$ that

$$a \wedge x = o \quad \text{or} \quad a \wedge x = a,$$

and

$$a \wedge y = o \quad \text{or} \quad a \wedge y = a.$$

The formulas $a \wedge x = o$ and $a \wedge y = o$ imply

$$a = a \wedge (x-y) = (a \wedge x)-(a \wedge y) = o-o = o.$$

Hence $a = o$, which contradicts the fact that a is an atom (Definition 2). Similarly, the formulas $a \wedge x = a$ and $a \wedge y = a$ imply

$$a = a \wedge (x-y) = (a \wedge x)-(a \wedge y) = a-a = o,$$

which again contradicts Definition 2.

Thus, either $a \wedge x = o$ and $a \wedge y = a$, or $a \wedge x = a$ and $a \wedge y = o$. In the former case we have a non $\leqslant x$ and $a \leqslant y$, in the latter $a \leqslant x$ and a non $\leqslant y$. Thus in both cases $\Phi(x) \neq \Phi(y)$.

Let $f \in K^T$. If $a \in \bigcup_{t \in T} \Phi(f_t)$, then there exists $t \in T$ such that $a \in \Phi(f_t)$. This implies $a \leqslant f_t \leqslant \bigvee_{t \in T} f_t$; thus $a \in \Phi(\bigvee_{t \in T} f_t)$.

Hence we have proved that

$$\bigcup_{t \in T} \Phi(f_t) \subset \Phi\Big(\bigvee_{t \in T} f_t\Big). \tag{5}$$

Now suppose that $a \in \Phi\big(\bigvee_{t \in T} f_t\big)$; that is, $a \leqslant \bigvee_{t \in T} f_t$. If $a \wedge f_t = o$ for every t, then we have $a = a-(a \wedge f_t) = a-f_t$; thus

$$a = \bigwedge_{t \in T} (a-f_t) = a-\bigvee_{t \in T} f_t$$

$$= a-\Big(a \wedge \bigvee_{t \in T} f_t\Big) = a-a = o,$$

because $a \wedge \bigvee_{t \in T} f_t = a$. But this conclusion contradicts the fact that $a \in A$. Hence there exists $t \in T$ such that $o \neq a \wedge f_t \leqslant a$. This means $a \wedge f_t = a$; thus $a \leqslant f_t$ and finally $a \in \Phi(f_t) \subset \bigcup_{t \in T} \Phi(f_t)$. We have thus proved the following inclusion

$$\Phi(\bigvee_{t\in T} f_t) \subset \bigcup_{t\in T} \Phi(f_t).$$

This by (5) implies (3).

Formula (4) can be proved even more simply. We have

$$\begin{aligned} a = \Phi(\bigwedge_{t\in T} f_t) &\equiv (a \leqslant \bigwedge_{t\in T} f_t) \\ &\equiv \bigwedge_{t\in T} (a \leqslant f_t) \\ &\equiv \bigwedge_{t\in T} (a \in \Phi(f_t)) \\ &\equiv a \in \bigcap_{t\in T} \Phi(f_t). \end{aligned}$$

It remains to be shown that every set $X \subset A$ can be represented as $\Phi(x)$ for some $x \in K$. For this purpose, let

$$T = X, \quad f_t = t, \quad x = \bigvee_{a\in X} f_a$$

(this union exists because we assumed that K is a complete Boolean algebra).

According to (3) we have

$$\Phi(x) = \bigcup_{a\in X} \Phi(a) = \bigcup_{a\in X} \{a\} = X,$$

because a is the unique atom included in a and hence $\Phi(a) = \{a\}$.

In this way Theorem 9 is completely proved.

THEOREM 10:[1]) *Every complete and atomic Boolean algebra K is distributive.*

PROOF. According to Theorem 9 there exists a function Φ which establishes the isomorphism between K and the field of subsets of some set A. By Theorem 4, § 1, formula (2) implies the formula

$$\bigcap_{u\in U} \bigcup_{t\in T_u} \Phi(f_t) = \bigcup_{Y\in K} \bigcap_{t\in Y} \Phi(f_t),$$

[1]) See Tarski [5].

which by (3) and (4) implies

$$\Phi(\bigwedge_{u\in U}\bigvee_{t\in T_u} f_t) = \Phi(\bigvee_{Y\in \boldsymbol{K}}\bigwedge_{t\in Y} f_t).$$

Since the function Φ is one-to-one, the formula above implies formula (1).

THEOREM 11: *If the Boolean algebra K is complete and distributive, then it is atomic*

PROOF. Suppose that K is complete and distributive but not atomic. Let a_0 be an element $\neq o$ which does not contain any atom. Let $T_u = \{u, -u\}$ for $u \in K$ and $f_t = t \wedge a_0$ for $t \in K$. Since $K = \bigcup_{u\in K} T_u$, we have equation (1), where $M = K$ and where $\boldsymbol{K}$ is defined by (2) for $M = K$. It follows from the definition of the set T_u that

$$\bigvee_{t\in T_u} f_t = f_u \vee f_{-u} = (a_0 \wedge u) \vee (a_0 \wedge -u) = a_0 \wedge (u \vee -u) = a_0,$$

hence

$$\bigwedge_{u\in U}\bigvee_{t\in T_u} f_t = a_0.$$

It follows from (1) that there exists a set $Y_0 \in \boldsymbol{K}$ such that

$$\bigwedge_{t\in Y_0} f_t \neq o.$$

Let

(6) $$b = \bigwedge_{t\in Y_0} f_t = a_0 \wedge \bigwedge_{t\in Y_0} t.$$

Since a_0 contains no atom, b is not an atom. This means that there exists an element c such that

(7) $$o \neq c \leqslant b \quad \text{and} \quad c \neq b.$$

According to the definition of $\boldsymbol{K}$ (see (2)), $Y_0 \cap T_c \neq \emptyset$; that is, either $c \in Y_0$ or $-c \in Y_0$. This implies by (6) that either $b \leqslant c$ or $b \leqslant -c$. In the former case we infer that $b = c$, in the latter case that $c \leqslant -c$. Thus either $c = b$ or $c = o$, which contradicts (7). This contradiction completes the proof.

Example. The Boolean algebra K of regular closed sets in the plane is complete but not distributive.

For we proved in Chapter I, p. 39 that $\boldsymbol{K}$ is a Boolean algebra with respect to the operations $\bigcup$, $\odot$ and $'$. Moreover, it was shown that each element of $\boldsymbol{K}$, different from $\emptyset$, contains a non-empty element distinct from itself. The Boolean algebra $\boldsymbol{K}$ is thus not atomic. From Examples 1.3 and 1.4 it follows that $\boldsymbol{K}$ is complete, from Theorem 11 it follows that $\boldsymbol{K}$ is not distributive.

This Boolean algebra $\boldsymbol{K}$ is an interesting example which shows that not all laws which hold for the algebra of sets can be transferred to the theory of Boolean algebras, even in the case of complete Boolean algebras.[1])

Exercises

1. Prove that the equation dual to (1) also holds in atomic Boolean algebras

$$\bigvee_{u\in U}\bigwedge_{t\in T_u} f_t = \bigwedge_{Y\in K}\bigvee_{t\in Y} f_t. \tag{8}$$

Furthermore, show that there exist complete Boolean algebras in which (8) does not hold.

2. Give an example of a Brouwerian lattice K such that for some set T, for some function $f\in K^T$ and for some element $a\in K$ we have

$$a\wedge\Big(\bigvee_{t\in T} f_t\Big) \neq \bigvee_{t\in T}(a\wedge f_t).$$

3. Give an example of a Brouwerian lattice K such that for some set T, for some function $f\in K^T$ and for some $a\in K$ the supremum $\bigvee_{t\in T} f_t$ exists but the supremum $\bigvee_t (a\wedge f_t)$ does not exist.

§11. Extensions of ordered sets to complete lattices

We shall prove that every ordered set can be treated as a subsystem of a complete lattice (that is, of a lattice in which most laws of the algebra of sets hold). We shall also solve a similar problem for Boolean

[1]) There are numerous other results concerning distributive Boolean algebras which we could not include in this book. The reader may found bibliographical references to these results in Sikorski [1].

In the whole Section 10 we dealt with the generalization of the operations of union and intersection. Other set-theoretical operations were also generalized for Boolean algebras. For instance the theory of cylindric algebras represents a generalization to Boolean algebras of the cartesian multiplication. See Henkin, Monk and Tarski [1].

algebras. For this purpose, we first introduce a general notion of embedding of one system in another.

DEFINITION 1: A relational system $\langle A, R\rangle$ is said to be a *subsystem* of $\langle B, S\rangle$ if $A \subset B$ and $\bigwedge_{x,y \in A} [xRy \equiv xSy]$ (that is, $R = (A \times A) \cap S$).

DEFINITION 2: A system $\langle B, S\rangle$ is said to be an *extension* of the system $\langle A, R\rangle$ if there exists a subsystem $\langle B_1, S_1\rangle$ of the system $\langle B, S\rangle$ isomorphic to $\langle A, R\rangle$.

In this case we also say that the system $\langle A, R\rangle$ is *embedded isomorphically* into the system $\langle B, S\rangle$ and that the function establishing the isomorphism *embeds* $\langle A, R\rangle$ in $\langle B, S\rangle$.

THEOREM 1: *Every ordered set A can be embedded isomorphically in the family of all subsets of some set* (*where the family is ordered by the inclusion relation*). *Consequently, A can be embedded in some complete and atomic Boolean algebra.*

PROOF. Let for $a \in A$

$$O(a) = \{x\colon x \leqslant a\}.$$

Because the relation $\leqslant$ is transitive, we have

$$a \leqslant b \to O(a) \subset O(b).$$

Since $a \in O(a)$, we have $O(a) \subset O(b) \to a \in O(b) \to a \leqslant b$. Hence

$$a \leqslant b \equiv O(a) \subset O(b),$$

and it follows directly that

$$O(a) = O(b) \to a = b.$$

Thus the function $a \to O(a)$ embeds A in the family of sets $O(a)$ ordered by the inclusion relation. Clearly, this family can be extended to the family $\boldsymbol{P}(X)$ where $X = \bigcup_{a \in A} O(a)$.

In general, the extension described in Theorem 1 preserves neither suprema nor infima; that is, the equation $a = b \vee c$ does not necessarily imply $O(a) = O(b) \cup O(c)$. We shall consider whether it is possible to extend the set A to a complete lattice preserving suprema and infima.

Let the function φ embed the ordered system $\langle A, \leqslant_A\rangle$ in the ordered system $\langle B, \leqslant_B\rangle$.

DEFINITION 3: The embedding φ *preserves suprema* if for every set T and for every function $f \in A^T$ such that the supremum $\bigvee_{t\in T} f_t$ exists, the supremum $\bigvee_{t\in T} \varphi(f_t)$ also exists and

$$\varphi\Big(\bigvee_{t\in T} f_t\Big) = \bigvee_{t\in T} \varphi(f_t).$$

We admit a similar definition for the infima.

We shall now consider a construction which will extend any ordered set to a complete lattice preserving suprema and infima. Let A be a set ordered by the relation $\leqslant$. For any set $X \subset A$, let

$$X^+ = \Big\{a \in A: \bigwedge_{x\in X} (x \leqslant a)\Big\}, \quad X^- = \Big\{a \in A: \bigwedge_{x\in X} (a \leqslant x)\Big\}.$$

The next statements follow from the definitions above:

(1) $$X \subset Y \to (X^+ \supset Y^+) \wedge (X^- \supset Y^-);$$

(2) *for any* $Z \subset A$ *we have* $Z \subset Z^{+-}$ *and* $Z \subset Z^{-+}$.

PROOF. By definition,

$$(z \in Z) \wedge (z' \in Z^+) \to (z \leqslant z'),$$

thus

$$(z \in Z) \to \bigwedge_{z'\in Z^+} (z \leqslant z') \to (z \in Z^{+-}).$$

The proof of the second part of (2) is similar.

(3) $$Z^{+-+} = Z^+ \quad \text{and} \quad Z^{-+-} = Z^-.$$

The inclusion $Z^+ \subset Z^{+-+}$ follows from (2). If $a \in Z^{+-+}$ then $\bigwedge_{z\in Z^{+-}} (z \leqslant a)$ and therefore $\bigwedge_{z\in Z} (z \leqslant a)$ since $Z \subset Z^{+-}$. Thus $a \in Z^+$.

The proof of the second part of (3) is similar.

We now introduce the notion of a cut for ordered sets.[1])

[1]) The notion of a cut originated with Dedekind [1] who used it in his construction of real numbers. The extension of Dedekind's theory to arbitrary linearly ordered sets is immediate. The much less obvious generalization to the case of arbitrary ordered sets is due to Mac Neille [1].

The pair $\langle X, Y\rangle$ is said to be a ***cut*** in the ordered set A if $X^+ = Y$ and $Y^- = X$. The set X is called the *lower section* and Y the *upper section* of the cut.

It follows from the definition that

(4) $$(x \in X) \wedge (y \in Y) \rightarrow (x \leqslant y).$$

In fact, every element of X^+ is in relation $\geqslant$ to every element of X.

It follows from (3) that

(5) *Pairs $\langle Z^-, Z^{-+}\rangle$ and $\langle Z^{+-}, Z^+\rangle$ are cuts. Moreover, every cut can be expressed in both of these forms.*

Finally, it follows from the definition of a cut that

(6) *If $a \in A$ then the pair $\langle \{a\}^-, \{a\}^+\rangle$ is a cut.*

We can introduce an order relation between cuts:

$$\langle X, Y\rangle \leqslant \langle U, V\rangle \equiv X \subset U.$$

Before we show that the relation $\leqslant$ is indeed an order relation, we first prove that

(7) $$\langle X, Y\rangle \leqslant \langle U, V\rangle \equiv V \subset Y.$$

PROOF. Suppose that $X \subset U$ and $v \in V$. If $x \in X$, then $x \in U = V^-$; thus $x \leqslant v$. Therefore $v \in X^+ = Y$, and finally $V \subset Y$. In a similar way we prove the opposite implication.

It is now immediate that the relation $\leqslant$ between cuts is reflexive, antisymmetric, and transitive, i.e. $\leqslant$ is an order relation.

Let $\mathfrak{P}$ denote the family of all cuts.

(8) *$\mathfrak{P}$ is a complete lattice.*

Let $\mathfrak{L} \subset \mathfrak{P}$ and

$$S = \bigcup_{\langle X, X^+\rangle \in \mathfrak{L}} X, \qquad T = \bigcup_{\langle X^-, X\rangle \in \mathfrak{L}} X.$$

The cut $\langle S^{+-}, S^+\rangle$ is the supremum of $\mathfrak{L}$. In fact, $S \subset S^{+-}$ and thus $\langle X, X^+\rangle \leqslant \langle S^{+-}, S^+\rangle$ for every cut $\langle X, X^+\rangle \in \mathfrak{L}$. If $\langle X, X^+\rangle \leqslant \langle U, V\rangle$ for every $\langle X, X^+\rangle \in \mathfrak{L}$, then $X \subset U$ and therefore $S \subset U$. This implies $S^+ \supset U^+ = V$, hence $\langle S^{+-}, S^+\rangle \leqslant \langle U, V\rangle$.

Similarly we can prove that the cut $\langle T^{-}, T^{-+}\rangle$ is the infimum of $\mathfrak{L}$.

(9) *The function* $f(x) = \langle\{x\}^{-}, \{x\}^{+}\rangle$ *embeds* A *in* $\mathfrak{P}$ *preserving suprema and infima.*

PROOF. By definition the following equivalences hold:

$$x \leqslant y \equiv \{x\}^{-} \subset \{y\}^{-} \equiv f(x) \leqslant f(y).$$

Suppose that $x = \bigvee_{t\in T} \varphi_t$ in the set A; then $\varphi_t \leqslant x$ and therefore $f(\varphi_t) \leqslant f(x)$ for $t \in T$.

Let $\langle X, Y\rangle$ be a cut such that $f(\varphi_t) \leqslant \langle X, Y\rangle$ for $t \in T$. Then $\{\varphi_t\}^{-} \subset X$, and since $\varphi_t \in \{\varphi_t\}^{-}$, we obtain $\varphi_t \in X$. For any $y \in Y$ we have $\varphi_t \leqslant y$, thus $x \leqslant y$. This implies $x \in Y^{-} = X$, hence $\{x\}^{-} \subset X$ and $f(x) \leqslant \langle X, Y\rangle$. Thus $f(x) = \bigvee_t f(\varphi_t)$ in the set $\mathfrak{P}$.

The proof for the infimum is similar.

The next theorem follows from (9).

THEOREM 2: *Every ordered set* A *can be extended to a complete lattice* $\mathfrak{P}$ *preserving suprema and infima.*

The lattice $\mathfrak{P}$ constructed above is called the *minimal extension* of the ordered set A.

We now consider the case where A is a Boolean algebra.

First, observe that *if* Z_1 *and* Z_2 *are arbitrary subsets of an ordered set* A, *then*

(10) $$(Z_1 \cup Z_2)^{+} = Z_1^{+} \cap Z_2^{+}, \quad (Z_1 \cup Z_2)^{-} = Z_1^{-} \cap Z_2^{-}.$$

Now let A be a lattice. We prove that *if* Y_1 *and* Y_2 *are upper sections of two cuts in* A, *then* $Y_1 \cap Y_2$ *is the set of all elements* $y_1 \vee y_2$ *where* $y_i \in Y_i$ *for* $i = 1, 2$. In fact, $y_1 \vee y_2 \geqslant y_i$; thus $y_1 \vee y_2 \in Y_i$ for $i = 1, 2$. Moreover, $y \in Y_1 \cap Y_2 \rightarrow y = y \vee y$, where the first component may be understood to be an element of Y_1 and the second to be an element of Y_2.

Likewise we can prove that *if* X_1 *and* X_2 *are lower sections of two cuts in* A, *then* $X_1 \cap X_2$ *is the set of all elements of the form* $x_1 \wedge x_2$ *where* $x_i \in X_i$ *for* $i = 1, 2$.

Combining the above with formula (10) we conclude that *if* A *is a lattice and* $\langle X_1, Y_1\rangle$, $\langle X_2, Y_2\rangle$ *are two cuts in* A, *then*

$$(11)\quad \begin{aligned}(X_1 \cup X_2)^+ &= Y_1 \cap Y_2 = \{y_1 \vee y_2 \colon (y_1 \in Y_1) \wedge (y_2 \in Y_2)\},\\ (Y_1 \cup Y_2)^- &= X_1 \cap X_2 = \{x_1 \wedge x_2 \colon (x_1 \in X_1) \wedge (x_2 \in X_2)\}.\end{aligned}$$

It follows from the definitions of supremum and infimum in $\mathfrak{P}$ that for any ordered set A and for any two cuts in A we have

$$(12)\quad \begin{aligned}\langle X_1, Y_1\rangle \vee \langle X_2, Y_2\rangle &= \langle (X_1 \cup X_2)^{+-}, (X_1 \cup X_2)^+\rangle\\ &= \langle (Y_1 \cap Y_2)^-, Y_1 \cap Y_2\rangle,\end{aligned}$$

$$(13)\quad \begin{aligned}\langle X_1, Y_1\rangle \wedge \langle X_2, Y_2\rangle &= \langle (Y_1 \cup Y_2)^-, (Y_1 \cup Y_2)^{-+}\rangle\\ &= \langle X_1 \cap X_2, (X_1 \cap X_2)^+\rangle.\end{aligned}$$

Finally, observe that *if A is a Boolean algebra and $Z^* = \{-z \colon z \in Z\}$ for any set $Z \subset A$, then*

(14) *if $\langle X, Y\rangle$ is a cut in A, then $\langle Y^*, X^*\rangle$ is also a cut in A.*

The proof of this lemma is left to the reader.

Now we shall prove the following

THEOREM 3: *The minimal extension of a Boolean algebra is a Boolean algebra.*

PROOF. Let $\mathfrak{P}$ be the minimal extension of a Boolean algebra A. It suffices to show that $\mathfrak{P}$ is a distributive lattice with zero O and with unit I and that for any cut $\langle X, Y\rangle \in \mathfrak{P}$ there exists a cut $\langle X_1, Y_1\rangle \in \mathfrak{P}$ such that

$$(15)\quad \langle X, Y\rangle \wedge \langle X_1, Y_1\rangle = O, \qquad \langle X, Y\rangle \vee \langle X_1, Y_1\rangle = I$$

(see p. 43, Theorem).

Clearly, the cut $O = \langle \{o\}, A\rangle$ is the zero of $\mathfrak{P}$ and $I = \langle A, \{i\}\rangle$ is the unit. The cut $\langle Y^*, X^*\rangle$ defined by (14) satisfies conditions (15). In fact, the lower section of $\langle X, Y\rangle \wedge \langle Y^*, X^*\rangle$ is $X \cap Y^*$ (see formula (13)). The only element contained in $X \cap Y^*$ is o, because $a \in X \cap$ $\cap Y^* \rightarrow (a \in X) \wedge (-a \in Y) \rightarrow (a \leqslant -a) \rightarrow a = o$. This proves the first part of (15); the second part can be proved similarly.

It remains to show that the distributive law holds. Since the inequality $(a \wedge c) \vee (b \wedge c) \leqslant (a \vee b) \wedge c$ holds in every lattice, it suffices to show that if $\langle X_1, Y_1\rangle$, $\langle X_2, Y_2\rangle$, and $\langle U, V\rangle$ are three cuts in A, then

$$\begin{aligned}&(\langle X_1, Y_1\rangle \vee \langle X_2, Y_2\rangle) \wedge \langle U, V\rangle\\ &\qquad\qquad \leqslant (\langle X_1, Y_1\rangle \wedge \langle U, V\rangle) \vee (\langle X_2, Y_2\rangle \wedge \langle U, V\rangle).\end{aligned}$$

Applying (12) and (13), this formula can be reduced to the form

$$(Y_1 \cap Y_2)^- \cap U \subset [(X_1 \cap U)^+ \cap (X_2 \cap U)^+]^-;$$

that is, by (11) and by the definitions of Z^+ and Z^-,

$$(16) \qquad [(a \in U) \wedge \bigwedge_{y_1 \in Y_1, y_2 \in Y_2} (a \leqslant y_1 \vee y_2) \wedge$$
$$\wedge \bigwedge_{x_1 \in X_1, x_2 \in X_2, u \in U} (b \geqslant x_1 \wedge u) \wedge (b \geqslant x_2 \wedge u)] \rightarrow (a \leqslant b).$$

Suppose now that the elements a and b satisfy the antecedent of the implication (16). For arbitrary x_1 in X_1 the inequality $b \geqslant x_1 \wedge a$ holds. This implies $-a \vee b \geqslant x_1$. Since x_1 is arbitrary, we conclude that the element $y = -a \vee b$ belongs to Y_1. Similarly we prove that $y \in \boldsymbol{Y_2}$. Because a satisfies the antecedent of (16), we obtain $a \leqslant y \vee y = y = -a \vee b$; hence $a \leqslant a \wedge (-a \vee b) = a \wedge b \leqslant b$. This proves (16) and thus the theorem is proved.

§ 12. Representation theory for distributive lattices

The notion of ideal is the basic concept involved in the representation theory for distributive lattices.

DEFINITION: A non-empty set $I \subset A$ is said to be an *ideal* of the distributive lattice A if

$$(1) \qquad (a \in I) \wedge (b \in I) \rightarrow (a \vee b \in I),$$

$$(2) \qquad (a \leqslant b) \wedge (b \in I) \rightarrow (a \in I).$$ [1]

An ideal I is said to be a *prime ideal* if $I \neq A$ and for $a, b \in A$,

$$(3) \qquad (a \wedge b \in I) \rightarrow [(a \in I) \vee (b \in I)].$$

Examples

1. Let A be a lattice of sets (for instance, the lattice of all subsets of an arbitrary set X) and let a be any element of the union $U(A)$. The family $\boldsymbol{I}$ of all sets $M \in \boldsymbol{A}$ which do not contain a is a prime ideal in the lattice $\boldsymbol{A}$.

[1] In all subsequent investigations ideals can be replaced by filters (see Chapter I. p. 17) simply by interchanging the symbols $\vee$ and $\wedge$, 0 and 1 and reversing the sign $\leqslant$.

2. The family of all finite sets $M \in A$ is an ideal in A.

3. If A is the family of all subsets of the set of real numbers, then the family of all sets of Lebesgue measure zero is an ideal in A.

The set $\{x: x \leqslant a\}$ is an ideal in any lattice. This ideal is called the *principal ideal generated by a*.

We shall make use of the following general theorems concerning ideals.

(4) $$(a \vee b \in I) \rightarrow (a \in I) \wedge (b \in I).$$

In fact, $a \leqslant a \vee b$ and $b \leqslant a \vee b$. If $a \vee b \in I$, then $a \in I$ and $b \in I$ by (2).

(5) $$a \in I \rightarrow a \wedge b \in I.$$

In fact, $a \wedge b \leqslant a$.

(6) *The set $I^*(b)$ of elements x such that $x \leqslant i \vee b$ for some $i \in I$ is an ideal and $I \subset I^*(b)$, $b \in I^*(b)$.*

In fact, if $x \leqslant i_1 \vee b$ and $y \leqslant i_2 \vee b$, then $x \vee y \leqslant (i_1 \vee i_2) \vee b$; therefore $x \vee y \in I^*(b)$, because $i_1 \vee i_2 \in I$. If $x \leqslant i \vee b$ and $y \leqslant x$, then $y \leqslant i \vee b$. Thus the set $I^*(b)$ is an ideal. The conditions $I \subset I^*(b)$ and $b \in I^*(b)$ are obvious.

(7) *Let the ideals I_t, for $t \in T$, constitute a monotone family of ideals such that $a \in I_t$ and $b \notin I_t$ for every t. Then the union $I = \bigcup_t I_t$ is an ideal and $a \in I$, $b \notin I$.*

In fact, if $x \in I_{t_1}$ and $y \in I_{t_2}$, then either both elements x, y belong to I_{t_1}, or both belong to I_{t_2}. In any case, $(x \vee y) \in I$.

If $y \in I_t$ and $x \leqslant y$, then $x \in I_t$; thus $x \in I$. Hence the set I is an ideal and $a \in I$ and $b \notin I$.

(8) *If it is not true that $b \leqslant a$, then there exists an ideal I such that $a \in I$ and $b \notin I$.*

In fact, the set $\{x: x \leqslant a\}$ is such an ideal.

Suppose that $b \leqslant a$ is not true and let $\mathbf{P}_{a,b}$ be the family of those ideals which contain a but do not contain b.

(9) *If the lattice A is distributive and $b \leqslant a$ is false, then every maximal element I of the family $\mathbf{P}_{a,b}$ is a prime ideal.*

In fact, suppose that $x \wedge y \in I$. If $x \notin I$ then the ideal I is a proper subset of the ideal $I^*(x)$. Thus $I^*(x)$ does not belong to the family $\boldsymbol{P}_{a,b}$. Since $a \in I^*(x)$, we have $b \in I^*(x)$. Thus $b \leqslant i_1 \vee x$ for some $i_1 \in I$. Similarly we show that if $y \notin I$, then $b \leqslant i_2 \vee y$ for some $i_2 \in I$. By the distributive law for the lattice A it follows that

(i) $\quad b = b \wedge b \leqslant (i_1 \vee x) \wedge (i_2 \vee y) = (i_1 \wedge i_2) \vee (i_1 \wedge y) \vee (x \wedge i_2) \vee (x \wedge y).$

By (5) the elements $i_1 \wedge i_2$, $i_1 \wedge y$, and $x \wedge i_2$ belong to I and by assumption $x \wedge y$ belongs to I. Thus the element denoted by the right-hand side of (i) belongs to I. This implies by (2) that $b \in I$, which contradicts $I \in \boldsymbol{P}_{a,b}$. Hence the hypothesis that neither x nor y belongs to I leads to a contradiction. Since $b \notin I$, we have $I \neq A$. Hence I is a prime ideal.

It will be shown in Chapter VII, p. 258, that the following theorem is a consequence of Theorems (7) and (8) (and of the axiom of choice).

°(10) *The family $\boldsymbol{P}_{a,b}$ has a maximal element, that is, there exists $I \in \boldsymbol{P}_{a,b}$ such that I is not a proper subset of any ideal belonging to $\boldsymbol{P}_{a,b}$.*

We have introduced the notions of distributive lattice, Boolean algebra, lattice of sets, and field of sets. The relations among these notions are shown in the following scheme:

distributive lattice
↙ ↘
lattice of sets Boolean algebra with 1
↘ ↙
field of sets

We shall prove that every distributive lattice is isomorphic to a lattice of sets and every Boolean algebra with 1 is isomorphic to a field of sets.

°THEOREM 1: *Every distributive lattice is isomorphic to a lattice of sets.*

PROOF. Let A be a distributive lattice. With each $a \in A$ we associate the family $\boldsymbol{R}(a)$ of prime ideals satisfying the condition $a \notin I$. This correspondence is one-to-one. In fact, if $a \neq b$, then either $a \leqslant b$ or $b \leqslant a$ is false. By (9) and (10) there exists a prime ideal I such that either $I \in \boldsymbol{P}_{a,b}$ or $I \in \boldsymbol{P}_{b,a}$. In other words, there exists an ideal such that either $a \in I$ and $b \notin I$ or $b \in I$ and $a \notin I$. In the first case we have $I \in \boldsymbol{R}(b)$ and $I \notin \boldsymbol{R}(a)$, in the second case $I \in \boldsymbol{R}(a)$ and $I \notin \boldsymbol{R}(b)$.

It follows from (1) and (4) that

$$\begin{aligned} I \in \boldsymbol{R}(a \vee b) &\equiv a \vee b \notin I \\ &\equiv [(a \notin I) \vee (b \notin I)] \\ &\equiv [I \in \boldsymbol{R}(a) \vee I \in \boldsymbol{R}(b)] \\ &\equiv [I \in \boldsymbol{R}(a) \cup \boldsymbol{R}(b)], \end{aligned}$$

and from (3) and (5)

$$\begin{aligned} I \in \boldsymbol{R}(a \wedge b) &\equiv a \wedge b \notin I \\ &\equiv (a \notin I) \wedge (b \notin I) \\ &\equiv [I \in \boldsymbol{R}(a)] \wedge [I \in \boldsymbol{R}(b)] \\ &\equiv I \in \boldsymbol{R}(a) \cap \boldsymbol{R}(b); \end{aligned}$$

thus

$$\boldsymbol{R}(a \vee b) = \boldsymbol{R}(a) \cup \boldsymbol{R}(b) \quad \text{and} \quad \boldsymbol{R}(a \wedge b) = \boldsymbol{R}(a) \cap \boldsymbol{R}(b).$$

These formulas show that the class of all families $\boldsymbol{R}(a)$ is a lattice of sets isomorphic to the lattice A.

°THEOREM 2: *Every Boolean algebra is isomorphic to a field of sets.*

PROOF. If the lattice A in Theorem 1 is a Boolean algebra, then it has a zero element o and a unit element i and, moreover, for every $a \in A$ there exists an element $-a \in A$ such that $a \wedge (-a) = o$ and $a \vee (-a) = i$. Under the correspondence $a \to \boldsymbol{R}(a)$ the element o corresponds to the empty set and the element i corresponds to the whole set A. Since $A = \boldsymbol{R}(a \vee -a) = \boldsymbol{R}(a) \cup \boldsymbol{R}(-a)$ and $o = \boldsymbol{R}(a \wedge (-a)) = \boldsymbol{R}(a) \cap \boldsymbol{R}(-a)$, we have $\boldsymbol{R}(-a) = A - \boldsymbol{R}(a)$. The set of all families $\boldsymbol{R}(a)$ is not only a lattice of sets but a field of sets as well. Q.E.D. [1])

We shall give a topological interpretation of Theorem 2. Let A be a Boolean algebra with the zero element o and the unit element i and let P be the set of all prime ideals of A. We assume each of the families

[1]) Theorems 1 and 2 are due to Stone [1]. Other related results dealing with σ-Boolean algebras may be found in Sikorski [1]. There exist also numerous papers devoted to representations of other types of lattices.

$\boldsymbol{R}(a)$ to be a neighborhood of every one of its elements. For any set $X \subset P$, let $I \in \overline{X}$ if every neighborhood of I contains elements of X.

°THEOREM 3. (i) *P is a compact topological space.* (ii) *The sets $\boldsymbol{R}(a)$ are both open and closed in P.* (iii) *Every set in P which is both open and closed is identical with one of the sets $\boldsymbol{R}(a)$.*

PROOF. The proof that the axioms of topology hold is left to the reader.

To prove that P is a compact space, we let $\boldsymbol{K}$ be a family of closed sets with the finite intersection property. We shall show that $\bigcap_{X \in \boldsymbol{K}} X \neq \emptyset$.

Let $\boldsymbol{K}^*$ be the the family of all finite intersections of the form $\bigcap_{j<n} X_j$, where n is an arbitrary natural number and $X_j \in \boldsymbol{K}$. The family $\boldsymbol{K}^*$ is thus a family of closed non-empty sets.

Let

$$I = \{a \in A: \bigvee_{X \in \boldsymbol{K}^*} (\boldsymbol{R}(a) \cap X = \emptyset)\}.$$

Clearly, $a_1 \leqslant a_2 \in I \to a_1 \in I$. If $a_1, a_2 \in I$ then for some $X_1, X_2 \in \boldsymbol{K}^*$ we have $\boldsymbol{R}(a_1) \cap X_1 = \emptyset = \boldsymbol{R}(a_2) \cap X_2$. Thus $[\boldsymbol{R}(a_1) \cup \boldsymbol{R}(a_2)] \cap (X_1 \cap X_2) = \emptyset$. Since $X_1 \cap X_2 \in \boldsymbol{K}^*$ and $\boldsymbol{R}(a_1) \cup \boldsymbol{R}(a_2) = \boldsymbol{R}(a_1 \vee a_2)$, we conclude that $a_1 \vee a_2 \in I$. Hence the set I is an ideal.

We show that $i \notin I$. Suppose the contrary. Then for some $X \in \boldsymbol{K}^*$ we have $\boldsymbol{R}(i) \cap X = \emptyset$. This implies $A \cap X = \emptyset$; that is, $X = \emptyset$, which contradicts the hypothesis that $\boldsymbol{K}$ has the finite intersection property.

From the formula just proved and from the fact that $i \notin I$ it follows by (10) that there exists a prime ideal $I_0 \supset I$. Thus this ideal is an element of P. We shall show that $I_0 \in \bigcap_{X \in \boldsymbol{K}} X$.

Let X be an arbitrary set belonging to $\boldsymbol{K}$ and let $\boldsymbol{R}(a)$ be a neighborhood of I_0. This implies that $a \notin I_0$. Therefore $a \notin I$, which by the definition of I shows that for every $Y \in \boldsymbol{K}^*$ we have $\boldsymbol{R}(a) \cap Y \neq \emptyset$. In particular, we have $\boldsymbol{R}(a) \cap X \neq \emptyset$. Hence every neighborhood of I_0 has a non-empty intersection with X, and $I_0 \in \overline{X} = X$. This shows that $I_0 \in \bigcap_{X \in \boldsymbol{K}} X$.

In this way part (i) of Theorem 3 is proved.

The proof of (ii) follows from the fact that the family $\boldsymbol{R}(a)$ being a neighborhood in P is open in P and its complement $P-\boldsymbol{R}(a)$ is also open, because it is equal to $\boldsymbol{R}(-a)$, which is also a neighborhood in P.

Finally, to prove (iii) suppose that X is both open and closed in P and let $\boldsymbol{L} = \{\boldsymbol{R}(a): \boldsymbol{R}(a) \subset X\}$. Since every point belonging to the open set X has at least one neighborhood $\boldsymbol{R}(a)$ which is contained in X, we have $\bigcup_{Y \in \boldsymbol{L}} Y = X$ and thus $X \cap \bigcap_{Y \in \boldsymbol{L}} (P-Y) = \emptyset$. Hence the intersection of the family composed of the set X and the sets $P-Y$, where $Y \in \boldsymbol{L}$, is empty. Since this family consists of closed sets, it does not have the finite intersection property. This means that there exists a finite subset $\{\boldsymbol{R}(a_0), \boldsymbol{R}(a_1), \ldots, \boldsymbol{R}(a_{n-1})\}$ of $\boldsymbol{L}$ such that $X \cap \bigcap_{j<n} (P-\boldsymbol{R}(a_j))$ $= \emptyset$. This implies $X = \bigcup_{j<n} \boldsymbol{R}(a_j) = \boldsymbol{R}(\bigvee_{j<n} a_j)$. Hence the set X is of the form $\boldsymbol{R}(a)$.

The space P constructed in Theorem 3 is called the *Stone space of A*.

The following corollary is a consequence of Theorem 3.

°COROLLARY 4: *Every Boolean algebra with a unit is isomorphic to the field of sets which are both open and closed in a compact space.*

Exercises

1. Construct a lattice of sets isomorphic to the lattice N ordered by the relation of divisibility.

2. Prove that if an ideal $I \neq A$ in a distributive lattice A is maximal (that is, if every ideal containing I is equal to either A or I), then I is a prime ideal. The converse theorem is false.

3. Prove that if A is a Boolean algebra with unit, then an ideal I is prime if and only if for arbitrary $a \in A$ either $a \in I$ or $-a \in I$.

4. Show that in a Boolean algebra with a unit the notions of prime ideal and maximal ideal are equivalent.

5. Show that the family of ideals in a distributive lattice A is, in turn, a distributive lattice with inclusion as the ordering relation.

CHAPTER V

THEORY OF CARDINAL NUMBERS

In this chapter and in the remainder of the book we shall use the axiom system Σ°[TR] (see p. 55) together with axiom VIII formulated on p. 88. As usual, theorems not marked with the sign $^\circ$ are proved without using the axiom of choice.

§ 1. Equipollence. Cardinal numbers

We now introduce the notion of equipollence, one of the most characteristic and important notions of set theory.[1])

DEFINITION: The set A is *equipollent* to the set B if there exists a one-to-one function f with domain A and range B. We write $A \sim B$, and we say that f establishes the *equipollence of A and B.*

Examples

1. If the set A is finite, i.e. the number of its elements is a certain natural number n, then the set B is equipollent to A if and only if B contains exactly n elements. Thus the notion of equipollence is a generalization to arbitrary sets of the notion, for finite sets, of having an equal number of elements.

2. Let A be the interval $a_1 < x < a_2$, B the interval $b_1 < x < b_2$. The function

$$f(x) = \frac{b_2 - b_1}{a_2 - a_1}(x - a_1) + b_1$$

[1]) The notion of equipollence was first defined and systematically investigated by Cantor. His first publication about this subject dates back to 1878. The notion was also known to Bolzano [1], Section 20, but his work did not have much influence.

is a one-to-one correspondence from A to B. By definition, $A \sim B$.

THEOREM 1: *For arbitrary sets A, B, and C the following formulas hold:*

(1) $A \sim A, \quad (A \sim B) \to (B \sim A), \quad (A \sim B) \wedge (B \sim C) \to (A \sim C).$

Thus, *equipollence is reflexive, symmetric and transitive.*

PROOF. The function I_A (see p. 71) establishes the equipollence of A with itself. If the function f establishes the equipollence of A and B then the function f^c establishes the equipollence of B and A (see p. 70). If f establishes the equipollence of A and B and g establishes the equipollence of B and C, then the composition $g \bigcirc f$ establishes the equipollence of A and C (see Theorem II.6.2).

The following formulas hold:

(2) $$(A \times B) \sim (B \times A),$$

(3) $$(A \times \{a\}) \sim A \sim A^{\{a\}}, \qquad \{a\}^A \sim \{a\},$$

(4) $$[A \times (B \times C)] \sim [(A \times B) \times C],$$

(5) $$(A_1 \sim B_1) \wedge (A_2 \sim B_2) \to [(A_1 \times A_2) \sim (B_1 \times B_2)],$$

(6) $$(A \sim B) \to (\boldsymbol{P}(A) \sim \boldsymbol{P}(B)),$$

(7) $$(A_1 \sim B_1) \wedge (A_2 \sim B_2) \wedge (A_1 \cap A_2 = \emptyset = B_1 \cap B_2) \to (A_1 \cup A_2 \sim B_1 \cup B_2),$$

(8) $$Y^{X \times T} \sim (Y^X)^T,$$

(9) $$(Y \times Z)^X \sim (Y^X \times Z^X),$$

(10) $$(A \cap B = \emptyset) \to (Y^{A \cup B} \sim Y^A \times Y^B).$$

We omit the proofs of (2)–(7), which are not difficult. On the other hand, we prove the important formulas (8)–(10).

Let $f \in Y^{X \times T}$; hence f is a function of two variables x and t, where x ranges through the set X and t through the set T and where f takes values in Y. For fixed t the function g_t (with one variable x) defined by $g_t(x) = f(x, t)$ is a function from X into Y, $g_t \in Y^X$. The function F defined by $F(t) = g_t$ associates with every $t \in T$ an element of the set Y^X, so $F \in (Y^X)^T$.

If f_1 and f_2 are distinct functions belonging to the set $Y^{X \times T}$, then the corresponding functions F_1 and F_2 are also distinct. In fact, if

$f_1(x_0, t_0) \neq f_2(x_0, t_0)$ then the elements $F_1(t_0)$ and $F_2(t_0)$ of the set Y^X are distinct.

Each function $F \in (Y^X)^T$ corresponds in the manner described above to some function $f \in Y^{X \times T}$, namely, to the function f defined by $f(x, t) = g_t(x)$, where $g_t = F(t)$.

It follows that the correspondence of the function $f \in Y^{X \times T}$ to the function $F \in (Y^X)^T$ establishes the equipollence of the set $Y^{X \times T}$ with the set $(Y^X)^T$, which proves formula (8).

For the proof of (9) notice that if $f \in (Y \times Z)^X$, then $f(x)$ is, for every $x \in X$, an ordered pair $\langle g(x), h(x) \rangle$, where $g(x) \in Y$ and $h(x) \in Z$. Thus $g \in Y^X$ and $h \in Z^X$. It is easy to show that this correspondence of the function f to the pair $\langle g, h \rangle$ determines a one-to-one mapping of the set $(Y \times Z)^X$ onto the set $Y^X \times Z^X$.

Finally, to prove (10) we associate with every function $f \in Y^{A \cup B}$ the ordered pair of restricted functions $\langle f|A, f|B \rangle$. Again it is not difficult to show that this correspondence is a one-to-one mapping of the set $Y^{A \cup B}$ onto the set $Y^A \times Y^B$.

Equations (2) and (8)–(10) are particular instances of the following theorems.

THEOREM 2: (COMMUTATIVE LAW) *Let $F \in (P(A))^X$. If φ is a permutation of the set X, then*

$$\prod_{x \in X} F_x \sim \prod_{x \in X} F_{\varphi(x)}. \tag{11}$$

PROOF. Associate with every function $f \in \prod_x F_x$ the composite function $g = f \circ \varphi$. If $f_1 \neq f_2$ then for some $x \in X$, $f_1(x) \neq f_2(x)$. Thus setting $y = \varphi^c(x)$ we obtain $f_1(\varphi(y)) \neq f_2(\varphi(y))$, that is, $g_1(y) \neq g_2(y)$. Hence the correspondence given by the equation $g = f \circ \varphi$ is one-to-one.

The function $g = f \circ \varphi$ belongs to the cartesian product $\prod_x F_{\varphi(x)}$. In fact, if $x \in X$ then $f(\varphi(x)) \in F_{\varphi(x)}$, that is, $g(x) \in F_{\varphi(x)}$.

Finally, every function belonging to the cartesian product $\prod_x F_{\varphi(x)}$ can be represented as $f \circ \varphi$ for some $f \in \prod_x F_x$. In fact, it suffices to take for f the function $g \circ \varphi^c$.

THEOREM 3: (ASSOCIATIVE LAW) *Let* $F \in (P(A))^X$. *If* $X = \bigcup_{y \in Y} T_y$, *where the sets* T_y *are pairwise disjoint; then*

(12) $$\prod_{x \in X} F_x \sim \prod_{y \in Y} \Big(\prod_{x \in T_y} F_x\Big),$$

(13) $$A^X \sim \prod_{y \in Y} (A^{T_y}).$$

PROOF. Let $G_y = \prod_{x \in T_y} F_x$. G_y is the set of all functions f with domain T_y such that $f(x) \in F_x$ for $x \in T_y$. $\prod_{y \in Y} G_y$ consists of the functions g with domain Y such that $g(y) \in G_y$ for $y \in Y$. We denote the value $g(y)$ of the function g by g_y; g_y is a function with domain T_y such that $g_y(x) \in F_x$.

We associate with the function $g \in \prod_{y \in Y} G_y$ the function f defined by the equation

(i) $$f(x) = g_y(x),$$

where y belongs to the set Y, and $x \in T_y$.

The domain of f is X; and for every $x \in X, f(x) \in F_x$. Thus $f \in \prod_{x \in T} F_x$.

The correspondence between the functions g and f is one-to-one. In fact, if $g^{(1)} \neq g^{(2)}$ then there exists $y \in Y$ such that $g_y^{(1)} \neq g_y^{(2)}$ and thus there exists $x \in T_y$ such that $g_y^{(1)}(x) \neq g_y^{(2)}(x)$. Hence, by (i), $f^{(1)}(x) \neq f^{(2)}(x)$.

It remains to be shown that to every function $f \in \prod_{x \in T} F_x$ there corresponds a function g. For this purpose it suffices to notice that the function g defined at $y \in Y$ by the equation

$$g_y = f | T_y$$

belongs to the cartesian product $\prod_{y} G_y$ and also satisfies (i).

To prove formula (13) it suffices to let $F_x = A$ in formula (12) for each x.

THEOREM 4: (LAW OF EXPONENTS FOR THE CARTESIAN PRODUCT) *Let* $F \in (P(A))^T$. *For every set* H

(14) $$\Big(\prod_{t \in T} F_t\Big)^H \sim \prod_{t \in T} (F_t^H).$$

PROOF. We define the function of two variables on the set $T \times H$ by the equation $G_{\langle t, h\rangle} = F_t$. We may represent the cartesian product in two ways as the union of disjoint sets

$$T \times H = \bigcup_{h \in H} T_h = \bigcup_{t \in T} H_t,$$

where T_h is the set of all pairs with the second coordinate equal to h and H_t is the set of all pairs with the first coordinate equal to t.

Applying Theorem 3 twice we obtain

$$\prod_{x \in T \times H} G_x \sim \prod_{t \in T} \Big(\prod_{x \in H_t} G_x \Big), \tag{15}$$

$$\prod_{x \in T \times H} G_x \sim \prod_{h \in H} \Big(\prod_{x \in T_h} G_x \Big). \tag{16}$$

For $x \in H_t$, $x = \langle t, h \rangle$. Thus $G_x = F_t$, which shows that

$$\prod_{x \in H_t} G_x = (F_t)^{H_t}.$$

On the other hand, since $H_t \sim H$,

$$(F_t)^{H_t} \sim F_t^H \quad \text{and} \quad \prod_{t \in T} (F_t)^{H_t} \sim \prod_{t \in T} (F_t^H).$$

Thus by (15) we get

$$\prod_{x \in T \times H} G_x \sim \prod_{t \in T} (F_t^H). \tag{17}$$

For a given h we associate the function $g \in \prod_{x \in T_h} G_x$ with the function f where $f(t) = g(t, h)$, obtaining

$$\prod_{x \in T_h} G_x \sim \prod_{t \in T} F_t.$$

Applying (16) we obtain

$$\prod_{x \in T \times H} G_x \sim \Big(\prod_{t \in T} F_t \Big)^H. \tag{18}$$

Finally (14) follows directly from (17) and (18).

We shall now introduce the cardinal numbers. First we note a theorem whose proof is immediate:

THEOREM 5: *For the sets A and B to be equipollent it is necessary and sufficient that the relational systems $\langle A, A \times A\rangle$ and $\langle B, B \times B\rangle$ be isomorphic.*

We shall denote by $\overline{\overline{A}}$ the relational type of the system $\langle A, A \times A\rangle$. We shall call $\overline{\overline{A}}$ the *cardinal number* or the *power* of the set A. From Theorem 5 and Axiom II.10.VIII we obtain the following:

THEOREM 6: *For arbitrary sets A, B the conditions $A \sim B$ and $\overline{\overline{A}} = \overline{\overline{B}}$ are equivalent.*

This theorem allows us to formulate statements about equipollence as equations involving cardinal numbers.

The notion of a cardinal number is not indispensable.

We may formulate all theorems of set theory so that they are statements, not about cardinal numbers or powers of sets, but rather about relationships between cardinal numbers, which can always be stated in terms of the notion of equipollence. On the other hand, many theorems can be stated more intuitively if they are formulated as theorems about cardinal numbers. For this reason it is convenient to introduce this notion.

§ 2. Countable sets

If X is a finite set containing exactly n elements (see III, §3), then Theorem 6, § 1 is satisfied if we set $\overline{\overline{X}} = n$ (see Theorem III.4.6). In the future we shall identify the cardinal number of a finite set of n elements with the natural number n.

The theory of cardinality for finite sets is not essentially richer than the arithmetic of the natural numbers. New notions appear when we turn to infinite sets.

DEFINITION: A set A is said to be *countable* (or *denumerable*) if it is either finite or equipollent with the set of all natural numbers.

Clearly, two arbitrary infinite countable sets are always equipollent (see Theorem 1.1). We shall denote the cardinal number of infinite countable sets by $\mathfrak{a}$.

In Chapter III, p. 92 we defined a sequence as a function with domain equal to the set of natural numbers. From this definition it follows that a set is countable if and only if it is the range of a sequence none of whose terms are equal. Speaking somewhat figuratively, a set A is countable if its elements can be "arranged" in an infinite sequence $a_0, a_1, a_2, a_3, \ldots$.

THEOREM 1: *Every countable non-empty set is the range of an infinite sequence. Conversely, the range of any infinite sequence is countable and non-empty.*

PROOF. The finite set whose elements are $a_0, a_1, a_2, a_3, \ldots, a_k$ is the range of the infinite sequence:

$$f(0) = a_0, \quad \ldots, \quad f(k) = a_k, \quad f(k+1) = a_k, \quad \ldots, \quad f(k+j) = a_k, \ldots$$

Every infinite countable set is by definition the set of terms of an infinite sequence.

To prove the converse we assume that X is the set of terms of an infinite sequence φ and that X is an infinite set. Let $\psi_0 = \varphi_0$ and

$$\psi_{n+1} = \varphi_m \quad \text{where} \quad m = \min_k \Big[\bigwedge_{j\leqslant n} (\varphi_k \neq \psi_j)\Big],$$

or

$$\psi_{n+1} = \varphi_0 \text{ if there exists no } k \text{ such that } \bigwedge_{j\leqslant n} (\psi_j \neq \varphi_k).$$

We prove by induction that for every n there exists a number k such that $\bigwedge_{j\leqslant n} (\psi_j \neq \varphi_k)$. It follows that $\bigwedge_{j\leqslant n} (\psi_{n+1} \neq \psi_j)$ and thus the sequence ψ has distinct terms. It remains to be shown that every element of the set X is a term of the sequence ψ.

For this purpose we assume that the set $\{k: \varphi_k \text{ is not a term of } \psi\}$ is non-empty and we let k_0 be the least element belonging to that set. Clearly $k_0 > 0$. If $i < k_0$ then φ_i is a term of ψ, say $\varphi_i = \psi_{m(i)}$. Let $m = \max_{i<k_0} m(i)$. The smallest number k such that $\bigwedge_{j\leqslant m} (\varphi_k \neq \psi_j)$ is then k_0 and thus from the definition of the sequence ψ we obtain $\psi_{m+1} = \varphi_{k_0}$, contradicting the definition of the number k_0.

In a similar manner we establish the following theorems.

THEOREM 2: *Every subset of a countable set is countable.*

THEOREM 3: *The union of two countable sets is countable.*

PROOF. Since the case where one of the given sets is empty causes no difficulty, we assume that A is the set of terms of the sequence

$$a_0, a_1, a_2, \ldots, a_n, \ldots$$

and that B is the set of terms of the sequence

$$b_0, b_1, b_2, \ldots, b_n, \ldots$$

The union of A and B is then the set of terms of the sequence

$$a_0, b_0, a_1, b_1, a_2, b_2, \ldots, a_n, b_n, \ldots,$$

and is therefore countable.

From Theorem 3 it follows by induction that the union of an arbitrary finite number of countable sets is countable.

A particular case of Theorem 3 is the following

THEOREM 4: *The union of a finite set with a countable set is countable.*

THEOREM 5: *The cartesian product of two countable sets is countable.*

PROOF. If A and B are infinite countable sets, then $A \sim N$ and $B \sim N$; and thus $A \times B \sim N \times N$. By Theorem 1.1 it follows that $N \times N \sim N$, and thus $A \times B \sim N$.

If one or both of the sets A and B are finite, then $A \times B$ is equipollent to a subset of $N \times N$, that is, to a subset of N and the theorem follows by Theorem 2.

THEOREM 6: *If the set A is countable, then the set of all finite sequences with terms in A is countable.*

PROOF. The theorem follows immediately from Theorem III.3.4.

THEOREM 7: *If ψ is an infinite sequence whose elements are also infinite sequences, then the set X of elements x which are terms of the sequences ψ_n is countable.*

PROOF. By definition, $X = \{x: \bigvee_{mn} (x = \psi_{mn})\} = \{x: \bigvee_{p} (x = \psi_{K(p), L(p)})\}$.

Thus X is the set of terms of the sequence φ defined by the equation $\varphi_p = \psi_{K(p), L(p)}$.

°THEOREM 8: *If A is a sequence whose elements are non-empty countable sets, then the union $\bigcup_n A_n$ is countable.*

PROOF. Let C_n be the set of sequences φ such that A_n is the set of terms of φ. By assumption $C_n \neq \emptyset$ for all $n \in N$. Therefore by the axiom of choice there exists a sequence ψ such that $\psi_n \in C_n$ for each n. Thus the union $\bigcup_n A_n$ is the set of those x for which there exist m, $n \in N$ such that $x = \psi_{mn}$, which proves the theorem on the basis of Theorem 7.

REMARK: The use of the axiom of choice is necessary for the proof of Theorem 8. For every countable set A there exists an infinite sequence containing all the elements of A among its terms. Yet there are infinitely many such sequences for a given set A and we have no way of distinguishing between them. In other words, we have no way of associating with every countable set an infinite sequence whose terms contain all the elements of the given set.

Examples of countable sets

1. *The set of integers is countable.*

In fact, it is the union $N \cup N'$ where N' is the set of integers $\leqslant 0$. Because $N \sim N'$, where the function $f(n) = -n$ establishes equipollence, both sets N and N' are countable. It follows that the set $N \cup N'$ is countable.

2. *The set of rational numbers is countable.*

Indeed, the sequence φ defined by $\varphi_p = K(p)/\big(L(p+1)\big)$ contains all the non-negative rational numbers among its terms and only such numbers. Thus the set of non-negative rational numbers is countable. From this we obtain that the set of negative rational numbers is countable (see Example 1). Hence the set of all rational numbers is countable.

3. *The set of polynomials of one variable with integral coefficients is countable.*

To every polynomial with integral coefficients there corresponds the unique sequence of its coefficients. By Theorem 6 the set of all finite sequences of integers is countable.

4. *The set of algebraic numbers is countable.*

In fact, with every polynomial we with rational coefficients may associate a finite sequence whose terms are all the roots of the polynomial. We let the first term of the sequence be that root which has the smallest modulus and among those of equal modulus that root which has the smallest argument. Similarly, we let the second term of the sequence be that root different from the first which has the smallest modulus and the smallest argument among the roots having the same modulus. In this way we define by induction the desired sequence. The countability of the set of all algebraic numbers follows now from Theorem 7.

We can obtain the same result from Theorem 8. But in this case we have to use the axiom of choice.

REMARK 1: Cantor used the notion of a cardinal number from 1878 on but defined it only in [5]. In free translation his definition reads: "A cardinal number of a set M is the notion which arises from M by abstraction from the nature of the elements of M and from their order". See Cantor [5], Section 1. The complicated symbol $\overline{\overline{M}}$ introduced by Cantor indicated the double process of abstraction which leads from a set M to its cardinal number. Many recent authors replace the Cantor symbol by $|M|$ which is more convenient to print.

Frege [1] had a similar conception as Cantor. He was concerned chiefly with natural numbers but mentioned the possibility of extending the notion of a number to arbitrary sets. See Frege [1], p. 96.

REMARK 2: It can be shown that the following statement cannot be proved in the system Σ[TR]: The union of a denumerable family of disjoint unordered pairs is denumerable. See Jech [2], p. 95. This result shows that Theorem 8 cannot be proved without the axiom of choice.

Exercises

1. Prove that the set of all intervals with rational endpoints (in the space of real numbers) is countable.

2. Prove that in 3-dimensional euclidean space (or more generally, in $\mathscr{E}^n$) the set of all spheres with radius of rational length and with center having rational coordinates is countable.

3. Let f be a function with field contained in the set of real numbers. We say that f has a *proper extremum* at the point a if there exists an interval P containing a such that $f(x) < f(a)$ for all $x \in P - \{a\}$ or else $f(x) > f(a)$ for all $x \in P - \{a\}$. Prove that the set of proper extrema of such a function f is at most countable.

Hint: Use Exercise 1.

Generalize the theorem to functions defined on the space $\mathscr{E}^n$ (replacing P by an n-dimensional sphere).

4. Prove that every disjoint family of intervals in the space of real numbers is countable.

Hint: Use Exercise 1.

Generalize the theorem to families of disjoint open sets in the space $\mathscr{E}^n$ using Exercise 2.

5. Let Z be a set of points in the plane. We call the point $p \in Z$ *isolated* if there exists an open circle K (i.e. without circumference) such that $\{p\} = Z \cap K$. Prove that the set of isolated points of a given set Z is countable.

Hint: Use Exercise 2.

Generalize the theorem to the space $\mathscr{E}^n$ (replacing circle by ball in the definition of isolated point).

6. Prove that every monotonic discontinuous function from the set of real numbers to the set of real numbers has a countable number of points of discontinuity.

Hint: Every monotonic function has both a limit from the right and a limit from the left at every point; at points of discontinuity those limits are unequal. Apply Exercise 4.

§ 3. The hierarchy of cardinal numbers

We shall prove that besides the finite cardinal numbers and the number $\mathfrak{a}$ there exists infinitely many other cardinal numbers.

For this purpose we prove the following very useful theorem.

THEOREM 1: (ON DIAGONALIZATION[1]) *If the domain T of function F is contained in a set A and if the values of F are subsets of A, then the set*

$$Z = \{t \in T: t \notin F(t)\}$$

is not a value of the function F.

PROOF. We have to show that for every $t \in T$, $F(t) \neq Z$. From the definition of the set Z it follows that if $t \in T$, then

$$[t \in Z] \equiv [t \notin F(t)].$$

Thus if $F(t) = Z$ we obtain the contradiction:

$$(t \in Z) \equiv (t \notin Z).$$

[1]) Theorem 1, one of the most important results in set theory, was proved by Cantor in [4].

For $A = T$ Theorem 1 has a geometrical interpretation. We may consider the set $A \times A$ represented as a square (see p. 62, Chapter II). We let $R = \{\langle x, y\rangle : y \in F(x)\}$. The set $F(x)$ is the projection onto the vertical axis of those pairs belonging to R which have first coordinate equal to x. The set Z is the projection onto the vertical axis of those points along the diagonal of the square which do not belong to R. It is then geometrically obvious that $Z \neq F(x)$ for any $x \in A$; for if $\langle x, x\rangle \in R$, then $x \in F(x)$ but $x \notin Z$, and if $\langle x, x\rangle \notin R$, then $x \notin F(x)$ but $x \in Z$. This interpretation motivates calling Theorem 1 the *Diagonalization Theorem.*

We apply Theorem 1 to prove that there exist distinct infinite cardinal numbers.

THEOREM 2: *The set* $\boldsymbol{P}(A)$ *is not equipollent to* A, *nor to any subset of* A.

For otherwise there would exist a one-to-one function whose domain is a subset of A and whose range is the family of all subsets of A. But this contradicts Theorem 1.

THEOREM 3: *No two of the sets*

(1) $$A, \boldsymbol{P}(A), \quad \boldsymbol{P}(\boldsymbol{P}(A)), \quad \boldsymbol{P}(\boldsymbol{P}(\boldsymbol{P}(A)))$$

are equipollent.

PROOF. Let P_k be the kth set in sequence (1) and suppose that there exist k and l such that $k > l$ and P_k is equipollent to a subset of P_l. The set P_{k-1} is clearly equipollent to a subset of P_k, namely to the subset of singletons $\{x\}$ where $x \in P_{k-1}$. Thus the set P_{k-1} is equipollent to a subset of P_l. Repeating this argument we conclude that each of the sets $P_{k-1}, P_{k-2}, \ldots, P_{l+1}$ is equipollent to some subset of P_l, but this contradicts Theorem 2 because $P_{l+1} = \boldsymbol{P}(P_l)$.

THEOREM 4: *Let the family* $\boldsymbol{A}$ *have the property*

(2) *For every* $X \in \boldsymbol{A}$ *there exists a set* $Y \in \boldsymbol{A}$ *which is not equipollent to any subset of* X.

Then the union $\bigcup(\boldsymbol{A})$ *is not equipollent to any* $X \in \boldsymbol{A}$ *nor to any subset of* $X \in \boldsymbol{A}$.

PROOF. Assume that $\bigcup(\boldsymbol{A}) \sim X_1 \subset X \in \boldsymbol{A}$. It follows that there exists a one-to-one function f such that $f^1\big(\bigcup(\boldsymbol{A})\big) = X_1$. By assumption (2) there is a set $Y \in \boldsymbol{A}$ which is not equipollent to any subset of X. As $Y \subset \bigcup(\boldsymbol{A})$, we have $f^1(Y) \subset f^1(\bigcup(\boldsymbol{A}))$; that is, $f^1(Y) \subset X_1$ and consequently $Y \sim f^1(Y) \subset X$. The contradiction shows that it is not the case that $\bigcup(\boldsymbol{A}) \sim X_1$.

Theorems 3 and 4 give us some ideas of how many distinct infinite cardinal numbers exist. Starting with the set N of natural numbers which has power $\mathfrak{a}$, we can construct the sets

$$N, \boldsymbol{P}(N)\, \boldsymbol{P}(\boldsymbol{P}(N)),\ \boldsymbol{P}(\boldsymbol{P}(\boldsymbol{P}(N))), \ldots \tag{3}$$

no two of which are equipollent by Theorem 3. In this way we obtain *infinitely many distinct cardinal numbers.*

By the axiom of replacement there exists the family $\boldsymbol{A}$ whose elements are exactly all the sets (3) (see p. 97). By Theorem 2 the family $\boldsymbol{A}$ satisfies condition (2); thus by Theorem 4 the union $S = \bigcup(\boldsymbol{A})$ has a cardinal number different from each of the sets (3) and from each of their subsets. Again applying Theorem 3 we obtain the sequence of sets

$$S, \boldsymbol{P}(S), \boldsymbol{P}(\boldsymbol{P}(S)), \boldsymbol{P}(\boldsymbol{P}(\boldsymbol{P}(S))), \ldots \tag{4}$$

no two of which are equipollent and none of which is equipollent to any of the sets (3). We obtain in this way a new infinite quantity of distinct cardinal numbers.

We obtain still other cardinals by constructing the family $\boldsymbol{B}$ consisting of all the sets (3) and (4) and by constructing a new sequence

$$Q = \bigcup(\boldsymbol{B}), \boldsymbol{P}(Q), \boldsymbol{P}(\boldsymbol{P}(Q)), \boldsymbol{P}(\boldsymbol{P}(\boldsymbol{P}(Q))).$$

We may continue this procedure indefinitely. We see that the hierarchy of distinct infinite cardinals obtained in this way is incomparably richer than the hierarchy of finite cardinals, which coincides with the natural numbers.

As a further consequence of Theorem 2 we note the following.

THEOREM 5: *There exists no family of sets* $\boldsymbol{U}$ *which, for every set* X, *contains an element* Y *equipollent to* X.

PROOF. By Theorem 2 the set $\boldsymbol{P}(\bigcup(\boldsymbol{U}))$ is not equipollent to any subset of the set $\bigcup(\boldsymbol{U})$ and hence it is not equipollent to any of the sets Y belonging to $\boldsymbol{U}$ because $Y \in \boldsymbol{U}$ implies $Y \subset \bigcup(\boldsymbol{U})$.

THEOREM 6: *There exists no set containing all sets.*

For otherwise this set would satisfy the conditions of Theorem 5.

Theorem II.3.6 shows again that we cannot state as an axiom consistent with our axiom system that there exists a set composed of all elements which satisfy an arbitrary formula.

Theorem 5 is also another indication of the vastness of the hierarchy of cardinal numbers, which is so "large" that it is impossible to construct a set containing at least one set of each power.

Exercises

1. Prove that the set N^N is uncountable.

Hint: If φ is a sequence of elements of N^N, then the sequence ϑ defined by $\vartheta_n = \varphi_n(n)+1$ is not a term of the sequence φ.

2. Let X be a compact space $\neq \emptyset$ having the property: for every finite set S and for every open set $G \neq \emptyset$ there exists an open non-empty set G^* such that $G^* \subset G$ and $\bar{G}^* \cap S = \emptyset$. Show that $\overline{\overline{X}} \neq \mathfrak{a}$. Apply this inequality to show that the Cantor set is uncountable.

Hint: Use the axiom of choice to associate with every open set $G \neq \emptyset$ and every finite set $S \subset X$ an open subet $G^* = G^*(G, S) \subset G$ such that $\bar{G}^* \cap S = \emptyset$. Assuming that φ is an infinite sequence of elements of X, let

$$G_0 = X, \quad G_{n+1} = G^*(G_n, \{\varphi_0, \dots, \varphi_n\});$$

prove that $\bigcap_n \bar{G}_n \neq \emptyset$.

3.[1]) We say that the sequence $b_1, b_2, \dots$ of natural numbers increases faster than the sequence $a_1, a_2, \dots$ if $\lim_{n=\infty} \frac{a_n}{b_n} = 0$. Prove the following statements:

(i) for every sequence there exists another sequence which increases faster;

(ii) let Z be a set of sequences such that for every sequence $a_1, a_2, \dots$ there exists a sequence $b_1, b_2, \dots$ belonging to Z which increases faster than $a_1, a_2, \dots$; then the set Z is uncountable.

Hint: Assume $\overline{\overline{Z}} = \overline{\overline{N}}$. We then may represent the elements of Z as the rows of the table

$$\begin{array}{l} a_{1,1}, a_{1,2}, \dots, a_{1,n}, \dots \\ a_{2,1}, a_{2,2}, \dots, a_{2,n}, \dots \\ \dots\dots\dots\dots\dots \\ a_{n,1}, a_{n,2}, \dots, a_{n,n}, \dots \\ \dots\dots\dots\dots\dots \end{array}$$

[1]) In connection with Exercise 3 see Hardy [1]. Topics dealt with in Section 4 are all due to Cantor [5], Sections 2, 3 and 4.

Using an appropriate diagonalization argument, define a sequence increasing faster than every sequence in the table.

§ 4. The arithmetic of cardinal numbers

We shall define operations of addition, multiplication, and exponentiation for cardinal numbers. The definitions will be chosen so that they will coincide with the ordinary definitions for finite cardinals (that is for the natural numbers).

DEFINITION 1: The cardinal number $\mathfrak{m}$ is the *sum* of the cardinals $\mathfrak{n}_1$ and $\mathfrak{n}_2$,

$$\mathfrak{m} = \mathfrak{n}_1 + \mathfrak{n}_2,$$

if every set of power $\mathfrak{m}$ is the union of two disjoint sets, one of which has power $\mathfrak{n}_1$ and the other of which has power $\mathfrak{n}_2$.

LEMMA 1: *Given two arbitrary sets A_1 and A_2, there exist sets B_1 and B_2 such that*

$$(0) \qquad A_1 \sim B_1, \quad A_2 \sim B_2, \quad B_1 \cap B_2 = \emptyset.$$

Choose a_1 and a_2 such that $a_1 \neq a_2$ (for instance, $a_1 = \emptyset$, $a_2 = \{\emptyset\}$). Then, by 1(4), the sets $B_1 = \{a_1\} \times A_1$ and $B_2 = \{a_2\} \times A_2$ satisfy the desired conditions.

THEOREM 2: *The sum $\mathfrak{n}_1 + \mathfrak{n}_2$ of any two given cardinals $\mathfrak{n}_1$ and $\mathfrak{n}_2$ always exists.*

PROOF. Assume that $\overline{\overline{A_1}} = \mathfrak{n}_1$ and $\overline{\overline{A_2}} = \mathfrak{n}_2$. Let B_1 and B_2 satisfy conditions (0). Then $B_1 \cup B_2$ is the union of two disjoint sets of powers $\mathfrak{n}_1$ and $\mathfrak{n}_2$. Clearly every set equipollent to $B_1 \cup B_2$ has the same property. Thus $\overline{\overline{B_1 \cup B_2}} = \mathfrak{n}_1 + \mathfrak{n}_2$.

Moreover, we have proved that

$$\overline{\overline{A}} + \overline{\overline{B}} = \overline{\overline{A \cup B}} \quad \text{if} \quad A \cap B = \emptyset.$$

THEOREM 3: *Addition of cardinal numbers is commutative and associative: for arbitrary cardinal numbers $\mathfrak{n}_1, \mathfrak{n}_2$ and $\mathfrak{n}_3$, we have*

$$(1) \qquad \mathfrak{n}_1 + \mathfrak{n}_2 = \mathfrak{n}_2 + \mathfrak{n}_1,$$

$$(2) \qquad \mathfrak{n}_1 + (\mathfrak{n}_2 + \mathfrak{n}_3) = (\mathfrak{n}_1 + \mathfrak{n}_2) + \mathfrak{n}_3.$$

PROOF. If $\overline{\overline{A}} = \mathfrak{n}_1 + \mathfrak{n}_2$ then $A = A_1 \cup A_2$, where $A_1 \cap A_2 = \emptyset$, $\overline{\overline{A_1}} = \mathfrak{n}_1$ and $\overline{\overline{A_2}} = \mathfrak{n}_2$. Thus $A = A_2 \cup A_1$ and $\overline{\overline{A}} = \mathfrak{n}_2 + \mathfrak{n}_1$, which proves (1); the proof of (2) is similar.

Example. By Theorems 2(3) and 2(4) we have

$$\mathfrak{a} + \mathfrak{a} = \mathfrak{a}, \quad n + \mathfrak{a} = \mathfrak{a}. \tag{3}$$

DEFINITION 2: The cardinal number $\mathfrak{m}$ is the *product* of $\mathfrak{n}_1$ and $\mathfrak{n}_2$, i.e.

$$\mathfrak{m} = \mathfrak{n}_1 \cdot \mathfrak{n}_2,$$

if every set of power $\mathfrak{m}$ is equipollent to the cartesian product $A_1 \times A_2$ where $\overline{\overline{A_1}} = \mathfrak{n}_1$ and $\overline{\overline{A_2}} = \mathfrak{n}_2$.

Thus

$$\overline{\overline{A_1}} \cdot \overline{\overline{A_2}} = \overline{\overline{A_1 \times A_2}}.$$

It is clear that, for arbitrary cardinals $\mathfrak{n}_1$ and $\mathfrak{n}_2$, the product $\mathfrak{n}_1 \cdot \mathfrak{n}_2$ always exists.

Definition 2 is a generalization to the case of arbitrary cardinal numbers of the usual notion of multiplication: for example, we consider the product $3 \cdot 4$ as the number of elements of a set which can be represented as three groups of four elements; that is, as the number of elements in the set $A \times B$, where A contains exactly three elements, and B four.

THEOREM 4: *Multiplication of cardinal numbers is commutative, associative, and distributive over addition*:

$$\mathfrak{n}_1 \cdot \mathfrak{n}_2 = \mathfrak{n}_2 \cdot \mathfrak{n}_1, \tag{4}$$

$$\mathfrak{n}_1 \cdot (\mathfrak{n}_2 \cdot \mathfrak{n}_3) = (\mathfrak{n}_1 \cdot \mathfrak{n}_2) \cdot \mathfrak{n}_3, \tag{5}$$

$$\mathfrak{n}_1 \cdot (\mathfrak{n}_2 + \mathfrak{n}_3) = \mathfrak{n}_1 \cdot \mathfrak{n}_2 + \mathfrak{n}_1 \cdot \mathfrak{n}_3. \tag{6}$$

PROOF. Equations (4) and (5) are immediate consequences of equations 1(2) and 1(4). Equation (6) follows from the equations (see Chapter II § 4):

$$A_1 \times (A_2 \cup A_3) = (A_1 \times A_2) \cup (A_1 \times A_3),$$

$$[A_2 \cap A_3 = \emptyset] \rightarrow [(A_1 \times A_2) \cap (A_1 \times A_3) = \emptyset].$$

THEOREM 5: 1 *is the unit for multiplication*; *namely,*

$$\mathfrak{n} \cdot 1 = \mathfrak{n}. \tag{7}$$

The proof follows from 1 (3).

Example. By Theorem 2.5

$$\mathfrak{a} \cdot \mathfrak{a} = \mathfrak{a}, \quad \mathfrak{a} \cdot n = \mathfrak{a}. \tag{8}$$

Denote the n-fold product $\mathfrak{m} \cdot \mathfrak{m} \cdot \ldots \cdot \mathfrak{m}$ by $\mathfrak{m}^n$. By Definition 2, $\mathfrak{m}^n$ is the power of the set of all sequences of n-elements $\langle a_1, a_2, \ldots, a_n \rangle$, where $a_1, \ldots, a_n$ are elements of a set A of power $\mathfrak{m}$. In other words (see Chapter II, § 6),

$$(\overline{\overline{A}})^n = \overline{\overline{A^n}},$$

Generalizing the example above, we obtain the following definition.

DEFINITION 3: The cardinal $\mathfrak{m}$ is the *cardinal* $\mathfrak{n}$ *raised to* $\mathfrak{p}$*-th power*,

$$\mathfrak{m} = \mathfrak{n}^{\mathfrak{p}},$$

if every set of power $\mathfrak{m}$ is equipollent to the set A^B, where $\overline{\overline{A}} = \mathfrak{n}$ and $\overline{\overline{B}} = \mathfrak{p}$.

Thus

$$(\overline{\overline{A}})^{\overline{\overline{B}}} = \overline{\overline{A^B}}.$$

It is clear that for every two cardinal numbers $\mathfrak{n}$ and $\mathfrak{p}$ the cardinal $\mathfrak{n}^{\mathfrak{p}}$ always exists.

THEOREM 6: *For arbitrary cardinals* $\mathfrak{n}, \mathfrak{p}$ *and* $\mathfrak{q}$:

$$\mathfrak{n}^{\mathfrak{p}+\mathfrak{q}} = \mathfrak{n}^{\mathfrak{p}} \cdot \mathfrak{n}^{\mathfrak{q}}, \tag{9}$$

$$(\mathfrak{n} \cdot \mathfrak{p})^{\mathfrak{q}} = \mathfrak{n}^{\mathfrak{q}} \cdot \mathfrak{p}^{\mathfrak{q}}, \tag{10}$$

$$(\mathfrak{n}^{\mathfrak{p}})^{\mathfrak{q}} = \mathfrak{n}^{\mathfrak{p} \cdot \mathfrak{q}}, \tag{11}$$

$$\mathfrak{n}^1 = \mathfrak{n}, \tag{12}$$

$$1^{\mathfrak{n}} = 1. \tag{13}$$

These equations follow directly from equations 1(4), 1(8)–1(10).

THEOREM 7: *If A has power* $\mathfrak{m}$, *then the set* $\boldsymbol{P}(A)$ (*which consists of all subsets of* A) *has power* $2^{\mathfrak{m}}$:

$$2^{\overline{\overline{A}}} = \overline{\overline{\boldsymbol{P}(A)}}.$$

PROOF. $2^{\mathfrak{m}}$ is the power of the set $\{0, 1\}^A$, consisting of all functions f whose values are the numbers 0 and 1 and whose domain is the set A.

Each such function is uniquely determined by the set X_f of those a for which $f(a) = 1$ (f is the characteristic function of this set, see p. 119). To distinct functions f_1 and f_2 correspond distinct sets X_{f_1} and X_{f_2}. Thus associating with the function $f \in \{0, 1\}^A$ the set $X_f \subset A$, we obtain a one-to-one correspondence between the sets $\{0, 1\}^A$ and $\boldsymbol{P}(A)$.

§ 5. Inequalities between cardinal numbers. The Cantor–Bernstein theorem and its generalizations

We obtain the "less than" relation between cardinal numbers from the following definition.

DEFINITION: The cardinal number $\mathfrak{m}$ is *not greater* than the cardinal number $\mathfrak{n}$,

$$\mathfrak{m} \leqslant \mathfrak{n},$$

if every set of power $\mathfrak{m}$ is equipollent to a subset of a set of power $\mathfrak{n}$.

If $\mathfrak{m} \leqslant \mathfrak{n}$ and $\mathfrak{m} \neq \mathfrak{n}$ we say that $\mathfrak{m}$ is *less* than $\mathfrak{n}$ or that $\mathfrak{n}$ is *greater* than $\mathfrak{m}$; we write $\mathfrak{m} < \mathfrak{n}$ or $\mathfrak{n} > \mathfrak{m}$.

For example,

$$n < \mathfrak{a}, \tag{1}$$

$$\mathfrak{m} < 2^{\mathfrak{m}}. \tag{2}$$

For the proof of (2) we notice that $\mathfrak{m} \leqslant 2^{\mathfrak{m}}$, because every set A of power $\mathfrak{m}$ is equipollent to the subset of $\boldsymbol{P}(A)$ consisting of all singletons of elements of A. Moreover, $\mathfrak{m} \neq 2^{\mathfrak{m}}$ by Theorem 3.2.

The following theorem is an interesting consequence of the definition of inequality.

°THEOREM 1: *If f is a function defined on the set X and $f^1(X) = Y$, then $\overline{\overline{Y}} \leqslant \overline{\overline{X}}$.*

PROOF. For any y in Y put $W_y = \{x \in X: f(x) = y\}$. Since $W_y \neq \emptyset$ and $W_y \cap W_{y'} = \emptyset$ for $y \neq y'$, there exists by the axiom of choice a set A containing exactly one element from every W_y. It follows that A is equipollent to the family of all sets W_y and thus to the set $f^1(X)$. Since A is a subset of X, we conclude that $\overline{\overline{Y}} \leqslant \overline{\overline{X}}$.

Example. The projection of a plane set Q onto an arbitrary straight line has power $\leqslant \overline{\overline{Q}}$. In this case the sets W_y are the intersections $Q \cap L$ where L is a straight line parallel to the direction of the projection.

REMARK: We write $\mathfrak{m} \leqslant^* \mathfrak{n}$ if $\mathfrak{m} = 0$ or if every set of power $\mathfrak{m}$ is the image of every set of power $\mathfrak{n}$. It is an easy consequence of Theorem 1 that $\{\mathfrak{m} \leqslant \mathfrak{n}\} \equiv \{\mathfrak{m} \leqslant^* \mathfrak{n}\}$; we saw that the proof of this equivalence uses the axiom of choice. Without using this axiom we are not even able to prove the intuitive proposition that the conditions $\mathfrak{m} \leqslant^* \mathfrak{n}$ and $\mathfrak{n} < \mathfrak{m}$ are incompatible.[1])

The relation $\leqslant$ possesses many properties of its arithmetical counterpart.

(3) $$(\mathfrak{m} \leqslant \mathfrak{n}) \wedge (\mathfrak{n} \leqslant \mathfrak{p}) \rightarrow (\mathfrak{m} \leqslant \mathfrak{p}),$$

(4) $$(\mathfrak{m} \leqslant \mathfrak{n}) \rightarrow (\mathfrak{m}+\mathfrak{p} \leqslant \mathfrak{n}+\mathfrak{p}),$$

(5) $$(\mathfrak{m} \leqslant \mathfrak{n}) \rightarrow (\mathfrak{m}\mathfrak{p} \leqslant \mathfrak{n}\mathfrak{p}),$$

(6) $$(\mathfrak{m} \leqslant \mathfrak{n}) \rightarrow (\mathfrak{m}^{\mathfrak{p}} \leqslant \mathfrak{n}^{\mathfrak{p}}),$$

(7) $$(\mathfrak{m} \leqslant \mathfrak{n}) \rightarrow (\mathfrak{p}^{\mathfrak{m}} \leqslant \mathfrak{p}^{\mathfrak{n}}).$$

Law (3) expresses the transitivity of the relation $\leqslant$. Laws (4)–(7) express the monotonicity of addition, multiplication and exponentiation with respect to $\leqslant$.

As an example we prove (3). Let A, B and C be sets of power $\mathfrak{m}$, $\mathfrak{n}$ and $\mathfrak{p}$. By hypothesis, A is equipollent to a subset B_1 of B and B to a subset C_1 of C. Let f and g establish the equipollences $A \sim B_1$ and $B \sim C_1$. The composition $g \circ f$ is one-to-one and maps A onto a subset of C_1. Thus $\mathfrak{m} \leqslant \mathfrak{p}$. Q.E.D.

The laws of monotonicity do not hold for the relation $<$: for instance, $2 < \mathfrak{a}$ but $2+\mathfrak{a} = \mathfrak{a}+\mathfrak{a} = \mathfrak{a}\cdot\mathfrak{a} = 2\cdot\mathfrak{a}$; similarly, $2 < 3$, but $2^{\mathfrak{a}} = 3^{\mathfrak{a}}$ as will be shown in § 6.

In the arithmetic of natural numbers the laws converse to (4)–(7) are called the cancellation laws for the relation $\leqslant$ with respect to

[1]) The fact that the proof of Theorem 1 rests on the existence of a choice set for the family of all the sets W_y was pointed out by Levi [1] before the axiom of choice was explicitly formulated.

The relation $\leqslant^*$ was introduced by Tarski and Lindenbaum [1].

Lévy [1] proved that Theorem 1 is not provable in the system Σ [TR] provided that this system is consistent. See Jech [2], p. 162, Problem 8.

Also the impossibility of proving the implication $\mathfrak{m} \leqslant^* \mathfrak{n} \rightarrow \mathfrak{m} \leqslant \mathfrak{n}$ without the axiom of choice was shown by Lévy [1]. See also Sierpiński [20].

the operations of addition, multiplication and exponentiation; these theorems hold in arithmetic provided that $\mathfrak{p} > 1$. In the arithmetic of arbitrary cardinal numbers all of the cancellation laws fail to hold: it suffices to let $\mathfrak{m} = 2$, $\mathfrak{n} = 3$ and $\mathfrak{p} = \mathfrak{a}$ to obtain a counterexample.

On the other hand, the cancellation laws with respect to addition, multiplication and exponentiation hold for the relation $<$. They follow without difficulty from the law of trichotomy which we now state but which we shall prove only in Chapter VIII.

°*For arbitrary cardinals* $\mathfrak{m}$ *and* $\mathfrak{n}$ *either* $\mathfrak{m} \leqslant \mathfrak{n}$ *or* $\mathfrak{n} \leqslant \mathfrak{m}$.

In the remainder of this section we shall treat the question of the asymmetry of the relation $<$. This problem was investigated already by Cantor but not completely solved by him.

The asymmetry of the relation $<$ is equivalent to the theorem:

(i) $$(\mathfrak{m} \leqslant \mathfrak{n}) \wedge (\mathfrak{n} \leqslant \mathfrak{m}) \rightarrow (\mathfrak{m} = \mathfrak{n}).$$

In fact, if (i) holds, then the formulas $\mathfrak{m} < \mathfrak{n}$ and $\mathfrak{n} < \mathfrak{m}$ never hold simultaneously; otherwise (i) would yield $\mathfrak{m} = \mathfrak{n}$. Conversely, if the relation $<$ is asymmetric and satisfies the antecedent of implication (i), then necessarily $\mathfrak{m} = \mathfrak{n}$, because otherwise the $\leqslant$ signs in the antecedent of the implication could be replaced by $<$, in contradiction to the assumption of the asymmetry of $<$.

To prove (i) we first prove the following more general proposition.

THEOREM 2:[1]) *If A and B are sets and f and g are one-to-one, where $f \in B^A$ and $g \in A^B$, then the sets A and B can be represented as unions of disjoint sets $A = A_1 \cup A_2$ and $B = B_1 \cup B_2$, where*

$$f^1(A_1) = B_1 \quad \textit{and} \quad g^1(B_2) = A_2.$$

PROOF. We call the element $a \in A$ *extendable* if $a \in g^1(B)$ and $g^c(a) \in f^1(A)$. For extendable a we set $a^* = f^c\big(g^c(a)\big)$ and call a^* the *extension* of a.

Now we construct a maximal sequence of successive extensions starting with the element a. By $n(a)$ we denote the largest natural number such that there exists a sequence of $n(a)$ terms constructed by starting with a and taking successive extensions, provided that such a maximal natural number exists. Otherwise, when for every natural number k

[1]) Theorem 2 is due to Banach [1].

there exists such a sequence of k terms, we let $n(a) = N$. The sequence defined by

$$\varphi_0(a) = a, \quad \varphi_{j+1}(a) = \varphi_j(a)^* \quad \text{for} \quad j \in n(a)$$

is the desired maximal sequence of successive extensions starting from a. For non-extendable a we put $n(a) = 1$ and $\varphi_0(a) = a$.

If $n(a)$ is finite then we let

$$s(a) = \varphi_{n(a)-1}(a).$$

We define

$$A_2 = \{a \in A\colon n(a) = N \vee (s(a) \in g^1(B)) \wedge (g^c(s(a)) \notin f^1(A))\}.$$

$$A_1 = A - A_2, \quad B_1 = f^1(A_1), \quad B_2 = B - B_1.$$

For the proof of the theorem it suffices to show that $g^1(B_2) = A_2$, i.e. that

(8) $$b \in B_2 \to g(b) \in A_2,$$

(9) $$A_2 \subset g^1(B_2).$$

Proof of (8). Assume $b \in B_2$, that is, $b \notin f^1(A_1)$ and let $a = g(b)$. If $b \notin f^1(A)$ then a is not extendable, $s(a) = a$, and by definition we have $a \in A_2$. If $b \in f^1(A)$ then $b \in f^1(A_2)$ and $b = f(a')$, where $a' \in A_2$. Clearly $a' = f^c(g^c(a))$, that is, $a' = a^*$; thus $a^* \in A_2$. If $n(a^*) = N$ then $n(a) = N$ and $a \in A_2$; otherwise $s(a^*) = s(a)$ and again $a \in A_2$.

Proof of (9). Assume that $a \in A_2$. If a is extendable then $a = g(f(a^*))$. If at the same time $n(a)$ is finite, then $s(a) = s(a^*)$ and $a^* \in A_2$. The same is true if $n(a) = N$, because then $n(a^*) = N$ as well. Thus in both cases $a^* \notin A_1$, $f(a^*) \notin f^1(A_1)$ and thus $f(a^*) \in B_2$. It follows $a = g(f(a^*)) \in g^1(B_2)$.

If a is not extendable, then $s(a) = a$ and by the assumption that $a \in A_2$ we obtain $a \in g^1(B)$. If a were an element of $g^1(B_1)$ then, by the definition of the set B_1, a would have the form $g(f(a'))$ and would be extendable contrary to the assumption. Thus $a \in g^1(B_2)$ Q.E.D.

As a corollary we have the *Cantor–Bernstein theorem*:[1])

[1]) The Cantor–Bernstein theorem was conjectured by Cantor but proved correctly first by Bernstein (see Borel [1]).

The proof given here is due to Banach [1] but based on an idea similar to that invented by Bernstein. The applications to problems of finite decompositions are due to Banach and Tarski [1].

THEOREM 3: *If* $\mathfrak{m} \leqslant \mathfrak{n}$ *and* $\mathfrak{n} \leqslant \mathfrak{m}$ *then* $\mathfrak{m} = \mathfrak{n}$.

PROOF. Let $\overline{\overline{A}} = \mathfrak{m}$ and $\overline{\overline{B}} = \mathfrak{n}$. Since $\mathfrak{m} \leqslant \mathfrak{n}$, there exists a one-to-one function f from A onto a subset of B. Since $\mathfrak{n} \leqslant \mathfrak{m}$, there exists, similarly, a one-to-one function g from B onto a subset of A. By Theorem 2, $A = A_1 \cup A_2$, where A_1 and A_2 are disjoint; and $B = B_1 \cup B_2$, where B_1 and B_2 are disjoint and where $f^1(A_1) = B_1$ and $g^1(B_2) = A_2$. Thus $A_1 \sim B_1$ and $A_2 \sim B_2$; hence $A \sim B$.

The Cantor–Bernstein theorem can be generalized as follows. Let R be an equivalence relation on the family $\boldsymbol{P}(A)$ and let R satisfy the following two conditions:

$$(10) \quad XRY \to \bigvee_{f \in Y^X} [(f \text{ is a one-to-one function}) \wedge \bigwedge_{Z \subset X} (ZRf^1(Z))],$$

$$(11) \quad (X_1 \cap X_2 = \emptyset = Y_1 \cap Y_2) \wedge (X_1 RY_1) \wedge (X_2 RY_2) \to (X_1 \cup X_2 RY_1 \cup Y_2).$$

THEOREM 4: *If the relation R with field $\boldsymbol{P}(A)$ satisfies for arbitrary subsets of A the conditions* (10) *and* (11), *and if X stands in the relation R to some subset of Y and Y stands in the relation R to some subset of X, then XRY.*

PROOF. Assume that XRY_1 where $Y_1 \subset Y$, and YRX_1 where $X_1 \subset X$. From (10) it follows that there exist one-to-one functions f and g such that f maps X into Y_1 and satisfies the condition that $ZRf^1(Z)$ for every $Z \subset X$; and similarly g maps Y into X_1 and satisfies $TRg^1(T)$ for all $T \subset Y$. By Theorem 2, $X = X' \cup X''$ and $Y = Y' \cup Y''$, where $X' \cap X'' = \emptyset$ and $Y' \cap Y'' = \emptyset$ and where $Y' = f^1(X')$ and $X'' = g^1(Y'')$. Since $X' \subset X$, we have $X'Rf^1(X')$, that is, $X'RY'$; similarly, $X''RY''$. It then follows by (11) that XRY. Q.E.D.

We give two examples of relations satisfying conditions (10) and (11):

1. *The relation of equipollence on subsets of A.* Theorem 4 for this relation is then identical with the Cantor–Bernstein theorem.

2. Let $A = \mathscr{E}^n$. We call two subsets X and Y of A *equivalent by finite decomposition* if there exist a natural number k and two sequences $X_0, \ldots, X_{k-1}$ and $Y_0, \ldots, Y_{k-1}$ such that

$$X = \bigcup_{j<k} X_j, \qquad Y = \bigcup_{j<k} Y_j,$$

$$X_i \cap X_j = \emptyset = Y_i \cap Y_j \quad \text{for} \quad 0 \leqslant i < j < k,$$

$$X_i \text{ and } Y_i \text{ are isometric for } i < k.$$

If X and Y are equivalent by finite decomposition we write $X \sim_{\text{fin}} Y$.

THEOREM 5: *The relation* $\sim_{\text{fin}}$ *satisfies conditions* (10) *and* (11) *and is an equivalence relation on* $\boldsymbol{P}(A)$.

We omit the proof.

We prove one more theorem about transformations. Like Theorem 2 it is a generalization of the Cantor–Bernstein theorem.

THEOREM 6: (MEAN-VALUE THEOREM) *Let* A, B, C, A' *and* B' *be sets such that* $A \supset C \supset B$, $A' \supset B'$, $A \sim A'$ *and* $B \sim B'$. *Then there exists a set* C' *such that* $A' \supset C' \supset B'$ *and* $C \sim C'$.[1]

PROOF. It suffices to prove the existence of a function h from A into A' satisfying the following conditions:

(12) *the restriction* $h|C$ *is one-to-one,*

(13) $$h^1(C) \supset B'.$$

In fact, if h satisfies (12) and (13), then the desired set C' is $h^1(C)$.

We define h as follows:

$$h(x) = \begin{cases} f(x) & \text{for} \quad x \in A - X, \\ g^c(x) & \text{for} \quad x \in X, \end{cases}$$

where—for the present—X is an arbitrary subset of B and where f and g are one-to-one functions such that $f^1(A) = A'$ and $g^1(B') = B$. The function h defined in this way certainly satisfies condition (12) provided that

(14) $$g^{-1}(X) \cap f^1(C-X) = \emptyset.$$

In fact, if h satisfies (14) then $h(x') = h(x'')$ holds for no x' and x'' such that $x' \in X$ and $x'' \in C-X$.

The function h satisfies (13) if besides satisfying (14) it also satisfies

(15) $$g^{-1}(X) \cup f^1(C-X) \supset B',$$

because $h^1(C) = h^1(X) \cup h^1(C-X) = g^{-1}(X) \cup f^1(C-X)$.

[1]) The mean value theorem is due to Tarski. See Lindenbaum and Tarski [1], Theorem 15, p. 303.

Both condition (14) and condition (15) hold if

$$g^1\big(B'-f^1(C-X)\big) = X. \tag{16}$$

To complete the proof it suffices to show that there exists a set $X \subset B$ satisfying (16). For this purpose we let $F(X) = g^1\big(B'-f^1(C-X)\big)$ and we notice that

$$X_1 \subset X_2 \subset B \to F(X_1) \subset F(X_2) \subset B.$$

A function F from $\boldsymbol{P}(B)$ into $\boldsymbol{P}(B)$ will be called a *monotonic function on the subsets* of B if it satisfies the condition above. Therefore it suffices to prove the following lemma:

For every monotonic function on the subsets of a given set B, there exists a set X such that $F(X) = X$.

We construct X as follows: let $\boldsymbol{K} = \{X \subset B\colon F(X) \subset X\}$. The family $\boldsymbol{K}$ is non-empty because $B \in \boldsymbol{K}$. We shall show that the condition $F(X) = X$ is satisfied by the set $X_0 = \bigcap_{X \in \boldsymbol{K}} X$.

In fact, $X \in \boldsymbol{K} \to X_0 \subset X$ and thus by monotonicity $X \in \boldsymbol{K} \to F(X_0) \subset F(X) \subset X$. It follows that $F(X_0) \subset X$ for every $X \in \boldsymbol{K}$, and hence $F(X_0) \subset \bigcap_{X \in \boldsymbol{K}} X = X_0$, so that $F(X_0) \subset X_0$. By monotonicity it follows moreover that $F\big(F(X_0)\big) \subset F(X_0)$, and thus $F(X_0) \in \boldsymbol{K}$, so $X_0 \subset F(X_0)$. We have $F(X_0) = X_0$. Q.E.D.

The Cantor–Bernstein theorem is a consequence of Theorem 6. In fact, if $\mathfrak{m} \leqslant \mathfrak{n}$ then there exist sets X, Y such that $\overline{\overline{X}} = \mathfrak{m}$, $\overline{\overline{Y}} = \mathfrak{n}$ and $X \subset Y$. If we assume moreover that $\mathfrak{n} \leqslant \mathfrak{m}$, then X contains a subset Z of power $\mathfrak{n}$. Putting in Theorem 6 $A' = B' = B = Z$, $A = Y$, and $C = X$, we conclude that there exists a set C' such that $A' \supset C' \supset B'$ and $C' \sim C$. Thus $C' = Z$ and $C' \sim X$, so $Z \sim X$, and consequently $\mathfrak{m} = \mathfrak{n}$.

Exercises

1. We say that a cardinal number $\mathfrak{n}$ *absorbs* $\mathfrak{m}$ if $\mathfrak{m}+\mathfrak{n} = \mathfrak{n}$. Show:
(a) $\mathfrak{n}$ absorbs $\mathfrak{m}$ if and only if $\mathfrak{a}\cdot\mathfrak{m} \leqslant \mathfrak{n}$;
(b) if $\mathfrak{n}$ absorbs $\mathfrak{m}$, then every cardinal larger than $\mathfrak{n}$ absorbs $\mathfrak{m}$;
(c) $\mathfrak{n}$ absorbs $\mathfrak{m}$ if and only if $\mathfrak{n}$ absorbs $k\cdot\mathfrak{m}$ $(k \in N)$;
(d) $\mathfrak{n}$ absorbs $\mathfrak{m}$ if and only if $\mathfrak{n}$ absorbs $\mathfrak{a}\cdot\mathfrak{m}$. [Tarski]

2. Without using the axiom of choice show that $\mathfrak{m} \leqslant \mathfrak{a} \equiv \mathfrak{m} \leqslant^* \mathfrak{a}$.

3. Without using the axiom of choice show that $\neg(2^{\mathfrak{m}} \leqslant^* \mathfrak{m})$. [Tarski]

4. Show that $2^{\mathfrak{m}} \geqslant \mathfrak{a} \rightarrow 2^{\mathfrak{m}} \geqslant 2^{\mathfrak{a}}$.

5. Show that the closed circle T is equivalent by finite decomposition to the union $T \cup F$, where F is an arbitrary line segment disjoint from T.[1]

§ 6. Properties of the cardinals $\mathfrak{a}$ and $\mathfrak{c}$

We introduce the following notation:

$$\mathfrak{c} = 2^{\mathfrak{a}}.$$

The cardinal $\mathfrak{c}$ is called the *power of the continuum.*

We often meet the cardinal $\mathfrak{c}$, as well as the cardinal $\mathfrak{a}$, in many parts of set theory and its applications. We shall prove several formulas concerning the numbers n (*natural numbers*), $\mathfrak{a}$ and $\mathfrak{c}$.

(1) $$\mathfrak{c} = \mathfrak{c}+\mathfrak{c}.$$

In fact (see p. 180), $\mathfrak{c}+\mathfrak{c} = 2\mathfrak{c} = 2\cdot 2^{\mathfrak{a}} = 2^{1+\mathfrak{a}} = 2^{\mathfrak{a}} = \mathfrak{c}$, because $1+\mathfrak{a} = \mathfrak{a}$.

(2) $$n < \mathfrak{a} < \mathfrak{c}.$$

The inequalities follow from equations 5(1) and 5(2).

(3) $$n+\mathfrak{c} = \mathfrak{a}+\mathfrak{c} = \mathfrak{c}.$$

In fact, by (2) we have (see 5 (4))

$$\mathfrak{c} \leqslant n+\mathfrak{c} \leqslant \mathfrak{a}+\mathfrak{c} \leqslant \mathfrak{c}+\mathfrak{c},$$

and by (1)

$$\mathfrak{c} \leqslant n+\mathfrak{c} \leqslant \mathfrak{a}+\mathfrak{c} \leqslant \mathfrak{c}.$$

Applying the Cantor–Bernstein theorem, we obtain (3).

(4) $$\mathfrak{c} = \mathfrak{c}\cdot\mathfrak{c}.$$

Indeed, $\mathfrak{c} = 2^{\mathfrak{a}} = 2^{\mathfrak{a}+\mathfrak{a}} = 2^{\mathfrak{a}}\cdot 2^{\mathfrak{a}} = \mathfrak{c}\cdot\mathfrak{c}$, because $\mathfrak{a}+\mathfrak{a} = \mathfrak{a}$.

(5) $$n\cdot\mathfrak{c} = \mathfrak{a}\cdot\mathfrak{c} = \mathfrak{c} \qquad (\text{for } n > 0).$$

By 4(2) and 4(5) we have the inequalities $\mathfrak{c} \leqslant n\cdot\mathfrak{c} \leqslant \mathfrak{a}\cdot\mathfrak{c} \leqslant \mathfrak{c}\cdot\mathfrak{c}$; hence in view of (4) we obtain (5) by applying the Cantor–Bernstein theorem.

[1]) For a deeper analysis of problems dealt with in Section 5 see Tarski [9].

By induction from (4) we obtain

(6) $$\mathfrak{c}^n = \mathfrak{c} \quad (\text{for } n > 0).$$

(7) $$n^{\mathfrak{a}} = \mathfrak{a}^{\mathfrak{a}} = \mathfrak{c}^{\mathfrak{a}} = \mathfrak{c} \quad (\text{for } n > 1).$$

In fact (see 4(10)),

$$\mathfrak{c} = 2^{\mathfrak{a}} \leqslant n^{\mathfrak{a}} \leqslant \mathfrak{a}^{\mathfrak{a}} \leqslant \mathfrak{c}^{\mathfrak{a}} = (2^{\mathfrak{a}})^{\mathfrak{a}} = 2^{\mathfrak{a}\cdot\mathfrak{a}} = 2^{\mathfrak{a}} = \mathfrak{c},$$

whence (7) follows by applying the Cantor–Bernstein theorem.

The following equations concerning the number $\mathfrak{f} = 2^{\mathfrak{c}}$ are proved similarly:

(8) $$\begin{aligned} n+\mathfrak{f} &= \mathfrak{a}+\mathfrak{f} = \mathfrak{c}+\mathfrak{f} = \mathfrak{f}+\mathfrak{f} = \mathfrak{f}, \\ n\cdot\mathfrak{f} &= \mathfrak{a}\cdot\mathfrak{f} = \mathfrak{c}\cdot\mathfrak{f} = \mathfrak{f}\cdot\mathfrak{f} = \mathfrak{f} \qquad (\text{for } n > 0), \\ n^{\mathfrak{c}} &= \mathfrak{a}^{\mathfrak{c}} = \mathfrak{c}^{\mathfrak{c}} = \mathfrak{f}^{\mathfrak{c}} = \mathfrak{f} \qquad (\text{for } n > 1). \end{aligned}$$

We shall give examples of sets of powers $\mathfrak{c}$ and $2^{\mathfrak{c}}$.

THEOREM 1: *The Cantor set C has power* $\mathfrak{c}$.

PROOF. $C = \{0, 1\}^N$, so $\overline{\overline{C}} = 2^{\mathfrak{a}}$ by Theorem 4.7.

THEOREM 2: *The set of all infinite sequences of natural numbers has power* $\mathfrak{c}$.

PROOF. This set is N^N. Therefore its power is $\mathfrak{a}^{\mathfrak{a}} = \mathfrak{c}$.

THEOREM 3: *The following sets have the power of the continuum*:

(a) *the set of irrational numbers in the interval* $(0, 1)$;

(b) *the set of all points in* $(0, 1)$;

(c) *the set $\mathscr{E}$ of all real numbers*;

(d) *the set of all points of the space $\mathscr{E}^n$, where n is a natural number.*

PROOF. (a) follows from Chapter IV, § 7 and from Theorem 2. (b) follows from the observation that the interval $(0, 1)$ is the union of the countable set of rationals in $(0, 1)$ and the set of irrationals in $(0, 1)$, which has power $\mathfrak{c}$. (c) holds because the function $y = 1/2 + \frac{1}{\pi}\operatorname{arctg} x$ is one-to-one and maps the set $\mathscr{E}$ onto the interval $(0, 1)$. (d) follows from (c) and equation (6).

THEOREM 4: *If $\overline{\overline{A}} = \mathfrak{c}$, $\overline{\overline{B}} = \mathfrak{a}$ and $B \subset A$, then $\overline{\overline{A-B}} = \mathfrak{c}$.*

PROOF. By (6) we have that $A\times A \sim A$; therefore it suffices to show that if M is a countable subset of $A\times A$, then the difference $A\times A - M$

has power $\mathfrak{c}$. The projection onto A of the points in M constitute at most a countable set, which implies that there exists an element of A which does not belong to the projection of M. The set $\{\langle a, y\rangle : y \in A\}$ is disjoint from M and has power $\mathfrak{c}$, thus the difference $A \times A - M$ has power $\geqslant \mathfrak{c}$. On the other hand, this set has power $\leqslant \mathfrak{c}$ as a subset of $A \times A$. Thus the difference $A \times A - M$ has power $\mathfrak{c}$ by the Cantor–Bernstein theorem.

COROLLARY 5: *The set of transcendental numbers has power* $\mathfrak{c}$.

PROOF. It suffices to apply Theorem 4 to the case where A is the set $\mathscr{E}$ of real numbers and B is the set of algebraic numbers.

This corollary, proved by Cantor in 1874, was one of the first applications of set theory to concrete mathematical problems.

THEOREM 6: *The set* $\mathscr{E}^N$ *of infinite sequences of real numbers has power* $\mathfrak{c}$.

PROOF. $\overline{\overline{\mathscr{E}^N}} = \mathfrak{c}^{\mathfrak{a}} = \mathfrak{c}$ by (7).

THEOREM 7: *The set of continuous functions of one real variable has power* $\mathfrak{c}$.

PROOF. Let $r_1, r_2, \ldots, r_n, \ldots$ be an enumeration of all rational numbers. With every continuous function f of one real variable we associate the sequence of real numbers

$$f(r_1), f(r_2), \ldots, f(r_n), \ldots \tag{9}$$

If f and g are distinct then the corresponding sequences

$$f(r_1), f(r_2), \ldots, f(r_n), \ldots, \qquad g(r_1), g(r_2), \ldots, g(r_n), \ldots$$

are also distinct. In fact, $f \neq g$ implies that $f(x) \neq g(x)$ for some x; so if r_{k_n} is a sequence of rationals converging to x, then it is not true that $f(r_{k_n}) = g(r_{k_n})$ for every n, because in that case, by the continuity of f and g, we would have

$$f(x) = \lim_{n=\infty} f(r_{k_n}) = \lim_{n=\infty} g(r_{k_n}) = g(x).$$

Thus the set of continuous functions of one real variable is equipollent to the set of sequences (9), which has power $\leqslant \mathfrak{c}$ by Theorem 4. On the other hand, the set of continuous functions has power $\geqslant \mathfrak{c}$ because it contains all constant functions. Thus by Cantor–Bernstein theorem we obtain Theorem 7.

THEOREM 8: *The set* $\mathscr{E}^{\mathscr{E}}$ *of all functions of one real variable has power* $2^{\mathfrak{c}}$.

PROOF. $\overline{\overline{\mathscr{E}^{\mathscr{E}}}} = \mathfrak{c}^{\mathfrak{c}} = 2^{\mathfrak{c}}$ by (8).

Exercise

Prove that the family $\boldsymbol{R}$ of closed sets of the space $\mathscr{E}$ has power $\mathfrak{c}$.

Hint: To prove that $\overline{\overline{\boldsymbol{R}}} \leqslant \mathfrak{c}$ associate with every $X \in \boldsymbol{R}$ a family of intervals with rational endpoints disjoint from X and show that the set of all such families has power $\mathfrak{c}$. The inequality $\overline{\overline{\boldsymbol{R}}} \geqslant \mathfrak{c}$ holds because all one-element sets belong to $\boldsymbol{R}$.

§ 7. The generalized sum of cardinal numbers

Let T be an arbitrary set, $\mathfrak{f}$ a function defined on T with cardinal numbers as values. Instead of $\mathfrak{f}(x)$ we shall also write $\mathfrak{f}_x$.

Assume that the function $\mathfrak{f}$ satisfies the following condition

(W) $\begin{cases} \textit{there exists a set-valued function } F^{(0)} \textit{ defined on } T \textit{ such that} \\ \overline{\overline{F_x^{(0)}}} = \mathfrak{f}_x \textit{ for all } x \in T. \end{cases}$

Condition (W) can easily be shown to hold for many functions $\mathfrak{f}$. Such is the case, for instance, when $\mathfrak{f}$ has only finitely many distinct values. We shall show in Chapter VII that every $\mathfrak{f}$ satisfies condition (W), so that condition (W) does not actually affect the generality of our treatment.

°THEOREM 1: *There exists a set-values function F defined on T such that*

(1) $$\overline{\overline{F_x}} = \mathfrak{f}_x \quad \textit{for} \quad x \in T,$$

(2) $$F_x \cap F_y = \emptyset \quad \textit{for} \quad x \neq y.$$

Moreover, if $F^{(1)}$ and $F^{(2)}$ both satisfy (1) *and* (2), *then*

$$\bigcup_x F_x^{(1)} \sim \bigcup_x F_x^{(2)}.$$

PROOF. For $x \in T$ let

$$F_x = F_x^{(0)} \times \{x\},$$

where $F^{(0)}$ is any function satisfying (W). If $x \neq y$ then $F_x \cap F_y = \emptyset$, because the set F_x consists of ordered pairs with x as the second component while F_y consists of ordered pairs having y as the second component. Moreover, $\overline{\overline{F_x}} = \overline{\overline{F_x^{(0)}}} = \mathfrak{f}_x$. Thus F satisfies conditions (1) and (2).

Assume now that functions $F^{(1)}$ and $F^{(2)}$ satisfy conditions (1) and (2). For every $x \in T$ the set Φ_x of one-to-one functions from $F_x^{(1)}$ onto $F_x^{(2)}$ is non-empty. If $x \neq y$ then $\Phi_x \cap \Phi_y = \emptyset$, because every function belonging to Φ_x has domain F_x and therefore is different from every function belonging to Φ_y.

By the axiom of choice it follows that there exists a set Ψ containing exactly one element in common with each of the sets Φ_x. Let φ_x be the only element of $\Psi \cap \Phi_x$; then φ_x is a one-to-one function from the set $F_x^{(1)}$ onto the set $F_x^{(2)}$.

It is now easy to show that the function $f = \bigcup_{x \in T} \varphi_x$ maps the union $\bigcup_x F_x^{(1)}$ onto $\bigcup_x F_x^{(2)}$ in a one-to-one manner. This completes the proof.

DEFINITION: The *sum of the cardinal numbers* $\mathfrak{f}_x$ for $x \in T$ is the cardinal $\overline{\overline{\bigcup_x F_x}}$, where F is any function satisfying (1) and (2).

We denote this sum by $\sum_{x \in T} \mathfrak{f}_x$ or by $\sum \mathfrak{f}_x$:

$$\sum_{x \in T} \mathfrak{f}_x = \overline{\overline{\bigcup_{x \in T} F_x}}. \tag{3}$$

The definition is correct since the number $\overline{\overline{\bigcup_x F_x}}$ does not depend on the choice of the function F satisfying conditions (1) and (2) and since such a function always exists. However, we cannot prove the existence of such a function without appealing to the axiom of choice, so that the definition of the sum of an arbitrary set of cardinal numbers is based upon the axiom of choice.[1])

If $T = \{1, 2\}$, then $\sum_{t \in T} \mathfrak{f}_t = \mathfrak{f}_1 + \mathfrak{f}_2$. If $T = N$, then we shall also write

$$\mathfrak{f}_0 + \mathfrak{f}_1 + \mathfrak{f}_2 + \mathfrak{f}_3 + \ldots \quad \text{or} \quad \sum_{n=0}^{\infty} \mathfrak{f}_n$$

and speak of the sum of a series of cardinal numbers.

[1]) The fact that the sum of an infinite sequence of cardinal numbers cannot be properly defined without assuming the axiom of choice was pointed out by Sierpiński in 1918. See Sierpiński [23], pp. 208–256.

°THEOREM 2: (GENERALIZED COMMUTATIVE LAW) *If* φ *is an arbitrary permutation of the set* T, *then* $\sum \mathfrak{f}_x = \sum \mathfrak{f}_{\varphi(x)}$.

For the proof it suffices to notice that

$$\sum_x \mathfrak{f}_x = \overline{\overline{\bigcup_x F_x}} = \overline{\overline{\bigcup_x F_{\varphi(x)}}} = \sum_x \mathfrak{f}_{\varphi(x)},$$

where F is any function satisfying conditions (1) and (2). The equations hold on the basis of Theorem IV.1.3 and formula (3).

°THEOREM 3: (GENERALIZED ASSOCIATIVE LAW) *If* $T = \bigcup_{y\in I} T_y$ *where the sets* T_y *are disjoint, then*

$$\sum_{x\in T} \mathfrak{f}_x = \sum_{y\in I} \Big(\sum_{x\in T_y} \mathfrak{f}_x\Big).$$

PROOF. For $y \in I$, let $\mathfrak{g}_y = \sum_{x\in T_y} \mathfrak{f}_x$, that is, $\mathfrak{g}_y = \overline{\overline{\bigcup_{x\in T_y} F_x}}$. Then

$$\sum_{y\in I} \mathfrak{g}_y = \overline{\overline{\bigcup_{y\in I}\Big(\bigcup_{x\in T_y} F_x\Big)}} = \overline{\overline{\bigcup_{x\in T} F_x}},$$

by Theorem IV.9.2. It follows that

$$\sum_{y\in J} \mathfrak{g}_y = \sum_{x\in T} \mathfrak{f}_x,$$

which proves Theorem 3.

°THEOREM 4: (GENERALIZED DISTRIBUTIVE LAW FOR MULTIPLICATION WITH RESPECT TO ADDITION) *The equation*

$$\Big(\sum_x \mathfrak{f}_x\Big) \cdot \mathfrak{m} = \sum_x (\mathfrak{f}_x \cdot \mathfrak{m})$$

holds for every cardinal $\mathfrak{m}$.

PROOF. Let $\overline{\overline{M}} = \mathfrak{m}$. We have

$$\Big(\sum_x \mathfrak{f}_x\Big)\mathfrak{m} = \overline{\overline{\Big(\bigcup_x F_x\Big) \times M}} \quad \text{and} \quad \sum_x (\mathfrak{f}_x \cdot \mathfrak{m}) = \overline{\overline{\bigcup_x (F_x \times M)}}.$$

At the same time (see Exercise IV.1.2)

$$\Big(\bigcup_x F_x\Big) \times M = \bigcup_x (F_x \times M).$$

°THEOREM 5: *If* $\mathfrak{g}_x \leqslant \mathfrak{f}_x$ *for* $x \in T$, *then* $\sum_x \mathfrak{g}_x \leqslant \sum_x \mathfrak{f}_x$.

PROOF. Let K_x be the family of those $X \subset F_x$ which have power $\mathfrak{g}_x$. By assumption $K_x \neq \emptyset$ for every $x \in T$. It follows that there exists a function G (see Theorem IV.6.2) such that

$$G \in \prod_{x \in T} K_x, \quad \text{that is,} \quad G_x \in K_x.$$

Therefore $G_x \subset F_x$ and $\overline{\overline{G_x}} = \mathfrak{g}_x$. This implies

$$\bigcup_x G_x \subset \bigcup_x F_x \quad \text{and} \quad \overline{\overline{\bigcup_x G_x}} = \sum_x \mathfrak{g}_x,$$

which proves that $\sum_x \mathfrak{g}_x \leqslant \sum_x \mathfrak{f}_x$.

°THEOREM 6: *If $S \subset T$ then*

$$\sum_{x \in S} \mathfrak{f}_x \leqslant \sum_{x \in T} \mathfrak{f}_x.$$

PROOF. Let

$$\mathfrak{g}_x = \begin{cases} \mathfrak{f}_x & \text{for} \quad x \in S, \\ 0 & \text{for} \quad x \in T-S. \end{cases}$$

By Theorem 5,

$$\sum_{x \in T} \mathfrak{g}_x \leqslant \sum_{x \in T} \mathfrak{f}_x, \quad \text{but} \quad \sum_{x \in T} \mathfrak{g}_x = \sum_{x \in S} \mathfrak{f}_x.$$

°THEOREM 7: *If $\mathfrak{f}_x = \mathfrak{m}$ for all $x \in T$, and $\mathfrak{n} = \overline{\overline{T}}$, then*

$$\sum_{x \in T} \mathfrak{f}_x = \mathfrak{m} \cdot \mathfrak{n}.$$

PROOF. Let $\overline{\overline{M}} = \mathfrak{m}$. For every x there exists a one-to-one mapping f of the set F_x onto M; let Φ_x be the set of all such functions. By the axiom of choice there exists a set Ψ containing exactly one element from each of the sets Φ_x. Let f_x be the unique element of $\Psi \cap \Phi_x$.

For $t \in \bigcup_x F_x$ we let

$$\varphi(t) = \langle f_x(t), x \rangle,$$

where x is the (unique) element of T such that $t \in F_x$.

The function φ maps the union $\bigcup F_x$ onto $M \times T$ and is one-to-one. In fact, if

$$\varphi(t_1) = \langle f_{x_1}(t_1), x_1 \rangle \quad \text{and} \quad \varphi(t_2) = \langle f_{x_2}(t_2), x_2 \rangle,$$

then $x_1 \neq x_2$ implies $t_1 \neq t_2$, because t_1 and t_2 belong to the disjoint sets F_{x_1} and F_{x_2}. On the other hand, if $x_1 = x_2 = x$ then, since f_x is one-to-one, $\varphi(t_1) \neq \varphi(t_2)$ implies that $t_1 \neq t_2$.

Therefore $\bigcup F_x \sim M \times T$, which proves Theorem 7.

°THEOREM 8: *If* $\mathfrak{f}_x \leqslant \mathfrak{m}$ *for* $x \in T$ *and* $\mathfrak{n} = \overline{\overline{T}}$, *then* $\sum \mathfrak{f}_x \leqslant \mathfrak{m} \cdot \mathfrak{n}$.

PROOF. Letting $\mathfrak{m}_x = \mathfrak{m}$ for $x \in T$, we have by Theorem 5 $\sum \mathfrak{f}_x \leqslant \sum \mathfrak{m}_x$, and by Theorem 7 $\sum \mathfrak{m}_x = \mathfrak{m} \cdot \mathfrak{n}$.

Examples

1. We shall calculate the sum $\sum k_n$, where k_n is a natural number and n runs through the set of natural numbers. For this purpose we notice that from Theorem 7 it follows that

$$(4) \qquad 1+1+1+ \ldots = 1 \cdot \mathfrak{a} = \mathfrak{a}, \qquad \mathfrak{a}+\mathfrak{a}+\mathfrak{a}+ \ldots = \mathfrak{a} \cdot \mathfrak{a} = \mathfrak{a};$$

since, by Theorem 5,

$$1+1+1+ \ldots \leqslant \sum_{n=1}^{\infty} k_n \leqslant \mathfrak{a}+\mathfrak{a}+\mathfrak{a}+ \ldots,$$

we have by the Cantor–Bernstein theorem

$$\sum_{n=1}^{\infty} k_n = \mathfrak{a}.$$

In particular

$$2+2+2+ \ldots = \mathfrak{a},$$
$$1+2+3+ \ldots = \mathfrak{a},$$
$$1!+2!+3!+ \ldots = \mathfrak{a}.$$

2. It follows from equation (4) that the union of a countable number of countable sets is countable (see Theorem 2.8).

3. By Theorem 7, $\mathfrak{c}+\mathfrak{c}+\mathfrak{c}+ \ldots = \mathfrak{c} \cdot \mathfrak{a} = \mathfrak{c}$. Similarly, the sum $\sum_{x \in T} \mathfrak{f}_x$ where the power of T is $\mathfrak{c}$ and where each $\mathfrak{f}_x$ equals $\mathfrak{c}$ is itself equal to $\mathfrak{c} \cdot \mathfrak{c}$, that is to $\mathfrak{c}$.

§ 8. The generalized product of cardinal numbers

As before, we shall use the axiom of choice and assume that $\mathfrak{f}$ is a function having cardinal numbers as values and satisfying condition (W) given on p. 191.

°THEOREM 1: *If the functions* $F^{(1)}$ *and* $F^{(2)}$ *satisfy the condition* $\overline{\overline{F_x^{(1)}}} = \mathfrak{f}_x = \overline{\overline{F_x^{(2)}}}$ *for* $x \in T$, *then*

$$\overline{\overline{\prod_{x \in T} F_x^{(1)}}} = \overline{\overline{\prod_{x \in T} F_x^{(2)}}}.$$

PROOF. As in the proof of Theorem 7.1, using the axiom of choice, we show that there exists a set which for every $x \in T$ contains exactly one function φ_x which is one-to-one and maps $F_x^{(1)}$ onto $F_x^{(2)}$. With each function $f_1 \in \prod_{x \in T} F_x^{(1)}$ we associate the function f_2, where

(1) $$f_2(t) = \varphi_t(f_1(t)) \quad \text{for} \quad t \in T.$$

Since $f_1(t) \in F_t^{(1)}$ for every $t \in T$, $\varphi_t(f_1(t)) \in \varphi_t(F_t^{(2)}) = F_t^{(2)}$, which proves that $f_2(t) \in F_t^{(2)}$, that is, $f_2 \in \prod_{x \in T} F_t^{(2)}$.

If $f_1' \neq f_2''$ then, for some $t, f_1'(t) \neq f_1''(t)$. Thus, since φ_t is one-to-one,

$$\varphi_t(f_1'(t)) \neq \varphi_t(f_1''(t)),$$

that is, $f_2'(t) \neq f_2''(t)$. Therefore the correspondence between f_1 and f_2 is one-to-one.

Finally, if $f_2 \in \prod F_x^{(2)}$ then the function f_1 defined by the equation

$$f_1(t) = \varphi_t^c(f_2(t))$$

belongs to $\prod F_x^{(1)}$ and satisfies condition (1). Thus every function belonging to $\prod F_x^{(2)}$ corresponds to some function belonging to $\prod F_x^{(1)}$.

Thus we have proved Theorem 1. This theorem leads to the following definition.

DEFINITION: The *product of the cardinal numbers* $\mathfrak{f}_x$ is the power of the cartesian product $\prod_x F_x$, where F is an arbitrary function such that $\overline{\overline{F_x}} = \mathfrak{f}_x$, that is,

$$\prod_{x \in T} \mathfrak{f}_x = \overline{\overline{\prod_{x \in T} F_x}}, \quad \text{where} \quad \overline{\overline{F_x}} = \mathfrak{f}_x.$$

Just as for generalized sum, the use of the notion of generalized product rests upon the axiom of choice, without which we cannot prove Theorem 1 which is the basis for the definition of product.

If $T = \{1, 2\}$, then $\prod_{t \in T} \mathfrak{f}_t = \mathfrak{f}_1 \cdot \mathfrak{f}_2$. For this reason we write $\mathfrak{f}_0 \cdot \mathfrak{f}_1 \cdot \ldots$, or $\prod_{n=0}^{\infty} \mathfrak{f}_n$ when $T = N$.

From Theorems 1.2–4 we obtain directly the *commutative, associative and distributive laws*:

$$\prod_x \mathfrak{f}_x = \prod_x \mathfrak{f}_{\varphi(x)}$$

(where φ is a permutation of T),

$$\prod_{y \in U} \Big(\prod_{x \in T_y} \mathfrak{f}_x\Big) = \prod_{x \in T} \mathfrak{f}_x$$

(where $T = \bigcup_{y \in U} T_y$ and $T_{y_1} \cap T_{y_2} = \emptyset$ for $y_1 \neq y_2$),

$$\Big(\prod_x \mathfrak{f}_x\Big)^{\mathfrak{g}} = \prod_x \mathfrak{f}_x^{\mathfrak{g}}.$$

If all of the values of the function $\mathfrak{f}$ are identical, then multiplication coincides with exponentiation; that is,

(2) $\quad$ if $\mathfrak{f}_x = \mathfrak{f}_0$ for $x \in T$ and $\mathfrak{t} = \overline{\overline{T}}$, then $\prod_x \mathfrak{f}_x = \mathfrak{f}_0^{\mathfrak{t}}$.

By equation (13), Theorem 1.3 we derive from the definition of sum that

(3) $$\mathfrak{f}^{\sum_y \mathfrak{t}_y} = \prod_y (\mathfrak{f}^{\mathfrak{t}_y}).$$

Finally we note without proof that

(4) $\quad$ *if* $\mathfrak{g}_x \leqslant \mathfrak{f}_x$, *then* $\prod_x \mathfrak{g}_x \leqslant \prod_x \mathfrak{f}_x$,

analogously to Theorem 7.5.

Examples

1. In (2) let $f_0 = \mathfrak{a}$ and $T = N$. Since $\mathfrak{a}^{\mathfrak{a}} = \mathfrak{c}$, we obtain

$$\mathfrak{a} \cdot \mathfrak{a} \cdot \mathfrak{a} \cdot \ldots = \mathfrak{c}.$$

2. In (2) let $f_0 = 2$ and $T = N$. From the equation $2^{\mathfrak{a}} = \mathfrak{c}$ we have

$$2 \cdot 2 \cdot 2 \cdot \ldots = \mathfrak{c}.$$

3. From (4) we have $2 \cdot 2 \cdot 2 \cdot \ldots \leqslant 2 \cdot 3 \cdot 4 \cdot 5 \cdot \ldots$ and $2 \cdot 3 \cdot 4 \cdot 5 \cdot \ldots \leqslant \mathfrak{a} \cdot \mathfrak{a} \cdot \mathfrak{a} \cdot \ldots$. Thus by Cantor–Bernstein theorem we conclude that

$$1 \cdot 2 \cdot 3 \cdot 4 \cdot \ldots = \mathfrak{c}.$$

In the same way it can be shown that $k_1 \cdot k_2 \cdot k_3 \cdot \ldots = \mathfrak{c}$ if for all n $k_n > 1$.

°THEOREM 2: (J. König [1]) *If* $\mathfrak{g}_x < \mathfrak{f}_x$ *for all* $x \in T$, *then*

$$\sum_x \mathfrak{g}_x < \prod_x \mathfrak{f}_x.$$

PROOF. From the assumption that (W) holds (p. 191) it follows that there exists a function F such that $\overline{\overline{F_x}} = \mathfrak{f}_x$. We may assume, moreover, that $F_x \cap F_y = \emptyset$ for $x \neq y$ (see Theorem 7.1).

Arguing as in the proof of Theorem 7.5 we conclude that there exists a function G such that $G_x \subset F_x$ and $\overline{\overline{G_x}} = \mathfrak{g}_x$. It follows that $G_x \cap G_y = \emptyset$ for $x \neq y$ and that $F_x - G_x \neq \emptyset$.

We show first that

$$\sum_x \mathfrak{g}_x \leqslant \prod_x \mathfrak{f}_x. \tag{5}$$

Let f be an arbitrary function belonging to $\prod_x (F_x - G_x)$. Such a function exists by Theorem IV.6.3. For every $a \in \bigcup_x G_x$ let

$$f_a(x) = \begin{cases} f(x) & \text{for} \quad a \notin G_x, \\ a & \text{for} \quad a \in G_x. \end{cases}$$

Clearly, $f_a \in \prod F_x$. If $a \neq b$ then $f_a \neq f_b$, because a and b either belong to different sets G_x, G_y and then $f_a(x) = a \in G_x$, $f_b(x) = f(x) \in F_x - G_x$, and thus $f_a(x) \neq f_b(x)$; or else a and b belong to the same set G_x and then $f_a(x) = a \neq b = f_b(x)$.

Thus the functions f_a constitute a subset of the cartesian product $\prod_x F_x$ equipollent to the union $\bigcup G_x$, which proves formula (5).

It remains to show that

$$\sum_x \mathfrak{g}_x \neq \prod_x \mathfrak{f}_x. \tag{6}$$

For this purpose we observe that every set S equipollent to the union $\bigcup_x G_x$ can be considered as a union of disjoint sets:

$$S = \bigcup_x H_x, \quad \text{where} \quad \overline{\overline{H}}_x = \mathfrak{g}_x .$$

Assume that $S \subset \prod_x F_x$; let $h \in H_x$. Thus for every t, $h(t) \in F_t$, and in particular $h(x) \in F_x$. Hence if h ranges over the set H_x, then the elements $h(x)$ form a set K_x contained in F_x, and so (see Theorem 5.1) $\overline{\overline{K}}_x \leqslant \overline{\overline{H}}_x = \mathfrak{g}_x < \overline{\overline{F}}_x$. It follows that $F_x - K_x \neq \emptyset$ for every x, and hence $\prod_x (F_x - K_x) \neq \emptyset$.

Let $\varphi \in \prod_x (F_x - K_x)$. Hence $\varphi(x) \notin K_x$, and $\varphi \notin H_x$, because $h(x) \in K_x$ for all h belonging to H_x. This implies that the function φ belongs to none of the sets H_x contained in the union S; that is, $\varphi \notin S$ and it follows that $S \neq \prod_x F_x$, which proves (6).

Taking in König's theorem $\mathfrak{g}_x = 1$, $\mathfrak{f}_x = 2$, and $\mathfrak{m} = \overline{\overline{T}}$, we obtain Cantor's inequality $\mathfrak{m} < 2^{\mathfrak{m}}$ (see 5(2)). Thus König's theorem is a generalization of this inequality.

COROLLARY 3: *If* $\mathfrak{m}_n < \mathfrak{m}_{n+1}$ *for* $n = 0, 1, 2, \ldots$ *and* $\mathfrak{m}_0 > 0$, *then*

$$\sum_{n=0}^{\infty} \mathfrak{m}_n < \prod_{n=0}^{\infty} \mathfrak{m}_n .$$

PROOF. By König's theorem,

$$\mathfrak{m}_0 + \mathfrak{m}_1 + \mathfrak{m}_2 + \ldots < \mathfrak{m}_1 \cdot \mathfrak{m}_2 \cdot \mathfrak{m}_3 \cdot \ldots,$$

whence

$$\mathfrak{m}_0 + \mathfrak{m}_1 + \mathfrak{m}_2 + \ldots < \mathfrak{m}_0 \cdot \mathfrak{m}_1 \cdot \mathfrak{m}_2 \cdot \ldots .$$

COROLLARY 4: *For no cardinal* $\mathfrak{n}$ *can* $\mathfrak{n}^{\mathfrak{a}}$ *be represented as the sum of an infinite strictly increasing sequence of cardinal numbers.*

PROOF. Let $\mathfrak{n}^{\mathfrak{a}} = \mathfrak{m}_0 + \mathfrak{m}_1 + \mathfrak{m}_2 + \ldots$. Hence $\mathfrak{m}_p \leqslant \mathfrak{n}^{\mathfrak{a}}$ and (see (4))

$$\prod_{p=0}^{\infty} \mathfrak{m}_p \leqslant (\mathfrak{n}^{\mathfrak{a}})^{\mathfrak{a}} = \mathfrak{n}^{\mathfrak{a}} = \sum_{p=0}^{\infty} \mathfrak{m}_p .$$

By Theorem 3, the sequence $\mathfrak{m}_0, \mathfrak{m}_1, \mathfrak{m}_2, \ldots$ cannot be increasing.

Thus in particular neither $\mathfrak{c}$ nor $2^{\mathfrak{c}}$ is the sum of an infinite increasing sequence. On the other hand, the number

$$\mathfrak{a}+2^{\mathfrak{a}}+2^{2^{\mathfrak{a}}}+\ldots,$$

and more generally the number

$$\mathfrak{n}+2^{\mathfrak{n}}+2^{2^{\mathfrak{n}}}+\ldots,$$

cannot be written as a cardinal raised to the power $\mathfrak{a}$.

CHAPTER VI

LINEARLY ORDERED SETS

§ 1. Introduction

The notion of a linearly ordered set was introduced in Chapter II, § 9, p. 81. A linear ordering is also called a *total, complete* or *simple* ordering.[1])

The types of relational systems $\langle A, R\rangle$, where R is a linear order relation, are called *order types*. The order type of the system $\langle A, R\rangle$ will usually be denoted by $\overline{A}$ (although it would be more proper to denote it by $\overline{R}$).

Examples

1. For φ, $\psi \in N^N$, let $\varphi \leqslant \psi$ if $\varphi = \psi$ or if the least n such that $\varphi_n \neq \psi_n$ satisfies the conditions $\varphi_n < \psi_n$. The relation $\leqslant$ linearly orders the set N^N. If $A \subset N^N$, then the relation $\leqslant$ restricted to A linearly orders the set A.

2. The set A consisting of all natural numbers of the form 2^n is linearly ordered by the divisibility relation, that is, by the relation

$$\Big\{\langle m, n\rangle \colon (m \in A) \wedge (n \in A) \wedge \bigvee_k (m = kn)\Big\}.$$

3. The set $N \times N$ is linearly ordered by the relation R which holds between pairs $\langle m, n\rangle$ and $\langle p, q\rangle$ if and only if $(2m+1)/2^n \leqslant (2p+1)/2^q$. This relation is isomorphic to the relation $\leqslant$ in the set of real numbers of the form $(2m+1)/2^n$.

4. Let $P(m)$ assert that m is an even number. The set N is linearly ordered by the relation

[1]) The notion of a linearly ordered set is due to Cantor [5]. Cantor's paper contains all the essential results which are given in the present Chapter.

$$\{\langle m, n\rangle\colon [P(m) \wedge P(n) \wedge (m \leqslant n)] \vee [P(m) \wedge \neg P(n)] \vee \\ \vee [\neg P(m) \wedge \neg P(n) \wedge (n \leqslant m)]\}.$$

In this ordering every even number precedes every odd number. Of two even numbers, the smaller precedes the larger; of two odd numbers, the larger precedes the smaller.

Schematically this ordering can be illustrated by the sequence

$$2, 4, 6, 8, 10, \ldots \quad \ldots, 9, 7, 5, 3, 1,$$

where every number preceding a number x is written to the left of x.

5. The set of complex numbers is linearly ordered by the relation

$$\{\langle x, y\rangle\colon [R(x) < R(y)] \vee [R(x) = R(y)] \wedge [I(x) \leqslant I(y)]\}.$$

In this ordering a number x precedes a number y if the real part $R(x)$ of the complex number x is less than the real part $R(y)$ of y. In the case where the real parts of x and y are equal, x precedes y if the imaginary part $I(x)$ of x is less than the imaginary part $I(y)$ of y.

6. Let $\alpha(n)$ be the number of distinct prime factors of the natural number n. The set of natural numbers is linearly ordered by the relation

$$\{\langle x, y\rangle\colon [\alpha(x) < \alpha(y)] \vee [\alpha(x) = \alpha(y)] \wedge (x \leqslant y)\}.$$

7. The set of concentric circles is linearly ordered by the inclusion relation.

DEFINITION: An element x is said to be a *first element* of the linearly ordered set A (with respect to the relation R) if xRy for all $y \in A$. On the other hand, if yRx for all y, then x is said to be a *last element* of A (with respect to R). Generally speaking, not every set has a first or last element; but if such an element exists, then it is uniquely determined.

THEOREM 1: *In a finite non-empty subset X of a linearly ordered set A there is a first element and a last element.*

PROOF. The proof is by induction on the number of elements of X. If X has only one element, then the theorem is obvious. Suppose that the theorem holds for subsets with n elements. Let $X = Y \cup \{a\}$ where $a \notin Y$ and Y has n elements. Let b_1 be the first and b_2 the last element of Y. Since A is linearly ordered, either a precedes b_1 or b_1

precedes a. That element which precedes the other is clearly the first element of Y. Similarly we show that one of the elements a and b_2 is the last element of X.

In the case of linear order relations we usually speak of *similarity* of relations instead of their isomorphism. The following theorem shows that in the case of similarity the definition of isomorphism can be simplified: instead of proving that two formulas xRy and $f(x)Sf(y)$ are equivalent it suffices to prove only the implication (1) below.

THEOREM 2: *In order that two sets A and B linearly ordered respectively by relations R and S be similar it is necessary and sufficient that there exists a one-to-one function f which maps the set A onto the set B so that*

$$xRy \to f(x)Sf(y) \tag{1}$$

for all $x, y \in A$.

PROOF. Clearly, it suffices to show that if $x, y \in A$ and $f(x)Sf(y)$, then xRy. Suppose the contrary: $\neg(xRy)$. Since the relation R is connected in A, we have either $x = y$ or yRx. In the first case, xRy as the relation R is reflexive in A, but this contradicts the hypothesis $\neg(xRy)$. In the second case, (1) implies $f(y)Sf(x)$ and therefore $f(x) = f(y)$ because S is antisymmetric. Since f is one-to-one, we infer that $x = y$, which again contradicts $\neg(xRy)$. Hence the theorem is proved.

Similar sets are clearly equipollent. The converse theorem holds for finite sets only.

THEOREM 3: *Two finite linearly ordered equipollent sets are similar.*

PROOF. Suppose that sets A and B, linearly ordered by relations R and S, respectively, have n elements. For $n = 0$ the empty function satisfies the conditions of Theorem 2, consequently it establishes the isomorphism between the relations R and S.

Now suppose that Theorem 3 holds for sets with n elements and let A and B have $n+1$ elements. Let a be the first element of A and b the first element of B. By assumption, there exists a function f_1 which establishes the similarity between the sets $A - \{a\}$ and $B - \{b\}$.

Let

$$f = f_1 \cup \{\langle a, b\rangle\}.$$

It is easy to check that f is a function which establishes the similarity between A and B. In this way Theorem 3 is proved by induction.

By means of a counterexample it can be shown that Theorem 3 is false for infinite sets. For example, it fails for the set of natural numbers (see pp. 201 and 202, see Examples 4 and 6).

It follows from Theorem 3 that for any linearly ordered set A of n elements we can put $\bar{A} = n$.

Now we shall introduce some terminology. We say that x *precedes* y if

$$xRy \quad \text{and} \quad x \neq y.$$

In this case we write $x \prec_R y$ (or $x \prec y$ if there is no confusion about the relation R). We also write $y \succ_R x$ or $y \succ x$.

We say that y *lies between* x *and* z if

$$x \prec y \prec z \quad \text{or} \quad x \succ y \succ z.$$

If $x \in A$ and the set $\{y\colon x \prec y\}$ has a first element, then this element is called a *direct successor* of x (with respect to R). The last element of $\{y\colon y \prec x\}$ (if one exists) is called a *direct predecessor* of x. Each element $x \in A$ possesses at most one direct successor and at most one direct predecessor.

A proper subset X of the set A is said to be an *initial segment* (a *final segment*) if $x \in X$ implies that every element preceding x belongs to X (every element after x belongs to X).

The set $X \subset A$ is said to be an *interval* if the condition $x, y \in X$ implies that every element lying between x and y belongs to X.

Let

$$O_R(x) = \{y\colon (yRx) \wedge (y \neq x)\} = \{y\colon y \prec x\}.$$

The subscript R will sometimes be omitted.

It is easily seen that $O_R(x)$ is an initial segment. However, not every initial segment is of the form $O_R(x)$.

We say that an interval X of a linearly ordered set A *precedes* an interval Y of A if

$$(x \in X) \wedge (y \in Y) \rightarrow x \prec y.$$

Every family of disjoint intervals is linearly ordered by the relation "X precedes Y or $X = Y$."

Exercises

1. Let $\boldsymbol{M}$ be a family of subsets of a set Z such that

(i) $\boldsymbol{M}$ is a monotonic family,

(ii) M is not included in any monotonic family of subsets of Z different from M. Prove that the relation defined by the equivalence

$$xRy \equiv \{(x = y) \vee \bigvee_{E \in M} [(x \notin E) \wedge (y \in E)]\}$$

linearly orders the set Z. The family of final segments of this set is identical with M.[1])

2. Let M be a monotonic family of subsets of a set Z. Prove that the family of all sets of the form $\bigcup_{X \in S} X$ and $\bigcap_{X \in S} X$, where $S \subset M$, is monotonic.

§ 2. Dense, scattered, and continuous sets

A set A is said to be *densely ordered* by an order relation $\prec$ if for any two elements $x, y \in A$ there exists an element $z \in A$ between x and y. We say then also that A is *dense*. In a densely ordered set no element has either a direct successor or a direct predecessor. This property is characteristic for densely ordered sets. In fact, if no element of the set A has a direct predecessor and $x, y \in A$, $x \prec y$, then x cannot be the last element of the set $\{z: z \prec y\}$, for then x would be a direct predecessor of y. Thus there exists z such that $x \prec z \prec y$. Hence the set A is densely ordered.

All one-element sets, as well as the empty set, are densely ordered. All other densely ordered sets contain infinitely many elements.

A set which is not densely ordered may contain a densely ordered subset. For instance, the set consisting of all positive real numbers and negative integers ordered by the relation $\leqslant$ is not densely ordered, because no element of this set lies between -2 and -1. However, this set does contain a densely ordered subset.

A linearly ordered set which contains no infinite densely ordered subset is said to be *scattered*. For instance, the set of integers and the set composed of all fractions $1/n$ ($n = \pm 1, \pm 2, \ldots$) are scattered if the order relation is $\leqslant$.

Every subset of a scattered set is scattered.

THEOREM 1: *If A and B are two scattered subsets of a linearly ordered set M, then the union $A \cup B$ is also scattered.*

[1]) The theorem stated in Exercise 1 shows that the theory of linearly ordered sets can be reduced to the theory of monotonic families of sets. See Kuratowski [4].

PROOF. Suppose that there exists an infinite densely ordered subset C of the set $A \cup B$. Since $C = (C \cap A) \cup (C \cap B)$, either $C \cap A$ or $C \cap B$ is infinite. Let $C \cap A$ be infinite. Since this set is not densely ordered (as a subset of the scattered set B), there exists a pair a_1, a_2 of elements of the set $C \cap A$ such that $a_1 \prec a_2$ and such that no element of $C \cap A$ lies between a_1 and a_2. This implies that for every $x \in C$

(1) $$(a_1 \prec x \prec a_2) \rightarrow x \in B.$$

Let $B_1 = C \cap \{x\colon a_1 \prec x \prec a_2\}$. This set is infinite, for there are infinitely many elements of C between a_1 and a_2. If $x_1, x_2 \in B_1$ and $x_1 \prec x_2$, then there exists $x \in C$ lying between x_1 and a_2; thus $x \in B_1$. This implies that the set B_1 is densely ordered. By (1), $B_1 \subset B$, which means that B_1 is an infinite densely ordered subset of B. This contradicts the assumption that B is a scattered set. Hence Theorem 1 is proved.

A set X contained in a linearly ordered set A is said to be *densely ordered* in A if, for every two elements x and y of the set A, there exists an element z of X lying between x and y. For example, the set of rational numbers is densely ordered in the set of real numbers, where the order relation is $\leqslant$.

It is clear that if a set X is densely ordered in A, then the sets A and X are both densely ordered.

Of course, *two sets X and Y densely ordered in A which have neither first nor last elements are always cofinal and coinitial. If a set X is cofinal with Y and Y is cofinal with Z, then the sets X and Z are also cofinal.* A similar law of transitivity holds for coinitial sets.

Let $\langle X, Y\rangle$ be a cut in a linearly ordered set A. The intersection $X \cap Y$ contains at most one element. In fact, if $x, y \in X \cap Y$, then $x \preceq y$ and $y \preceq x$; thus $x = y$. If $X \cap Y = \emptyset$, then we say that the cut $\langle X, Y\rangle$ determines a *gap* in the set A. If $X \cap Y = \{a\}$, then we say that the element *a lies in the cut* $\langle X, Y\rangle$. It can easily be shown that in this case $X = \{a\}^-$, $Y = \{a\}^+$ and $a = \bigvee_{x \in X} x = \bigwedge_{y \in Y} y$. A cut $\langle X, Y\rangle$ is said to be *proper* if $X \neq \emptyset \neq Y$.

A set A is said to be *continuously ordered* if no proper cut in A determines a gap in A. We also say that A is *continuous*.

If $\langle X_1, Y_1\rangle$ and $\langle X_2, Y_2\rangle$ are cuts in A, then either $X_1 \subset X_2$ or $X_2 \subset X_1$. In fact, suppose that $a \in X_1 - X_2$ and let $b \in X_2$. By connectedness, it follows that b precedes a, for in the opposite case we would have $a \in X_2$. Thus $b \in X_1$ and we obtain $X_2 \subset X_1$. This implies the following theorem.

THEOREM 2: *The minimal extension* $\mathfrak{P}$ (see p. 155) *of a linearly ordered set is continuously ordered.*

In fact, $\mathfrak{P}$ is a complete lattice and, as has been shown before, the ordering in $\mathfrak{P}$ is connected; thus it is linear. The complete linearly ordered lattice $\mathfrak{P}$ is continuously ordered, because if $\langle \mathfrak{O}_1, \mathfrak{O}_2 \rangle$ is a proper cut in $\mathfrak{P}$, then the supremum of the set $\mathfrak{O}_1$ lies in this cut.

COROLLARY 3:[1]) *Every linearly ordered set can be extended to a continuously ordered set (preserving suprema and infima).*

To conclude this section, we shall prove a theorem showing that the study of any order type can be reduced to the study of dense and scattered order types. For this purpose we need the notion of the ordered union of linearly ordered sets.

Let T be a set linearly ordered by the relation Q and let F and R be functions defined for $x \in T$ and such that R_x is a relation linearly ordering the set F_x. Suppose that $F_{x_1} \cap F_{x_2} = \emptyset$ for $x_1 \neq x_2$.

THEOREM 4: *Let S be the relation which holds between two elements a and b of the union $\bigcup_x F_x$ if and only if either*

$$a \text{ and } b \text{ belong to the same component } F_x \text{ and } aR_x b,$$

or

$$a \text{ and } b \text{ belong to different components } F_{x_1} \text{ and } F_{x_2} \text{ and } x_1 Q x_2.$$

Then the relation S linearly orders the union $\bigcup_x F_x$.

PROOF. The reflexivity of S is obvious.

If a and b belong to different components of the union $\bigcup_x F_x$, then either aSb or bSa, because the relation Q orders the set of indices.

On the order hand, if a and b belong to the same component F_x, then either aSb or bSa, because the relation R_x is connected in F_x. Thus the relation S is also connected.

If $a \in F_{x_1}$, $b \in F_{x_2}$, aSb and bSa, then $x_1 Q x_2$ and $x_2 Q x_1$. This implies that $x_1 = x_2$, since the relation Q is antisymmetric.

[1]) Corollary 3 goes back to Dedekind [1].

Finally, suppose that $a \in F_x$, $b \in F_y$, $c \in F_z$, aSb and bSc. These conditions imply xQy and yQz; thus xQz. If $x \neq z$, then aSc. On the other hand, if $x = z$, then we have $x = y = z$, because xQy, yQz and the relation Q is antisymmetric. By the definition of S we obtain $aR_x b$ and $bR_x c$. Since the relation R_x is transitive, we obtain $aR_x c$ and thus aSc. Hence the relation S is also transitive.

Thus the relation S linearly orders the set $\bigcup F_x$.

In the following, by the *ordered union* of linearly ordered disjoint sets F_x we shall always understand the union $\bigcup_x F_x$ ordered by the relation S defined in Theorem 4. In this case the relation Q ordering the set of indices is assumed to be fixed.

We say also that $\bigcup_x F_x$ is the union of sets F_x *over the indexing set* T.

Let A be an arbitrary set, linearly ordered by the relation R. For $x, y \in A$ let $[x, y]$ denote the set of those z which equal either x or y, or which satisfy one of the conditions $x \prec z \prec y$ or $y \prec z \prec x$.

Clearly, $[x, y] = [y, x]$. We shall prove that for arbitrary $x, y, z \in A$

$$[x, y] \subset [x, z] \cup [z, y]. \tag{2}$$

In fact, if $t \in [x, y]$ and $t = x$ or $t = y$, then clearly $t \in [x, z] \cup [z, y]$. If $x \prec t \prec y$ and $t = z$ or $t \prec z$, then $t \in [x, z]$. On the other hand, if $z \prec t$, then $t \in [z, y]$. Similarly, if $y \prec t \prec x$, then $t \in [x, z] \cup [z, y]$.

Let V_x be the set of all y such that the set $[x, y]$ is scattered.

Clearly, $V_x \neq \emptyset$, because $x \in V_x$.

We shall prove that the set V_x is also scattered.

Suppose that on the contrary $C \subset V_x$ and that the set C is infinite and densely ordered.

For any $c_1, c_2 \in C$ such that $c_1 \prec c_2$ we have by (2)

$$[c_1, c_2] \subset [c_1, x] \cup [c_2, x].$$

Thus the set $[c_1, c_2]$ is contained in the union of two scattered sets. By Theorem 1 this union is scattered. But this is impossible, for between any two distinct elements of the set C there always lies at least one other element of C. Thus the assumption that V_x is not scattered leads to a contradiction.

The set V_x is an *interval in the set* A. In fact, suppose that $y, z \in V_x$ and $y \prec t \prec z$. If $x = t$ or $x \prec t$, then $[x, t] \subset [x, z]$. Thus, as a subset of the scattered set $[x, z]$, the set $[x, t]$ is also scattered. On the other hand, if $t \prec x$, then $[t, x] \subset [y, x]$. This means that the set $[t, x]$ is scattered as a subset of the scattered set $[y, x]$. It follows now that $t \in V_x$. Thus the set V_x is an interval.

If $x \neq y$, then either $V_x \cap V_y = \emptyset$ or $V_x = V_y$. In fact, if $z \in V_x \cap V_y$, then the set $[x, y]$, as a subset of the union $[x, z] \cup [y, z]$, is scattered. It follows that if $u \in V_x$, then the set $[u, y]$ is scattered because $[u, y] \subset [u, x] \cup [x, y]$. Similarly, from $u \in V_y$ it follows that $u \in V_x$. Thus $V_x = V_y$.

Let $\mathbf{A}$ be the family of all the sets V_x. This family consists of disjoint non-empty subsets of A and is linearly ordered by the relation ϱ which holds between V_x and V_y if and only if $V_x = V_y$ or V_x precedes V_y (see p. 204).

The set A is the ordered union

$$A = \bigcup_{P \in \mathbf{A}} P,$$

where the family $\mathbf{A}$ is linearly ordered by the relation ϱ and each interval P is linearly ordered by the relation R. In fact the union $\bigcup(\mathbf{A})$ is contained in A and every element x of the set A belongs to V_x; thus every x belongs to one component of the union $\bigcup(\mathbf{A})$.

We now prove that *the relation ϱ is a dense ordering of the family $\mathbf{A}$*. Suppose that $V_x \varrho V_y$ and $V_x \neq V_y$; that is, $V_x \cap V_y = \emptyset$. This implies that the interval $[x, y]$ is not scattered, that is, for some z lying between x and y one of the sets $[x, z]$ and $[z, y]$, say the first, contains an infinite densely ordered set M. If m, n, p are elements of the set M such that $x \prec m \prec n \prec p \prec z$, then the sets $[x, n]$ and $[n, y]$ contain infinite densely ordered subsets. This shows that $V_x \varrho V_n \varrho V_y$ and $V_x \neq V_n \neq V_y$.

It this way we obtain the following theorem.

THEOREM 5:[1]) *Every ordered set is the union of scattered sets over a densely ordered indexing set.*

Exercises

1. Give an example of an infinite set which has a first element, has no last element, and in which every element except the first has a direct predecessor. Moreover, this set should not be similar to the set of natural numbers.

[1]) Theorem 5 is due to Schönflies [1], p. 184. A detailed analysis of countable scattered sets is given in Erdös and Hajnal [1].

2. Show that if the set X is densely ordered and if sets X_1 and X_2 are continuously ordered and contain subsets dense in themselves and similar to X, then the sets X_1 and X_2 are similar.

3. Prove that the set $\mathfrak{G}$ of those relations $R \in \boldsymbol{P}(N\times N)$ which densely order their fields is a G_δ-set in the space $\boldsymbol{P}(N\times N)$.

4. Show that the union of scattered sets over a scattered indexing set is itself scattered.

5. Prove that if the sets F_x contain neither first nor last elements, and if they are infinite and densely ordered, then the union $\bigcup_{x\in T} F_x$ is densely ordered (for any ordered set T).

6. Prove that if a set T is infinite, densely ordered and $F_x \neq 0$ for each x, then the set $\bigcup_{x\in T} F_x$ contains a densely ordered subset.

7. Prove that if the union $\bigcup_{x\in T} F_x$ is continuous, then the set T has no gaps.

8. Prove that if a set T is continuously ordered and contains first and last elements and if the sets F_x are continuously ordered, then the union $\bigcup_{x\in T} F_x$ is continuously ordered. [Hausdorff]

9. Give an example of a linearly ordered set A which contains a densely ordered subset X which is not dense in A.

§ 3. Order types ω, η, and λ

We shall illustrate the notion of order type by means of examples.

Order type ω. The order type ω is the order type of the set N ordered by the relation $\leqslant$.

THEOREM 1: *A linearly ordered set A is of type ω if and only if*

(i) *A has a first element a_0,*

(ii) *every element x of the set A has a direct successor x^*,*

(iii) *if $a_0 \in X \subset A$ and if the set X contains the direct successor of every element of X, then $X = A$.*

PROOF. Conditions (i), (ii), (iii) are invariant under any transformation which preserves order. Since they are satisfied by the set of natural numbers ordered by the relation $\leqslant$, they are necessary conditions for a set A to be of type ω.

Suppose now that the set A linearly ordered by the relation R satisfies conditions (i), (ii), (iii). Let us define a function f which establishes the similarity between A and the set of natural numbers as follows:

$$(1) \qquad f(0) = a_0, \quad f(n+1) = [f(n)]^*.$$

These formulas define by induction the function f. It follows from (1) that the range of f contains a_0, and that it contains the direct successor of every of its element. Condition (iii) implies that the range of f coincides with A.

Let us prove that

$$(2) \qquad m < n \rightarrow f(m) \prec f(n).$$

It follows from (1) that formula (2) holds for $n = m+1$. If we suppose that (2) holds for some n, then (2) also holds for $n+1$. Indeed, if $f(m) \prec f(n)$, then $f(m) \prec [f(n)]^*$, because $f(n) \prec [f(n)]^*$.

It follows directly from (2) that

$$m \neq n \rightarrow f(m) \neq f(n), \qquad m \leqslant n \rightarrow f(m) R f(n).$$

The first of these formulas shows that the function f is one-to-one. Together these formulas show by Theorem 1.2 that the function f establishes a similarity between A and the set of natural numbers ordered by $\leqslant$.

It should be stressed that the formula expressing the condition (iii) above contains a variable X ranging over $\boldsymbol{P}(A)$. We say that such formulas are of *second order*. The use of second order formulas in the characterization of the order type ω is essential. We can easily show that elementary formulas, i.e. ones in which all quantifiers are limited to A, cannot characterize the order type ω. This results from the following two observations: (1) If a first order formula is valid in a relational system $\langle A, R\rangle$, it is also valid in every reduced product of N copies of $\langle A, R\rangle$ reduced modulo any prime ideal (see Corollary IV.9.5); (2) If $\mathfrak{A} = \langle A, R\rangle$ has the order type ω and $\mathfrak{A}_n = \mathfrak{A}$ for each $n \in N$, then the reduced product of the systems $\mathfrak{A}_n$ modulo a prime ideal $\boldsymbol{I} \subset \boldsymbol{P}(N)$ which contains all finite sets is not isomorphic to $\mathfrak{A}$.

Order type η. Before defining this type, we prove the following important theorem.

THEOREM 2: *Every two non-empty, denumerable, linearly ordered and dense sets which have neither first nor last elements are similar.*

PROOF. Let A and B be sets satisfying the assumptions of the theorem. To simplify notation, we shall use the same symbol to denote the relations ordering both sets.

It follows from the assumptions that the sets A and B are infinite. Thus there exist one-to-one sequences $a \in A^N$ and $b \in B^N$ such that $a^1(N) = A$ and $b^1(N) = B$.

Let us define by induction two permutations φ and ψ of the set N such that the mapping $f: a_{\varphi(n)} \to b_{\psi(n)}$ establishes a similarity between the sets A and B. For this purpose, let $\varphi(0) = \psi(0) = 0$. Now we consider two cases depending upon whether n is even or odd.

Case 1: *n even.*[1]) Let

$$\varphi(n+1) = \min_k \Big[\bigwedge_{j \leqslant n} (a_k \neq a_{\varphi(j)})\Big],$$

$$\psi(n+1) = \min_k \Big(\bigwedge_{j \leqslant n} \{(b_k \neq b_{\psi(j)}) \wedge [(b_{\psi(j)} \prec b_k) \equiv (a_{\varphi(j)} \prec a_{\varphi(n+1)})]\}\Big).$$

Case 2: *n odd.* The definition is similar, but the roles of φ and ψ are interchanged.

$$\psi(n+1) = \min_k \Big[\bigwedge_{j \leqslant n} (b_k \neq b_{\psi(j)})\Big].$$

$$\varphi(n+1) = \min_k \Big(\bigwedge_{j \leqslant n} \{(a_k \neq a_{\varphi(j)}) \wedge [(a_{\varphi(j)} \prec a_k) \equiv (b_{\psi(j)} \prec b_{\psi(n+1)})]\}\Big).$$

We shall prove by induction that if $n \in N$ and $j < n$, then

$$\varphi(n) \neq \varphi(j), \tag{3}$$

$$\psi(n) \neq \psi(j), \tag{4}$$

$$a_{\varphi(n)} \prec a_{\varphi(j)} \equiv b_{\psi(n)} \prec b_{\psi(j)}, \qquad a_{\varphi(n)} \succ a_{\varphi(j)} \equiv b_{\psi(n)} \succ b_{\psi(j)}. \tag{5}$$

It is clear that these formulas hold for $n = 0$. Suppose that $n_0 > 0$ and that (3)–(5) hold for $n < n_0$. Let $n_0 = n'+1$. The proof now splits into two cases according as n' is even or n' is odd. We shall consider only the first case.

Since the set A is infinite, there exist numbers k such that $a_k \neq a_{\varphi(j)}$ for $j \leqslant n'$. By definition, $\varphi(n'+1)$ is one of these numbers k. This proves (3), for $n = n'+1 = n_0$.

[1]) If Φ is a formula and there exists no n such that $\Phi(n)$, then the symbol $\min\limits_k \Phi(k)$ shall denote the number 0.

To prove the remaining formulas, let

$$P = \{j \leqslant n' : a_{\varphi(j)} \prec a_{\varphi(n_0)}\}, \qquad Q = \{j \leqslant n' : a_{\varphi(n_0)} \prec a_{\varphi(j)}\}.$$

Thus

$$(p \in P) \wedge (q \in Q) \rightarrow (a_{\varphi(p)} \prec a_{\varphi(q)}),$$

and since (5) holds by assumption for $n \leqslant n'$, we obtain

$$(p \in P) \wedge (q \in Q) \rightarrow (b_{\psi(p)} \prec b_{\psi(q)}).$$

Since the set B is densely ordered, the formula above shows that there exist numbers k such that $b_{\psi(p)} \prec b_k$ for every p in P and $b_k \prec b_{\psi(q)}$ for every q in Q. It follows from the definition of ψ that $\psi(n_0)$ is one of these numbers k. Thus $b_{\psi(n_0)} \neq b_{\psi(j)}$ for $j \in P \cup Q$. Moreover,

$$b_{\psi(n_0)} \prec b_{\psi(j)} \equiv (j \in Q) \equiv a_{\varphi(n_0)} \prec a_{\varphi(j)}$$

and similarly for $\succ$ $(j \in P \cup Q)$. In this way formulas (4) and (5) are proved.

Formulas (3)–(5) show that the function $f: a_{\varphi(n)} \rightarrow b_{\psi(n)}$ establishes similarity between the sets $\{a_{\varphi(n)}: n \in N\}$ and $\{b_{\psi(n)}: n \in N\}$. It remains to be shown that these sets are identical with A and B respectively. In other words, we have to show that every natural number occurs in the sequences φ and ψ. We consider only the sequence φ.

Suppose on the contrary that $N - \varphi^1(N) \neq \emptyset$ and let k_0 be the least number in this set. Clearly, $k_0 > 0$. For $h < k_0$, let n_h denote the unique number such that $\varphi(n_h) = h$ and let n be an even number greater than all numbers n_h, $h < k_0$. Since $a_{k_0} \neq a_{\varphi(j)}$ for all $j \leqslant n$, and for every $h < k_0$ there exists $j \leqslant n$ such that $a_n = a_{\varphi(j)}$, namely $j = n_h$, we obtain

$$k_0 = \min_k \bigwedge_{j \leqslant n} (a_k \neq a_{\varphi(j)}).$$

This implies $k_0 = \varphi(n+1)$, which contradicts $k_0 \notin \varphi^1(N)$.

Theorem 2 is proved.

Theorem 2 shows that there exists only one type of set which is simultaneously $\neq \emptyset$, densely ordered, countable and without first and last elements. This type is denoted by η.

An example of an ordered set of type η is the set of rational numbers ordered by the relation $\leqslant$. Another example is given on p. 201 (Example 3).

Sets of type η have the following property of universality:

THEOREM 3: *If* $\overline{B} = \eta$ *and* A *is any denumerable linearly ordered set, then there exists a set* $C \subset B$ *such that* A *and* C *are similar.*

PROOF. We may assume that the set A is infinite. Using the notions introduced in the proof of Theorem 2, we define the sequences φ and ψ as in Case 1. However, this time we do not confine n to even numbers only, but we let n be any natural number. Then formulas (3)–(5) are satisfied. We can prove in the same way as in Theorem 2 that $\varphi^1(N) = N$ and that the sets $\{a_{\varphi(n)}: n \in N\}$ and $\{b_{\psi(n)}: n \in N\}$ are similar. The first of these sets is equal to A and the second is contained in B. This proves the theorem.

The order type λ. We precede the definition by a theorem.

THEOREM 4: *Let* A *and* B *be sets satisfying the following conditions:*

(i) A *and* B *are linearly and continuously ordered.*

(ii) *There exists subsets* $A_1 \subset A$ *and* $B_1 \subset B$ *dense in* A *and* B *respectively which are both coinitial and cofinal with* A *and* B *respectively.*

(iii) *The sets* A_1 *and* B_1 *are of type* η.

Then A *and* B *are similar.*

We shall only outline the proof of this theorem. By Theorem 2 there exists a function f_1 which maps A_1 onto B_1 and preserves order. It can easily be shown that the sets $X(a) = A_1 \cap \{a\}^-$ and $Y(a) = A_1 \cap \{a\}^+$ determine a proper cut in the set A_1. Hence the pair $\langle P, Q\rangle = \langle f_1^1(X(a)), f_1^1(Y(a))\rangle$ is a proper cut in the set B_1.

Let

$$\tilde{X}(a) = \Big\{b \in B: \bigwedge_{y \in Q} (b \leqslant y)\Big\},$$

$$\tilde{Y}(a) = \Big\{b \in B: \bigwedge_{x \in P} (x \leqslant b)\Big\}.$$

It can easily be shown that the pair $\langle \tilde{X}(a), \tilde{Y}(a)\rangle$ is a proper cut in the set B. Since B is continuous, there exists an element $f(a)$ lying in this cut: it is the last element of the set $\tilde{X}(a)$ and simultaneously the first element of the set $\tilde{Y}(a)$.

The mapping f satisfies the condition $a' \leqslant a'' \to f(a') \leqslant f(a'')$. In fact, if $a' \leqslant a''$, then $X(a') \subset X(a'')$. Thus $f_1^1(X(a')) \subset f_1^1(X(a''))$, which proves that $\tilde{Y}(a') \supset \tilde{Y}(a'')$. Hence $f(a') \leqslant f(a'')$.

It remains to be shown that the function f is one-to-one and that it maps the set A onto the set B. For this purpose, we repeat the previous construction interchanging the roles of the sets A and B and obtain a function g mapping B into A. One can show that $f(g(b)) = b$ for every $b \in B$, which proves the theorem.[1])

Theorem 4 enables us to admit the following definition.

A linearly ordered set A is of *type* λ if it is continuous, contains a subset A_1 of type η which is dense in A, and is coinitial and cofinal with A.

An example of a set of order type λ is provided by

THEOREM 5: *If A is a set of order type η then the set X obtained from the minimal extension of A by removing the first and the last elements has the order type λ.*

The continuity results from Theorem 2.2. The set A' consisting of all cuts $\langle \{a\}^-, \{a\}^+ \rangle$ where a ranges over A is dense in X. Finally if $\langle P, Q \rangle$ is a proper cut of A then for any x in P not lying in the cut $\langle P, Q \rangle$ the element $\langle \{x\}^-, \{x\}^+ \rangle$ precedes $\langle P, Q \rangle$ in X and so A' is coinitial with X; we can show similarly that A' is cofinal with X and so X has the order type λ.

Taking for A the set of all rational numbers, we obtain that the set $\mathscr{E}$ of real numbers is ordered by the relation $\leqslant$ in the type λ.

As an interesting application of Theorem 3 we prove

THEOREM 6: *Each linearly ordered set A which possesses an infinite countable subset X dense in A is similar to a set of real numbers ordered by the relation $\leqslant$.*

To prove this theorem we use Theorem 3 and infer that X can be

[1]) Generalizations of Theorems 2 and 3 for sets of higher cardinalities are discussed below in Chapter IX, Section 2.

Theorems 2 and 3 which were proved in Section 9 of Cantor's 1895 paper [5] assert that the order type η is in a sense universal for all denumerable linear orderings. In spite of their simplicity these theorems had a profound impact on further development of set theory and model theory; see Chang–Keisler [1], Chapter V. The method used in the proofs of these theorems is often called the "back-and-forth" method.

Theorem 4 is also due to Cantor, l.c., Section 11. Cantor used the letter "θ" for the order type of real numbers but later the letter "λ" became universally accepted.

mapped into the set Q of rational numbers < 1 so that $x \prec x'$ implies $(x) < f(x')$ for arbitrary x, x' in X. Putting $F(a) = \sup\{f(x): (x \in X) \wedge \wedge (x \prec a \vee x = a)\}$ for each a in A, we obtain a mapping of A into real numbers and verify immediately that $a \prec b$ implies $F(a) < F(b)$ for each a, b in A. This proves that A is similar to the set $F^1(A) \subset \mathscr{E}$, Q.E.D.

We also add a remark concerning formulas expressing the characteristic properties of the order type λ. One of these formulas (expressing the continuity of A) is not elementary because it contains a variable ranging over arbitrary subsets of A. Similarly as in the case of the order type ω one can show that it is not possible to characterize the order type λ by elementary sentences alone. This is an immediate corollary of a basic theorem of logic called the Skolem–Löwenheim theorem which says that for every uncountable relational system $\langle A, R\rangle$ there is a countable subsystem $\langle A', R'\rangle$ such that exactly the same elementary sentences are valid in $\langle A, R\rangle$ as in $\langle A', R'\rangle$.

Also properties characterizing the order type η are not elementary because the denumerability of a set is not expressible by elementary sentences. Modifying slightly the proof given for the case of the order type ω, we can prove that no set of elementary sentences characterizes the order type η.

Sets of types ω, η, and λ which we discussed in this section are either denumerable or have the power $\mathfrak{c}$. In the next chapter we shall give other examples of order types and of ordered sets whose cardinalities are different from $\mathfrak{a}$ and $\mathfrak{c}$. In this connection it is worthwhile to mention the following metamathematical fact: In the system Σ[TR] which, as we remember, does not contain the axiom of choice it is impossible to prove that for every cardinal number there exists a linearly ordered set whose power is this cardinal number. Thus for instance the theorem: *There exists a relation which linearly orders the set* $\boldsymbol{P}(\boldsymbol{P}(N))$ cannot be proved in the system Σ[TR].

Exercises

1. Show that every dense and infinite set contains a subset of type η.

2. Show that every infinite continuous set is of power $\geqslant \mathfrak{c}$.

3. Let $r_0, r_1, \ldots$ be an infinite sequence without repetitions consisting of all rational numbers. For $c = \sum_{j=0}^{\infty} c_j/3^j$ where $c_j = 0$ or $c_j = 2$, let $M_c = \{r_j : c_j = 2\}$ and let $\bar{c}$ denote the type of the set M_c ordered by the relation $\leqslant$. Prove that every denumerable order type can be represented in the form $\bar{c}$.[1])

4. Let $C_\tau = \{c : \bar{c} = \tau\}$. Show that the Cantor set C is the union $\bigcup_\tau C_\tau$ where the union is over all denumerable order types and where the components of this union are pairwise disjoint.

5. Show that the set C_η is a G_δ-set in the space C.

6. Let $\boldsymbol{M}$ be a monotonic family of open subsets of the real line (or generally of the space $\mathscr{E}^n$). Prove that this family is similar to a set of real numbers (ordered by $\leqslant$).

Hint: Let $P_1, P_2, \ldots$ be a sequence of all intervals with rational endpoints. Assume that each interval occurs in this sequence infinitely many times. For a given set $G \in \boldsymbol{M}$, let $k_1, k_2, \ldots$ be a sequence of all natural numbers such that $P_{k_n} \subset G$. The function

$$t(G) = \sum_{n=1}^{\infty} \frac{1}{2^{k_n}}, \qquad t(0) = 0$$

establishes the required similarity.

7. Prove that any two denumerable Boolean algebras without atoms are isomorphic. Each at most denumerable Boolean algebra can be isomorphically embedded in a denumerable Boolean algebra without atoms.

Hint: Use the "back-and-forth" argument.

§ 4. Arithmetic of order types

We can define operations on order types which are similar to certain operations in ordinary arithmetic just as we did in the case of cardinal numbers. This arithmetic of order types allows us to simplify arguments concerning linearly ordered sets.

Inverse types. It is easy to show that *if a relation R linearly orders the set A, then so does the inverse relation* R^{i} (see p. 64). Of course, *the isomorphism of R and S implies the isomorphism of the inverse relations* R^{i} *and* S^{i}.

The order type of the set A ordered by R^{i} is said to be the *inverse* of the order type of the set A ordered by R. If the order type of A

[1]) In connection with Exercises 3, 4, and 5 see Kuratowski [12]; see also Scott [2].

ordered by R is α, then the order type of A ordered by R^1 is denoted by α^*.

It follows from the equivalence $x(R^1)^1 y \equiv xRy$ that

$$\alpha^{**} = \alpha. \tag{1}$$

Examples

1. If n is a finite order type then $n^* = n$, because every two finite equipollent sets are similar.

2. Likewise $\eta^* = \eta$ and $\lambda^* = \lambda$. On the other hand, $\omega^* \neq \omega$ because a set of type ω^* (for example, the set of negative integers) possesses a last element whereas a set of type ω has no last element.

The sum of order types. Let α and β be two order types and let A and B be two sets linearly ordered by R and S such that $\overline{A} = \alpha$, $\overline{B} = \beta$. We assume that $A \cap B = \emptyset$. This assumption can always be satisfied, for if A and B are not disjoint, then we can replace them by the sets $A \times \{1\}$ and $B \times \{2\}$ which are similar to A and B and disjoint.

The *sum* $\alpha+\beta$ is defined by

$$\alpha+\beta = \overline{A \cup B}$$

where the set $A \cup B$ is ordered as follows: all elements of A precede all elements of B and the order in each of the sets A and B is preserved.

In particular, if α and β are finite order types, the definition of the sum $\alpha+\beta$ coincides with the definition of the sum of natural numbers.

It is easy to see that the sum $\alpha+\beta$ does not depend on the sets A and B but only upon their order types. Moreover, the following formulas clearly hold:

$$(\alpha+\beta)+\gamma = \alpha+(\beta+\gamma), \qquad \alpha+0 = \alpha = 0+\alpha.$$

On the other hand, the commutative law does not hold: for example, $\omega+1 \neq 1+\omega$. In fact, $1+\omega = \omega$ (thus $1+\omega$ is equal to the type of the set of natural numbers), whereas $\omega+1$ is the type of a set with a last element.

Product of order types. Let $\alpha = \overline{A}$, $\beta = \overline{B}$. The *product* $\alpha \cdot \beta$ of the order types α and β is defined by the formula

$$\alpha \cdot \beta = \overline{A \times B},$$

where the set $A \times B$ is ordered as follows. Let $\langle x, y\rangle$ and $\langle x_1, y_1\rangle$ be two elements of $A \times B$. Then $\langle x, y\rangle \prec \langle x_1, y_1\rangle$ if $y \prec y_1$. If $y = y_1$, then $\langle x, y\rangle \prec \langle x_1, y_1\rangle$ if $x \prec x_1$.

For example $\lambda \cdot \lambda$ or λ^2 is the order type of the set of points in the plane ordered as above.

It is easy to check that, just as for the sum, the product $\alpha \cdot \beta$ depends only upon α and β.

For finite order types, the definition given above coincides with that of multiplication of natural numbers. Moreover, we have for arbitrary order types α, β, γ the formulas:

$$(\alpha\beta)\gamma = \alpha(\beta\gamma), \qquad \alpha 1 = 1\alpha = \alpha, \qquad \alpha 0 = 0\alpha = 0.$$

Similarly as for addition, multiplication is not commutative. For example, $\omega 2 \neq 2\omega$. In fact,

$$2\omega = \overline{\{1, 2\} \times N} = \omega, \qquad \omega 2 = \overline{N \times \{1, 2\}} = \omega + \omega.$$

The distributive law is satisfied only in the form

$$\alpha(\beta+\gamma) = \alpha\beta+\alpha\gamma.$$

In fact, let

$$\alpha = \overline{A}, \qquad \beta = \overline{B}, \qquad \gamma = \overline{C}, \qquad B \cap C = \emptyset.$$

Then we have (see p. 62)

$$\begin{aligned}\alpha(\beta+\gamma) &= \overline{A \times (B \cup C)} = \overline{A \times B \cup A \times C} \\ &= \overline{A \times B} \cup \overline{A \times C} = \alpha\beta+\alpha\gamma,\end{aligned}$$

because

$$(A \times B) \cap (A \times C) = \emptyset.$$

Exponentiation of order types in the case of a finite exponent can be defined by induction:

$$\alpha^0 = 1, \qquad \alpha^{n+1} = \alpha^n \cdot \alpha.$$

No obvious counterpart of the arithmetical "less than" relation exists for order types. The most natural is the following relation of embeddability: We say that an order type α is *embeddable in the order type* β if any set of type α is similar to a subset of any set of order type β.

The relation of embeddability is transistive and reflexive but not symmetric: for instance η is embeddable in $\eta+1$ and conversely, although $\eta \neq \eta+1$. It is easy to give examples of types which are incomparable with respect to the relation of embeddability (e.g. $\omega+\omega$ and $\omega^*+\omega$).[1])

Exercises

1. Prove that $\eta+\eta = \eta$, $\lambda+1+\lambda = \lambda$, $\lambda+\lambda \neq \lambda$.

2. Using operations on the order type ω, give an example of an infinite linearly ordered set which possesses first and last elements such that every element except the first has a direct predecessor and every element except the last has a direct successor.

3. Prove that $(\alpha+\beta)^* = \beta^*+\alpha^*$.

4. Prove that $\eta^2 = \eta$.

5. Prove that $(\omega\eta)^2 = (\omega\eta+\omega)^2$, but $\omega\eta \neq \omega\eta+\omega$. [A. Davis–W. Sierpiński]

6. Prove that ω^2 is the type of the set of natural numbers ordered by the following relation: m precedes n if either m has fewer prime factors than n, or m has the same number of prime factors as n and $m \leqslant n$.

7. Prove that a set A of type λ^2 does not contain a subset which is denumerable and dense in A.

§ 5. Lexicographical ordering

The product of order types is related to lexicographical ordering. In order to define this ordering, let us assume that T is a set linearly ordered by the relation Q and let each $x \in T$ be associated with a set F_x ordered by the relation R_x. We do not assume that the sets F_x are disjoint. Let

$$P = \prod_{x \in T} F_x .$$

Thus P is the set of functions f whose domain is T such that $f(x) \in F_x$ for all $x \in T$.

Two arbitrary functions f and g belonging to P determine the set

$$D(f, g) = \{x \in T\colon f(x) \neq g(x)\} .$$

Clearly, $D(f, g) = \emptyset$ if and only if $f = g$.

We now define a relation S in P in the following way: fSg holds if

[1]) For deeper results concerning the relation of embeddability in the cases of special classes of order types compare Laver [1] and [2].

and only if either $f = g$ or the set $D(f, g)$ possesses a first element x_0 and $f(x_0)\,R_{x_0}\,g(x_0)$.

This definition can be written in symbols as follows

$$fSg \equiv (f = g) \vee \bigvee_x \Big\{ \big(f(x) <_{R_x} g(x)\big) \wedge \bigwedge_y \big[(y <_Q x) \rightarrow (f(y) = g(y))\big]\Big\}.$$

If the relation S linearly orders the cartesian product P, then this product is said to be *lexicographically ordered* (or *ordered according to the principle of first differences.*[1])

We shall investigate the conditions under which the relation S linearly orders P.

THEOREM 1: *The relation S is reflexive, antisymmetric and transitive in P.*

PROOF. The reflexivity of S is obvious.

Suppose that both fSg and gSf. If the functions f and g are distinct, then the set $D(f, g)$ has a first element x and this element satisfies the conditions $f(x)\,R_x\,g(x)$ and $g(x)\,R_x\,f(x)$. Since, by assumption, the relation R_x orders F_x, we have $f(x) = g(x)$, which is incompatible with $x \in D(f, g)$. Thus fSg and gSf imply $f = g$. This shows that the relation S is antisymmetric.

Suppose that fSg and gSh. If $f = g$ or $g = h$ then clearly fSh. Thus we may assume that $f \neq g$ and $g \neq h$. Hence the sets $D(f, g)$ and $D(g, h)$ have first elements x and y respectively and these elements satisfy the conditions:

$$f(x) <_{R_x} g(x), \qquad g(y) <_{R_y} h(y).$$

If z precedes x and y, then $f(z) = g(z) = h(z)$. On the other hand, if z_0 is that one of the elements x and y which precedes the other, then we have either $f(z_0) <_{R_{z_0}} g(z_0)$ and $g(z_0) = h(z_0)$ (if $x <_Q y$), or $f(z_0) = g(z_0)$ and $g(z_0) <_{R_{z_0}} h(z_0)$ (if $y <_Q x$), or finally $f(z_0) <_{R_{z_0}} h(z_0)$ and $g(z_0) <_{R_{z_0}} h(z_0)$ (if $x = y$). In any case $f(z_0) <_{R_{z_0}} h(z_0)$, which shows that z_0 is the first element of $D(f, h)$. Hence fSh and it follows that S is transitive.

REMARK: The following example shows that the relation S need not be connected.

[1]) The lexicographical and anti-lexicographical orderings were first defined by Hausdorff. See his book [1].

Let T be a set of type ω^* (for instance, the set of negative integers), $F_x = \{0, 1\}$ and let R_x be the relation $\leqslant$. Let f be the function whose value is 0 for even numbers and 1 for odd numbers and let $g(x) = 1-f(x)$. Then the set $D(f, g)$ is equal to T and therefore it has no first element. Thus neither fSg nor gSf.

THEOREM 2: *If $\overline{T} = n$ or $\overline{T} = \omega$, then the relation S linearly orders the set P.*[1])

For the proof, it suffices to show that S is connected in P. For this purpose assume that $f, g \in P$ and $f \neq g$. The set $D(f, g)$ is non-empty and therefore it has a first element x. Since R_x is connected in F_x, we infer that $f(x) R_x g(x)$ or $g(x) R_x f(x)$. This implies that fSg or gSf.

THEOREM 3: *If the sets $A_1, \ldots, A_n$ are of types $\alpha_1, \ldots, \alpha_n$, then the set $A_1 \times A_2 \times \ldots \times A_n$ lexicographically ordered is of type $\alpha_n \alpha_{n-1} \ldots \alpha_1$.*

PROOF. The proof is by induction on n. For $n = 1$ the theorem is obvious. Suppose that it holds for n and consider the product $P = A_1 \times \times A_2 \times \ldots \times A_{n+1}$ of $(n+1)$ sets, with lexicographical ordering. Let $B = A_2 \times A_3 \times \ldots \times A_{n+1}$. Ordering the set $A_1 \times B$ lexicographically, we obtain a set similar to P. Hence it suffices to show that $A_1 \times B$ is of type $\alpha_{n+1} \cdot \alpha_n \cdot \ldots \cdot \alpha_2 \cdot \alpha_1$. But this follows directly from the definition of the product of types.

The definition of anti-lexicographical ordering (ordering by the principle of last differences) is similar to that of lexicographical ordering. The notion of anti-lexicographical ordering rather than that of the lexicographical ordering lies at the basis of the notion of the product of types.

Examples

1. The product $\lambda\lambda$ is the type of the set of complex numbers ordered lexicographically (where the complex number $x+iy$ is identified with the ordered pair $\langle x, y\rangle$).

2. The product $\eta\lambda$ is the type of the set of complex numbers of the form $r+iy$, where r is a rational number and y is a real number, ordered anti-lexicographically. On the other hand, the product $\lambda\eta$ is the

[1]) In general, Theorem 2 holds when T is an arbitrary well-ordered set. See Chapter VII.

type of the same set ordered lexicographically. These types are distinct, for a set of type $\lambda\eta$ contains continuous intervals whereas a set of type $\eta\lambda$ does not.

3. Let T be the set of natural numbers with the usual ordering, let $F_x = \{0, 1\}$ for $x \in T$, and let R_x be the relation $\leqslant$. The lexicographical ordering S is isomorphic in this case to the relation $\leqslant$ in the Cantor set C (understood as the set of real numbers of the form $\sum_{n=1}^{\infty} c_n/3^n$ where $c_n = 0$ or $c_n = 2$ for $n \in N$). In fact, associating with each function $f \in \prod_x F_x$ the number $c_f = \sum_{n=0}^{\infty} 2f(n)/3^{n+1}$, we see that $c_f < c_g$ if and only if $f \neq g$ and, moreover, the smallest number n_0 such that $f(n_0) \neq g(n_0)$ satisfies the inequality $f(n_0) < g(n_0)$.

4. Again let $T = N$ (where the order relation is $\leqslant$) and let $F_n = N$ for $n \in N$. With each function $f \in N^N$ we associate the real number

$$r_f = \sum_{n=0}^{\infty} 2^{-(f(0)+f(1)+\ldots+f(n)+n+1)}.$$

Clearly, $0 < r_f \leqslant 1$ and each real number x, $0 < x \leqslant 1$, can be represented in this form in exactly one way. In fact, if $x = \sum_{n=0}^{\infty} 2^{-\varphi(n)}$ is the binary representation of x with infinitely many digits different from 0, then the sequence φ is strictly increasing and $\varphi(0) > 0$. Assuming $f(0) = \varphi(0)-1$ and $f(n) = \varphi(n)-\varphi(n-1)-1$ for $n > 0$, we obtain $x = r_f$.

In order that $r_f < r_g$ it is necessary and sufficient that $f \neq g$ and that the smallest number n_0 such that $f(n_0) \neq g(n_0)$ satisfies the inequality $f(n_0) > g(n_0)$.

In this case the relation S of lexicographical ordering is similar to the relation $\geqslant$ in the set of numbers x, $0 < x \leqslant 1$. Hence the type of this relation is $1+\lambda$.

CHAPTER VII

WELL-ORDERED SETS

§ 1. Definitions. Principle of transfinite induction

We say that a relation R *well orders* a set X if R linearly orders X and every non-empty subset of X contains a first element (with respect to the relation R).[1])

Examples

1. Every set of type ω is well ordered.

2. The set consisting of the number 1 and of all numbers of the form $1-1/n$, $n = 1, 2, \ldots$ is well ordered by the relation $\leqslant$. The type of this set is $\omega+1$.

3. Let $\alpha(n)$ be the number of distinct prime factors of the number n. The relation

$$\{\langle x, y\rangle\colon\ [\alpha(x) < \alpha(y)] \vee ([\alpha(x) = \alpha(y)] \wedge [x \leqslant y])\}$$

well orders the set of natural numbers. The type of this set is ω^2 (see Exercise VI.4.6).

4. The ordered union (see p. 208) $S = \bigcup_{x\in T} F_x$, where the set T and the component sets F_x are well ordered, is also well ordered.

For, let Y be a non-empty subset of S. The set of all x such that $Y \cap F_x \neq \emptyset$ is a non-empty subset of T, therefore it contains a first element x_0. Thus the intersection $Y \cap F_{x_0}$ is a non-empty subset of the well-ordered set F_{x_0} and therefore it has a first element. This element is clearly the first element of Y.

[1]) The notion of a well-ordered set, one of the basic notions of set theory, is due to Cantor [3]. It is interesting that the logically simpler notion of a linearly ordered set was introduced by Cantor much later (in 1895) evidently in the course of systematizing his results in the theory of well-ordered sets.

5. The cartesian product of any finite number of well-ordered sets is itself well ordered by the relation of lexicographical ordering.

6. Every subset of a well-ordered set is also well ordered.

The following two theorems are simple consequences of the definition.

THEOREM 1: *In every well-ordered set there exists a first element. Every element except the last element (if such exists) has a direct successor.*

THEOREM 2′: *No subset of a well-ordered set is of type* ω^*.

°THEOREM 2″: *If a linearly ordered set* A *is not well ordered, then it contains a subset of type* ω^*.

PROOF. Let P be a non-empty subset of A which contains no first element. Let

$$Q(x) = \{y\colon (y \prec x) \wedge (y \in P)\}.$$

We have $Q(x) \neq \emptyset$ for every $x \in P$. By Theorem II.6.8 there exists a function f defined for every $x \in P$ such that $f(x) \in Q(x)$.

Let p_0 be any element of P. We define by induction a sequence p_1, $p_2, \ldots, p_n, \ldots$ letting $p_n = f(p_{n-1})$ for $n > 0$.

Since $f(x) \prec x$, this sequence is of type ω^*.

Theorems 2′ and 2″ imply the following.

°THEOREM 2: *In order that a linearly ordered set be well ordered it is necessary and sufficient that it contain no subset of type* ω^*.

THEOREM 3: *Each initial segment of a well-ordered set* A *is of the form* $O(x)$ *for some* $x \in A$.

In fact, if x is the first element of the difference $A-X$, then $O(x) = X$ (see p. 204).

THEOREM 4: (PRINCIPLE OF TRANSFINITE INDUCTION[1])) *If a set* A *is well ordered,* $B \subset A$ *and if for every* $x \in A$ *the set* B *satisfies the condition*

$$[O(x) \subset B] \rightarrow (x \in B), \tag{1}$$

then $B = A$.

PROOF. Suppose that $A-B \neq \emptyset$. Then there exists a first element x in $A-B$. This means that if $y \prec x$ then $y \notin A-B$, that is, $y \in B$. This

[1]) The principle of transfinite induction was implicit in Cantor [6], pp. 336–339. A first explicit formulation is due to Hessenberg [1], p. 53.

shows that $O(x) \subset B$. Now it follows from (1) that $x \in B$, which contradicts the hypothesis that $x \notin B$.

Theorem 4 can be reformulated as follows. Let a subset B of a well-ordered set A be called *hereditary* if it satisfies condition (1). Then Theorem 4 asserts that the only hereditary subset of A is the set A itself.

For many formulas Φ it can be proved that the set $\{x \in A: \Phi(x)\}$ is hereditary; consequently, for such formulas the theorem $\bigwedge_{x \in A} \Phi(x)$ holds. This method of proving theorems of the form $\bigwedge_{x \in A} \Phi(x)$ is called the *method of transfinite induction.* Of course, this method and the method of proof by induction in ordinary arithmetic (which consists in showing that the set $\{n \in N: \Phi(n)\}$ is inductive) are analogous.

We shall use transfinite induction to prove several theorems about the similarity of well-ordered sets.

Let A be a set linearly ordered by the relation R. A function f which establishes similarity between A and the set $f^1(A)$ contained in A, is said to be an *increasing function.* Such functions satisfy the condition

$$x \prec y \to f(x) \prec f(y). \tag{2}$$

THEOREM 5: *If a function f defined on a well-ordered set A is increasing, then for every x we have $xRf(x)$ (that is, $x \prec f(x)$ or $x = f(x)$).*

PROOF. Let $B = \{x: xRf(x)\}$. Let $O(x) \subset B$. We show that $x \in B$. In fact, let $y \in O(x)$, that is $y \prec x$. By (2) it follows that $f(y) \prec f(x)$. Since $y \in B$, we have $yRf(y)$ and thus $y \prec f(x)$. This shows that the element $f(x)$ occurs after every element y of $O(x)$, that is, $f(x) \in A - O(x)$. Since x is the first element of $A - O(x)$, we have $xRf(x)$ and finally $x \in B$. Hence the set B is hereditary. Q.E.D.

COROLLARY 6: *If the well-ordered sets A and B are similar, then there exists only one function which establishes their similarity.*

PROOF. Suppose that the sets A and B are well ordered by R and S and that there exist two functions f and g establishing similarity between A and B.

The function $g^1 \circ f$ is clearly increasing in A (see p. 203). By Theorem 5 we thus have $xRg^1(f(x))$ for every $x \in A$. Hence $g(x) S f(x)$. Considering $f^c O g$ instead of $g^c O f$ we have by the same argument $f(x) S g(x)$. This implies $f(x) = g(x)$, because S is antisymmetric.

COROLLARY 7: *No well-ordered set is similar to any of its initial segments.*

In fact, if the sets A and $O(x)$ were similar, then the function f establishing similarity would be increasing and would satisfy $f(x) \in O(x)$, that is, $f(x) \prec x$. But this contradicts Theorem 5.

COROLLARY 8: *No two distinct initial segments of a well-ordered set are similar.*

For the proof, it suffices to apply Corollary 7 and to observe that, given two distinct initial segments, one is always an initial segment of the other.

THEOREM 9:[1]) *Let A and B be two well-ordered sets. Then either*

(i) *A and B are similar, or*

(ii) *the set A is similar to a segment of B, or*

(iii) *the set B is similar to a segment of A.*

PROOF. Let R and S be relations which well order A and B respectively and let

$$Z = \left\{x \in A: \bigvee_{y \in B} \overline{O_R(x)} = \overline{O_S(y)}\right\}.$$

In other words (see notation on p. 204):

$x \in Z \equiv$ {the initial segment $O_R(x)$ of A is similar to some initial segment $O_S(y)$ of B}.

By virtue of Corollary 8, for given $x \in Z$ there exists only one such segment. Thus there exists a function f defined on Z such that this segment is of the form $O_S(f(x))$.

First we show that either $Z = A$ or Z is a segment of A, that is, there exists an a in A such that $Z = O_R(a)$. In fact, let $x \prec x' \in Z$. Since $O_R(x)$ is a segment of $O_R(x')$, the function establishing similarity between $O_R(x')$ and $O_S(f(x'))$ also maps $O_R(x)$ onto a segment of B. Hence $x \in Z$.

Similarly: either $f^1(Z) = B$ or else $f^1(Z)$ is a segment of B: $f^1(Z) = O_S(b)$. To show this it suffices to observe that

[1]) Theorem 9 is due to Cantor [6], p. 216.

$$f^1(Z) = \{y \in B: \bigvee_{x \in Z} [y = f(x)]\} = \{y \in B: \bigvee_{x \in Z} \overline{O_S(y)} = \overline{O_R(x)}\}.$$

In fact, if $O_S(y)$ is similar to a segment of A then y is of the form $f(x)$ where $x \in Z$.

Finally, observe that f establishes the similarity between Z and $f^1(Z)$. Indeed, we have just shown that $x \prec x' \in Z$ implies that $O_S(f(x))$ is a segment of $O_S(f(x'))$, therefore $f(x) \prec f(x')$.

A priori we have one of the following four possibilities:

(i) $Z = A$ and $f^1(Z) = B$,

(ii) $Z = A$ and $f^1(Z) = O_S(b)$,

(iii) $Z = O_R(a)$ and $f^1(Z) = B$,

(iv) $Z = O_R(a)$ and $f^1(Z) = O_S(b)$.

The first three possibilities correspond to those stated in the theorem. Case (iv) is impossible, because then $\overline{O_R(a)} = \overline{O_S(b)}$ and thus, by the definition of Z, $a \in Z$; that is, $a \in O_R(a)$, which contradicts the definition of a segment. Q.E.D.

COROLLARY 10: *If A and B are well ordered, then either $\overline{\overline{A}} \leqslant \overline{\overline{B}}$ or $\overline{\overline{B}} \leqslant \overline{\overline{A}}$. That is, powers of well-ordered sets obey the law of trichotomy.*

§ 2. Ordinal numbers

By *ordinal numbers* (or *ordinals*) we shall understand the order types of well-ordered sets. Theorem 1.9 allows us to define a "less than" relation for ordinals.

DEFINITION 1: We say that an ordinal α is *less* than an ordinal β if any set of type α is similar to a segment of a set of type β. We denote this relation by $\alpha < \beta$ or $\beta > \alpha$.

We write "$\alpha \leqslant \beta$" instead of "$\alpha < \beta$ or $\alpha = \beta$."

THEOREM 1: *For any ordinals α and β one and only one of the formulas $\alpha < \beta$, $\alpha = \beta$, $\alpha > \beta$ holds.*

This theorem is a direct consequence of Theorem 1.9.

THEOREM 2: *If α, β and γ are ordinals and if $a < \beta$ and $\beta < \gamma$, then $\alpha < \gamma$.*

PROOF. Let A, B and C be sets of types α, β and γ, respectively. By assumption, the set A is similar to a segment of the set B and B is similar to a segment of C. Thus A is similar to a segment of C. Q.E.D.

The following formulas can be proved without difficulty:

$$(\alpha \leqslant \beta) \wedge (\beta \leqslant \alpha) \rightarrow (\alpha = \beta), \qquad (\alpha \leqslant \beta) \wedge (\beta \leqslant \gamma) \rightarrow (\alpha \leqslant \gamma).$$

THEOREM 3: *If the well-ordered sets A and B are of types α and β and if the set A is similar to a subset B_1 of the set B, then $\alpha \leqslant \beta$.*

PROOF. If this were not so, then we would have $\beta < \alpha$ and then B would be similar to a segment of B_1. This contradicts Theorem 5.

We now examine sets of ordinal numbers.

THEOREM 4: *The set $W(\alpha)$ consisting of all ordinals less than α is well ordered by the relation $\leqslant$. Moreover, the type of $W(\alpha)$ is α.*

PROOF. Let A be a well-ordered set of type α. Associating the type of the segment $O(a)$ with the element $a \in A$ we infer (by the axiom of replacement) that the set $W(\alpha)$ exists and simultaneously we obtain a one-to-one mapping of A onto $W(\alpha)$. It is easily seen that following conditions are equivalent:

(i) a_1 precedes a_2 or $a_1 = a_2$,
(ii) $O(a_1)$ is a segment of $O(a_2)$ or $O(a_1) = O(a_2)$,
(iii) the type of $O(a_1)$ is not greater than the type of $O(a_2)$.

This shows that the relation $\leqslant$ indeed orders $W(\alpha)$ in type α.

THEOREM 5: *Every set of ordinals is well ordered by the relation $\leqslant$. In other words, in any non-empty set Z of ordinals there exists a smallest ordinal.*

PROOF. Let $\alpha \in Z$. If α is not the smallest ordinal of Z, then $Z \cap W(\alpha) \neq \emptyset$. Then in the set $Z \cap W(\alpha)$ there exists a smallest number β, as the set $W(\alpha)$ is well ordered (see Theorem 4). At the same time β is the smallest ordinal in Z. In fact, if $\xi \in Z - W(\alpha)$ then $\xi \geqslant \alpha$; thus $\xi > \beta$.

THEOREM 6: *For every set Z of ordinals there exists an ordinal greater than all ordinals belonging to Z.*

PROOF. By the axiom of replacement there exists a set K whose elements are all the sets $W(\alpha)$ corresponding to the ordinals α belonging to Z:

$$W(\alpha) \in \boldsymbol{K} \equiv \alpha \in Z.$$

Consider the union of all sets belonging to $\boldsymbol{K}$

$$S = \bigcup_{X \in \boldsymbol{K}} X.$$

By Theorem 5 the set S is well ordered by $\leqslant$. Let σ be its order type.

For $\alpha \in Z$ the set $W(\alpha)$ is either a segment of S or identical with S. In any case, $\alpha \leqslant \sigma$. This implies that $\alpha < \sigma+1$ for every $\alpha \in Z$. Thus the ordinal $\sigma+1$ is greater than every ordinal of Z.

COROLLARY 7: *There exists no set of all ordinals.*[1]

COROLLARY 8: *There exists a smallest ordinal not belonging to a given set Z.*

Let $\alpha \notin Z$ (such an ordinal exists by Corollary 7). If α is not the smallest ordinal not belonging to Z, then the set $W(\alpha)-Z$ is non-empty. The smallest number in this set (see Theorem 5) is simultaneously the smallest ordinal not belonging to Z.

COROLLARY 9: *If a set Z of ordinals has the property $(\gamma < \xi \in Z) \to (\gamma \in Z)$, then there exists an ordinal α such that $Z = W(\alpha)$.*

Namely, this ordinal α is the smallest ordinal among all ordinals not belonging to Z.

In fact, if $\xi \in Z$ then $\xi < \alpha$, because $\alpha \leqslant \xi$ would imply $\alpha \in Z$. Hence $Z \subset W(\alpha)$.

On the other hand if $\beta \in W(\alpha)$ then $\beta < \alpha$ and, by the definition of α, $\beta \in Z$. Therefore $W(\alpha) \subset Z$.

§ 3. Transfinite sequences

An ordinal is said to be a *limit ordinal* if it has no direct predecessor. Thus 0 is a limit ordinal.

THEOREM 1: *Each ordinal can be represented in the form $\lambda+n$ where λ is a limit ordinal and n is a finite ordinal (natural number).*

[1]) Before set theory was axiomatized, Corollary 7 was considered to be an antinomy. It was discovered by Burali-Forti [1].

PROOF. Let α be an ordinal, A a set of type α. Every set of the form $A-O(a)$ is said to be a *remainder* of A. Clearly,

$$A-O(a_1) \subset A-O(a_2) \equiv (a_1 \succ a_2) \vee (a_1 = a_2).$$

This implies that there exists no infinite increasing sequence of distinct remainders. Therefore there exists only a finite number of $m \in N$ such that there exists a remainder of power m. If n is the greatest such number and if $A-O(a)$ is a remainder of power n, then the segment $O(a)$ has no last element. Thus $\overline{O(a)}$ is the limit ordinal λ. This implies that $\alpha = \lambda+n$. Q.E.D.

By a *transfinite sequence of type* α or by an α-*sequence* we understand a function φ whose domain is $W(\alpha)$. If the values of this function (also called the *terms* of the α-sequence) are ordinals and if $\gamma < \beta < \alpha$ implies $\varphi(\gamma) < \varphi(\beta)$, then we say that this α-sequence is *increasing*.

Let φ be a λ-sequence of ordinals where λ is a limit ordinal. By Theorem 2.6 there exist ordinals greater than all the ordinals $\varphi(\gamma)$ where $\gamma < \lambda$. The smallest such ordinal (see Corollary 2.8) is called the *limit* of the λ-sequence $\varphi(\gamma)$ for $\gamma < \lambda$ and is denoted by $\lim\limits_{\gamma<\lambda} \varphi(\gamma)$.

For example,

$$\omega = \lim_{n<\omega} n = \lim_{n<\omega} n^2.$$

We say that an ordinal λ is *cofinal* with a limit ordinal α if λ is the limit of an increasing α-sequence:

$$\lambda = \lim_{\xi<\alpha} \varphi(\xi). \tag{1}$$

An ordinal cofinal with a limit ordinal is clearly itself a limit ordinal.

The connection between this notion and the notion of cofinality for sets is established by the following theorem.

THEOREM 2: *An ordinal λ is cofinal with the limit ordinal α if and only if $W(\lambda)$ contains a subset of type α cofinal with $W(\lambda)$.*

PROOF. Let A be a subset of $W(\lambda)$ cofinal with $W(\lambda)$ and such that $\overline{A} = \alpha$. For every ordinal $\xi < \alpha$ there exists an ordinal $\varphi(\xi)$ in A such that the set $\{\eta: (\eta \in A) \wedge (\eta < \varphi(\xi))\}$ is of type ξ. The sequence $\varphi(\xi)$ is clearly increasing and $\varphi(\xi) < \lambda$ for $\xi < \alpha$, because $\varphi(\xi) \in A \subset W(\lambda)$.

If $\mu < \lambda$ then there exists an ordinal $\xi \in A$ such that $\mu < \xi$, because the sets A and $W(\lambda)$ are cofinal. Thus $\mu < \xi \leqslant \varphi(\xi)$ (see p. 228), which proves that λ is the least ordinal greater than all the ordinals $\varphi(\xi)$. This proves (1).

Suppose in turn that (1) holds. Let A be the set of all terms of φ. We have $\eta < \lambda$ for $\eta \in A$ and consequently, since λ is a limit ordinal, there exists $\xi \in W(\lambda)$ such that $\eta < \xi$. Conversely, if $\xi \in W(\lambda)$ then $\xi < \lambda$ and by the definition of limit there exists $\xi' < \alpha$ such that $\xi < \varphi(\xi')$. This means that some ordinal in A is greater than ξ. Hence the sets A and $W(\lambda)$ are cofinal.

It follows directly from the definition of limit that

(2) $$\lim_{\gamma<\lambda} \varphi(\gamma) > \varphi(\gamma) \quad \text{for} \quad \gamma < \lambda,$$

(3) $$\Big\{\bigwedge_{\gamma<\lambda} [\mu > \varphi(\gamma)]\Big\} \equiv \{\mu \geqslant \lim_{\gamma<\lambda} \varphi(\gamma)\}.$$

THEOREM 3: *If φ and ψ are two increasing transfinite sequences, λ is a limit ordinal and $\xi = \lim_{\gamma<\lambda} \psi(\gamma)$, then*

$$\lim_{\delta<\xi} \varphi(\delta) = \lim_{\gamma<\lambda} \varphi\big(\psi(\gamma)\big).$$

PROOF. If $\gamma < \lambda$ then by (2) $\psi(\gamma) < \xi$ and again by (2) $\varphi\big(\psi(\gamma)\big) < \lim_{\delta<\xi} \varphi(\delta)$. Applying (3) we obtain

(4) $$\lim_{\gamma<\lambda} \varphi\big(\psi(\gamma)\big) \leqslant \lim_{\delta<\xi} \varphi(\delta).$$

If $\delta < \xi$ then by (3) we infer that for some ordinal $\gamma < \lambda$ we have $\psi(\gamma) \geqslant \delta$. Since the sequence φ is increasing, $\varphi\big(\psi(\gamma)\big) \geqslant \varphi(\delta)$, and by (3) it follows that $\lim_{\gamma<\lambda} \varphi\big(\psi(\gamma)\big) > \varphi(\delta)$.

Applying (3) again we obtain

$$\lim_{\gamma<\lambda} \varphi\big(\psi(\gamma)\big) \geqslant \lim_{\delta>\xi} \varphi(\delta).$$

This together with (4) proves Theorem 3.

It follows from Theorem 3 that if a limit ordinal η is cofinal with a limit ordinal ξ and ξ is cofinal with a limit ordinal λ, then η is cofinal with λ.

We say that an α-sequence φ is *continuous* if for every limit ordinal $\lambda < \alpha$ we have

$$\varphi(\lambda) = \lim_{\gamma<\lambda} \varphi(\gamma).$$

THEOREM 4: *Let φ be an increasing continuous α-sequence. For a given ordinal $\gamma < \alpha$, let*

(5) $$\varkappa_\gamma = \lim_{n<\omega} \gamma_n, \quad \textit{where} \quad \gamma_0 = \gamma \quad \textit{and} \quad \gamma_{n+1} = \varphi(\gamma_n).$$

Then $\varphi(\varkappa_\gamma) = \varkappa_\gamma$ (when $\varkappa_\gamma < \alpha$ and $\gamma_n < \alpha$ for $n = 1, 2, \ldots$).

PROOF. By (5) we have $\varkappa_\gamma > \gamma_{n+1} = \varphi(\gamma_n)$, and it follows by (3) that

$$\varkappa_\gamma \geqslant \lim_{n<\omega} \varphi(\gamma_n) = \varphi(\lim_{n<\omega} \gamma_n) = \varphi(\varkappa_\gamma).$$

On the other hand, $\varkappa_\gamma \leqslant \varphi(\varkappa_\gamma)$, because the sequence φ is increasing (see p. 226).

Each ordinal ξ satisfying the equation $\varphi(\xi) = \xi$ is said to be a *critical ordinal* of the sequence φ. Thus Theorem 4 states that if γ belongs to the domain of an increasing continuous sequence φ, then there exists a critical ordinal of this sequence greater than γ, provided that this sequence is defined for sufficiently large ordinals.

§ 4. Definitions by transfinite induction

The theory discussed in this section is similar to the theory of inductive definitions in arithmetic of natural numbers.

THEOREM 1: (ON DEFINITIONS BY TRANSFINITE INDUCTION[1])) *Given a set Z and an ordinal α, let Φ denote the set of all ξ-sequences for $\xi < \alpha$ with values belonging to Z. For each function $h \in Z^\Phi$ there exists one and only one transfinite sequence f defined on $\xi \leqslant \alpha$ and such that*

(1) $$f(\xi) = h[f|W(\xi)] \quad \textit{for every } \xi \leqslant \alpha.$$

PROOF. We show first that there exists at most one sequence f satisfying condition (1). Suppose that g is a sequence defined on the set $W(\alpha+1)$ and satisfying the condition

(2) $$g(\xi) = h[g|W(\xi)] \quad \text{for every } \xi \leqslant \alpha.$$

[1]) The theorem on definitions by transfinite induction was known already to Cantor [6], p. 231, although not in the full generality.

Let $B = \{\xi: f(\xi) = g(\xi)\}$. If $\xi \leqslant \alpha$ and $W(\xi) \subset B$, then $f|W(\xi) = g|W(\xi)$, and by (1) and (2) $f(\xi) = g(\xi)$, that is, $\xi \in B$. Thus we have shown that the condition $W(\xi) \subset B$ implies $\xi \in B$. By the principle of transfinite induction (p. 225) we infer that $W(\alpha+1) \subset B$; that is, for every $\xi \leqslant \alpha$, we have $f(\xi) = g(\xi)$. This means that the functions f and g are identical.

We now prove that there exists a function f satisfying (1).

Suppose by way of contradiction that for given α such a function does not exist. Clearly we may assume that α is the smallest ordinal with this property; otherwise we can find the smallest ordinal with this property in the set $W(\alpha)$. Thus for each $\xi < \alpha$ there exists a function f_ξ satisfying the condition

(3) $$f_\xi(\gamma) = h[f_\xi|W(\gamma)] \quad \text{for every } \gamma \leqslant \xi.$$

It follows from the part of the theorem already proved that for a given ξ there exists exactly one function f_ξ satisfying (3). We now infer that if $\gamma \leqslant \xi$ then $f_\xi|W(\gamma+1) = f_\gamma$. Hence $f_\gamma(\zeta) = f_\xi(\zeta)$ provided that $\zeta < \gamma$. This implies:

(4) $$[f_\gamma|W(\gamma)] = [f_\xi|W(\gamma)] \quad \text{for} \quad \gamma \leqslant \xi.$$

Let

(5) $$f(\xi) = f_\xi(\xi) \text{ for } \xi < \alpha \quad \text{and} \quad f(\alpha) = h(C_\alpha),$$

where C_α denotes the α-sequence such that $C_\alpha(\xi) = f_\xi(\xi)$ for $\xi < \alpha$. The function f satisfies condition (1). In fact, if $\gamma \leqslant \xi < \alpha$ then by (3), (4) and (5)

(6) $$f(\gamma) = h[f_\gamma|W(\gamma)] = h[f_\xi|W(\gamma)] = f_\xi(\gamma),$$

whence $f|W(\xi) = f_\xi|W(\xi)$ and by (3)

$$f(\xi) = h[f_\xi|W(\xi)] = h[f|W(\xi)] \quad \text{for} \quad \xi < \alpha.$$

Finally, $f(\alpha) = h[f|W(\alpha)]$, because by (5) $f|W(\alpha) = C_\alpha$.

In applications of the theorem, the function h is often defined by three formulas: the first gives the value $h(\varphi)$ for the void sequence φ (i.e. the value $h(\emptyset)$), the second the value $h(\varphi)$ for sequences $\varphi \in Z^\Phi$ whose type is not a limit ordinal (i.e. is of the form $\xi+1$), the third gives the value $h(\varphi)$ for sequences whose type λ is a limit ordinal. For instance, the first formula may be of the form

$$h(\varnothing) = A,$$

the second of the form

$$h(\varphi) = F(\varphi(\xi)),$$

and the third of the form

$$h(\varphi) = G\Big(\bigcup_{\eta<\lambda} \varphi(\eta)\Big) \quad \text{or} \quad h(\varphi) = G\Big(\bigcap_{\eta<\lambda} \varphi(\eta)\Big),$$

where F and G are given functions and A is a given set.

Then the sequence f, which exists by Theorem 1, satisfies the conditions

$$\begin{cases} f(\varnothing) = A, \\ f(\xi+1) = F(f(\xi)), \\ f(\lambda) = G\Big(\bigcup_{\eta<\lambda} f(\eta)\Big) \text{ or } f(\lambda) = G\Big(\bigcap_{\eta<\lambda} f(\eta)\Big). \end{cases}$$

Usually when we apply Theorem 1 to prove the existence of f, we give only these three formulas.

Examples

1. *Derivatives of order* α.[1]) Let A be a subset of the real line (or, more generally, $A \subset \mathscr{E}^n$). The derivative of order α of the set A is defined by transfinite induction as follows

$$A^{(0)} = \bar{A}, \quad A^{(\xi+1)} = A^{(\xi)\text{`}}, \quad A^{(\lambda)} = \bigcap_{\gamma<\lambda} A^{(\gamma)},$$

where λ is a limit ordinal $\leqslant \alpha$.

In this case we have $Z = \boldsymbol{P}(\bar{A})$, $h(\varnothing) = \bar{A}$, $h(\varphi) = [\varphi(\xi)]^{\text{`}}$ if φ is of type $\xi+1$, $h(\varphi) = \bigcap_{\gamma<\lambda} \varphi(\gamma)$ if φ is of type λ.

2. *Borel sets of type* α.[2]) The family $\boldsymbol{F}_\alpha$ of Borel sets of type α is defined by transfinite induction:

(i) $\boldsymbol{F}_0$ is the family of all closed subsets (of a given space),

(ii) $\boldsymbol{F}_\xi = \big(\bigcup_{\gamma<\xi} \boldsymbol{F}_\gamma\big)_\sigma$ or $\boldsymbol{F}_\xi = \big(\bigcup_{\gamma<\xi} \boldsymbol{F}_\gamma\big)_\delta$ for $0 < \xi \leqslant \alpha$,

[1]) We recall that the derivative of order 1, i.e. A`, is the set of limit points of the set A.

[2]) Borel sets were introduced by Borel [1].

depending on whether ξ is an even or an odd ordinal (ordinals of the form $\lambda+n$, where λ is a limit ordinal, are said to be *even* if n is even and *odd* if n is odd). See page 125 for the definitions of σ and δ.

In this case we have

$$h(\emptyset) = \boldsymbol{F}_0 \quad \text{and} \quad h(\varphi) = \Big(\bigcup_{\gamma<\xi} \varphi(\gamma)\Big)_\sigma \quad \text{or} \quad h(\varphi) = \Big(\bigcup_{\gamma<\xi} \varphi(\gamma)\Big)_\delta$$

depending on whether the type ξ of φ is an even or an odd ordinal.

Similarly we define the family $\boldsymbol{G}_\alpha$ by the conditions:

(iii) $\boldsymbol{G}_0$ is the family of open sets,

(iv) $\boldsymbol{G}_\xi = \big(\bigcup_{\gamma<\xi} \boldsymbol{G}_\gamma\big)_\sigma$ or $\boldsymbol{G}_\xi = \big(\bigcup_{\gamma<\xi} \boldsymbol{G}_\gamma\big)_\delta$ for $0 < \xi \leqslant \alpha$,

depending on the character of ξ (even or odd).

3. *Analytically representable functions of class* α.[1]) The set of these functions is denoted by Φ_α and is defined by transfinite induction as follows:

(a) Φ_0 is the set of all real continuous functions (of a real variable),

(b) $\Phi_\xi = \big(\bigcup_{\gamma<\xi} \Phi_\gamma\big)_\lambda$ for $0 < \xi \leqslant \alpha$,

where in general Δ_λ denotes the set of all functions which are limits of convergent sequences of functions belonging to the class Δ.

In this example:

$$Z = \boldsymbol{P}(\mathscr{E}^{\mathscr{E}}), \quad h(\emptyset) = \Phi_0, \quad h(\varphi) = \Big(\bigcup_{\gamma<\xi} \varphi(\gamma)\Big)_\lambda,$$

where ξ is the type of the sequence φ.

As another example of an application of Theorem 1 we prove the following theorem.

[1]) Baire functions were first studied in the book Baire [1]. The notion of a rank of a set originated with Russell who introduced a classification of sets into "types". He called objects which are not sets "*objects of type* 0". Sets whose elements are of type n are called *objects of type* $n+1$. Russell allowed only objects of a finite type and maintained that other sets are meaningless. This was the basic principle of his "simple theory of types". See Russell and Whitehead [1].

The aim of the theories of types was to eliminate antinomies from set theory. Both theories are now obsolete but their role in the development of set theory cannot be neglected. See Quine [1], Chapter XI.

THEOREM 2: *Every limit ordinal of the form* $\lambda = \lim_{\xi<\alpha} \varphi(\xi)$ *is cofinal with some ordinal* $\gamma \leqslant \alpha$.[1])

PROOF. According to Theorem 1, there exists an α-sequence ψ such that

(i) $\psi(\xi)$ is the smallest ordinal ζ such that $\varphi(\xi) < \zeta < \lambda$ and $\bigwedge_{\eta<\xi} \psi(\eta) < \zeta$, if such an ordinal exists;

(ii) $\psi(\xi) = \lambda$ otherwise.

We consider two cases:

I. There exists no ordinal $\xi < \alpha$ such that $\psi(\xi) = \lambda$. In this case the sequence ψ is increasing by (i) and (by induction) $\psi(\xi) > \varphi(\xi)$ for all ξ. Since λ is the smallest ordinal greater than all the $\varphi(\xi)$, we infer that $\lim_{\xi<\alpha} \psi(\xi) = \lambda$. Thus λ is cofinal with α.

II. There exist ordinals $\xi < \alpha$ such that $\psi(\xi) = \lambda$. Let γ be the smallest such ordinal, that is, $\psi(\gamma) = \lambda$ and $\psi(\eta) < \lambda$ for $\eta < \gamma$. This implies that γ is a limit ordinal. In fact, $\gamma = \delta+1$ then $\psi(\delta) < \lambda$ and therefore there exists an ordinal μ such that $\psi(\delta) < \mu < \lambda$ and $\varphi(\delta) < \mu < \lambda$, which contradicts $\psi(\gamma) = \lambda$. It follows from (i) that for $\xi_1 < \xi_2 < \gamma$ we have $\psi(\xi_1) < \psi(\xi_2)$. This means that $\psi|W(\gamma)$ is increasing. Let $\varrho = \lim_{\xi<\gamma} \psi(\xi)$. We shall show that $\varrho = \lambda$. For suppose that $\varrho < \lambda$. The ordinal ϱ is greater than all the $\psi(\eta)$ for $\eta < \gamma$ and therefore there exist ordinals $< \lambda$ greater than $\varphi(\gamma)$ and greater than $\psi(\eta)$ for $\eta < \gamma$ (namely any ordinal greater than both ϱ and $\varphi(\gamma)$ and less than λ satisfies these conditions). But this contradicts the assumption $\psi(\gamma) = \lambda$. Hence $\varrho \geqslant \lambda$. We have $\varrho \leqslant \lambda$, because $\psi(\xi) < \lambda$ for $\xi < \gamma$. Thus $\lambda = \varrho = \lim_{\xi<\gamma} \psi(\xi)$. This means that λ is cofinal with γ. Q.E.D.

Although Theorem 1 is very general, it is still insufficient. The reason is that in Theorem 1 we proved the existence of a function f whose range is contained in a set Z given in advance, whereas in several cases we have to construct a function whose range is not a subset of any set known to us before. As an example we prove the following theorem:

[1]) We do not suppose that the function φ is increasing.

THEOREM 3: *For every set A and every ordinal number α there exists exactly one function with domain $D(f) = W(\alpha+1)$ which satisfies the conditions*:

(C$_0$) $$f(\emptyset) = A,$$

(C$_1$) $$\xi+1 \in D(f) \to f(\xi+1) = \boldsymbol{P}(f(\xi)),$$

(C$_2$) $$(\lambda \text{ is a limit number}) \wedge (\lambda \in D(f)) \to f(\lambda) = \bigcup_{\xi<\lambda} f(\xi).$$

PROOF. We first show by an easy induction that for every α there is at most one function f with domain $W(\alpha)$ satisfying (C$_0$)–(C$_2$).

Assume that there is an ordinal α such that for no function f with domain $W(\alpha+1)$ conditions (C$_0$)–(C$_2$) hold. We can assume that α is the least such ordinal. Hence if $\xi < \alpha$ then there exists exactly one function f with domain $W(\xi+1)$ satisfying (C$_0$)–(C$_2$).

From the axiom of replacement it follows that there is a set S consisting of all these functions f_ξ. Since $f_\xi \subset f_\eta$ for $\xi < \eta < \alpha$, we easily infer that the set $F = \bigcup(S)$ is a function with domain $W(\alpha)$ which satisfies (C$_0$)–(C$_2$). Now we shall extend F by adding to it one more pair of the form $\langle \alpha, X \rangle$ so as to obtain a function with domain $W(\alpha+1)$ which also satisfies (C$_0$)–(C$_2$). It is easy to verify that it is sufficient to take as X the set $\boldsymbol{P}(\bigcup(\mathrm{Rg}(F)))$ if α is a successor ordinal and the set $\bigcup(\mathrm{Rg}(F))$ if α is a limit number.

Thus the assumption that there are ordinals α for which an f satisfying (C$_0$)–(C$_2$) and having $W(\alpha+1)$ as its domain does not exist results in a contradiction. Theorem 3 is thus proved.

Theorem 3 allows us to introduce the notion of a rank of a set. We denote by $R_\alpha(A)$ the set $f(\alpha)$ where f is a unique function satisfying Theorem 3. If $A = \emptyset$ then we write R_α instead of $R_\alpha(\emptyset)$.

DEFINITION: The *rank of a set X* is the least ordinal α such that $X \in R_\alpha$. We denote the rank of X by $\varrho(X)$.

Since $R_\lambda = \bigcup_{\xi<\lambda} R_\xi$ whenever λ is a limit number, it is clear that $\varrho(X)$ is always a successor ordinal. If $\varrho(X) = \alpha+1$, then $X \in R_{\alpha+1} - R_\alpha$. This property is characteristic for the rank of X. It can also be easily proved that $X \in Y$ implies the inequality $\varrho(X) < \varrho(Y)$ and that $\varrho(Y)$ is the least successor ordinal greater than $\varrho(X)$ for each X in Y.

We shall use the sets R_α in later chapters (see X § 2, p. 285).

Theorem 3 above illustrates a general method of proving the existence of functions satisfying recursive conditions. Other examples of such functions are given in exercises below.

Exercises

1. Prove that for each pair of ordinals (α, β) there exists a function f with domain $W(\alpha)\times W(\beta)$ such that $f(\xi, 0) = 1$, $f(\xi, \eta+1) = f(\xi, \eta)\cdot\xi$, $f(\xi, \lambda) = \lim_{\sigma<\lambda} f(\xi, \sigma)$ where λ is a limit number and $(\xi, \eta+1)$, (ξ, λ) belong to the domain of f.

Prove moreover that if $\alpha \leqslant \alpha'$ and $\beta \leqslant \beta'$ then the function f' corresponding to the pair (α', β') is an extension of the function f corresponding to the pair (α, β).

REMARK: The value $f(\xi, \eta)$ is usually denoted by ξ^η.

2. Define in a similar manner a function f with domain $W(\alpha)$ such that for every $\xi < \alpha$ the following equations hold: $f(0) = \mathfrak{a}$, $f(\xi+1) = \overline{\overline{P(f(\xi))}}$, $\lambda < \alpha \rightarrow f(\lambda) = \sum_{\xi<\lambda} f(\xi)$ (α and λ denote here limit numbers).

REMARK: The cardinal number $f(\xi)$ is usually called the *ξ-th beth* where "beth" is the second letter of the Hebrew alphabet; see also p. 285.

§ 5. Ordinal arithmetic[1]

The following theorem is a direct consequence of the definitions given in Chapter VI, pp. 218–219

THEOREM 1: *The sum and the product of two ordinal numbers are ordinal numbers.*

By means of Example 1.4 we obtain the following

THEOREM 2: *The ordered sum of ordinal numbers, where the indexing set is well ordered, is itself an ordinal number.*

The first part of Theorem 1 follows from Theorem 2, by letting the indexing set be a two-element set. Similarly, assuming that the indexing set is of type β and that all components are equal to α, we conclude that $\alpha\beta$ is an ordinal. We shall prove some arithmetic laws for the ordinal addition and multiplication.

THE FIRST MONOTONIC LAW FOR ADDITION:

(1) $$(\alpha < \beta) \rightarrow (\gamma+\alpha < \gamma+\beta).$$

PROOF. Let $\overline{C} = \gamma$, $\overline{B} = \beta$ and $B \cap C = \emptyset$. Since $\alpha < \beta$, the set B

[1]) Ordinal arithmetic discussed in Sections 5–7 is due to Cantor [3].

contains a segment A of type α. The ordered sum $C \cup A$, which is of type $\gamma+\alpha$, is a segment of $C \cup B$, which is of type $\gamma+\beta$. Thus $\gamma+\alpha < \gamma+\beta$.

It follows from (1) (for $\alpha = 0$) that

(2) $$(\beta > 0) \to (\gamma+\beta > \gamma).$$

Thus the sum of two ordinals different from 0 is greater than the first component.

THE SECOND MONOTONIC LAW FOR ADDITION:

(3) $$(\alpha \leqslant \beta) \to (\alpha+\gamma \leqslant \beta+\gamma).$$

In fact, assuming that $\overline{A} = \alpha$, $\overline{B} = \beta$, $\overline{C} = \gamma$, $A \subset B$, $B \cap C = \emptyset$ and applying Theorem 2.3 to the ordered unions $A \cup C$ and $B \cup C$, we obtain (3).

In particular, it follows from (3) that

(4) $$\beta+\gamma \geqslant \gamma.$$

Thus the sum of two ordinals is not less than the second component. On the other hand, from $1+\omega = \omega$ we see that the sum does not need to be greater than the second component (although the first component is not zero).

THEOREM 3: *If $\alpha \geqslant \beta$ then there exists exactly one ordinal γ such that $\alpha = \beta+\gamma$.*

PROOF. Let $\overline{A} = \alpha$, let B be a segment of A of type β and let $\gamma = \overline{A-B}$. Clearly, $\alpha = \beta+\gamma$. To prove the uniqueness of γ suppose that $\beta+\gamma_1 = \beta+\gamma_2$. By (1) this implies that $\gamma_1 \nless \gamma_2$ and $\gamma_2 \nless \gamma_1$. Thus $\gamma_1 = \gamma_2$ by Theorem 2.1.

It follows from Theorem 3 and formula (2) that the inequality $\alpha \geqslant \beta$ is a necessary and sufficient condition for the equation $\alpha = \beta+x$ to be solvable.

On the other hand, the equation $\alpha = x+\beta$ is not always solvable, for example we have $\omega \neq x+2$ for every ordinal x.

In connection with Theorem 3 we introduce the following definition.

The *difference of the ordinals α and β* ($\alpha \geqslant \beta$) is defined to be the unique ordinal γ such that $\alpha = \beta+\gamma$. This ordinal is denoted by $\alpha-\beta$. Thus

(5) $$\alpha = \beta+(\alpha-\beta).$$

For instance, $\omega - n = \omega$, because $n + \omega = \omega$. Similarly $\omega^2 - \omega = \omega^2$, because $\omega + \omega^2 = \omega^2$.

THE MONOTONIC LAWS FOR ORDINAL SUBTRACTION:

$$(\alpha > \alpha_1) \to (\alpha - \beta > \alpha_1 - \beta), \tag{6}$$

$$(\beta > \beta_1) \to (\alpha - \beta \leqslant \alpha - \beta_1). \tag{7}$$

In order for the subtraction to be possible we assume in (6) that $\alpha_1 \geqslant \beta$ and in (7) that $\alpha \geqslant \beta$.

PROOF. For the purpose of obtaining a contradiction, suppose that $\alpha - \beta \leqslant \alpha_1 - \beta$. From formulas (5) and (1) it follows that $\alpha = \beta + (\alpha - \beta) \leqslant \beta + (\alpha_1 - \beta) = \alpha_1$, which contradicts the assumption $\alpha > \alpha_1$.

Similarly, assume that $\alpha - \beta > \alpha - \beta_1$; then it follows from (5) and (1) that $\alpha = \beta + (\alpha - \beta) > \beta + (\alpha - \beta_1)$. This contradicts $\beta > \beta_1$, because $\beta > \beta_1$ implies by (3) that $\beta + (\alpha - \beta_1) \geqslant \beta_1 + (\alpha - \beta_1) = \alpha$ and consequently we would obtain $\alpha > \alpha$.

The identity $\omega - 2 = \omega - 3 = \omega$ shows that the symbol $\leqslant$ in (7) cannot in general be replaced by $<$.

THE FIRST MONOTONIC LAW FOR ORDINAL MULTIPILCATION:

$$(\alpha < \beta) \to (\gamma\alpha < \gamma\beta) \quad \textit{for} \quad \gamma > 0. \tag{8}$$

In fact, $\gamma\beta$ is the type of the cartesian product $C \times B$ ordered anti-lexicographically, where $\overline{B} = \beta$ and $\overline{C} = \gamma$. Let $\overline{A} = \alpha$ and let A be a segment of B. The cartesian product $C \times A$ ordered anti-lexicographically is a segment of $C \times B$. Q.E.D.

THE SECOND MONOTONIC LAW FOR ORDINAL MULTIPLICATION:

$$(\alpha \leqslant \beta) \to (\alpha\gamma \leqslant \beta\gamma). \tag{9}$$

To prove (9), we consider sets A, B, C of types α, β, γ respectively, and we assume that $A \subset B$. Thus $A \times C \subset B \times C$ which proves (9).

It follows from the identity $1 \cdot \omega = 2 \cdot \omega$ that the symbol $\leqslant$ in (9) cannot be replaced by $<$.

LEFT DISTRIBUTIVITY OF ORDINAL MULTIPLICATION WITH RESPECT TO ORDINAL SUBTRACTION:

$$\alpha(\beta - \gamma) = \alpha\beta - \alpha\gamma \quad \textit{for} \quad \beta \geqslant \gamma. \tag{10}$$

PROOF. Since multiplication is left-distributive over addition (see p. 219), we have by (5) $\alpha\beta = \alpha[\gamma+(\beta-\gamma)] = \alpha\gamma+\alpha(\beta-\gamma)$. This implies (10) by the definition of subtraction.

We now prove a theorem about division of ordinal numbers.

THEOREM 4: *If β is an ordinal > 0, then for each ordinal α there exist ordinals γ and ϱ such that*

$$\alpha = \beta\gamma+\varrho \quad \text{and} \quad \varrho < \beta. \tag{11}$$

The ordinals γ and ϱ are then uniquely determined.

PROOF. Since $1 \leqslant \beta$, we have by (9) $\alpha \leqslant \beta\alpha$. If $\alpha = \beta\alpha$, then it suffices to let $\gamma = \alpha$ and $\varrho = 0$. Therefore suppose that $\alpha < \beta\alpha$. The product $\beta\alpha$ is the type of the set $B\times A$ where $\overline{B} = \beta$, $\overline{A} = \alpha$. It follows from the hypothesis $\alpha < \beta\alpha$ that the set $B\times A$ contains a segment $O(\langle b, a\rangle)$ of type α. Since

$$\langle y, x\rangle \in O(\langle b, a\rangle) \equiv \{(x \prec a)\vee[(x = a)\wedge(y \prec b)]\},$$

we have

$$O(\langle b, a\rangle) = \big(B\times O_A(a)\big)\cup\big(O_B(b)\times\{a\}\big),$$

where every element belonging to the first component precedes every element belonging to the second component. Since the first component is of type $\beta\cdot\overline{O_A(a)} = \beta\gamma$ and the second of type $\overline{O_B(b)} = \varrho$, we have $\alpha = \beta\gamma+\varrho$ where $\gamma < \alpha$, $\varrho < \beta$. In this way we have proved that there exist ordinals γ and ϱ satisfying conditions (11).

To prove uniqueness, suppose that

$$\beta\gamma+\varrho = \beta\gamma_1+\varrho_1, \quad \varrho < \beta, \quad \varrho_1 < \beta. \tag{12}$$

Let $\gamma > \gamma_1$. Then $\gamma = \gamma_1+(\gamma-\gamma_1)$ and

$$\beta\gamma+\varrho = \beta[\gamma_1+(\gamma-\gamma_1)]+\varrho = \beta\gamma_1+\beta(\gamma-\gamma_1)+\varrho. \tag{13}$$

Since $\gamma-\gamma_1 \geqslant 1$, we have by (8)

$$\beta(\gamma-\gamma_1) \geqslant \beta. \tag{14}$$

It follows from (12) and (13) that

$$\beta\gamma_1+\varrho_1 \geqslant \beta\gamma_1+\beta(\gamma-\gamma_1),$$

whence by (14)

$$\beta\gamma_1+\varrho_1 \geqslant \beta\gamma_1+\beta. \tag{15}$$

As $\beta > \varrho_1$, we obtain by (1) $\beta\gamma_1+\beta > \beta\gamma_1+\varrho_1$. But this is impossible, because (15) implies $\beta\gamma_1+\varrho_1 > \beta\gamma_1+\varrho_1$. Thus the hypothesis $\gamma > \gamma_1$ leads to a contradiction. Similarly it can be shown that $\gamma_1 > \gamma$ does not hold. Hence $\gamma = \gamma_1$.

This formula and (12) imply that $\beta\gamma+\varrho = \beta\gamma+\varrho_1$; by (1) this formula implies $\varrho = \varrho_1$.

Thus Theorem 4 is completely proved.

The ordinal γ in (11) is called the *quotient* and ϱ the *remainder.*

The following theorem is a consequence of Theorem 4.

THEOREM 5: (THE EUCLIDEAN ALGORITHM FOR ORDINAL NUMBERS) *For any two ordinals α_0 and α_1 different from* 0, *there exist a natural number n and sequences $\alpha_2, \ldots, \alpha_n, \beta_1, \ldots, \beta_n$ such that $\alpha_1 > \alpha_2 > \ldots > \alpha_n > 0$ and*

$$\alpha_0 = \alpha_1\beta_1+\alpha_2, \quad \alpha_1 = \alpha_2\beta_2+\alpha_3, \quad \ldots,$$
$$\alpha_{n-2} = \alpha_{n-1}\beta_{n-1}+\alpha_n, \quad \alpha_{n-1} = \alpha_n\beta_n.$$

PROOF. According to Theorem 4 and to the theorem on inductive definitions, there exist infinite sequences φ and ψ such that

$$\varphi_0 = \alpha_0, \quad \varphi_1 = \alpha_1, \quad \psi_0 = \psi_1 = 1$$

and, for $j > 1$,

$$\varphi_{j-1} = \varphi_j\psi_{j+1}+\varphi_{j+1}$$

and

$$\varphi_{j+1} < \varphi_j \quad \text{if} \quad \varphi_j \neq 0,$$
$$\varphi_{j+1} = \psi_{j+1} = 0 \quad \text{if} \quad \varphi_j = 0.$$

Since there exists no infinite decreasing sequence of ordinals, all terms of φ from a certain term on are equal to 0. Let $n' = \min_j(\varphi_j = 0)$. Clearly, $n' > 0$, because by assumption φ_0 and φ_1 are $\neq 0$. To prove the theorem it suffices to let $n = n'-1$, $\alpha_j = \varphi_j$ for $2 \leqslant j < n'$ and $\beta_j = \psi_{j+1}$ for $1 \leqslant j < n'$.

Now we shall prove several formulas concerning the operation of taking limits (p. 231). We assume that λ is a limit ordinal and that φ is an increasing λ-sequence. Then we have

(16) $$\lim_{\xi<\lambda}[\alpha+\varphi(\xi)] = \alpha+\lim_{\xi<\lambda}\varphi(\xi).$$

PROOF. Let $\beta = \lim_{\xi<\lambda}\varphi(\xi)$. If $\xi < \lambda$, then $\varphi(\xi) < \beta$ and therefore $\alpha+\varphi(\xi) < \alpha+\beta$. Let $\zeta < \alpha+\beta$; we shall prove that there exists $\xi < \lambda$ such that $\zeta < \alpha+\varphi(\xi)$. If $\zeta < \alpha$, then $\zeta < \alpha+\varphi(0)$; on the other hand, if $\zeta \geqslant \alpha$, then $\zeta = \alpha+(\zeta-\alpha)$ and $\zeta-\alpha < (\alpha+\beta)-\alpha \leqslant \beta$. It follows that for some $\xi < \lambda$ we have $\zeta-\alpha < \varphi(\xi)$, thus $\zeta < \alpha+\varphi(\xi)$. Hence the ordinal $\alpha+\beta$ is the smallest ordinal greater than all ordinals $\alpha+\varphi(\xi)$ for $\xi < \lambda$. This proves (16).

(17) $$\lim_{\xi<\lambda}[\alpha\cdot\varphi(\xi)] = \alpha\cdot\lim_{\xi<\lambda}\varphi(\xi).$$

PROOF. We may clearly assume that $\alpha \neq 0$. Let $\beta = \lim_{\xi<\lambda}\varphi(\xi)$. For $\xi < \lambda$ we have $\varphi(\xi) < \beta$, thus $\alpha\cdot\varphi(\xi) < \alpha\cdot\beta$. Let $\zeta < \alpha\cdot\beta$. By Theorem 4 there exist ordinals γ and ϱ such that $\zeta = \alpha\gamma+\varrho < \alpha\beta$ and $\varrho < \alpha$. If $\gamma \geqslant \beta$ then we have $\zeta \geqslant \alpha\beta+\varrho \geqslant \alpha\beta$, which contradicts the hypothesis. Thus $\gamma < \beta$, which implies that for some $\xi < \lambda$ we have $\gamma \leqslant \varphi(\xi)$. Hence

$$\zeta \leqslant \alpha\cdot\varphi(\xi)+\varrho \leqslant \alpha\cdot\varphi(\xi)+\alpha = \alpha\cdot[\varphi(\xi)+1],$$

and

$$\zeta \leqslant \alpha\cdot\varphi(\xi+1),$$

because the function φ is increasing by assumption. Since λ is a limit ordinal, we have $\xi+1 < \lambda$ and the formula $\zeta \leqslant \alpha\cdot\varphi(\xi+1)$ shows that ζ is not greater than some ordinal of the form $\alpha\cdot\varphi(\eta)$ where $\eta \leqslant \lambda$. Thus the ordinal $\alpha\beta$ is the smallest ordinal greater than all ordinals $\alpha\cdot\varphi(\eta)$ for $\eta < \lambda$. This proves (17).

REMARKS. Assuming $\varphi(\xi) = \xi$, we infer from (1) and (16) that *the function* $s(\xi) = \alpha+\xi$ *is increasing and continuous* (*on the set* $W(\beta)$ *for every* β). It follows from (8) and (17) that the function $p(\xi) = \alpha\cdot\xi$ possesses the same property. Theorem 3.4 implies that there exist critical ordinals for the function s and for the function p. A critical ordinal for the function s is $\xi = \alpha\cdot\omega$, and, more generally, every ordinal of the form $\alpha\cdot\omega+\varrho$ where ϱ is an arbitary ordinal. In fact,

$$s(\alpha\cdot\omega+\varrho) = \alpha+\alpha\cdot\omega+\varrho = \alpha(1+\omega)+\varrho = \alpha\cdot\omega+\varrho.$$

One of the critical ordinals for the function p is $\lim_{n<\omega}\alpha^n$, where α^n

$= \underbrace{\alpha \cdot \alpha \ldots \alpha}_{n}$. All critical ordinals for the function p will be determined in the following section.

Exercises[1])

1. Determine whether $\lim[\varphi(\xi)+\psi(\xi)] = \lim\varphi(\xi)+\lim\psi(\xi)$.

2. Determine whether $\lim[\varphi(\xi)]\cdot\alpha = [\lim\varphi(\xi)\cdot\alpha]$.

3. Prove that if $\alpha_1+\beta_1 = \alpha+\beta$ and $\beta_1 > \beta$, then $\alpha_1 < \alpha$.

4. Show that for every ordinal α there exists a finite number of ordinals β such that the equation $\alpha = \xi+\beta$ is solvable for ξ (each such ordinal is called a *remainder* of α).

5. Show that if a sequence φ is increasing, $\alpha = \lim\limits_{\xi<\lambda}\varphi(\xi)$ and if $\alpha = \varphi(\xi)+\varrho(\xi)$, then there exists an ordinal $\mu < \lambda$ such that $\varrho(\xi)$ is constant for $\mu < \xi < \lambda$ and this constant is equal to the smallest remainder of α.

§ 6. Ordinal exponentiation

The operation of *ordinal exponentiation* is defined by transfinite induction as follows:

(1) $$\gamma^0 = 1,$$

(2) $$\gamma^{\xi+1} = \gamma^\xi\cdot\gamma,$$

(3) $$\gamma^\lambda = \lim_{\xi<\lambda}\gamma^\xi,$$

where λ is a limit ordinal.

We say that γ^α is the *power* of γ, γ is the *base* and α the *exponent*. It follows from the definition that if $\gamma > 1$, then

(4) $$\alpha < \beta \rightarrow \gamma^\alpha < \gamma^\beta.$$

We shall prove that

(5) $$\gamma^{\xi+\eta} = \gamma^\xi\cdot\gamma^\eta.$$

PROOF. Given an ordinal ξ, let B denote the set of those $\zeta \in W(\eta+1)$ for which $\gamma^{\xi+\zeta} = \gamma^\xi\cdot\gamma^\zeta$. We shall show that if $\zeta \leqslant \eta$, then

(*) $$W(\zeta) \subset B \rightarrow \zeta \in B.$$

[1]) More material for exercises on ordinal arithmetic can be found in Sierpiński [1]. The theorem in exercise 5 is due to Hoborski [1].

In fact, the following three cases are possible: (i) $\zeta = 0$; (ii) ζ is not a limit ordinal; (iii) ζ is a limit ordinal > 0. In case (i), $\zeta \in B$, because $\gamma^{\xi+0} = \gamma^{\xi} = \gamma^{\xi}\cdot 1 = \gamma^{\xi}\cdot\gamma^{0}$. In case (ii), $\zeta = \zeta_1+1$ where $\zeta_1 \in W(\zeta)$; thus, by assumption, $\zeta_1 \in B$. Hence we have $\gamma^{\xi+\zeta_1} = \gamma^{\xi}\cdot\gamma^{\zeta_1}$ and therefore

$$\gamma^{\xi+\zeta} = \gamma^{\xi+(\zeta_1+1)} = \gamma^{(\xi+\zeta_1)+1} = \gamma^{\xi+\zeta_1}\cdot\gamma = (\gamma^{\xi}\cdot\gamma^{\zeta_1})\gamma$$
$$= \gamma^{\xi}(\gamma^{\zeta_1}\cdot\gamma) = \gamma^{\xi}\cdot\gamma^{\zeta_1+1} = \gamma^{\xi}\cdot\gamma^{\zeta},$$

which shows that $\zeta \in B$. Finally, in case (iii), $\xi+\zeta$ is a limit ordinal, thus

$$\gamma^{\xi+\zeta} = \lim_{\alpha<\xi+\zeta}\gamma^{\alpha}.$$

Applying Theorem 3.3 to the functions $\varphi(\alpha) = \gamma^{\alpha}$ and $\psi(\alpha) = \xi+\alpha$ we infer that

$$\lim_{\alpha<\xi+\zeta}\gamma^{\alpha} = \lim_{\alpha<\zeta}\gamma^{\xi+\alpha}, \qquad \gamma^{\xi+\zeta} = \lim_{\alpha<\zeta}\gamma^{\xi+\alpha}.$$

Since for $\alpha < \zeta$ we have $\alpha \in W(\zeta)$, it follows that $\alpha \in B$, i.e. $\gamma^{\xi+\alpha} = \gamma^{\xi}\cdot\gamma^{\alpha}$. Thus

$$\gamma^{\xi+\zeta} = \lim_{\alpha<\zeta}(\gamma^{\xi}\cdot\gamma^{\alpha}) = \gamma^{\xi}\lim_{\alpha<\zeta}\gamma^{\alpha} = \gamma^{\xi}\cdot\gamma^{\zeta},$$

which implies that $\zeta \in B$.

Hence implication $(*)$ is proved. By induction it follows that $B = W(\eta+1)$. Thus $\eta \in B$, which proves (5).

(6) $$(\gamma^{\xi})^{\eta} = \gamma^{\xi\eta}.$$

The proof is analogous to that of (5). Let B denote the set of those $\zeta \in W(\eta+1)$ for which $(\gamma^{\xi})^{\zeta} = \gamma^{\xi\zeta}$. It suffices to show that implication $(*)$ holds. As previously we consider cases (i), (ii), and (iii). In case (i) we have $\zeta \in B$, because $(\gamma^{\xi})^{0} = 1 = \gamma^{0} = \gamma^{\xi 0}$. In case (ii) we have $\zeta = \zeta_1+1$, where ζ_1 satisfies the condition $(\gamma^{\xi})^{\zeta_1} = \gamma^{\xi\zeta_1}$. This formula in turn implies

$$(\gamma^{\xi})^{\zeta} = (\gamma^{\xi})^{\zeta_1+1} = (\gamma^{\xi})^{\zeta_1}\gamma^{\xi} = \gamma^{\xi\zeta_1}\gamma^{\xi} = \gamma^{\xi\zeta_1+\xi} = \gamma^{\xi(\zeta_1+1)} = \gamma^{\xi\zeta},$$

which shows that $\zeta \in B$. Finally, if ζ is a limit ordinal > 0, then

$$(\gamma^{\xi})^{\zeta} = \lim_{\alpha<\zeta}(\gamma^{\xi})^{\alpha} = \lim_{\alpha<\zeta}\gamma^{\xi\alpha}.$$

Since Theorem 3.3 implies that

$$\lim_{\alpha<\zeta}\gamma^{\xi\alpha} = \lim_{\eta<\xi\zeta}\gamma^{\eta} = \gamma^{\xi\zeta},$$

we have $\zeta \in B$. Hence formula (6) is proved.

(7) $$\gamma > 1 \to \gamma^{\xi} \geqslant \xi .$$

This formula follows from (4) by Theorem 1.5.

The operation of ordinal exponentiation allows us to find all critical ordinals of the function $p(\xi) = \alpha\cdot\xi$ (compare p. 244). Namely these critical ordinals are all ordinals of the form $\alpha^{\omega+\sigma}$ where σ is any ordinal. In fact,

$$p(\alpha^{\omega+\sigma}) = \alpha\cdot\alpha^{\omega+\sigma} = \alpha^{1+(\omega+\sigma)} = \alpha^{(1+\omega)+\sigma} = \alpha^{\omega+\sigma},$$

The function $f(\xi) = \gamma^{\xi}$ is—according to (3) and (4)—increasing and continuous (on every set $W(\alpha)$). Thus this function possesses critical ordinals by Theorem 3.5. According to this theorem, those critical ordinals can be obtained as the limits of the sequences α_n where α_0 is an arbitrary ordinal and $\alpha^{n+1} = \gamma^{\alpha_n}$.

For example, assuming that $\gamma = \omega$, $\alpha_0 = 1$ we have

$$\alpha_1 = \omega, \quad \alpha_2 = \omega^{\omega}, \quad \alpha_3 = \omega^{\omega^{\omega}}, \ \ldots$$

The limit of this sequence,

$$\varepsilon = \lim_{n<\omega}\alpha_n,$$

is the smallest critical ordinal of the function ω^{ξ}, i.e. the smallest ordinal satisfying the equality

(8) $$\omega^{\varepsilon} = \varepsilon .$$

Such ordinals ε are called *epsilon-ordinals*.

Exercises

1. Let λ be a limit ordinal. Show that if the function φ is continuous on the set $W(\lambda)$ and satisfies the conditions $\varphi(0) = 1$, $\varphi(1) = \gamma$, $\varphi(\alpha+\beta) = \varphi(\alpha)\cdot\varphi(\beta)$ for $\alpha, \beta, \alpha+\beta < \lambda$, then $\varphi(\xi) = \gamma^{\xi}$ for $\xi < \lambda$.

2. Show that if $\omega^{\xi} = \alpha+\beta$ and $\beta \neq 0$, then $\beta = \omega^{\xi}$.

3. Show that for every ordinal α there exists an epsilon-ordinal greater than α.

§ 7. Expansions of ordinal numbers for an arbitrary base

The operation of ordinal exponentiation can be used in order to represent ordinal numbers in the form similar to decimal expansion of natural numbers. For this purpose, we shall first prove the following theorem.

THEOREM 1: *If* $\gamma > 1$ *and* $1 \leqslant \alpha < \gamma^{\xi}$, *then there exist ordinals* η, β *and* ϱ *such that*

$$\alpha = \gamma^{\eta} \cdot \beta + \varrho, \quad 0 \leqslant \eta < \xi, \quad \beta < \gamma \quad \textit{and} \quad \varrho < \gamma^{\eta}.$$

PROOF. Let ζ be the smallest ordinal such that $\alpha < \gamma^{\zeta}$. Clearly, $0 < \zeta \leqslant \xi$. If ζ were a limit ordinal, then we would have $\gamma^{\zeta} = \lim_{\lambda < \zeta} \gamma^{\lambda}$ and $\gamma^{\lambda} \leqslant \alpha$ for $\lambda < \zeta$. This implies $\gamma^{\zeta} \leqslant \alpha$, which contradicts the definition of ζ. Thus $\zeta = \eta + 1$, where

$$0 \leqslant \eta < \xi, \quad \gamma^{\eta} \leqslant \alpha < \gamma^{\eta+1}.$$

By virtue of Theorem 5.4 there exist ordinals β and ϱ such that

$$\alpha = \gamma^{\eta} \cdot \beta + \varrho, \quad \varrho < \gamma^{\eta}.$$

If $\beta \geqslant \gamma$ then we would have $\alpha \geqslant \gamma^{\eta} \cdot \gamma + \varrho \geqslant \gamma^{\eta+1}$. Therefore $\beta < \gamma$ and the ordinals β, η and ϱ satisfy the theorem.

THEOREM 2: *If* $\gamma > 1$ *and* $1 \leqslant \alpha < \gamma^{\eta}$, *then there exist a natural number* n *and sequences* $\beta_1, \beta_2, \ldots, \beta_n$ *and* $\eta_1, \eta_2, \ldots, \eta_n$ *such that*

$$\alpha = \gamma^{\eta_1}\beta_1 + \gamma^{\eta_2}\beta_2 + \ldots + \gamma^{\eta_n}\beta_n, \tag{1}$$

$$\eta > \eta_1 > \eta_2 > \ldots > \eta_n, \quad 0 \leqslant \beta_i < \gamma \quad \textit{for} \quad i = 1, 2, \ldots, n. \tag{2}$$

The proof is almost a repetition of that of Theorem 5.5. Namely, we define by induction three sequences φ, ψ, ϑ such that

$$\varphi_0 = \alpha, \quad \psi_0 = \min_{\xi}(\alpha < \gamma^{\xi}), \quad \vartheta_0 = 0,$$

$$\varphi_j = \gamma^{\psi_{j+1}} \vartheta_{j+1} + \varphi_{j+1} \quad \text{and} \quad \varphi_{j+1} < \gamma^{\psi_{j+1}}, \quad \psi_{j+1} < \psi_j, \quad \vartheta_{j+1} < \gamma$$

if $\varphi_j \neq 0$, and

$$\varphi_{j+1} = \psi_{j+1} = \vartheta_{j+1} = 0$$

if $\varphi_j = 0$.

The existence of these sequences follows from the theorem on definition by induction (see p. 233) and from Theorem 1. Clearly, $\varphi_j = 0$ from a j on.

Now we let $n^* = \min_j(\varphi_j = 0)$, $n = n^*-1$ and $\eta_j = \psi_j$, $\beta_j = \vartheta_j$ for $1 \leqslant j < n^*$.

Formula (1) for ordinals β_j and η_j satisfying condition (2) is called the *expansion of an ordinal number* α *for the base* γ. The ordinals β_j are called *digits* and the ordinals η_i *exponents* of this expansion. If $\gamma = \omega$, then the digits are natural numbers.

Examples

$\alpha = \omega^2+\omega\cdot 5+9$ is the expansion of the ordinal α for the base ω. To expand the same ordinal for the base 2 it suffices to notice that $\omega = \lim_{n<\omega} 2^n = 2^\omega$, thus

$$\omega^2 = (2^\omega)^2 = 2^{\omega\cdot 2} \quad \text{and} \quad \omega\cdot 5 = 2^{\omega+2}+2^\omega.$$

Therefore

$$\omega^2+\omega\cdot 5+9 = 2^{\omega\cdot 2}+2^{\omega+2}+2^\omega+2^3+2^0.$$

In a similar way we obtain

$$\omega^\omega = 2^{\omega^2}.$$

For epsilon-ordinals, the expansion for the base ω is $\varepsilon = \omega^\varepsilon$. Thus an epsilon-ordinal ε cannot be represented in the form (1) for $\gamma = \omega$ with exponents smaller than ε.

Two ordinals represented in the form (1) can be compared with respect to their magnitude by means of the following theorem.

THEOREM 3: *If* $\eta > \xi_1 > \ldots > \xi_p$ *and* $\gamma > \vartheta_n$ *for* $n \leqslant p$, *then*

$$\gamma^\eta > \gamma^{\xi_1}\vartheta_1+\gamma^{\xi_2}\vartheta_2+\ldots+\gamma^{\xi_p}\vartheta_p.$$

PROOF. From the assumption it follows that $\eta \geqslant \xi_1+1$ and $\gamma-\vartheta_1 > 0$. This implies

$$\gamma^\eta \geqslant \gamma^{\xi_1}\gamma = \gamma^{\xi_1}\vartheta_1+\gamma^{\xi_1}(\gamma-\vartheta_1) \geqslant \gamma^{\xi_1}\vartheta_1+\gamma^{\xi_1}.$$

Since $\gamma^{\xi_1} \geqslant \gamma^{\xi_2}\gamma$, we have $\gamma^{\xi_1} \geqslant \gamma^{\xi_2}\vartheta_2+\gamma^{\xi_2}$. Thus

$$\gamma^\eta \geqslant \gamma^{\xi_1}\vartheta_1+\gamma^{\xi_2}\vartheta_2+\gamma^{\xi_2}.$$

Repeating this operation, after p steps we obtain the inequality

$$\gamma^\eta \geqslant \gamma^{\xi_1}\vartheta_1+\gamma^{\xi_2}\vartheta_2+\ldots+\gamma^{\xi_p}\vartheta_p+\gamma^{\xi_p},$$

from which the required inequality follows directly (since $\gamma^{\xi_p} > 0$).

THEOREM 4: *If*

$$\alpha = \gamma^{\eta_1}\beta_1 + \ldots + \gamma^{\eta_{i-1}}\beta_{i-1} + \gamma^{\eta_i}\beta_i + \ldots + \gamma^{\eta_n}\beta_n,$$
$$\zeta = \gamma^{\eta_1}\beta_1 + \ldots + \gamma^{\eta_{i-1}}\beta_{i-1} + \gamma^{\xi_i}\vartheta_i + \ldots + \gamma^{\xi_p}\vartheta_p,$$

where

$$\eta_1 > \eta_2 > \ldots > \eta_n, \quad \eta_{i-1} > \xi_i > \ldots > \xi_p,$$
$$0 < \beta_1, \ldots, \beta_n < \gamma, \quad 0 < \vartheta_i, \ldots, \vartheta_p < \gamma,$$

and $\gamma^{\eta_i}\beta_i \neq \gamma^{\xi_i}\vartheta_i$, *then*

$$(\alpha > \zeta) \equiv [(\eta_i > \xi_i) \vee (\eta_i = \xi_i) \wedge (\beta_i > \vartheta_i)].$$

PROOF. Suppose that $\eta_i > \xi_i$. It follows from Theorem 3 that

$$\gamma^{\eta_i}\beta_i + \ldots + \gamma^{\eta_n}\beta \geqslant \gamma^{\eta_i} > \gamma^{\xi_i}\vartheta_i + \gamma^{\xi_{i+1}}\vartheta_{i+1} + \ldots + \gamma^{\xi_p}\vartheta_p,$$

therefore $\alpha > \zeta$, because by 5 (1) we have

$$\alpha = \gamma^{\eta_1}\beta_1 + \ldots + \gamma^{\eta_{i-1}}\beta_{i-1} + \gamma^{\eta_i}\beta_i + \ldots + \gamma^{\eta_n}\beta_n$$
$$> \gamma^{\eta_1}\beta_1 + \ldots + \gamma^{\eta_{i-1}}\beta_{i-1} + \gamma^{\xi_i}\vartheta_i + \ldots + \gamma^{\xi_p}\vartheta_p = \zeta.$$

In turn, suppose that $\eta_i = \xi_i$ and $\beta_i > \vartheta_i$. It follows from Theorem 3 that

$$\gamma^{\eta_i}(\beta_i - \vartheta_i) \geqslant \gamma^{\eta_i} > \gamma^{\xi_{i+1}}\vartheta_{i+1} + \ldots + \gamma^{\xi_p}\vartheta_p.$$

This implies

$$\gamma^{\eta_i}\vartheta_i + \gamma^{\eta_i}(\beta_i - \vartheta_i) > \gamma^{\xi_i}\vartheta_i + \gamma^{\xi_{i+1}}\vartheta_{i+1} + \ldots + \gamma^{\xi_p}\vartheta_p,$$

or

$$\gamma^{\eta_i}\beta_i > \gamma^{\xi_i}\vartheta_i + \gamma^{\xi_{i+1}}\vartheta_{i+1} + \ldots + \gamma^{\xi_p}\vartheta_p.$$

Therefore we have

$$\gamma^{\eta_i}\beta_i + \ldots + \gamma^{\eta_n}\beta_n > \gamma^{\xi_i}\vartheta_i + \ldots + \gamma^{\xi_p}\vartheta_p$$

which shows that $\alpha > \zeta$.

Finally, if either $\eta_i < \xi_i$ or $\eta_i = \xi_i$ and $\beta_i < \vartheta_i$, then applying an analogous reasoning we obtain the inequality $\alpha < \zeta$. This concludes the proof of Theorem 4.

Theorem 4 shows that expansions of ordinals possess properties analogous to those of expansions of natural numbers. In both cases, comparing the magnitudes of two numbers, we consider the first non-identical components of their expansions and we compare the exponents. If the exponents are equal, we compare their coefficients.

THEOREM 5: *For a given base, every ordinal number can be represented in the form* (1) *in exactly one way.*

PROOF. In fact, by virtue of Theorem 4, if

$$\gamma^{\eta_1}\beta_1 + \ldots + \gamma^{\eta_n}\beta_n = \gamma^{\xi_1}\vartheta_1 + \ldots + \gamma^{\xi_p}\vartheta_p$$

(where $\eta_1 > \eta_2 > \ldots > \eta_n$, $\xi_1 > \xi_2 > \ldots > \xi_p$, $0 < \beta_1, \ldots, \beta_n < \gamma$, $0 < \vartheta_1, \ldots, \vartheta_p < \gamma$), then $n = p$ and $\eta_k = \xi_k$, $\beta_k = \vartheta_k$ for $k \leqslant n$.

Theorem 4 allows us to establish a connection between the notion of power of ordinal numbers defined in § 6, and the notion of lexicographical ordering introduced in § 5, Chapter VI.

THEOREM 6: *The power* γ^η *is the order type of the set of those functions belonging to the lexicographically ordered set* $W(\gamma)^{W(\eta)}$ *which have values* $\neq 0$ *only for a finite number of arguments.*

PROOF. According to Theorem 2.4 $\gamma^\eta = \overline{W(\gamma^\eta)}$. To each ordinal $\alpha \in W(\gamma^\eta)$ there corresponds a unique expansion

$$\gamma^{\eta_1}\beta_1 + \gamma^{\eta_2}\beta_2 + \ldots + \gamma^{\eta_n}\beta_n,$$

where $\eta > \eta_1 > \eta_2 > \ldots > \eta_n$ and $0 < \beta_1, \ldots, \beta_n < \gamma$. In turn, this expansion can be associated with the function $f_\alpha \in W(\gamma)^{W(\eta)}$ defined as follows

$$f_\alpha(\eta_i) = \beta_i \quad \text{for} \quad i = 1, 2, \ldots, n,$$
$$f_\alpha(\xi) = 0 \quad \text{for} \quad \xi \neq \eta_1, \eta_2, \ldots, \eta_n.$$

Conversely, each function $g \in W(\gamma)^{W(\eta)}$ which assumes values different from 0 only for a finite number of arguments can be associated with the ordinal $\alpha \in W(\gamma^\eta)$ such that $g = f_\alpha$.

Finally, it follows from Theorem 4 that f_α precedes f_ζ in the lexicographical ordering of the set $W(\gamma)^{W(\eta)}$ if and only if $\alpha < \zeta$.

As another application of expansion (1), we shall establish a characteristic property of powers of the ordinal ω.

DEFINITION: An ordinal ϱ is said to be a *remainder* of an ordinal α if $\varrho \neq 0$ and there exists an ordinal σ such that $\alpha = \sigma + \varrho$.

THEOREM 7: *In order that every remainder of an ordinal α be equal to α it is necessary and sufficient that the ordinal α be a power of the ordinal ω.*

PROOF. If every remainder of the ordinal α equals α, then in the expansion

$$\alpha = \omega^{\eta_1}\beta_1+\omega^{\eta_2}\beta_2+ \ldots +\omega^{\eta_n}\beta_n$$

we have $n = 1$ and $\beta_1 = 1$, i.e. $\alpha = \omega^{\eta_1}$.

Suppose that $\alpha = \omega^\beta$ and let ϱ be a remainder of α. Thus for a certain σ we have

(3) $$\alpha = \sigma+\varrho \quad \text{and} \quad \sigma < \alpha, \quad \text{i.e.} \quad \sigma < \omega^\beta.$$

Let us expand σ for the base ω:

$$\sigma = \omega^\eta n+\omega^{\eta_1}n_1+ \ldots +\omega^{\eta_k}n_k.$$

We have $\eta < \beta$, i.e. $\eta+1 \leqslant \beta$. Let $\tau = \omega^\beta-\omega^{\eta+1}$. We obtain

$$\begin{aligned}\omega^\beta \leqslant \sigma+\omega^\beta \leqslant \omega^\eta(n+1)+\omega^\beta &= \omega^\eta(n+1)+\omega^{\eta+1}+\tau \\ = \omega^\eta(n+1+\omega)+\tau = \omega^\eta\cdot\omega+\tau &= \omega^{\eta+1}+\tau = \omega^\beta\end{aligned}$$

and we infer that $\omega^\beta = \sigma+\omega^\beta$, that is, $\alpha = \sigma+\alpha$. According to (3), $\sigma+\varrho = \sigma+\alpha$, which shows that $\varrho = \alpha$. Q.E.D.

Using expansions for the base ω we can define two operations on ordinals, called *natural addition* and *natural multiplication*.[1]) These operations have more properties in common with operations of addition and multiplication of natural numbers than the operations of ordinal addition and multiplication considered before.

In order to define these operations let us consider two ordinals

$$\alpha = \omega^{\eta_1}n_1+\omega^{\eta_2}n_2+ \ldots +\omega^{\eta_k}n_k,$$
$$\beta = \omega^{\zeta_1}m_1+\omega^{\zeta_2}m_2+ \ldots +\omega^{\zeta_l}m_l.$$

Upon completing these expansions by powers of ω with the coefficients 0, we obtain expansions with the same powers of ω:

(4) $$\alpha = \omega^{\xi_1}p_1+\omega^{\xi_2}p_2+ \ldots +\omega^{\xi_h}p_h,$$

(5) $$\beta = \omega^{\xi_1}q_1+\omega^{\xi_2}q_2+ \ldots +\omega^{\xi_h}q_h.$$

[1]) Natural addition and multiplication were discovered by Hessenberg [1], pp. 591–594.

The *natural sum* of α and β is defined to be

$$\alpha\,(+)\,\beta = \omega^{\xi_1}(p_1+q_1)+\omega^{\xi_2}(p_2+q_2)+\ \dots\ +\omega^{\xi_h}(p_h+q_h).$$

The *natural product* $\alpha\,(\cdot)\,\beta$ is defined to be the ordinal arising by formal multiplication of the expansions (4) and (5) as though they were polynomials in ω: multiplying two powers of ω we take the natural sum of the exponents and the terms obtained in this way are ordered according to their magnitude.

Natural addition and natural multiplication are *commutative.*

Examples

1. A natural sum may be different from the ordinary sum, for instance:

$$[\omega^2+\omega+1]\,(+)\,[\omega^3+\omega] = \omega^3+\omega^2+\omega\cdot 2+1,$$

$$[\omega^2+\omega+1]+[\omega^3+\omega] = \omega^3+\omega.$$

2. $[\omega^2+\omega+1]\,(\cdot)\,[\omega^3+\omega] = \omega^5+\omega^4+\omega^3\cdot 2+\omega^2+\omega.$

3. $[\omega^{\omega+1}+\omega^{\omega}+1]\,(\cdot)\,[\omega^{\omega+1}+\omega^{\omega}+\omega]$

$$= \omega^{\omega\cdot 2+2}+\omega^{\omega\cdot 2+1}\cdot 2+\omega^{\omega\cdot 2}+\omega^{\omega+2}+\omega^{\omega+1}\cdot 2+\omega^{\omega}+\omega.$$

4. Expansions (4) and (5) can be rewritten as

$$\alpha = \omega^{\xi_1}p_1(+)\,\omega^{\xi_2}p_2(+)\ \dots\ (+)\,\omega^{\xi_h}p_h,$$

$$\beta = \omega^{\xi_1}q_1(+)\,\omega^{\xi_2}q_2(+)\ \dots\ (+)\,\omega^{\xi_h}q_h.$$

Exercises

1. Show that the sum $\alpha\,(+)\,\beta$ is an increasing function with respect to α as well as with respect to β.

2. Show that for every ordinal γ there exist at most finitely many pairs α, β such that $\alpha\,(+)\,\beta = \gamma$.

3. Prove that if $\xi < \omega^{\omega^\alpha}$ and $\eta < \omega^{\omega^\alpha}$, then $\xi\cdot\eta < \omega^{\omega^\alpha}$.

Conversely, if an ordinal ζ satisfies the condition

(6) $$(\xi < \zeta)\wedge(\eta < \zeta) \rightarrow (\xi\cdot\eta < \zeta),$$

then there exists an α such that $\zeta = \omega^{\omega^\alpha}$.

REMARK: The ordinals ζ satisfying (6) for all ξ, η are called the *principal ordinals of multiplication.*[1])

[1]) Principal ordinals for other operations were defined and investigated by Jacobstahl [1], p. 149.

§ 8. The well-ordering theorem

Well-ordered sets owe their importance mainly to the fact that for each set there exists a relation which well orders it. This theorem, called the *well-ordering theorem* or *Zermelo's theorem*[1]) is equivalent to the axiom of choice on the basis of the axioms Σ[TR]. In this section we shall prove this equivalence and formulate several theorems equivalent to Zermelo's theorem. Some applications of these theorems will also be given.

THEOREM 1: *If A is any set such that there exists a choice function for the family $\boldsymbol{P}(A)-\{\emptyset\}$, then there exists a relation well ordering A.*

PROOF. Let f be a choice function for the family $\boldsymbol{P}(A)-\{\emptyset\}$; we can extend this function to the whole family $\boldsymbol{P}(A)$ letting $f(\emptyset) = p$ where p is any fixed element which does not belong to A.

Now let C denote the family of relations $R \subset A \times A$ well ordering their field. In virtue of the axiom of replacement, there exists a set consisting of all ordinals $\overline{R}$ where $R \in C$. Let α be smallest ordinal greater than every ordinal $\overline{R}$ of this set.

According to the theorem on definition by induction, there exists a transfinite sequence φ of type α such that

$$\varphi_\xi = f(A-\{\varphi_\eta\colon \eta < \xi\}).$$

If $\varphi_\xi \neq p$ then $\varphi_\xi \in A-\{\varphi_\eta\colon \eta < \xi\}$ and $\varphi_\xi \neq \varphi_\eta$ for $\eta < \xi$. If for all $\xi < \alpha$ we had $\varphi_\xi \neq p$, then there would exist a transfinite sequence of type α with distinct terms belonging to A. This implies that there exists a relation well ordering a subset of A into type α. But this contradicts the definition of α. Therefore there exists a smallest ordinal β such that $\varphi_\beta = p$. This implies that $A = \{\varphi_\eta\colon \eta < \beta\}$, thus A is the set of all terms of a transfinite sequence of type β whose terms are all distinct. Consequently there exists a relation well ordering the set A into type β. Q.E.D.

REMARK: In the proof above we used only ordinals of the form $\overline{R}$ where $R \subset A \times A$ and the ordinal α. It is therefore easy to reformulate the proof in such a way as to eliminate the notion of an ordinal. We

[1]) The well-ordering theorem was first proved by Zermelo [1].

simply replace $\vec{R}$ by the family of all relations $S \subset A \times A$ which are similar to R and α by the family of all well-ordering relations $R \subset A \times A$. The proof thus modified can be based on axioms Σ and is independent of the axiom of replacement.[1])

Another method of eliminating ordinal numbers consists in replacing them by the so-called *von Neumann's ordinals* to be discussed in § 9. The proof of Theorem 1 obtained by this modification is also based only on the axioms Σ but this time with the axiom of replacement.

The converse to Theorem 1 is also true:

THEOREM 2: *If there exists a relation well ordering a set A, then there exists a choice function for the family $\boldsymbol{P}(A) - \{\emptyset\}$.*

In fact, for $X \in \boldsymbol{P}(A) - \{\emptyset\}$ we define $f(X)$ to be the first element of X in the given well-ordering.

The following corollary is a direct consequence of Theorem 1.

°COROLLARY 3: *For every set A there exists a relation well ordering A.*

Now we shall formulate the so-called *maximum principle*, which is often used in place of the well-ordering theorem.

We shall use the terminology introduced in Chapter IV. If A is an ordered set, then we call a *chain of A* any linearly ordered subset of A. An upper bound of a subset B of A is any element x of A such that $x \geqslant b$ for every b in B. An element a of A is called *maximal* if there is no x such that $x > a$. We say that A is *closed* if every chain $B \subset A$ has a least upper bound in A.

THEOREM 4: *If A is an ordered set in which every chain has an upper bound and if there exists a choice function f for the family $\boldsymbol{P}(A) - \{\emptyset\}$, then A has a maximal element.*

PROOF. As in the proof of Theorem 1 we extend f by putting $f(\emptyset) = p$, where p is a fixed element not in A. Let α be the ordinal defined in the proof of Theorem 1. By the theorem on definitions by transfinite induction there exists a sequence φ of type α such that for each $\xi < \alpha$

$$\varphi_\xi = f\Big(\Big\{x \in A : \bigwedge_{\eta<\xi} [(\varphi_\eta \in A) \wedge (x > \varphi_\eta)]\Big\}\Big).$$

[1]) A proof of the well-ordering theorem on the basis of the system Σ° can be found in numerous books, e.g. Fraenkel [2].

If $\varphi_\xi \neq p$ then $\varphi_\eta \neq p$ and $\varphi_\xi > \varphi_\eta$ for every $\eta < \xi$. If we had $\varphi_\xi \neq p$ for each $\xi < \alpha$ we would obtain a transfinite sequence of type α with distinct terms belonging to A which would contradict the definition of α. Consequently there exists an ordinal ξ such that $\varphi_\xi = p$. If β is the smallest such ordinal, then there is no element x satisfying the inequalities $x > \varphi_\eta$ for each $\eta < \beta$. Hence every upper bound of the set $\{\varphi_\eta: \eta < \beta\}$ is a maximal element of A.

°COROLLARY 5: (The so-called ZORN MAXIMUM PRINCIPLE) *In every closed, ordered set there exists a maximal element.*

The following theorem is a special case of Theorem 4.

THEOREM 6: *If A is a family of sets with the following property*

$$(*) \qquad \Big(\bigcup_{X\in \boldsymbol{B}} X\Big) \in A \text{ for every monotonic family } \boldsymbol{B} \subset A$$

and if there exists a choice function for the family $\boldsymbol{P}(A)-\{\emptyset\}$, then there exists a maximal element in A.[1])

For the proof it suffices to notice that the family A is ordered by the inclusion relation and the union $\bigcup_{X\in \boldsymbol{B}} X$ is the least upper bound of $\boldsymbol{B}$.

Theorems 4 and 6 show that the existence of maximal elements follows from the axiom of choice. We now show that, conversely, the axiom of choice follows from the existence of maximal elements.

THEOREM 7: *If for every family of sets A satisfying condition $(*)$ there exists a maximal element, then for every family $\boldsymbol{Z}$ of non-empty sets there exists a choice function.*

PROOF. Let A be a family of functions f such that

(i) The domain of f is a family $\boldsymbol{C}_f \subset \boldsymbol{Z}$.

(ii) $f(X) \in X$ for all $X \in \boldsymbol{C}_f$.

Let us recall that a function f with domain $\boldsymbol{C}_f$ is the set of pairs $\langle X, f(X)\rangle$ where $X \in \boldsymbol{C}_f$. Thus the formula $f_1 \subset f_2$ means that $\boldsymbol{C}_{f_1} \subset \boldsymbol{C}_{f_2}$

[1]) Theorem 6 was proved by Kuratowski in [5]. The maximum principle was formulated by Zorn [1]. The main merit of Zorn's paper were examples which showed how easy it is to establish existential statements with the use of the maximum principle. Before Zorn's paper such proofs used transfinite induction and were much more cumbersome.

and $f_1(X) = f_2(X)$ for all $X \in \boldsymbol{C}_{f_1}$:

(1) $$f_1 \subset f_2 \equiv \{(\boldsymbol{C}_{f_1} \subset \boldsymbol{C}_{f_2}) \wedge \bigwedge_{X \in \boldsymbol{C}_{f_1}} [f_1(X) = f_2(X)]\},$$

that is, the function f_2 is an extension of the function f_1.

We shall show that the family $\boldsymbol{A}$ satisfies condition (*).

For this purpose, suppose that $\boldsymbol{B}$ is a chain included in $\boldsymbol{A}$ and let F denote the union $\bigcup(\boldsymbol{B})$. The elements of the set F are pairs of the form $\langle X, y\rangle$ where $X \in \boldsymbol{Z}$. In fact, each component f of the union F is a set of such pairs. If $\langle X, y_1\rangle \in F$ and $\langle X, y_2\rangle \in F$, then there exist f_1 and f_2 such that $\langle X, y_1\rangle \in f_1 \in \boldsymbol{B}$ and $\langle X, y_2\rangle \in f_2 \in \boldsymbol{B}$. This implies that $y_1 = f_1(X)$ and $y_2 = f_2(X)$.

Since the family $\boldsymbol{B}$ is monotonic, we have either $f_1 \subset f_2$ or $f_2 \subset f_1$. By (1) we infer that in both cases $y_1 = f_1(X) = f_2(X) = y_2$.

Thus the set F satisfies the condition

$$[\langle X, y_1\rangle \in F] \wedge [\langle X, y_2\rangle \in F] \rightarrow (y_1 = y_2),$$

i.e. F is a function. The domain of this function is $\bigcup_{f \in \boldsymbol{B}} \boldsymbol{C}_f$, i.e. a family included in $\boldsymbol{Z}$.

If $\langle X, y\rangle \in F$ then there exists a function $f \in \boldsymbol{B}$ such that $\langle X, y\rangle \in f$. This implies that $y = f(X)$; hence $y = F(X) = f(X) \in X$. Therefore the function F belongs to the family $\boldsymbol{A}$. This shows that this family satisfies condition (*).

By assumption there exists a maximal element f_0 in the family $\boldsymbol{A}$. We shall show that $\boldsymbol{C}_{f_0} = \boldsymbol{Z}$. Suppose the contrary. Then there exists an element X of the difference $\boldsymbol{Z} - \boldsymbol{C}_{f_0}$ which is a non-empty set. Thus there also exists an element $x \in X$. Letting $f = f_0 \cup \{\langle X, x\rangle\}$, we obtain $f_0 \subset f$, $f_0 \neq f$ and $f \in \boldsymbol{A}$. But this contradicts the hypothesis that f_0 is a maximal element in $\boldsymbol{A}$.

Hence the function f_0 satisfies the condition $f_0(X) \in X$ for all $X \in \boldsymbol{Z}$. Q.E.D.

Applications

1. *Extension of an order to a linear order.*

°THEOREM 8: *For every relation R_0 ordering the set A there exists a relation linearly ordering A which contains R_0.*

PROOF. Let K be the family of relations ordering A and containing R_0. It can easily be shown that this family satisfies condition $(*)$ of Theorem 6, therefore it contains a maximal element R.

We shall show that R is the required extension of R_0 to a linear order. Since R is by definition an order, it suffices to show that the relation R is connected. Suppose on the contrary that there exist elements $a, b \in A$ such that $\neg(aRb) \wedge \neg(bRa)$. We shall show that the relation

$$R' = R \cup \{\langle x, y \rangle : (xRa) \wedge (bRy)\} = R \cup S$$

orders A. This will provide a contradiction since $R' \supset R$ and $R' \neq R$.

Clearly, $xR'x$ for every $x \in A$. To show that the relation R' is transitive, suppose that $xR'y$ and $yR'z$. One of the following cases holds:

(i) $(xRy) \wedge (yRz)$;

(ii) $(xRy) \wedge (yRa) \wedge (bRz)$;

(iii) $(xRa) \wedge (bRy) \wedge (yRz)$;

(iv) $(xRa) \wedge (bRy) \wedge (yRa) \wedge (bRz)$.

Case (iv) is impossible: it implies bRa, which contradicts the hypothesis. In case (i) we obtain xRz since R is transitive. Therefore we also have $xR'z$. In cases (ii) and (iii) we obtain $(xRa) \wedge (bRz)$ by the transitivity of R, thus xSz and, consequently, $xR'z$. This shows that R' is transitive.

Finally, in order to prove that R' is antisymmetric, suppose that $xR'y$ and $yR'x$. We now have the cases analogous to (i)–(iv) where z is replaced by x. Cases (ii)–(iv) are impossible, case (i) implies $x = y$. Thus Theorem 8 is proved.

°2. Let A be a distributive lattice and I_0 its ideal not containing an element b (see p. 158). The family of all ideals $I \supset I_0$ of the lattice A not containing the element b satisfies condition $(*)$ of Theorem 6. Therefore there exist maximal elements of this family. In particular, there exist maximal elements in the family $P_{a,b}$ of those ideals which contain a but which do not contain b (provided that a non $\geqslant b$, (see p. 159).

If, in particular, A is a lattice with unit i and I is an ideal different from A (hence I does not contain i), then there exists at least one maximal ideal different from A and containing I. Such an ideal is prime (see Exercise IV.12.2).

°3. Let $\overline{\overline{A}} = \mathfrak{m} \geqslant \mathfrak{a}_0$ and let $\boldsymbol{M}$ be a family of a power $\leqslant \mathfrak{m}$ such that each element of $\boldsymbol{M}$ is a subset of A and has the power $\mathfrak{m}$. Then there is a set Z such that

$$(2) \quad Z \subset A, \quad \overline{\overline{Z}} = \mathfrak{m}, \quad \overline{\overline{A-Z}} = \mathfrak{m}$$

$$\text{and} \quad Z \cap X \neq \emptyset \neq X - Z \quad \text{for every } X \in \boldsymbol{M}.$$

Proof. In virtue of the well-ordering theorem there is a smallest ordinal α such that there exists a sequence of type α without repetitions composed of all the elements of A. Let x_ξ denote the ξth term of this sequence and let M_ξ be the ξth term of a sequence of type α (not necessarily without repetitions) which contains all the elements of $\boldsymbol{M}$.

We define by transfinite induction two sequences p and q of type α. Namely p_ξ is the first term x_ν belonging to $M_\xi - S_\xi$, and q_ξ is the first term x_μ belonging to $(M_\xi - \{p_\xi\}) - S_\xi$ where

$$S_\xi = \{p_\eta \colon \eta < \xi\} \cup \{q_\eta \colon \eta < \xi\}.$$

The elements p_ξ and q_ξ exist, because the set S_ξ is of power $< \mathfrak{m}$ and M_ξ is of power $\mathfrak{m}$.

The set Z consisting of all p_ξ where $\xi < \alpha$ satisfies condition (2).

The theorem above has an interesting topological application.[1]) Let $A = \mathscr{E}$ (thus $\mathfrak{m} = \mathfrak{c}$) and let $\boldsymbol{M}$ be a family of non-empty perfect sets (a *perfect set* is a set identical with its derivative). Such sets are of power $\mathfrak{c}$.[2]) Hence there is a set Z which has a point in common with every perfect subset of the set $\mathscr{E}$ and whose complement possesses the same property.

It can be proved that such a set is non-measurable in the sense of Lebesgue.

4. Let X be an arbitrary set, $\boldsymbol{R}^*$ a family contained in $\boldsymbol{P}(X)$. We shall prove that *the set A of all families $\boldsymbol{R}$ with the finite intersection*

[1]) The theorem given in Example 3 and its application are due to Bernstein [2].
[2]) See also Kuratowski [1], p. 514.

property satisfying the condition $\boldsymbol{R}^* \subset \boldsymbol{R} \subset \boldsymbol{P}(X)$ *possesses property* (∗) (see p. 256).

Suppose that $B \subset A$ and that the set B is linearly ordered by inclusion. We shall show that $\bigcup_{R \in B} \boldsymbol{R} \in A$. Clearly, it suffices to show that this union possesses the finite intersection property. Let $n \in N$ and $X_i \in B$ for $i < n$. For each i there is a family $\boldsymbol{R}_i \in B$ such that $X_i \in \boldsymbol{R}_i$. Since B is linearly ordered, one of these families, say $\boldsymbol{R}_0$, contains all the remaining families. This implies $X_i \in \boldsymbol{R}_0$ for $i < n$ and since $\boldsymbol{R}_0$ has the finite intersection property, we obtain $\bigcap_i X_i \neq \emptyset$. Q.E.D.

° It follows from Theorem 6 that *for every family* $\boldsymbol{R}_0 \subset \boldsymbol{P}(X)$ *with the finite intersection property there is a maximal family* $\boldsymbol{R}$ *with the finite intersection property such that* $\boldsymbol{R} \subset \boldsymbol{P}(X)$ *and* $\boldsymbol{R}$ *contains* $\boldsymbol{R}_0$.

5. *Hamel's basis.*[1]) A set $X \subset \mathscr{E}$ is said to be *independent* if for any finite sequence $x_0, x_1, \ldots, x_{n-1}$ of distinct elements of X the equation $r_0 x_0 + r_1 x_1 + \ldots + r_{n-1} x_{n-1} = 0$ is satisfied by rational numbers $r_0, \ldots, r_{n-1}$ if and only if all these numbers are equal to 0. An example of an independent set is $\{\sqrt{2}, \sqrt{3}\}$.

It is easy to show that if $\boldsymbol{B}$ is a monotonic family of independent sets, then $\bigcup_{X \in B} X$ is also an independent set. By Theorem 6 this implies

°THEOREM 9: *There exists a maximal independent set.*

Such a set is called a *Hamel basis* for $\mathscr{E}$.

If H is a Hamel basis, then every number $x \neq 0$ can be uniquely represented in the form

$$x = \sum_{i<n} r_i b_i, \tag{3}$$

where $n \in N$, b_i are distinct elements of the basis and r_i rational coefficients different from 0. For if there existed a number x not having such a representation, then the set $H \cup \{x\}$ would be independent, contrary to the assumption that H is maximal. On the other hand, if there were two representations $\sum_{i<n} r_i' b_i' = \sum_{j<m} r_j'' b_j''$, then the elements of the set $\{b_0', \ldots, b_{n-1}', b_0'', \ldots, b_{m-1}''\}$ would not be independent.

[1]) Hamel's basis was first defined in Hamel [1].

°COROLLARY 10: *There exist non-continuous functions of the real variable x satisfying for all x, y the equation*

$$f(x+y) = f(x)+f(y).$$

In fact, let H be a Hamel basis and let $x_0 \in H$. Denoting by $f(x)$ the number r_0 such that in the expansion (3) x_0 occurs with the coefficient r_0, we obtain the required function. This function is not continuous, because it takes only rational values and is not constant.

The theory given in this section enables us to prove the theorem mentioned on p. 191.

°THEOREM 11: *If $\mathfrak{f}$ is a function defined on the set T whose values are cardinal numbers, then there exists a function F defined on T such that $\overline{\overline{F_t}} = \mathfrak{f}_t$ for every $t \in T$.*

PROOF. Let $t \in T$. Since $\mathfrak{f}_t$ is a cardinal, there exists a set X such that $\mathfrak{f}_t = \overline{\overline{X}}$. According to Corollary 3 there exist an ordinal α and a relation R such that R orders X into type α. Let α_t be the smallest ordinal with this property. Now we define the function F by $F_t = W(\alpha_t)$.

6. *Chains in pseudo-trees.* As the last application of the maximal principle we shall prove the following theorem which will be needed later in Chapter IX.

THEOREM 12: *Each chain C in a pseudo-tree can be extended to a branch.*

PROOF. The family of all chains which contain C satisfies the assumptions of the maximum principle and so contains a maximal element.

Exercises

1. A family A of sets is said to be *inductive* if it has the following properties:

(i) if $X \in A$, then every finite set $Y \subset X$ belongs to A,

(ii) if every finite set $Y \subset X$ belongs to A, then $X \in A$.

Show (without the axiom of choice) that the maximum principle is equivalent to the theorem: every inductive family possesses a maximal element.[1])

2. Show (without the axiom of choice) that the maximum principle is equivalent to the following theorem: every linearly ordered subset Z of an ordered set A (that is, a set with the property $x, y \in Z \rightarrow [(x \leqslant y) \vee (y \leqslant x)]$ is contained in the maximal linearly ordered set included in A.[2])

[1]) The formulation of the maximum principle given in Exercise 1 is due to Teichmüller [1], this paper contains also other useful forms of the maximum principle.

[2]) The result in Exercise 2 is due to Birkhoff [1], p. 42.

3. Show (without using the axiom of choice) that the maximum principle is equivalent to the following theorem: for every family $\boldsymbol{F}$ of non-empty sets there exists a maximal family of disjoint sets contained in $\boldsymbol{F}$.[1])

§ 9. Von Neumann's method of elimination of ordinal numbers

In this section our exposition is based exclusively on the axioms of Σ.

We shall show that it is possible to define sets possessing exactly the same properties as ordinal numbers. We shall establish a one-to-one correspondence between those sets and types of well-ordering relations.

DEFINITION:[2]) A set A is said to be an *ordinal number in the sense of von Neumann* (briefly: a *VN ordinal*) if it has the following properties:

1. *Every element of A is a set.*
2. *If $X \in A$ then $X \subset A$.*
3. *If $X, Y \in A$ then $X = Y$ or $X \in Y$ or $Y \in X$.*
4. *If $\emptyset \neq B \subset A$ then there exists an X such that $X \in B$ and $X \cap B = \emptyset$.*

Examples of VN ordinals:

(i) the empty set $N_0 = \emptyset$,
(ii) the set $N_1 = \{\emptyset\}$,
(iii) the set $N_2 = \{\emptyset, \{\emptyset\}\} = \{N_0, N_1\}$,
(iv) the set $N_3 = \{N_0, N_1, N_2\}$,
(v) the set $N_\omega = \{N_0, N_1, N_2, \ldots\}$,
(vi) the set $N_{\omega+1} = N_\omega \cup \{N_\omega\}$.

We prove several properties of VN ordinals.

5. *If A is a VN ordinal then there exists no finite sequence of sets $X_1, \ldots, X_k$ such that $X_k \in X_1 \in X_2 \in \ldots \in X_{k-1} \in X_k \in A$.*

PROOF. Suppose that there exist sets $X_1, \ldots, X_k$ with this property. Let $B = \{X_1, \ldots, X_k\}$. Since $X_k \in A$, we have $X_k \subset A$ by 2 and thus

[1]) For the exercise 3 see Vaught [1]. Many similar results dealing with the maximum principle can be found in the books: Sierpiński [1], Rubin and Rubin [1].

[2]) The definition of VN ordinals is due to von Neumann [1]. In the recent literature on abstract set theory it is customary to call the VN ordinals simply ordinals and to identify cardinal numbers with initial ordinals. In this way it is possible to eliminate from set theory the primitive notion of a relational type.

$X_{k-1} \in A$. By the same argument, $X_{k-2} \in A$ and so forth. Therefore all the sets $X_1, \ldots, X_k$ belong to A and consequently $B \subset A$. None of the sets X_i satisfies condition 3. In fact, for $i > 1$ we have $X_{i-1} \in X_i \cap B$ and for $i = 1$ we have $X_k \in X_1 \cap B$. Q.E.D.

6. *If A is a VN ordinal and $M \in A$, then M is also a VN ordinal.*

PROOF. We prove that M satisfies 1–4.

(i) If $X \in M$, then we also have $X \in A$ because 2 implies $M \subset A$, and thus X is a set.

(ii) Suppose that $X \in M$ and $Y \in X$. We have

$$Y \in X \in M \in A,$$

which in view of the inclusion $M \subset A$ implies that $Y \in X \in A$; consequently $Y \in A$ by 2. According to 3 we have either $Y = M$ or $M \in Y$ or $Y \in M$. In the first case we obtain

$$M \in X \in M \in A,$$

and in the second

$$M \in Y \in X \in M \in A,$$

which contradicts Theorem 5. Thus $Y \in M$. As Y is arbitrary we infer that $X \subset M$. Therefore the set M satisfies condition 2.

(iii) If $X, Y \in M$, then $X, Y \in A$ because $M \subset A$. Thus 3 implies that either $X = Y$ or $X \in Y$ or $Y \in X$.

(iv) Suppose that $\emptyset \neq B \subset M$. The set B is thus a non-empty subset of A and in view of 4 it contains an element X such that $X \cap B = \emptyset$. This shows that M itself satisfies condition 4.

7. *If A and B are VN ordinals, then*

$$(A \in B) \equiv (A \subset B) \wedge (A \neq B).$$

PROOF. If $A \in B$, then $A \subset B$ by 2 and $A \neq B$ since otherwise we would have $B \in B$, contrary to 5. Suppose that $A \neq B$ and $A \subset B$. The set $B - A$ is therefore a non-empty subset of B and according to 4 there exists a set $X \in B - A$ such that $X \cap (B - A) = \emptyset$. Now it suffices to show that $X = A$, because $X = A$ together with $X \in B$ imply $A \in B$, which proves the theorem.

The condition $X \in B$ implies $X \subset B$. Since $X \cap (B - A) = \emptyset$, we get $X - A = \emptyset$, that is, $X \subset A$. Suppose that $A - X \neq \emptyset$. Hence there exists a set Y such that $Y \in A - X$ and $Y \cap A - X = \emptyset$. Since $A - X \subset B$, we infer that $Y \in B$, and according to 3 we have either $Y \in X$ or $X \in Y$ or $X = Y$. But $Y \in X$ is impossible, for $Y \in A - X$; similarly $X \in Y$ is not the case, because it would imply $X \notin A - X$, that is (in view of $X \in A$) $X \in X \in A$, contrary to 5; finally, we cannot have $X = Y$ since $Y \in A$ and $X \in B - A$.

Hence we have proved that $A - X = \emptyset$, i.e. $A \subset X$, which shows that $A = X$.

8. *Each VN ordinal is well ordered by the inclusion relation.*

PROOF. It suffices to show that if A is a VN ordinal, then

(a) the inclusion relation is connected in A;

(b) every non-empty subset of A has a first element.

Now (a) follows from 3, 6 and 7; and (b) follows from 4 and 7.

9. *If A and B are VN ordinals, then $A \cap B$ is also a VN ordinal.*

PROOF. We show that $A \cap B$ satisfies conditions 1–4.

(i) From $X \in A \cap B$ it follows that $X \in A$; thus X is a set.

(ii) From $X \in A \cap B$ it follows that $X \in A$ and $X \in B$; hence $X \subset A$; $X \subset B$ and consequently $X \subset A \cap B$.

(iii) From $X, Y \in A \cap B$ it follows that $X, Y \in A$; thus $X = Y$ or $X \in Y$ or $Y \in X$.

(iv) From $\emptyset \neq M \subset A \cap B$ it follows that $\emptyset \neq M \subset A$; hence there exists an X such that $X \in M$ and $X \cap M = \emptyset$.

10. *If A and B are VN ordinals, then either $A \subset B$ or $B \subset A$.*

PROOF. Suppose that $A \neq A \cap B \neq B$. From 9 and 7 it follows that $A \cap B \in A$ and $A \cap B \in B$. This implies $A \cap B \in A \cap B$, which contradicts 5 because $A \cap B$ is a VN ordinal. Hence $A = A \cap B$ or $B = A \cap B$.

11. *If A and B are distinct VN ordinals, then either A is a segment of B or B is a segment of A. Consequently, these sets are not similar (with respect to the inclusion relation).*

PROOF. Suppose that $A \neq B$; by 10 we have either $A \subset B$ or $B \subset A$. Suppose that the former holds. It follows from 7 that $A \in B$, which shows that the elements of A precede (in the set B) the element A. Hence the set A is a segment of B, consequently it cannot be similar to B (see Corollary 1.7).

12. *For every relation R well ordering its field there exists exactly one VN ordinal ordered by the inclusion relation similarly to R.*

PROOF. Let Z be the field of the relation R and H the set of those $z \in Z$ for which there exists exactly one VN ordinal N_z satisfying the condition: N_z is ordered by the inclusion relation similarly to the segment $O(z)$ of the set Z; in this case we write $N_z \sim O(z)$.

Suppose that $O(x) \subset H$. We shall show that $x \in H$. By 11 there exists at most one VN ordinal similar to $O(x)$. Hence it suffices to show that there exists at least one such VN ordinal.

Let

$$N_x = \{X: \bigvee_z (z \prec x) \wedge (X = N_z)\}.$$

We are going to show that N_x is a VN ordinal. In fact, condition 1 clearly holds.

If $N_z \in N_x$ and $Y \in N_z$, then Y is a VN ordinal. Moreover, since $Y \in N_z \sim O(z)$, we see by 11 that Y is similar to a segment of N_z; hence Y is similar to a segment $O(t)$ of the set $O(z)$. Since $t \prec z \prec x$, we obtain $t \in O(x)$ and, consequently, $t \in H$ and $Y \sim N_t$. In view of 11 this implies $Y = N_t$ and finally $Y \in N_x$, because $N_t \in N_x$. Thus $N_z \subset N_x$; that is, N_x satisfies condition 2.

Condition 3 follows directly from 10 and 7.

Now let B be a non-empty set contained in N_x and let z_0 be the smallest element of Z such that $N_{z_0} \in B$. If there were $Y \in N_{z_0}$ such that $Y \in B$, then we would have $Y = N_z$ where $z \prec z_0$, which contradicts the definition of z_0.

Hence the set N_x is a VN ordinal. If $z_1 \prec z_2 \prec x$, then $N_{z_1} \sim O(z_1)$ and $N_{z_2} \sim O(z_2)$. Thus N_{z_1} is similar to a segment of N_{z_1}. This implies by 11 that $N_{z_1} \subset N_{z_2}$ and $N_{z_1} \neq N_{z_2}$. Therefore, $N_x \sim O(x)$, which proves that $x \in H$.

By induction we now infer that $H = Z$. The set

$$N^* = \{X: \bigvee_x (x \in Z) \wedge (X = N_x)\}$$

is a VN ordinal ordered similarly to Z. The proof is analogous to that carried out for the set N_x.

It follows from properties 12, 11, 8 that VN ordinals indeed satisfy all the requirements for ordinal numbers.

In connection with the reasoning just given it is worth mentioning that in the proof of property 12 we made essential use of the axiom of replacement. Without this axiom the existence of the sets N_x and N^* could not be proved.

CHAPTER VIII

ALEPHS AND RELATED TOPICS

In this chapter we shall discuss applications of the theory of well-ordering to the arithmetic of cardinal numbers.

§ 1. Ordinal numbers of power $\mathfrak{a}$ [1]

The cardinal number of an ordinal ξ is the power of any set ordered in type ξ. We denote this cardinal number by $\overline{\xi}$.

Thus

$$\overline{\xi} = \overline{\overline{W(\xi)}}.$$

Ordinals of power $\mathfrak{a}$ can be treated as the types of well-ordered sets of natural numbers. This fact implies the following theorem.

THEOREM 1: *All ordinals of power* $\mathfrak{a}$ *form a set.*

DEFINITION 1: The smallest ordinal greater than every ordinal of power $\mathfrak{a}$ will be denoted by ω_1.

The existence of the ordinal ω_1 follows from Corollary VII.2.8.

THEOREM 2: $(\xi < \omega_1) \equiv (\overline{\xi} \leqslant \mathfrak{a})$.

PROOF. If $\overline{\xi} \leqslant \mathfrak{a}$, then $\xi < \omega_1$ by Definition 1. Conversely, if $\xi < \omega_1$, then there exists an ordinal ζ such that $\xi \leqslant \zeta$ and $\overline{\zeta} = \mathfrak{a}$, thus $\overline{\xi} \leqslant \overline{\zeta} = \mathfrak{a}$.

DEFINITION 2:[2] $\aleph_1 = \overline{\omega}_1$, i.e. $\aleph_1 = \overline{\overline{W(\omega_1)}}$.

In Theorem 2, letting $\xi = \omega_1$ we obtain $\aleph_1 \not\leqslant \mathfrak{a}$. On the other hand, $\mathfrak{a} \leqslant \aleph_1$ and thus we have the following corollary.

COROLLARY 3: $\aleph_1 > \mathfrak{a}$.

[1]) Called by Cantor *numbers of the second class*. According to Cantor, the first class consists of finite numbers.

[2]) $\aleph$ is the Hebrew letter aleph.

It follows that the set of ordinals ξ such that $\overline{\xi} \leqslant \mathfrak{a}$, that is, the set $W(\omega_1)$, is uncountable.

THEOREM 4: *If* $\mathfrak{m} < \aleph_1$ *then* $\mathfrak{m} \leqslant \mathfrak{a}$.

In other words, *there is no cardinal which lies between* $\mathfrak{a}$ *and* $\aleph_1$.

PROOF. Let $\mathfrak{m} < \aleph_1$. Then there exists a set $M \subset W(\omega_1)$ such that $\overline{\overline{M}} = \mathfrak{m}$. Let $\overline{M} = \xi$. Thus $\xi \leqslant \omega_1$. Moreover, $\xi \neq \omega_1$ as otherwise $\overline{\xi} = \overline{\omega}_1$ and $\mathfrak{m} = \aleph_1$. Therefore, $\xi < \omega_1$ and thus, by Theorem 2, $\overline{\xi} \leqslant \mathfrak{a}$ and thus $\mathfrak{m} \leqslant \mathfrak{a}$. Q.E.D.

The following form of the induction principle holds for ordinals $\xi < \omega_1$.

THEOREM 5: *Let the set A of ordinals satisfy the conditions*:

(1) $$0 \in A,$$

(2) $$\xi \in A \rightarrow (\xi+1) \in A,$$

(3) *if* φ *is an increasing sequence and if* $\varphi(n) \in A$ *for* $n \in N$, *then*

$$[\lim_n \varphi(n)] \in A.$$

Then $W(\omega_1) \subset A$.

PROOF. Suppose that the theorem does not hold. Let α be the least ordinal such that $\alpha < \omega_1$ and $\alpha \notin A$. From (1) it follows that $\alpha \neq 0$. If α is not a limit ordinal, then $\alpha = \xi+1$ for a ξ in A, whence by (2) $\xi+1 \in A$, and $\alpha \in A$, contrary to the definition of α. It remains to examine the case in which α is a limit ordinal.

Since $\overline{\alpha} \leqslant \mathfrak{a}$, there exists a relation R which well orders the set N of natural numbers into type α.

We define a sequence $k_0, k_1, \ldots$ of natural numbers by induction: let k_0 be the first element of the set N with respect to the relation R and let k_{n+1} be the least number $k > k_n$ such that $\overline{O_R(k_n)} < \overline{O_R(k)}$ (where $O_R(k)$ denotes, as usual, the segment of N determined by k). Such a number k exists because α is a limit ordinal.

Let $\varphi(n) = \overline{O_R(k_n)}$. Thus $\varphi(n) < \varphi(n+1)$. Moreover, $\varphi(n) < \alpha$ for $n = 0, 1, 2, \ldots$, because $O_R(k_n)$ is a segment of a set of type α. If $\xi < \alpha$ then there exists a number m such that $\overline{O_R(m)} = \xi$. Since the sequence $k_0, k_1, \ldots$ is increasing, it follows that for some n, $O_R(m) \subset O_R(k_n)$, whence $\xi \leqslant \varphi(n)$. Hence $\alpha = \lim_{n<\omega} \varphi(n)$. On the other hand, $\varphi(n) < \alpha$ implies $\varphi(n) \in A$ for $n = 0, 1, 2, \ldots$ and thus, by (3), $\alpha \in A$. But this contradicts the definition of α.

Clearly the set $W(\omega_1)$ satisfies conditions (1) and (2). In fact, a more general theorem holds for $W(\omega_1)$.

THEOREM 6: *If $\xi < \omega_1$ and $\eta < \omega_1$, then $\xi+\eta < \omega_1$ and $\xi \cdot \eta < \omega_1$.*

For $\overline{\overline{\xi+\eta}} = \bar{\xi}+\bar{\eta}$ and $\overline{\overline{\xi \cdot \eta}} = \bar{\xi} \cdot \bar{\eta}$; since $\bar{\xi} \leqslant \mathfrak{a}$ and $\bar{\eta} \leqslant \mathfrak{a}$, $\mathfrak{a}+\mathfrak{a} = \mathfrak{a} = \mathfrak{a} \cdot \mathfrak{a}$, it follows that $\overline{\overline{\xi+\eta}} \leqslant \mathfrak{a}$ and $\overline{\overline{\xi \cdot \eta}} \leqslant \mathfrak{a}$.

Using the axiom of choice we shall show that the set $W(\omega_1)$ satisfies condition (3).

°THEOREM 7: *If $\varphi(0) < \varphi(1) < \ldots$ and $\varphi(n) < \omega_1$, then $\lim_{n<\omega} \varphi(n) < \omega_1$.*

PROOF. Let $\alpha = \lim_{n<\omega} \varphi(n)$. Then $W(\alpha) = \bigcup_n W(\varphi(n))$. The set $W(\alpha)$, being a countable union of countable sets, is countable (see p. 171). Thus $\alpha < \omega_1$.

Application. Let F_α be the set defined on p. 235.

°THEOREM 8: *The family F_{ω_1} is identical with the family* Bor *of Borel sets (i.e. the least σ-additive and δ-multiplicative family containing all closed sets).*

PROOF. By transfinite induction with respect to α it is easy to show that for every α (in particular for $\alpha = \omega_1$), $F_\alpha \subset$ Bor. It remains to show that the family F_{ω_1} is σ-additive and σ-multiplicative. For this purpose let X be a sequence of sets such that $X_n \in F_{\omega_1}$ for all n. Thus for all n there is an ordinal α_n such that $X_n \in F_{\alpha_n}$; we may assume that α_n is the least ordinal greater than α_{n-1} such that $X_n \in F_{\alpha_n}$. By Theorem 7 it follows that there exists β such that $\alpha_n < \beta < \omega_1$ for every n, moreover, we may assume that β is, for instance, odd. Then $\bigcup X_n \in \big(\bigcup_{\gamma<\beta} F_\gamma\big)_\sigma = F_{\beta+1} \subset F_{\omega_1}$ and $\bigcap X_n \in \big(\bigcup_{\gamma<\beta+1} F_\gamma\big)_\delta = F_{\beta+2} \subset F_{\omega_1}$.

Exercises

1. Prove (without using the axiom of choice) that if $\alpha < \omega_1$ and $\beta < \omega_1$, then $\alpha^\beta < \omega_1$.

2. Let Φ_α be the family of analytically representable functions of class α (see p. 236). Prove that the union $\bigcup_{\alpha<\omega_1} \Phi_\alpha$ is the least family of real functions such that

(i) every continuous function belongs to the family,

(ii) if f_n belongs to the family for $n \in N$ and if $f(x) = \lim_{n=\infty} f_n(x)$ for every x, then f belongs to the family.

Discuss the role played by the axiom of choice in the proof of this theorem.

3. If $\{X_\alpha\}_{\alpha<\omega_1}$ is a sequence of type ω_1 of closed subsets of the space $\mathscr{E}$ or N^N and if $X_\alpha \supset X_{\alpha+1}$ for all $\alpha < \omega_1$, then there exists an ordinal $\beta < \omega_1$ such that $X_\alpha = X_\beta$ for all $\alpha \geqslant \beta$.

Hint: Denote by $N_0, N_1, N_2, \ldots$ a sequence whose terms constitute an open subbase of the space. Associate with the ordinal α where $X_\alpha \neq X_{\alpha+1}$ the least number m such that $N_m \cap X_\alpha \neq \emptyset = N_m - X_{\alpha+1}$ and show that with distinct α are associated distinct natural numbers.

4. Prove that if A is an arbitrary subset of $\mathscr{E}$ (or of N^N), then for every transfinite sequence of derivatives of A (see p. 235) there exists a term $A^{(\beta)}$ such that $A^{(\beta)} = A^{(\alpha)}$ for all $\alpha \geqslant \beta$, where $\beta < \omega_1$

Hint: Notice that all derivatives $A^{(\alpha)}$ are closed sets and apply Exercise 3.

5. Prove the following theorem of Cantor–Bendixson: Every closed set $A \subset \mathscr{E}$ (or $A \subset N^N$) is the union of a perfect set and of a countable set.

Hint: Show that the difference $A - A'$ is a countable set (see Exercise V.2.5).

6. Show that $2^{2^{\mathfrak{a}}} \geqslant \aleph_1$ without using the axiom of choice.

Hint: The set $\boldsymbol{P}(N \times N)$ is the union $Z \cup \bigcup_{\xi<\omega_1} Z_\xi$, where Z_ξ is the set of relations well-ordering their fields in type ξ and where Z is the set of relations which are not well-orderings.

§ 2. The cardinal $\aleph(\mathfrak{m})$. Hartogs' aleph[1])

We now generalize the construction carried out in § 1 for the cardinal $\mathfrak{a}$ to the case of an arbitrary cardinal $\mathfrak{m}$.

THEOREM 1: *For every cardinal* $\mathfrak{m}$ *there exists a set*

$$Z(\mathfrak{m}) = \{\xi: \bar{\xi} \leqslant \mathfrak{m}\}.$$

PROOF. Let $\bar{\bar{A}} = \mathfrak{m}$. Every relation R whose field is contained in A is a subset of $A \times A$, that is, $R \subset \boldsymbol{P}(A \times A)$. Therefore there exists a set $\boldsymbol{R}$ of all relations $R \subset \boldsymbol{P}(A \times A)$ which well order their fields.

Associate with every relation $R \in \boldsymbol{R}$ its type. By the axiom of replacement we obtain the set $Z(\mathfrak{m})$ of ordinals such that

$$\xi \in Z(\mathfrak{m}) \to \bar{\xi} \leqslant \mathfrak{m}.$$

[1]) The construction of $\aleph(\mathfrak{m})$ which generalizes the construction given in Section 1 is due to Hartogs [1]. The letter "aleph" was first used by Cantor.

Conversely, if $\overline{\xi} \leqslant \mathfrak{m}$, then there exists a relation R ordering a subset of A in type ξ. Thus $\xi \in Z(\mathfrak{m})$. Q.E.D.

THEOREM 2: *If* $\xi \in Z(\mathfrak{m})$, *then* $W(\xi) \subset Z(\mathfrak{m})$.

For $\eta < \xi$ implies that $\overline{\eta} \leqslant \overline{\xi}$.

DEFINITION: $\aleph(\mathfrak{m}) = \overline{\overline{Z(\mathfrak{m})}}$.

By this definition we have correlated with every cardinal $\mathfrak{m}$ an aleph $\overline{\overline{Z(\mathfrak{m})}}$. This operation is referred to as Hartog's aleph function.

THEOREM 3: $\aleph(\mathfrak{m}) \not\leqslant \mathfrak{m}$.

PROOF. The set $Z(\mathfrak{m})$, as a set of ordinals, is well ordered by the relation $\leqslant$ (see p. 229). Let $\xi = \overline{Z(\mathfrak{m})}$. Suppose that $\aleph(\mathfrak{m}) \leqslant \mathfrak{m}$, that is, $\overline{\xi} \leqslant \mathfrak{m}$. Then $\xi \in Z(\mathfrak{m})$ and thus by Theorem 2 the set $W(\xi)$ is a segment of $Z(\mathfrak{m})$. But this is impossible because by Theorem VII.2.3 $\overline{W(\xi)} = \xi = \overline{Z(\mathfrak{m})}$, and no set is similar to its segment (p. 227).

COROLLARY 4: $\mathfrak{m} < \mathfrak{m} + \aleph(\mathfrak{m})$.

The inequality $\leqslant$ is obvious and the equation is impossible as it implies that $\aleph(\mathfrak{m}) \leqslant \mathfrak{m}$.

THEOREM 5: *If there exists an ordinal* ξ *of power* $\mathfrak{m}$, *then* $\aleph(\mathfrak{m}) > \mathfrak{m}$.

PROOF. By Theorem 2, $W(\xi) \subset Z(\mathfrak{m})$ and thus $\aleph(\mathfrak{m}) \geqslant \overline{\xi} = \mathfrak{m}$; hence, by Theorem 3, $\aleph(\mathfrak{m}) > \mathfrak{m}$.

THEOREM 6: *For every set X of ordinal numbers there exists an ordinal* α *such that* $\overline{\xi} < \overline{\alpha}$ *for every* $\xi \in X$.

PROOF. Let $S = \bigcup_{\xi \in X} W(\xi)$, $\mathfrak{m} = \overline{\overline{S}}$ and $\alpha = \overline{Z(\mathfrak{m})}$. Since $\xi = \overline{W(\xi)}$ and $W(\xi) \subset S$ for every $\xi \in X$, we have by Theorem 5

$$\overline{\xi} = \overline{\overline{W(\xi)}} \leqslant \overline{\overline{S}} = \mathfrak{m} < \aleph(\mathfrak{m}) = \overline{\overline{Z(\mathfrak{m})}} = \overline{\alpha}.$$

THEOREM 7: $\aleph(\mathfrak{m}) < 2^{\aleph(\mathfrak{m})} \leqslant 2^{2^{\mathfrak{m}^2}}$.

PROOF. Let $\overline{\overline{A}} = \mathfrak{m}$ and let X be the family of those relations $R \subset A \times A$ which well order their fields. Clearly, $X \subset \boldsymbol{P}(A \times A)$. The set X is the disjoint union

(2) $$X = \bigcup_{\alpha \in Z(\mathfrak{m})} X_\alpha$$

where X_α is the subfamily of X consisting of relations of type α. To every subset $Y \subset Z(\mathfrak{m})$ there corresponds in a one-to-one manner the union $\bigcup_{\alpha \in Y} X_\alpha = F(Y) \subset X$, therefore the family of subsets Y is of power $\leqslant 2^{\overline{\overline{X}}} = 2^{2\mathfrak{m}^2}$. Thus $2^{\aleph(\mathfrak{m})} \leqslant 2^{2\mathfrak{m}^2}$. Q.E.D.

Exercises

1. Using the axiom of choice, show that

(3) $$\aleph(\mathfrak{m}) \leqslant 2\mathfrak{m}^2.$$

Hint: From the axiom of choice it follows that there exists a set T containing exactly one element from each of the sets X_α (see formula (2)).

From the identity $\mathfrak{m}^2 = \mathfrak{m}$ which we shall prove in § 11, p. 309, we conclude a stronger inequality, namely

$$\aleph(\mathfrak{m}) \leqslant 2\mathfrak{m}.$$

2. In the definition of the set $Z(\mathfrak{m})$ and of the cardinal $\aleph(\mathfrak{m})$ replace the relation $\leqslant$ by $\leqslant^*$ (see p. 182) and prove for so defined $\aleph^*(\mathfrak{m})$ the theorems analogous to Theorems 3–5. [Lindenbaum]

§ 3. Initial ordinals

The ordinal φ is said to be an *initial ordinal* if φ is the least ordinal ξ such that $\overline{\xi} = \overline{\varphi}$; that is:

(1) $$\gamma < \varphi \rightarrow \overline{\gamma} < \overline{\varphi}.$$

For example, the ordinals ω and ω_1 are initial ordinals. We shall also denote ω by ω_0 in agreement with the notation for initial ordinals which we shall introduce in this section.

THEOREM 1: *For every infinite cardinal* $\mathfrak{m}$, *the type* φ *of the set* $Z(\mathfrak{m})$ (*that is, of the set of all ordinals of power* $\leqslant \mathfrak{m}$) *is an initial ordinal.*

Since $\xi < \gamma \in Z(\mathfrak{m})$ implies $\xi \in Z(\mathfrak{m})$, it follows that (see p. 230) $Z(\mathfrak{m}) = W(\varphi)$ for an ordinal φ. Thus the order type of $Z(\mathfrak{m})$ is φ. We shall now prove (1). Let γ be an ordinal such that $\gamma < \varphi$. Obviously $\overline{\gamma} \leqslant \overline{\varphi}$. Now we remark that $\overline{\gamma} \leqslant \mathfrak{m}$ because $\gamma < \varphi$ implies $\gamma \in W(\varphi) = Z(\mathfrak{m})$. Since $\varphi \notin W(\varphi)$, we also have $\overline{\varphi} \nleqslant \mathfrak{m}$ and hence it follows that $\overline{\varphi} \neq \overline{\gamma}$ which proves that $\overline{\gamma} < \overline{\varphi}$. Thus (1) holds.

The proof of the following more general theorem is similar.

THEOREM 2: *If* $\mathfrak{m}$ *is a function defined in a set* X *and* $\mathfrak{m}_x \geqslant \mathfrak{a}$ *for every* $x \in X$, *then the ordinal* $\overline{\bigcup_{x \in X} Z(\mathfrak{m}_x)}$ *is an initial ordinal.*

For a given initial ordinal φ we shall denote by $P(\varphi)$ the set of all initial ordinals $\psi < \varphi$ and we let

$$\iota(\varphi) = \overline{\overline{P(\varphi)}}. \tag{2}$$

DEFINITION 1: The ordinal $\iota(\varphi)$ is said to be the *index* of the initial ordinal φ.

Clearly, $\iota(\omega) = 0$, and $\iota(\omega_1) = 1$.

THEOREM 3: *If ψ and φ are initial ordinals and if $\psi < \varphi$, then $\iota(\psi) < \iota(\varphi)$.*

PROOF. By assumption, $\psi \in P(\varphi)$, and it follows that $P(\psi)$ is a segment of the set $P(\varphi)$; thus $\overline{\overline{P(\psi)}} < \overline{\overline{P(\varphi)}}$.

THEOREM 4: *To distinct initial ordinals there correspond distinct indices.*

Theorem 4 follows from Theorem 3.

THEOREM 5: *Every ordinal α is the index of some initial ordinal.*

PROOF. Suppose that α is not the index of any initial ordinal. Assume, moreover, that α is the least ordinal having this property. We shall show that these assumptions lead to a contradiction with Theorems 1 and 2.

In fact, if $\alpha = \beta+1$ then let ψ be such that $\iota(\psi) = \beta$, and let $\varphi = \overline{\overline{Z(\overline{\psi})}}$. By Theorem 1, φ is an initial ordinal; moreover:

$$W(\varphi) = Z(\overline{\psi}) = W(\psi) \cup \{\gamma\colon (\overline{\gamma} = \overline{\psi})\}, \text{ whence } P(\varphi) = P(\psi) \cup \{\psi\}.$$

Thus $\iota(\varphi) = \iota(\psi)+1 = \alpha$.

It remains to consider the case where α is a limit ordinal > 0. By assumption, to every $\xi < \alpha$ there corresponds exactly one ordinal ψ_ξ (by Theorem 4) such that $\iota(\psi_\xi) = \xi$. Let $\varphi = \overline{\overline{\bigcup_{\xi<\alpha} Z(\overline{\psi}_\xi)}}$. By Theorem 2, φ is an initial ordinal. Moreover, $\psi_\xi \in P(\varphi)$ for $\xi < \alpha$. Finally, if $\psi \in P(\varphi)$, then $\psi \in Z(\overline{\psi}_\xi)$ for some $\xi < \alpha$ and thus $\psi < \psi_{\xi+1}$, which by Theorem 3 implies $\iota(\psi) < \xi+1 < \alpha$; therefore ψ coincides with one of the ordinals ψ_ξ, $\xi < \alpha$. Thus $\overline{\overline{P(\varphi)}} = \alpha$, that is, $\iota(\varphi) = \alpha$.

On the basis of Theorem 5 we assume the following definition.

DEFINITION 2: Let ω_α denote the initial ordinal whose index is equal to α; that is, $\iota(\omega_\alpha) = \alpha$.

In this way we have associated with every ordinal an initial ordinal. Every initial ordinal is associated with its index and distinct initial ordinals are associated with distinct indices.

It is easy to check that the following theorems hold.

THEOREM 6: $(\alpha < \beta) \equiv (\omega_\alpha < \omega_\beta) \equiv (\overline{\omega}_\alpha < \overline{\omega}_\beta)$.

THEOREM 7: $Z(\overline{\omega}_\alpha) = W(\omega_{\alpha+1})$, *in other words*, $\aleph(\overline{\omega}_\alpha) = \overline{\omega}_{\alpha+1}$, *and* $\bigcup_{\xi<\lambda} Z(\overline{\omega}_\xi) = W(\omega_\lambda)$, *when* λ *is a limit ordinal.*

THEOREM 8: *If* λ *is a limit ordinal, then* $\omega^\gamma = \lim_{\xi<\lambda} \omega_\xi$.

Thus the function ω_ξ is continuous, and it follows that the ordinal ω_λ is cofinal with λ.

THEOREM 9: *Every initial ordinal is of the form* ω^γ *for some* γ.

PROOF. Assume that $\alpha > 0$ and that

$$\omega_\alpha = \omega^{\gamma_1} \cdot n_1 + \omega^{\gamma_2} \cdot n_2 + \ldots + \omega^{\gamma_k} \cdot n_k, \text{ where } \gamma_1 > \gamma_2 > \ldots > \gamma_k.$$

By Theorem VII.7.4 it follows that

$$\omega^{\gamma_1} \leqslant \omega_\alpha < \omega^{\gamma_1}(n_1+1). \tag{3}$$

Since $\gamma_1 \neq 0$, $\gamma_1 = 1+\delta$ for some δ; thus

$$\omega^{\gamma_1} \cdot (n_1+1) = \omega^{1+\delta} \cdot (n_1+1) = \omega \cdot \omega^\delta (n_1+1).$$

The power of the ordinal $\xi\eta$ equals the power of $\eta\xi$, because one is the power of the set $W(\xi) \times W(\eta)$ and the other is the power of the set $W(\eta) \times W(\xi)$. Therefore

$$\overline{\omega^{\gamma_1}(n_1+1)} = \overline{(n_1+1)\omega \cdot \omega^\delta} = \overline{\omega \cdot \omega^\delta} = \overline{\omega^{\gamma_1}},$$

and by (3) we have $\overline{\omega^{\gamma_1}} = \overline{\omega_\alpha}$. Since $\omega^{\gamma_1} \leqslant \omega_\alpha$, we hawe by (1)

$$\omega_\alpha = \omega^{\gamma_1},$$

since ω_α is initial.

THEOREM 10: *Every limit ordinal* $\lambda \neq 0$ *is cofinal with an initial or dinal* α, *where* α *is the least ordinal cofinal with* λ.

PROOF. We shall prove that the least ordinal α cofinal with λ is an initial ordinal. Let $\gamma < \alpha$. It suffices to show that $\overline{\gamma} \neq \overline{\alpha}$ (see formula (1)). Suppose on the contrary that $\overline{\gamma} = \overline{\alpha}$. Then there exists a sequence φ of type γ whose values are exactly all ordinals $\xi < \alpha$; thus $\alpha = \lim_{\zeta<\gamma} \varphi(\zeta)$.

It follows then by Theorem VII.4.2 that the ordinal α is cofinal with some $\beta \leqslant \gamma$. Thus λ is cofined with β, which contradicts the definition of α.

For every limit ordinal α we shall denote by $cf(\alpha)$ the least ξ such that α is cofinal with ω_ξ. Thus, for example,

$$cf(\omega) = 0, \quad cf(\omega_1) = 1, \quad cf(\omega_\omega) = 0, \quad cf(\omega_{\omega_1}) = 1.$$

Clearly, for every limit ordinal α we have

(4) $$cf(\omega_\alpha) \leqslant \alpha \quad \text{and} \quad cf(\alpha) \leqslant \alpha,$$

since $\alpha \leqslant \omega_\alpha$ and thus α is cofinal with ω_ξ for some $\xi \leqslant \alpha$. In §4 we shall prove that, for every ordinal α,

°(5) $$cf(\omega_{\alpha+1}) = \alpha+1$$

(see p. 278).

If $cf(\omega_\gamma) = \gamma$, we call ω_γ a *regular initial ordinal;* otherwise it is called a *singular initial ordinal.* We use the same terminology for alephs which we shall introduce in the next section. Note that according to (5) $\omega_{\delta+1}$ is a regular initial ordinal for each δ.

From the definitions given above we obtain the following simple result which we shall need in the next section: If λ is a limit ordinal and $cf(\lambda) < \lambda$, then ω_λ is $\lim_{\xi<\alpha} \varphi(\xi)$ of an increasing sequence φ of type $\alpha = \omega_{cf(\lambda)}$.

§4. Alephs and their arithmetic

Every cardinal number of an infinite well-ordered set, that is, any power of an ordinal $\geqslant \omega$ (see § 1, p. 267) is called an *aleph.*

The axiom of choice implies that every infinite cardinal is an aleph (see Corollary VII.8.3). However, many theorems about alephs do not require the use of the axiom of choice; in particular, the law of trichotomy (see p. 228) holds for alephs.

For every ordinal α, let

$$\aleph_\alpha = \overline{\omega}_\alpha, \quad \text{i.e.} \quad \aleph_\alpha = \overline{\overline{W(\omega_\alpha)}}.$$

In particular, $\aleph_0 = \mathfrak{a}$, $\aleph_1 = \overline{\omega}_1$ (see Definition 1.2). By Theorem

3.7 we have

$$\aleph(\aleph_\alpha) = \aleph_{\alpha+1}. \tag{0}$$

THEOREM 1: *If* $\alpha < \beta$ *then* $\aleph_\alpha < \aleph_\beta$.

Theorem 1 is an immediate consequence of Theorem 3.6.

THEOREM 2: *If* $\mathfrak{m} \leqslant \aleph_\alpha$, *then* $\mathfrak{m}$ *is an aleph.*

For $\mathfrak{m}$ is then the power of a subset of a well-ordered set.

THEOREM 3: *The cardinal* $\aleph_{\alpha+1}$ *is the direct successor of* $\aleph_\alpha$; *that is, there exists no cardinal* $\mathfrak{m}$ *such that* $\aleph_\alpha < \mathfrak{m} < \aleph_{\alpha+1}$.

PROOF. If $\mathfrak{m} < \aleph_{\alpha+1}$ then $\mathfrak{m}$ is the power of a subset of the set $W(\omega_{\alpha+1})$, and thus is the power of well-ordered set. Thus $\mathfrak{m}$ is an aleph: $\mathfrak{m} = \aleph_\beta$ and by Theorem 1, $\aleph_\alpha < \mathfrak{m} < \aleph_{\alpha+1}$ implies $\alpha < \beta < \alpha+1$, which is impossible.

The proof of Theorem 4 is similar.

THEOREM 4: *If* α *is a limit ordinal and if* $\aleph_\xi < \mathfrak{m}$ *for every* $\xi < \alpha$, *then* $\mathfrak{m} \nless \aleph_\alpha$.

Theorems 3 and 4 imply that the hierarchy of alephs is in a certain sense complete: it cannot be enriched by the introduction of new cardinals comparable with the alephs.

THEOREM 5:[1]) $\aleph_\alpha^2 = \aleph_\alpha$ *for every* α.

PROOF. Let $\xi < \omega_\alpha$ and $\eta < \omega_\alpha$ where $\xi \neq 0$ or $\eta \neq 0$. Expanding the numbers ξ and η at the base ω (see Theorem VII.7.5) and adding whenever necessary terms with coefficient 0, we can represent ξ and η uniquely in the form:

$$\begin{aligned} \xi &= \omega^{\gamma_1} \cdot m_1 + \omega^{\gamma_2} \cdot m_2 + \ldots + \omega^{\gamma_k} \cdot m_k, \\ \eta &= \omega^{\gamma_1} \cdot n_1 + \omega^{\gamma_2} \cdot n_2 + \ldots + \omega^{\gamma_k} \cdot n_k, \end{aligned} \tag{1}$$

where $\gamma_1 > \gamma_2 > \ldots > \gamma_k$ and for all $i \leqslant k$ either $m_i > 0$ or $n_i > 0$.

By Theorem 3.9, ω_α is of the form ω^λ. Since $\xi < \omega_\alpha$ and $\eta < \omega_\alpha$, it follows by (1) that (see p. 250) $\lambda > \gamma_1$.

Let $\Phi(0, 0) = 0$ and let

$$\Phi(\xi, \eta) = \omega^{\gamma_1} \cdot J(m_1, n_1) + \ldots + \omega^{\gamma_k} \cdot J(m_k, n_k). \tag{2}$$

[1]) Theorem 5 was found by Hessenberg [1].

In this way we have associated with every pair of ordinals ξ, η smaller than ω_α an ordinal $\Phi(\xi, \eta)$ which is smaller than ω_α (since $\gamma_1 < \lambda$). We prove that

$$[\Phi(\xi, \eta) = \Phi(\zeta, \tau)] \to (\xi = \zeta) \wedge (\eta = \tau). \tag{3}$$

This formula is obvious in the case in which $\Phi(\xi, \eta) = 0 = \Phi(\zeta, \tau)$, for then $\xi = \eta = 0 = \zeta = \tau$, so suppose that $\Phi(\xi, \eta) = \Phi(\zeta, \tau) > 0$. Then $\xi > 0$ or $\eta > 0$ and $\zeta > 0$ or $\tau > 0$.

Let ζ and τ be represented as

$$\begin{aligned} \zeta &= \omega^{\delta_1} \cdot p_1 + \omega^{\delta_2} \cdot p_2 + \ldots + \omega^{\delta_h} \cdot p_h, \\ \tau &= \omega^{\delta_1} \cdot q_1 + \omega^{\delta_2} \cdot q_2 + \ldots + \omega^{\delta_h} \cdot q_h, \end{aligned} \tag{4}$$

where for every $i \leqslant h$ either $p_i > 0$ or $q_i > 0$. Then

$$\Phi(\zeta, \tau) = \omega^{\delta_1} \cdot J(p_1, q_1) + \ldots + \omega^{\delta_h} \cdot J(p_h, q_h). \tag{5}$$

Since all coefficients in the expansions (2) and (5) are positive and since expansion at the base ω is unique (see p. 251), it follows that

$$k = h, \quad \gamma_i = \delta_i \quad \text{and} \quad J(m_i, n_i) = J(p_i, q_i) \quad \text{for} \quad i = 1, 2, \ldots, k.$$

Thus $m_i = p_i$ and $n_i = q_i$ for $i \leqslant k$ and hence, by (1) and (4), $\xi = \zeta$ and $\eta = \tau$. Thus (3) holds.

Finally, every $\vartheta < \omega_\alpha$ is a value of the function Φ. For assume $\vartheta = \omega^{\gamma_1} \cdot r_1 + \omega^{\gamma_2} \cdot r_2 + \ldots + \omega^{\gamma_k} \cdot r_k$, where $r_1, r_2, \ldots, r_k$ are positive; it suffices to take as ξ and η the ordinals (1) where $m_i = K(r_i)$ and $n_i = L(r_i)$ for $i = 1, 2, \ldots, k$. If $\vartheta = 0$ it suffices to let $\xi = \eta = 0$.

Therefore the function Φ establishes a one-to-one correspondence between elements of the set $W(\omega_\alpha)$ and ordered pairs of elements of $W(\omega_\alpha)$. Thus the sets $W(\omega_\alpha)$ and $W(\omega_\alpha) \times W(\omega_\alpha)$ are equipollent. Q.E.D.

COROLLARY 6: $\aleph_\alpha + \aleph_\beta = \aleph_{\max(\alpha,\beta)} = \aleph_\alpha \cdot \aleph_\beta$, *where* $\max(\alpha, \beta) = \alpha$ *if* $\alpha \geqslant \beta$ *and* $\max(\alpha, \beta) = \beta$ *if* $\beta \geqslant \alpha$.

PROOF. Assume that $\max(\alpha, \beta) = \alpha$. Then $\aleph_\beta \leqslant \aleph_\alpha$ and

$$\aleph_\alpha \leqslant \aleph_\alpha + \aleph_\beta \leqslant \aleph_\alpha + \aleph_\alpha = 2\aleph_\alpha \leqslant \aleph_\alpha \cdot \aleph_\alpha = \aleph_\alpha,$$

$$\aleph_\alpha \leqslant \aleph_\alpha \cdot \aleph_\beta \leqslant \aleph_\alpha \cdot \aleph_\alpha = \aleph_\alpha,$$

whence the desired identities follow by the Cantor–Bernstein Theorem (p. 185).

THEOREM 7: $\overline{\overline{W(\omega_{\alpha+1})-W(\omega_\alpha)}} = \aleph_{\alpha+1}$.

PROOF. Let $W(\omega_{\alpha+1}) = A$ and $W(\omega_\alpha) = B$. The difference $A-B$ is a well-ordered set and thus its cardinal number is an aleph, which we shall denote by $\aleph_\gamma$. The identity $A = (A-B)\cup B$ implies that $\aleph_{\alpha+1} = \aleph_\gamma+\aleph_\alpha = \aleph_{\max(\gamma,\alpha)}$, thus $\alpha+1 = \max(\gamma, \alpha)$. As $\alpha < \alpha+1$, it follows that $\gamma = \alpha+1$.

°LEMMA 8: $\overline{\overline{\bigcup_{t\in T} F_t}} \leqslant \sum_{t\in T} \overline{\overline{F_t}}$, *where F is any function whose values are sets.*

For

$$\sum_{t\in T} \overline{\overline{F_t}} = \overline{\overline{\{\langle t, x\rangle\colon (t\in T)\wedge(x\in F_t)\}}}; \tag{6}$$

on the other hand, the set $\bigcup_{t\in T} F_t$ can be obtained from the set $\{\langle t, x\rangle\colon (t\in T)\wedge(x\in F_t)\}$ by the mapping f defined by the formula $f(\langle t, x\rangle) = x$, and thus it has power less than or equal to that of the latter (see Theorem V.5.1).

°THEOREM 9: $cf(\omega_{\alpha+1}) = \alpha+1$.

Let $\beta = cf(\omega_{\alpha+1})$. Then the ordinal ω_β is cofinal with $\omega_{\alpha+1}$ and there exists a transfinite sequence φ of type ω_β whose limit is $\omega_{\alpha+1}$. It follows that if $\xi < \omega_{\alpha+1}$, then there exists an ordinal $\zeta < \omega_\beta$ such that $\xi < \varphi(\zeta)$ or, in other words,

$$W(\omega_{\alpha+1}) \subset \bigcup_{\zeta<\omega_\beta} W(\varphi(\zeta)). \tag{7}$$

Because $\overline{\overline{W(\omega_\beta)}} = \aleph_\beta$, and because for all $\zeta < \omega_\beta$ we have

$$\overline{\overline{W(\varphi(\zeta))}} < \aleph_{\alpha+1}, \quad \text{that is,} \quad \overline{\overline{W(\varphi(\zeta))}} \leqslant \aleph_\alpha,$$

it follows by (7), the lemma and Theorem 7.8 that

$$\aleph_{\alpha+1} = \overline{\overline{W(\omega_{\alpha+1})}} \leqslant \overline{\overline{\bigcup_{\zeta<\omega_\beta} W(\varphi(\zeta))}} \leqslant \sum_{\zeta<\omega_\beta} \overline{\overline{W(\varphi(\zeta))}} \leqslant \aleph_\beta\cdot\aleph_\alpha = \aleph_{\max(\alpha,\beta)}.$$

Thus $\alpha+1 \leqslant \max(\alpha, \beta)$ and $\alpha+1 \leqslant \beta$. At the same time (see 3(4)) $cf(\omega_{\alpha+1}) \leqslant \alpha+1$, that is, $\beta \leqslant \alpha+1$, and thus $\beta = \alpha+1$.

°THEOREM 10: *For arbitrary* α,

$$\sum_{\xi\leqslant\alpha} \aleph_\xi = \aleph_\alpha. \tag{8}$$

PROOF. Since $(\xi \leqslant \alpha) \to (\aleph_\xi \leqslant \aleph_\alpha)$ and

$$\overline{\overline{\{\xi \colon \xi \leqslant \alpha\}}} \leqslant \aleph_\alpha,$$

it follows by Theorem 8, p. 195 that

$$\sum_{\xi \leqslant \alpha} \aleph_\xi \leqslant \aleph_\alpha \cdot \aleph_\alpha = \aleph_\alpha,$$

which implies (8), because

$$\aleph_\alpha \leqslant \sum_{\xi \leqslant \alpha} \aleph_\xi.$$

°THEOREM 11: *If* $\alpha = \beta+1$, *then*

(9) $$\sum_{\xi < \alpha} \aleph_\xi = \aleph_\beta.$$

On the other hand, if α *is a limit ordinal* > 0, *then*

(10) $$\sum_{\xi < \alpha} \aleph_\xi = \aleph_\alpha.$$

PROOF. Let $\alpha = \beta+1$. Then $(\xi < \alpha) \equiv (\xi \leqslant \beta)$ and by (8)

$$\sum_{\xi < \alpha} \aleph_\xi = \sum_{\xi \leqslant \beta} \aleph_\xi = \aleph_\beta.$$

Suppose that α is a limit ordinal; then

$$W(\omega_\alpha) = \bigcup_{\xi < \alpha} W(\omega_\xi);$$

hence (by Lemma 8)

$$\aleph_\alpha = \overline{\overline{W(\omega_\alpha)}} = \overline{\overline{\bigcup_{\xi < \alpha} W(\omega_\xi)}} \leqslant \sum_{\xi < \alpha} \overline{\overline{W(\omega_\xi)}} = \sum_{\xi < \alpha} \aleph_\xi,$$

which implies (10), because by (8)

$$\sum_{\xi < \alpha} \aleph_\xi \leqslant \aleph_\alpha.$$

The following more general theorem can be proved in a similar manner.

°THEOREM 12. *If* α *is a limit ordinal,* φ *is a transfinite increasing sequence of type* α *and* $\lambda = \lim_{\xi < \alpha} \varphi(\xi)$, *then*

$$\sum_{\xi < \alpha} \aleph_{\varphi(\xi)} = \aleph_\lambda.$$

Remark. It is interesting to note that the aleph $\aleph_1$ and more generally, the aleph $\aleph_n$ can be defined without appealing to the notions of ordinal or cardinal numbers. In order to formulate the appropriate theorem, we introduce the following definition: set A contained in the cartesian product $X^n = X \times X \times \ldots \times X$ is *finite in the direction of the k-th axis* provided that for every element $(x_1, \ldots, x_{k-1}, x_{k+1}, \ldots, x_n)$ belonging to X^{n-1} the set

$$\{x_k: (x_1, \ldots, x_{k-1}, x_k, x_{k+1}, \ldots, x_n) \in A\}$$

is finite (more figuratively: every straight line parallel to the kth axis intersects A in a finite number of points).

The following theorem holds:

A necessary and sufficient condition for the set X to be of power $\leqslant \aleph_n$ (n finite $\geqslant 0$) is that the set X^{n+2} be representable as a union $A_1 \cup \ldots \cup A_{n+2}$, where A_k is finite in the direction of the k-th axis.[1]

§ 5. The exponentiation of alephs

First of all we note the following elementary theorem.

THEOREM 1: *If $\alpha \leqslant \beta$, then* $\aleph_\alpha^{\aleph_\beta} = 2^{\aleph_\beta}$.

PROOF. From the inequality $2 < \aleph_\alpha < 2^{\aleph_\alpha}$ and from the laws of exponentiation for cardinal numbers (p. 181) we obtain

$$2^{\aleph_\beta} \leqslant \aleph_\alpha^{\aleph_\beta} \leqslant 2^{\aleph_\alpha \cdot \aleph_\beta} = 2^{\aleph_\beta},$$

which implies the desired equality.

°THEOREM 2: (THE HAUSDORFF RECURSION FORMULA[2]),

$$\aleph_{\alpha+1}^{\aleph_\beta} = \aleph_\alpha^{\aleph_\beta} \cdot \aleph_{\alpha+1}.$$

PROOF. We shall examine two cases.

Case I: $\alpha+1 \leqslant \beta$. Then by Theorem 1 we have

$$\aleph_{\alpha+1}^{\aleph_\beta} = 2^{\aleph_\beta} = \aleph_\alpha^{\aleph_\beta}. \tag{2}$$

[1]) The characterization of $\aleph_n$ given here is due to Sierpiński [22], Kuratowski [14], and Sikorski [2].

[2]) The Hausdorff recursion formula is due to Hausdorff [1].

On the other hand, since $\aleph_{\alpha+1} \leqslant \aleph_\beta < 2^{\aleph_\beta}$, we also have

(3) $$\aleph_\alpha^{\aleph_\beta} \cdot \aleph_{\alpha+1} = 2^{\aleph_\beta} \cdot \aleph_{\alpha+1} \leqslant 2^{\aleph_\beta} \cdot 2^{\aleph_\beta} = 2^{\aleph_\beta + \aleph_\beta} = 2^{\aleph_\beta}.$$

By (2) and (3) it follows that

$$\aleph_{\alpha+1}^{\aleph_\beta} = 2^{\aleph_\beta} = \aleph_\alpha^{\aleph_\beta} \leqslant \aleph_\alpha^{\aleph_\beta} \cdot \aleph_{\alpha+1} \leqslant 2^{\aleph_\beta} = \aleph_{\alpha+1}^{\aleph_\beta},$$

which implies (1).

Case II: $\beta < \alpha+1$. In this case,

(4) $$\aleph_\alpha^{\aleph_\beta} \cdot \aleph_{\alpha+1} \leqslant \aleph_{\alpha+1}^{\aleph_\beta} \cdot \aleph_{\alpha+1} = \aleph_{\alpha+1}^{\aleph_\beta+1} = \aleph_{\alpha+1}^{\aleph_\beta}.$$

It remains to show that the opposite inequality also holds. For this purpose we consider the set $W(\omega_{\alpha+1})^{W(\omega_\beta)}$, that is, the set of all transfinite sequences of type ω_β whose terms are less than $\omega_{\alpha+1}$. We shall show that

(5) $$W(\omega_{\alpha+1})^{W(\omega_\beta)} \subset \bigcup_{\xi < \omega_{\alpha+1}} W(\xi)^{W(\omega_\beta)}.$$

For if φ is a transfinite sequence of type $\omega_\beta < \omega_{\alpha+1}$, then (see Theorem 4.9) the set of its term is not cofinal with $W(\omega_{\alpha+1})$. Thus there exists an ordinal $\xi < \omega_{\alpha+1}$ for which $W(\xi)$ contains all the terms of the sequence φ; hence $\varphi \in W(\xi)^{W(\omega_\beta)}$. Thus inclusion (5) holds.

As $\overline{\overline{W(\xi)}} = \overline{\overline{\xi}} \leqslant \aleph_\alpha$ for $\xi < \omega_{\alpha+1}$, it follows that $\overline{\overline{W(\xi)^{W(\omega_\beta)}}} \leqslant \aleph_\alpha^{\aleph_\beta}$. Therefore from (5), by Lemma 4.8 and by Theorem V.7.8 we obtain

$$\aleph_{\alpha+1}^{\aleph_\beta} = \overline{\overline{W(\omega_{\alpha+1})^{W(\omega_\beta)}}} \leqslant \overline{\overline{\bigcup_{\xi < \omega_{\alpha+1}} W(\xi)^{W(\omega_\beta)}}} \leqslant \sum_{\xi < \omega_{\alpha+1}} \overline{\overline{W(\xi)^{W(\omega_\beta)}}} \leqslant \aleph_\alpha^{\aleph_\beta} \cdot \aleph_{\alpha+1},$$

which together with (4) proves (1).

°THEOREM 3: (THE TARSKI RECURSION FORMULA[1])) *If φ is an increasing sequence of a limit type α and $\lambda = \lim_{\xi<\alpha} \varphi(\xi)$ and $\beta < cf(\alpha)$, then $\aleph_\lambda^{\aleph_\beta} = \sum_{\xi<\alpha} \aleph_{\varphi(\xi)}^{\aleph_\beta}$, in particular, $\aleph_\alpha^{\aleph_\beta} = \sum_{\xi<\alpha} \aleph_\xi^{\aleph_\beta}$ for every limit ordinal α and for every $\beta < cf(\alpha)$.*

PROOF. By assumption, $cf(\lambda) = cf(\alpha)$. Thus no transfinite sequence of type ω_β with values in $W(\lambda)$ is convergent to λ, hence for every such

[1]) Tarski's recursion formula is due to Tarski [2].

sequence there exists $\xi < \alpha$ such that this sequence belongs to $W(\varphi(\xi))^{W(\omega_\beta)}$. Therefore

$$W(\lambda)^{W(\omega_\beta)} \subset \bigcup_{\xi<\alpha} W(\varphi(\xi))^{W(\omega_\beta)},$$

which implies that

$$\aleph_\lambda^{\aleph_\beta} \leqslant \sum_{\xi<\alpha} \aleph_{\varphi(\xi)}^{\aleph_\beta}.$$

The opposite inequality follows from the remark that $\aleph_\lambda \geqslant \bar{\alpha}$, which implies that

$$\aleph_\lambda^{\aleph_\beta} = \aleph_\lambda \cdot \aleph_\lambda^{\aleph_\beta} \geqslant \bar{\alpha} \cdot \aleph_\lambda^{\aleph_\beta} \geqslant \aleph_\lambda^{\aleph_\beta} \geqslant \sum_{\xi<\alpha} \aleph_{\varphi(\xi)}^{\aleph_\beta}.$$

°THEOREM 4: (THE GENERALIZED HAUSDORFF FORMULA) *For finite n,*

$$\aleph_{\alpha+n}^{\aleph_\beta} = \aleph_\alpha^{\aleph_\beta} \cdot \aleph_{\alpha+n}. \tag{8}$$

PROOF. For $n = 1$, (8) coincides with (1). Assume that (8) holds for a particular n. Replacing in (1) α by $\alpha+n$ we obtain

$$\aleph_{\alpha+n+1}^{\aleph_\beta} = \aleph_{\alpha+n}^{\aleph_\beta} \cdot \aleph_{\alpha+n+1},$$

thus by the induction hypothesis and by Corollary 4.6 it follows that

$$\aleph_{\alpha+n+1}^{\aleph_\beta} = \aleph_\alpha^{\aleph_\beta} \cdot \aleph_{\alpha+n} \cdot \aleph_{\alpha+n+1} = \aleph_\alpha^{\aleph_\beta} \cdot \aleph_{\alpha+n+1}.$$

Thus (8) holds for $n+1$, and hence it holds for arbitrary finite n.

Setting $\alpha = 0$ in (8) and using Theorem 1, we obtain the following theorem.

°THEOREM 5: (THE BERNSTEIN FORMULA[1])) *For finite n,*

$$\aleph_n^{\aleph_\beta} = 2^{\aleph_\beta} \cdot \aleph_n. \tag{9}$$

Examples

1. (9) implies that

$$\aleph_1^{\aleph_0} = 2^{\aleph_0} \cdot \aleph_1.$$

2. Theorem 1 implies that

$$\aleph_0^{\aleph_1} = 2^{\aleph_1}.$$

[1]) See Bernstein [1]. For further formulas of the kind given in Theorems 2–4 see Bachmann [1].

3. In formula (6) let $\alpha = \omega_1$, $\beta = 0$. Since $cf(\omega_1) = 1$, we have by Theorem 4

$$\aleph_\Omega^{\aleph_0} = \sum_{\xi < \Omega} \aleph_\xi^{\aleph_0}.$$

We now give a theorem in the proof of which we shall use König's theorem (p. 198):

°THEOREM 6: *If α is a limit ordinal and $\beta \geqslant cf(\alpha)$, then for every cardinal $\mathfrak{m}$*

(10) $$\aleph_\alpha \neq \mathfrak{m}^{\aleph_\beta}.$$

PROOF. Let $\xi = cf(\alpha)$. Then the ordinal ω_ξ is cofinal with α; that is, there exists an increasing sequence φ of type ω_ξ such that

$$\lim_{\zeta < \omega_\xi} \varphi(\zeta) = \alpha.$$

By applying Theorem 4.12 it follows that

(11) $$\aleph_\alpha = \sum_{\zeta < \omega_\xi} \aleph_{\varphi(\zeta)}.$$

By König's theorem we have

(12) $$\sum_{\zeta < \omega_\xi} \aleph_{\varphi(\zeta)} < \prod_{\zeta < \omega_\xi} \aleph_{\varphi(\zeta)}.$$

On the other hand, since $\aleph_{\varphi(\zeta)} < \aleph_\alpha$ and $\overline{\omega_\xi} = \aleph_\xi \leqslant \aleph_\beta$, we obtain

(13) $$\prod_{\zeta < \omega_\xi} \aleph_{\varphi(\zeta)} \leqslant \aleph_\alpha^{\aleph_\beta}.$$

Formulas (11), (12) and (13) imply that

(14) $$\aleph_\alpha < \aleph_\alpha^{\aleph_\beta}.$$

If there existed a cardinal $\mathfrak{m}$ such that $\aleph_\alpha = \mathfrak{m}^{\aleph_\beta}$, then we would have

$$\aleph_\alpha^{\aleph_\beta} = \mathfrak{m}^{\aleph_\beta \aleph_\beta} = \mathfrak{m}^{\aleph_\beta} = \aleph_\alpha,$$

contrary to (14).

We conclude that (10) holds.

°COROLLARY 7: *If $\beta \geqslant cf(\alpha)$, then $\aleph_\alpha < \aleph_\alpha^{\aleph_\beta}$.*

From Theorem 6 it follows, in particular, that for no $\mathfrak{m}$ and β do the equations $\aleph_\omega = \mathfrak{m}^{\aleph_\beta}$, $\aleph_{\omega_\omega} = \mathfrak{m}^{\aleph_\beta}$, $\aleph_\varepsilon = \mathfrak{m}^{\aleph_\beta}$ hold (see p. 247 for the definition of ε); for (see p. 275) $cf(\omega) = cf(\omega_\omega) = cf(\varepsilon) = 0$.

It also follows from Theorem 6 that if $\aleph_{\omega_n} = \mathfrak{m}^{\aleph_\beta}$, then β can assume only one of the values $0, 1, \ldots, n-1$. In fact, $cf(\omega_n) = n$, and thus for $\beta \geqslant n$ the equation $\aleph_{\omega_n} = \mathfrak{m}^{\aleph_\beta}$ does not hold.

COROLLARY 8: $\aleph_\omega \neq 2^{\aleph_0}$.

For $\aleph_\omega = 2^{\aleph_0}$ implies that $\aleph_\omega^{\aleph_0} = 2^{\aleph_0 \cdot \aleph_0} = 2^{\aleph_0} = \aleph_\omega$, whereas by Corollary 7 we have $\aleph_\omega^{\aleph_0} > \aleph_\omega$.

We shall conclude this section by evaluating the power of the set

$$P_\mathfrak{n}(M) = \{X \subset M\colon \overline{\overline{X}} = \mathfrak{n}\}, \quad \text{where} \quad \overline{\overline{M}} = \mathfrak{m}.$$

For $\mathfrak{m}$ and $\mathfrak{n}$ finite the power of $P_\mathfrak{n}(M)$ is $\binom{\mathfrak{m}}{\mathfrak{n}}$.

°THEOREM 9: *If M is an infinite set of power $\mathfrak{m}$ and if $\mathfrak{n} \leqslant \overline{\overline{M}}$, then $P_\mathfrak{n}(M)$ has power $\mathfrak{m}^\mathfrak{n}$.*

PROOF. Let Z be a fixed set of power $\mathfrak{n}$. To every $X \in P_\mathfrak{n}(M)$ there corresponds in a one-to-one manner a non-empty family $C(X)$ of functions $f \in M^Z$ such that $f^1(Z) = X$; clearly, $C(X') \cap C(X'') = \emptyset$ for $X' \neq X''$. By applying the axiom of choice we conclude that the power of $P_\mathfrak{n}(M)$ is not greater that $\overline{\overline{M^Z}} = \mathfrak{m}^\mathfrak{n}$.

Conversely, to every function $f \in M^Z$ there corresponds the set $A_f = \{\langle z, f(z)\rangle\colon (z \in Z)\}$. This set has power $\mathfrak{n}$, because the set Z has power $\mathfrak{n}$. Moreover, $A_f \subset Z \times M$ and the set $Z \times M$ has power $\mathfrak{n} \cdot \mathfrak{m} = \mathfrak{m}$. If $f' \neq f''$ and, for example, $f'(z_0) \neq f''(z_0)$, then $A_{f'} \neq A_{f''}$, as $\langle z_0, f'(z_0)\rangle \in A_{f'} - A_{f''}$. It follows that the set $P_\mathfrak{n}(Z \times M)$ has power $\geqslant \mathfrak{m}^\mathfrak{n}$ and thus $P_\mathfrak{n}(M)$ has power $\geqslant \mathfrak{m}^\mathfrak{n}$ for the sets $P_\mathfrak{n}(Z \times M)$ and $P_\mathfrak{n}(M)$ are equipollent.

°COROLLARY 10: *If $\overline{\overline{M}} = \mathfrak{m} \geqslant \mathfrak{a}$, then the set $\{X \subset M\colon \overline{\overline{X}} < \mathfrak{m}\}$ has power $\sum_{\mathfrak{r} < \mathfrak{m}} \mathfrak{m}^\mathfrak{r}$.*

§ 6. The exponential hierarchy of cardinal numbers

In Section 4 we saw that the sum and the product of two alephs coincide with the greatest of them. Thus addition and multiplication of alephs are extremely simple operations. On the other hand the operation of exponentiation is by no means simple: given two ordinals α, β, it is in general impossible to evaluate the ordinal ξ satisfying the equation $\aleph_\alpha^{\aleph_\beta} = \aleph_\xi$.

In the present section we shall consider a class of cardinal numbers for which the laws of exponentiation are relatively simple. In the definition of this class we shall make use of sets R_α which we defined on p. 238. These sets satisfy the following recursive equations

$$R_0 = \emptyset, \qquad R_{\alpha+1} = \boldsymbol{P}(R_\alpha), \qquad R_\lambda = \bigcup\{R_\xi\colon \xi < \lambda\}$$

where λ is a limit number.

Let $\mathfrak{a}_\xi = \overline{\overline{R}}_{\omega+\xi}$. The cardinals $\mathfrak{a}_\xi$ constitute the exponential hierarchy of cardinals.[1])

In order to derive arithmetic laws for these cardinals we must first establish several properties of the sets R_α.

THEOREM 1: (a) $R_\alpha \subset R_\beta$ *for* $\alpha < \beta$; (b) $X \in R_\beta \to X \subset R_\beta$.

PROOF. The proof is by induction with respect to β. It suffices to show that if $\beta_0 \geqslant 0$, and if the theorem holds for all $\beta < \beta_0$, then it also holds for β_0.

The case $\beta_0 = 0$ is obvious. If $\beta_0 = \beta+1$ and $\alpha < \beta_0$, then $\alpha \leqslant \beta$ and thus by assumption $R_\alpha \subset R_\beta$ and from (b) we obtain $R_\beta \subset R_{\beta+1}$. If β_0 is a limit ordinal and $\alpha < \beta_0$, then $R_\alpha \subset R_{\beta_0}$ because R_{β_0} is the union of all R_ξ for $\xi < \beta_0$. Part (a) is proved.

Assume now that $X \in R_{\beta_0}$; then $\beta_0 \neq 0$ and β_0 either has the form $\beta+1$ or else it is a limit ordinal. In the first case $X \in \boldsymbol{P}(R_\beta)$, that is, $X \subset R_\beta$ and thus $X \subset R_{\beta_0}$ by (a) proved above. If β_0 is a limit ordinal, then there exists $\beta < \beta_0$ such that $X \in R_\beta$ and thus, by assumption, $X \subset R_\beta$ and by (a), $X \in R_{\beta_0}$. Thus part (b) holds.

THEOREM 2: $R_\alpha \times R_\alpha \subset R_{\alpha+2}$.

PROOF. The elements of $R_\alpha \times R_\alpha$ are pairs $\langle X, Y\rangle$ where $X, Y \in R_\alpha$. As $\{X, Y\} \subset R_\alpha$ and $\{X\} \subset R_\alpha$, thus $\{X, Y\} \in R_{\alpha+1}$ and $\{X\} \in R_{\alpha+1}$. It now follows that $\big\{\{X\}, \{X, Y\}\big\} \subset R_{\alpha+1}$, and thus $\big\{\{X\}, \{X, Y\}\big\} \in R_{\alpha+2}$. Q.E.D.

THEOREM 3: *If* α *is a limit ordinal then* $R_\alpha \times R_\alpha \subset R_\alpha$.

The proof follows from Theorem 2.

[1]) In the current literature the cardinals $\mathfrak{a}_\xi$ are usually denoted by the Hebrew letter "beth" with the index ξ.

THEOREM 4: *$R_\alpha \supset N_\alpha$, where N_α is the α-th VN ordinal.*

PROOF. For $\alpha = 0$, $R_\alpha = N_\alpha = \emptyset$. If $R_\alpha \supset N_\alpha$ then $N_\alpha \in \boldsymbol{P}(R) = R_{\alpha+1}$ and thus $\{N_\alpha\} \subset R_{\alpha+1}$. As $N_\alpha \subset R_\alpha$, it follows that $N_\alpha \cup \{N_\alpha\} \subset R_{\alpha+1}$, whence $N_{\alpha+1} \subset R_{\alpha+1}$. If α is a limit ordinal, and if the theorem holds for $\xi < \alpha$, then since $N_\alpha \subset \bigcup_{\xi<\alpha} N_\xi$, $N_\alpha \subset \bigcup_{\xi<\alpha} R_\xi = R_\alpha$.

THEOREM 5: *$\overline{\overline{R}}_n = 2^n$ for $n < \omega$; $\mathfrak{a}_0 = \aleph_0$.*

PROOF. The first part of the theorem follows from the definitions by induction.

To prove the second part of the theorem it suffices to prove that there exists a one-to-one mapping ν_n of the set R_n into N where ν_{n+1} is an extension of ν_n.

Put $\nu_0 = 0$; for $X \in R_{n+1}$ let $\nu_{n+1}(X) = \nu_n(X)$ if $X \in R_n$; if $X \in R_{n+1} - R_n$ then let $\nu_{n+1}(X) = \sum_{j<k} 2^{\nu_n(Y_j)}$, where $Y_0, Y_1, \ldots, Y_{k-1}$ are the elements of X ordered so that $\nu_n(Y_{j-1}) < \nu_n(Y_j)$ for $1 < j < k$. It is easy to show that the functions ν_n are one-to-one and that ν_{n+1} is an extension of ν_n.

THEOREM 6: *$\mathfrak{a}_{\alpha+1} = 2^{\mathfrak{a}_\alpha}$; $\mathfrak{a}_\alpha < \mathfrak{a}_\beta$ for $\alpha < \beta$.*

The theorem follows from the definitions and from Theorem 1.

THEOREM 7: *$\mathfrak{a}_\alpha^2 = \mathfrak{a}_\alpha$.*

PROOF. The theorem is obvious for $\alpha = 0$. If it holds for the ordinal β, and if $\alpha = \beta+1$, then it also holds for the ordinal α, because $\mathfrak{a}_\alpha = \overline{\overline{R_{\beta+1}}} = 2^{\mathfrak{a}_\beta}$; and thus $\mathfrak{a}_\alpha^2 = 2^{\mathfrak{a}_\beta+\mathfrak{a}_\beta}$, which implies that $\mathfrak{a}_\alpha^2 = \mathfrak{a}_\alpha$ as $\mathfrak{a}_\beta \leqslant \mathfrak{a}_\beta+\mathfrak{a}_\beta \leqslant \mathfrak{a}_\beta^2 = \mathfrak{a}_\beta$. Finally, if α is a limit ordinal, then by Theorem 3 it follows that $\mathfrak{a}_\alpha^2 \leqslant \mathfrak{a}_\alpha$; and because $\mathfrak{a}_\alpha^2 \geqslant \mathfrak{a}_\alpha$, we conclude that $\mathfrak{a}_\alpha^2 = \mathfrak{a}_\alpha$.

COROLLARY 8. $\mathfrak{a}_\alpha+\mathfrak{a}_\beta = \mathfrak{a}_\alpha\mathfrak{a}_\beta = \mathfrak{a}_{\max(\alpha,\beta)}$.

The corollary follows from Theorem 7 with a proof similar to that of the analogous theorem for alephs (see p. 277).

°THEOREM 9: *If α is a limit ordinal then*

$$\mathfrak{a}_\alpha = \sum_{\xi<\alpha} \mathfrak{a}_\xi .$$

PROOF. On the one hand,

$$\overline{\overline{\bigcup_{\xi<\alpha} R_\xi}} \leqslant \sum_{\xi<\alpha} \overline{\overline{R_\xi}} = \sum_{\xi<\alpha} \mathfrak{a}_\xi .$$

On the other, $\mathfrak{a}_\alpha = \mathfrak{a}_\alpha \cdot \mathfrak{a}_\alpha \geqslant \mathfrak{a}_\alpha \cdot \bar{\alpha}$ by Theorem 4, since $N_\alpha = \bar{\alpha}$. But since $\mathfrak{a}_\alpha \cdot \bar{\alpha} \geqslant \sum_{\xi<\alpha} \mathfrak{a}_\xi$, the desired equation follows by the Cantor–Bernstein theorem.

REMARK. The axiom of choice is used in this theorem because the very definition of the infinite sum of cardinal numbers demands the use of this axiom.

We shall now establish several laws concerning exponentiation of the cardinals $\mathfrak{a}_\xi$.

THEOREM 10: *If* $\alpha \leqslant \beta$ *then* $\mathfrak{a}_\alpha^{\mathfrak{a}_\beta} = \mathfrak{a}_{\beta+1}$.

PROOF. $\mathfrak{a}_{\beta+1} = 2^{\mathfrak{a}_\beta} \leqslant \mathfrak{a}_\alpha^{\mathfrak{a}_\beta} < (2^{\mathfrak{a}_\alpha})^{\mathfrak{a}_\beta} = 2^{\mathfrak{a}_\alpha \mathfrak{a}_\beta} = 2^{\mathfrak{a}_\beta} = \mathfrak{a}_{\beta+1}$.

THEOREM 11: *If* $\alpha+1 > \beta$ *then* $\mathfrak{a}_{\alpha+1}^{\mathfrak{a}_\beta} = \mathfrak{a}_{\alpha+1}$.

PROOF. $\mathfrak{a}_{\alpha+1}^{\mathfrak{a}_\beta} = 2^{\mathfrak{a}_\alpha \mathfrak{a}_\beta} = 2^{\mathfrak{a}_\alpha} = \mathfrak{a}_{\alpha+1}$.

It is not possible to prove any simple formula for the power $\mathfrak{a}_\alpha^{\mathfrak{a}_\beta}$, where α is a limit ordinal and $\beta < \alpha$. Certain fragmentary results in this area are collected in Theorem 18 below. First we shall establish certain relations between the hierarchy of alephs and the exponential hierarchy.

°THEOREM 12: *For all* α

$$\aleph_\alpha \leqslant \mathfrak{a}_\alpha . \tag{1}$$

PROOF. Let us proceed by induction.

By definition, $\aleph_0 = \mathfrak{a}_0$.

Assume that (1) holds for all $\beta < \alpha$. If $\alpha = \beta_1 + 1$, then using Theorem 3, p. 276 and Cantor's theorem (p. 181(2)) we infer that

$$\aleph_\alpha = \aleph_{\beta_1+1} \leqslant 2^{\aleph_{\beta_1}} \leqslant 2^{\mathfrak{a}_{\beta_1}} = \mathfrak{a}_{\beta_1+1} = \mathfrak{a}_\alpha .$$

If α is a limit ordinal, then (1) follows immediately from Theorem 9, p. 286.

Theorem 12 provides an estimation of alephs "from above" by means of the exponential hierarchy. No estimation of the numbers $\mathfrak{a}_\xi$ by means of alephs is possible, even for $\xi = 1$.

The well-ordering theorem implies that for every ordinal α there exists an ordinal $\pi(\alpha)$ such that $\mathfrak{a}_\alpha = \aleph_{\pi(\alpha)}$. The axioms of set theory yield only fragmentary information about the ordinals $\pi(\alpha)$.[1]) The following theorem is obvious.

°THEOREM 13: $\alpha < \beta \rightarrow \pi(\alpha) < \pi(\beta)$.

°THEOREM 14: *The function* π *is continuous (on every set of the form* $W(\alpha)$*).*

PROOF. Let α be a limit ordinal and let $\lambda = \lim_{\xi<\alpha} \pi(\xi)$. Applying Theorem 4.12 we obtain

$$\aleph_\lambda = \sum_{\xi<\alpha} \aleph_{\pi(\xi)} = \sum_{\xi<\alpha} \mathfrak{a}_\xi = \mathfrak{a}_\alpha,$$

and thus $\lambda = \pi(\alpha)$.

The following is an easy corollary of Theorem 14.

°COROLLARY 15: *If* α *is a non-zero limit ordinal, then* $cf(\alpha) \leqslant \pi(\alpha)$.

PROOF. Let $cf(\alpha) = \delta$ and let φ be an increasing sequence of type ω_δ such that $\alpha = \lim_{\xi<\omega_\delta} \varphi(\xi)$. Then

$$\pi(\alpha) = \lim_{\xi<\omega_\delta} \pi\big(\varphi(\xi)\big) = \lim_{\xi<\omega_\delta} \psi(\xi),$$

where the composite sequence $\psi = \pi \mathrm{O} \varphi$ is increasing. Thus $\pi(\alpha) \geqslant \omega_\delta \geqslant \delta$.

°THEOREM 16: *If* $\pi(\gamma+1)$ *is not a limit ordinal, then* $\pi(\gamma+1) > \gamma$; *if* $\pi(\gamma+1)$ *is a limit ordinal, then*

$$cf\big(\pi(\gamma+1)\big) > \gamma.$$

PROOF. The first part of the theorem is obvious since the function π is increasing.

Assume that $\pi(\gamma+1)$ is a limit ordinal and that $\delta = cf\big(\pi(\gamma+1)\big)$. If $\delta \leqslant \gamma$ were true then by Corollary 5.7 $\aleph_{\pi(\gamma+1)} < \aleph_{\pi(\gamma+1)}^{\aleph_\gamma}$ would also be true, that is,

$$\mathfrak{a}_{\gamma+1} < \mathfrak{a}_{\gamma+1}^{\aleph_\gamma} = 2^{\mathfrak{a}_\gamma \aleph_\gamma} = 2^{\mathfrak{a}_\gamma} = \mathfrak{a}_{\gamma+1},$$

which is impossible. Thus $\delta > \gamma$. Q.E.D.

[1]) The question how much information about $\pi(\alpha)$ can be deduced from the axioms of set theory has been discussed by Lake [1].

Using the ordinals $\pi(\alpha)$, we are now able to evaluate the power $\mathfrak{a}_\alpha^{\aleph_\beta}$ where α is a limit ordinal.

THEOREM 17: *If* $\alpha > 0$ *is a limit ordinal, then*

$$\mathfrak{a}_\alpha^{\aleph_\beta} = \begin{cases} \mathfrak{a}_\alpha & \textit{if} \quad \beta < cf(\alpha), \\ \mathfrak{a}_{\alpha+1} & \textit{if} \quad cf(\alpha) \leqslant \beta \leqslant \pi(\alpha). \end{cases}$$

PROOF. Assume that $\beta < cf(\alpha)$. Clearly it suffices to show that $\mathfrak{a}_\alpha^{\aleph_\beta} \leqslant \mathfrak{a}_\alpha$. By Theorem 5.3,

$$\aleph_{\pi(\alpha)}^{\aleph_\beta} = \sum_{\xi<\alpha} \aleph_{\pi(\xi)}^{\aleph_\beta} \leqslant \sum_{\xi<\alpha} \aleph_{\pi(\xi+1)}^{\aleph_\beta}.$$

The last sum can be separated into two sums $\sum'$ and $\sum''$ where in the first sum ξ is such that $\pi(\xi+1) \leqslant \beta$ and in the second such that $\beta < \pi(\xi+1)$. From Theorem 5.1 it follows that the ξth component of the sum $\sum'$ is equal to $2^{\aleph_\beta}$; and as $\aleph_\beta \leqslant \mathfrak{a}_\beta < \mathfrak{a}_{cf(\alpha)} \leqslant \mathfrak{a}_\alpha$, we have

$$\textstyle\sum' \leqslant \overline{\beta} \cdot \mathfrak{a}_\alpha \leqslant \mathfrak{a}_\beta \cdot \mathfrak{a}_\alpha = \mathfrak{a}_\alpha.$$

The ξth component of the second sum $= \mathfrak{a}_{\xi+1}^{\aleph_\beta} = 2^{\mathfrak{a}_\xi \aleph_\beta} \leqslant 2^{\mathfrak{a}_\xi \mathfrak{a}_{\xi+1}} = \mathfrak{a}_{\xi+2}$, because $\aleph_\beta < \aleph_{\pi(\xi+1)} = \mathfrak{a}_{\xi+1}$. Thus

$$\textstyle\sum'' \leqslant \displaystyle\sum_{\xi<\alpha} \mathfrak{a}_{\xi+2} \leqslant \mathfrak{a}_\alpha \overline{\alpha} = \mathfrak{a}_\alpha.$$

It follows that

$$\mathfrak{a}_\alpha^{\aleph_\beta} \leqslant \mathfrak{a}_\alpha + \mathfrak{a}_\alpha = \mathfrak{a}_\alpha.$$

Assume now that φ is an increasing sequence of type $\omega_{cf(\alpha)} = \gamma$ convergent to α and that $cf(\alpha) \leqslant \beta \leqslant \pi(\alpha)$. Then

$$\mathfrak{a}_\alpha^{\aleph_\beta} \leqslant \mathfrak{a}_\alpha^{\aleph_{\pi(\alpha)}} = \mathfrak{a}_\alpha^{\mathfrak{a}_\alpha} \leqslant 2^{\mathfrak{a}_\alpha^2} = \mathfrak{a}_{\alpha+1}.$$

On the other hand,

$$\mathfrak{a}_{\alpha+1} = 2^{\mathfrak{a}_\alpha} = 2^{\aleph_{\pi(\alpha)}} = 2^{\sum_{\xi<\gamma} \aleph_{\pi(\varphi_\xi)}},$$

$$\prod_{\xi<\gamma} 2^{\aleph_{\pi(\varphi(\xi))}} = \prod_{\xi<\gamma} \mathfrak{a}_{\varphi_\xi+1} \leqslant \mathfrak{a}_\alpha^{\overline{\gamma}} = \mathfrak{a}_\alpha^{\aleph_{cf(\alpha)}} \leqslant \mathfrak{a}_\alpha^{\aleph_\beta},$$

which proves Theorem 17.

°COROLLARY 18: *If* α *is a limit ordinal then*

$$\mathfrak{a}_\alpha^{\mathfrak{a}_\beta} = \begin{cases} \mathfrak{a}_\alpha & \textit{if} \quad \pi(\beta) < cf(\alpha), \\ \mathfrak{a}_{\alpha+1} & \textit{if} \quad cf(\alpha) \leqslant \pi(\beta) \leqslant \pi(\alpha), \\ \mathfrak{a}_{\beta+1} & \textit{if} \quad \beta \leqslant \alpha. \end{cases}$$

Corollary 18 shows that the difficulty in giving general formulas for the powers $\mathfrak{a}_\alpha^{\mathfrak{a}_\beta}$ is caused by the lack of knowledge about the values of the function π.

§ 7. The continuum hypothesis

Cantor conjectured[1]) that $\aleph_1 = \mathfrak{a}_1$. This hypothetical equation is called the *continuum hypothesis*, or CH for short. A more general hypothesis is that the hierarchy of alephs is identical with the exponential hierarchy, i.e., that $\aleph_\xi = \mathfrak{a}_\xi$ for each ordinal ξ. This hypothesis is called the *generalized continuum hypothesis* or GCH.

Obviously we can simplify essentially the laws of exponentiation of cardinals if we assume GCH.

It has been shown that CH can neither be proved nor disproved on the basis of the axioms Σ°[TR] provided that these axioms are consistent. These logical results do not settle the question originally asked by Cantor whether CH or GCH are true or false statements. However it must be said that these seemingly obvious questions are not very clear: the concepts of truth and falsity (as opposed to the concept of the derivability from axioms) do not have a clear meaning in abstract set theory. Thus we cannot rule out the possibility that Cantor's original questions will turn out to be simply meaningless.

In the present section we shall derive some theorems which illustrate the role played by CH or GCH in establishing mathematical theorems.

THEOREM 1: *In the system consisting of axioms* Σ°[TR] *and* VIII, *the hypothesis*

(1) $$\aleph_\alpha = \mathfrak{a}_\alpha$$

for each ordinal α (*i.e., the hypothesis GCH*) *is equivalent to*

[1]) Cantor's conjecture was first formulated by him in [1], Section 8. The consistency of GCH with other axioms of set theory was established by Gödel in 1939 in [1] and its independence by Cohen in 1964. The proofs of these results are easily available, see e.g. Cohen [1]. Easton [1] showed that there are arbitrarily many functions $f(\xi)$ such that the assumption: "$\mathfrak{a}_\xi = \aleph_{f(\xi)}$ for each regular $\mathfrak{a}_\xi$" is consistent with the axioms of set theory. For singular cardinals $\mathfrak{a}_\xi$ the problem is not yet completely solved.

(C) *If* $\mathfrak{m}$ *is any cardinal number then there exists no cardinal* $\mathfrak{r}$ *such that* $\mathfrak{m} < \mathfrak{r} < 2^{\mathfrak{m}}$.

PROOF. If $\mathfrak{m} = \aleph_\alpha$ then GCH implies that

$$2^{\mathfrak{m}} = 2^{\aleph_\alpha} = 2^{\mathfrak{a}_\alpha} = \mathfrak{a}_{\alpha+1} = \aleph_{\alpha+1},$$

and thus no cardinal lies between $\mathfrak{m}$ and $2^{\mathfrak{m}}$. Suppose now that there is no cardinal between $\mathfrak{m}$ and $2^{\mathfrak{m}}$. We shall prove by induction that (1) holds. For $\alpha = 0$, (1) is obvious. If (1) holds for an α, then it also holds for $\alpha+1$ since

$$\mathfrak{a}_{\alpha+1} = 2^{\mathfrak{a}_\alpha} \geqslant \aleph_{\alpha+1} > \aleph_\alpha = \mathfrak{a}_\alpha;$$

and if $\mathfrak{a}_{\alpha+1} \neq \aleph_{\alpha+1}$, then $\aleph_{\alpha+1}$ would lie between $2^{\mathfrak{a}_\alpha}$ and $\mathfrak{a}_\alpha$. Finally, if (1) holds for all $\xi < \lambda$ where λ is a limit ordinal, then

$$\mathfrak{a}_\lambda = \sum_{\alpha<\lambda} \mathfrak{a}_\alpha = \sum_{\alpha<\lambda} \aleph_\alpha = \aleph_\lambda.$$

The generalized continuum hypothesis implies a simplification of the laws of exponentiation of cardinal numbers.

°THEOREM 2: *GCH implies*:

(a) $\pi(\alpha) = \alpha$ *for all* α,

(b) *if* α *is a limit ordinal and* $\beta < \alpha$, *then*

$$\mathfrak{a}_\alpha^{\mathfrak{a}_\beta} = \begin{cases} \mathfrak{a}_\alpha & \text{for} \quad \beta < cf(\alpha), \\ \mathfrak{a}_{\alpha+1} & \text{for} \quad cf(\alpha) \leqslant \beta \leqslant \alpha. \end{cases}$$

PROOF. Clearly, $\pi(0) = 0$. If $\pi(\alpha) = \alpha$, then $\aleph_{\pi(\alpha+1)} = \mathfrak{a}_{\alpha+1}$ by definition and thus by GCH $\aleph_{\pi(\alpha+1)} = \aleph_{\alpha+1}$, which implies that $\pi(\alpha+1) = \alpha+1$. Finally, if λ is a limit ordinal and if $\pi(\alpha) = \alpha$ for all $\alpha < \lambda$, then

$$\aleph_{\pi(\lambda)} = \sum_{\xi<\lambda} \aleph_{\pi(\xi)} = \sum_{\xi<\lambda} \aleph_\xi = \aleph_\lambda$$

and thus $\pi(\lambda) = \lambda$.

Formula (b) is a consequence of (a) and of Theorem 6.17.

°THEOREM 3: *GCH implies that if*

$$s_{\alpha,\beta} = \sum_{\xi<\beta} \mathfrak{a}_\alpha^{\mathfrak{a}_\xi},$$

then

(i) $s_{\alpha,\delta+1} = \mathfrak{a}_\alpha^{\mathfrak{a}_\delta}$,

(ii) $s_{\gamma+1,\beta} = \mathfrak{a}_{\gamma+1}$ *if* $\beta < \gamma+1$, β *is a limit ordinal,*

(iii) $s_{\gamma+1,\beta} = \mathfrak{a}_\beta$ *if* $\beta > \gamma+1$, β *is a limit ordinal,*

(iv) $s_{\alpha,\beta} = \mathfrak{a}_\alpha$ *if* α *and* β *are limit ordinals and* $\beta < cf(\alpha)$,

(v) $s_{\alpha,\beta} = \mathfrak{a}_{\alpha+1}$ *if* α *and* β *are limit ordinals and* $cf(\alpha) < \beta \leqslant \alpha$,

(vi) $s_{\alpha,\beta} = \mathfrak{a}_\beta$ *if* α *and* β *are limit ordinals and* $\beta > \alpha$.

PROOF. (i) $\bar{\delta} \leqslant \mathfrak{a}_\delta$ implies that

$$\mathfrak{a}_\alpha^{\mathfrak{a}_\delta} \leqslant \sum_{\xi<\delta+1} \mathfrak{a}_\alpha^{\mathfrak{a}_\xi} \leqslant \mathfrak{a}_\alpha^{\mathfrak{a}_\delta} \cdot \bar{\delta} = \mathfrak{a}_\alpha^{\mathfrak{a}_\delta}.$$

(ii) By definition,

$$s_{\gamma+1,\beta} = \sum_{\xi<\beta} 2^{\mathfrak{a}_\gamma \cdot \mathfrak{a}_\xi} = \sum_{\xi<\beta} 2^{\mathfrak{a}_\gamma} = \sum_{\xi<\beta} \mathfrak{a}_{\gamma+1} = \bar{\beta} \cdot \mathfrak{a}_{\gamma+1} = \mathfrak{a}_{\gamma+1}.$$

(iii) Similarly we have

$$s_{\gamma+1,\beta} = \mathfrak{a}_{\gamma+1} + \sum_{\gamma+1<\xi<\beta} \mathfrak{a}_{\xi+1} = \mathfrak{a}_\beta.$$

(iv) From Theorem 2 we obtain

$$s_{\alpha,\beta} = \sum_{\xi<\beta} \mathfrak{a}_\alpha^{\mathfrak{a}_\xi} = \sum_{\xi<\beta} \mathfrak{a}_\alpha = \mathfrak{a}_\alpha \cdot \bar{\beta} = \mathfrak{a}_\alpha.$$

(v) Similarly,

$$s_{\alpha,\beta} = \mathfrak{a}_\alpha + \sum_{cf(\alpha)\leqslant\xi<\beta} \mathfrak{a}_\alpha^{\mathfrak{a}_\xi} = \mathfrak{a}_\alpha + \overline{\beta - cf(\alpha)} \cdot \mathfrak{a}_{\alpha+1} = \mathfrak{a}_{\alpha+1}.$$

(vi) From the elementary formula $\mathfrak{a}_\alpha^{\mathfrak{a}_\xi} = \mathfrak{a}_{\xi+1}$ for $\alpha < \xi$ it follows that

$$s_{\alpha,\beta} \geqslant \sum_{\alpha\leqslant\xi<\beta} \mathfrak{a}_\alpha^{\mathfrak{a}_\xi} = \sum_{\alpha\leqslant\xi<\beta} \mathfrak{a}_{\xi+1} = \mathfrak{a}_\beta;$$

on the other hand, it is clear that

$$\sum_{\xi<\beta} \mathfrak{a}_\alpha^{\mathfrak{a}_\xi} \leqslant \sum_{\xi<\alpha} \mathfrak{a}_\alpha^{\mathfrak{a}_\xi} + \sum_{\alpha\leqslant\xi<\beta} \mathfrak{a}_\alpha^{\mathfrak{a}_\xi} \leqslant \bar{\alpha} \cdot \mathfrak{a}_{\alpha+1} + \bar{\beta} \sum_{\xi<\beta} \mathfrak{a}_{\xi+1} \leqslant \mathfrak{a}_\beta.$$

The hypothesis GCH was used only to prove (iv) and (v). Using Theorem 2 we can further obtain explicit values for $s_{\alpha,\delta+1}$.

In a similar way we can also calculate the sum $t_{\alpha,\beta} = \sum_{\xi<\alpha} \mathfrak{a}_\xi^{\mathfrak{a}_\beta}$. We shall give this result without proof.

°THEOREM 4:

(i) $t_{\gamma+1,\beta} = \mathfrak{a}_\gamma^{\mathfrak{a}_\beta}$,

(ii) $t_{\alpha,\beta} = \mathfrak{a}_{\beta+1}$ *if* α *is a limit ordinal and* $\beta \geqslant \alpha$,

(iii) $t_{\alpha,\beta} = \mathfrak{a}_\alpha$ *if* α *is a limit ordinal and* $\beta < \alpha$.

Formula (i) reduces the evaluation of $t_{\gamma+1,\beta}$ to Theorem 2.

There are many simple properties of cardinal numbers which are equivalent to the hypothesis GCH. We shall give several examples:[1])

°THEOREM 5: *The hypothesis GCH is equivalent in the axiom system* Σ°[TR] *to each of the following formulas*:

(1) $$\bigwedge_\alpha (\aleph_{\alpha+1}^{\aleph_\alpha} = \aleph_{\alpha+1}),$$

(2) $$\bigwedge_\alpha (\aleph_{\alpha+1}^{\aleph_\alpha} < \aleph_{\alpha+2}^{\aleph_\alpha}),$$

(3) $$\bigwedge_\alpha \Big(\sum_{\xi\leqslant\alpha} \aleph_{\alpha+1}^{\aleph_\xi} = \aleph_{\alpha+1}\Big).$$

REMARK: Formula (3) states that for every set X of power $\aleph_{\alpha+1}$ the family of all subsets $Y \subset X$ which are not equipollent with X has the same power as X. For sets X whose power is an aleph with a limit index such a formula is not true (provided we accept GCH; see Theorem 3 (v)).

PROOF. GCH $\equiv$ (1), since

$$\aleph_\alpha < \aleph_{\alpha+1} \leqslant \aleph_{\alpha+1}^{\aleph_\alpha} \leqslant 2^{\aleph_\alpha\cdot\aleph_\alpha} = 2^{\aleph_\alpha}.$$

GCH $\rightarrow$ (2), because

$$\aleph_{\alpha+1}^{\aleph_\alpha} = 2^{\aleph_\alpha\cdot\aleph_\alpha} = 2^{\aleph_\alpha} = \aleph_{\alpha+1}$$

and

$$\aleph_{\alpha+2}^{\aleph_\alpha} = (2^{\aleph_{\alpha+1}})^{\aleph_\alpha} = 2^{\aleph_{\alpha+1}} = \aleph_{\alpha+2}.$$

[1]) Several equivalences similar to those given in Theorem 5 can be found in Bachmann [1]. See also Sierpiński [1].

$\neg$GCH $\to$ $\neg$(2). Assume that $\aleph_{\alpha+1} < 2^{\aleph_\alpha}$. Then

$$\aleph_{\alpha+1} < \aleph_{\alpha+2} \leqslant 2^{\aleph_\alpha}$$

and thus

$$2^{\aleph_\alpha} \leqslant \aleph_{\alpha+1}^{\aleph_\alpha} \leqslant \aleph_{\alpha+2}^{\aleph_\alpha} \leqslant (2^{\aleph_\alpha})^{\aleph_\alpha} = 2^{\aleph_\alpha},$$

which implies $\neg$(2).

GCH $\to$ (3). Indeed, by GCH $\sum_{\xi\leqslant\alpha} \aleph_{\alpha+1}^{\aleph_\xi} = s_{\alpha+1,\alpha+1}$ and by Theorem 3 (i) we easily obtain that

$$s_{\alpha+1,\alpha+1} = \mathfrak{a}_{\alpha+1} = \aleph_{\alpha+1}.$$

(3) $\to$ GCH. Indeed, (3) implies $\aleph_{\alpha+1}^{\aleph_\alpha} \leqslant \aleph_{\alpha+1}$ and thus

$$2^{\aleph_\alpha} \leqslant \aleph_{\alpha+1}^{\aleph_\alpha} \leqslant \aleph_{\alpha+1} \leqslant 2^{\aleph_\alpha},$$

whence $\aleph_{\alpha+1} = 2^{\aleph_\alpha}$.

°THEOREM 6: *The hypothesis GCH is equivalent in the system* Σ° [TR] *to the theorem*:

(T) *For every infinite set X, the set* $\boldsymbol{P}(X)$ *can be represented as the union* $\bigcup_{\xi<\alpha} M_\xi$ *of a strictly increasing sequence of sets* M_ξ *equipollent with X.*

PROOF. GCH $\to$ (T). Let $\overline{\overline{X}} = \aleph_\gamma$; from GCH it follows that the set $\boldsymbol{P}(X)$ is equipollent with $\bigcup_{\omega_\gamma<\xi<\omega_{\gamma+1}} W(\omega_\xi)$. Choosing for M_ξ the image of the set $W(\omega_\xi)$ we obtain the desired representation of $\boldsymbol{P}(X)$.

(T) $\to$ GCH. Assume that we have a representation of $\boldsymbol{P}(X)$ as described in (T) and let $\overline{\overline{X}} = \aleph_\gamma$. Then $2^{\aleph_\gamma} = \aleph_\gamma \cdot \overline{\alpha}$ and thus $\overline{\alpha} \geqslant \aleph_{\gamma+1}$ and $\alpha \geqslant \omega_{\gamma+1}$.

Let $S = \bigcup_{\xi<\omega_{\gamma+1}} M_\xi$. We shall show that $\boldsymbol{P}(X) = S$. Indeed, $\overline{\overline{S}} = \aleph_{\gamma+1}$ by the assumption that $M_\xi \subset M_\eta$, $M_\xi \neq M_\eta$ and that $\overline{\overline{M_\xi}} = \aleph_\gamma$ for $\xi < \eta < \alpha$. If $Y \in \boldsymbol{P}(X)$ then there exists $\eta < \alpha$ such that $Y \in M_\eta$. Because $\overline{\overline{M_\eta}} = \aleph_\gamma$, it follows that the set M_η cannot contain the entire union S and thus there exists $\xi < \omega_{\gamma+1}$ such that $M_\xi - M_\eta \neq \emptyset$; hence $\eta < \xi$ and $\eta < \omega_{\gamma+1}$. Therefore, $\boldsymbol{P}(X) \subset S$; the opposite inclusion is obvious. From the equation $\boldsymbol{P}(X) = S$ we infer that $2^{\aleph_\gamma} = \overline{\overline{S}} = \aleph_{\gamma+1}$. Q.E.D.

The hypothesis GCH and even its particular case $2^{\aleph_0} = \aleph_1$ have many consequences in various areas of mathematics and particularly in the theory of real functions. We shall give only one characteristic example.

°THEOREM 7:[1]) *There exists a non-countable subset Z of the set of real numbers $\mathscr{E}$ such that the intersection of Z with every nowhere-dense subset of $\mathscr{E}$ is countable.*

PROOF. Let $\boldsymbol{R}$ be the family of closed nowhere-dense subsets of $\mathscr{E}$ (see p. 32). Since $\overline{\overline{\boldsymbol{R}}} = \mathfrak{c}$ (see the exercise on p. 191), there exists a sequence F of type ω_1 whose set of terms coincides with the family $\boldsymbol{R}$.

Let

$$E_\xi = F_\xi - \bigcup_{\eta<\xi} F_\eta .$$

There exist uncountably many indices ξ such that $E_\xi \neq \emptyset$; otherwise there would exist an ordinal $\alpha < \omega_1$ such that

$$\bigcup_{\eta<\alpha} F_\eta = \bigcup_{\eta<\beta} F_\eta \quad \text{for} \quad \alpha \leqslant \beta < \omega_1$$

and then we would have $\bigcup_{\eta<\alpha} F_\eta = \mathscr{E}$ since every one-element set belongs to $\boldsymbol{R}$. But then the set $\mathscr{E}$ would be the countable union of nowhere-dense sets, in contradiction to one of the basic theorems of topology due to Baire.

Let $\boldsymbol{S}$ denote the family of non-empty sets E_ξ. By the axiom of choice there exists a set Z containing exactly one point from each of these sets. Therefore $\overline{\overline{Z}} > \aleph_0$. Let H be a nowhere-dense subset of E. Because the closure $\overline{H}$ of H is also nowhere-dense, it also belongs to $\boldsymbol{R}$ and thus $\overline{H} = F_\xi$ for some $\xi < \omega_1$. Since $\overline{\overline{Z \cap F_\xi}} \leqslant \aleph_0$, it follows that $\overline{\overline{Z \cap H}} \leqslant \aleph_0$.

Exercises

1. Derive from CH the following proposition: There exists a sequence $\{f_n\}_{n<\omega}$ of real functions such that $\mathscr{E}\times\mathscr{E}$ is the union $\bigcup\{f_n^1 : n < \omega\}$.

Hint: It follows from CH that there exists a mapping $\varphi\colon W(\omega_1) \xrightarrow[\text{onto}]{} \mathscr{E}$. For $\xi < \omega_1$ let $\alpha_\xi\colon N \xrightarrow[\text{onto}]{} W(\xi)$. For $x = \varphi(\xi)$ put $f_n(x) = \varphi(\alpha_\xi(n))$. [Sierpiński]

2. Derive from CH the following proposition: (M) the union of less than $2^{\aleph_0}$ linear sets of measure 0 has measure 0.

[1]) Theorem 7 is due to Lusin [1].

3. Derive from CH the following proposition: (S) there exists a set X of power $2^{\aleph_0}$, $X \subset \mathscr{E}$, which has at most denumerably many points in common with every set of measure 0.

Hint: Use the same method as in the proof of Theorem 7 and notice that a set of measure 0 is a subset of a G_δ-set of measure 0.

4. Derive CH from the conjunction of (M) and (S).

Hint: Let X be a set satisfying (S) and assume that $\aleph_1 < 2^{\aleph_0}$. If S is a subset of X of power $\aleph_1$, then X has measure 0 by (M) and hence $S \cap X = S$ is at most countable by (S). [Sierpiński]

§ 8. The number of prime ideals in the algebra $\boldsymbol{P}(A)$

In Sections 8, 9, and 10 we shall present some applications of the cardinal arithmetic which we developed in the present chapter. In Section 8 we shall calculate the cardinal number of the set of prime ideals of the algebra $\boldsymbol{P}(A)$ where A is an infinite set. In Section 9 we shall present a theorem on the decomposition of an infinite set A into a union of subsets any two of which intersect in less than $\bar{\bar{A}}$ elements. Finally in Section 10 we shall evaluate the cardinal numbers of families consisting of mutually disjoint open subsets of certain topological spaces.

The reasons why we present these rather unrelated results here are twofold: First they provide interesting applications of the theory of alephs. Secondly these results have found important applications in recent set-theoretical investigations. The interested reader will find references to the pertinent works in the notes at the end of this chapter.

Let A be an infinite set of power $\mathfrak{m}$.

°THEOREM 1: *The field of sets* $\boldsymbol{P}(A)$ *contains* $2^{2^{\mathfrak{m}}}$ *prime ideals.*[1])

First we reduce the proof to the following lemma:

°LEMMA 1: *There exists a family* $\boldsymbol{S} \subset \boldsymbol{P}(A)$ *of power* $2^{\mathfrak{m}}$ *such that every finite subfamily* $\boldsymbol{S}_1 \subset \boldsymbol{S}$ *is independent* (see p. 22).

With every function $f \subset \{0, 1\}^{\boldsymbol{S}}$ we associate the family

$$\boldsymbol{S}(f) = \{Z: [[(Z \in \boldsymbol{S}) \wedge (f(Z) = 0)] \vee [(A - Z \in \boldsymbol{S}) \wedge (f(Z) = 1)]]\}.$$

It is clear that distinct functions are associated with distinct families.

[1]) For $\mathfrak{m} = \aleph_0$ Theorem 1 was proved by Fichtenholz–Kantorovitch [1]; see also Hausdorff [6]. Tarski's papers [4] and [7] contain the full proof of Theorem 1 as well as many related results.

If k is a natural number > 0, and if $Z_i \in S(f)$ for $i < k$, then $\bigcup_{i<k} Z_i \neq A$. Indeed, assume for example that $Z_i \in S$ for $i < p$ and that $A - Z_i \in S$ for $p \leqslant i < k$. If the equation $\bigcup_{i<k} Z_i = A$ were true, then letting $W_i = A - Z_i$ for $p \leqslant i < k$ we would have $\bigcap_{i<p} (A - Z_i) \cap \cap \bigcap_{p \leqslant i \leqslant k} W_i = \emptyset$, contradicting the fact that the family composed of the sets $Z_0, \ldots, Z_{p-1}$ and $W_p, \ldots, W_{k-1}$ is independent.

Let $I(f)$ be the family composed of sets $Z \subset A$ having the following property: there exists a finite number of sets $Z_1, \ldots, Z_n \in S(f)$ such that $Z \subset Z_1 \cup \ldots \cup Z_n$. It is clear that

$$Z' \subset Z \in I(f) \rightarrow Z' \in I(f)$$

and that

$$(Z \in I(f)) \wedge (Z' \in I(f)) \rightarrow (Z \cup Z' \in I(f)).$$

Thus the family $I(f)$ is an ideal. Since no finite sum $Z_1 \cup \ldots \cup Z_n$ of elements of $S(f)$ is equal to A, it follows that $A \notin I(f)$. Finally by definition we have $S(f) \subset I(f)$.

By VII.8.2 there exists a maximal (and thus a prime) ideal $J(f)$ containing $I(f)$ (see Exercise IV.12.2).

If $f_1 \neq f_2$, then $J(f_1) \neq J(f_2)$. If for example $f_1(Z) = 0$ and $f_2(Z) = 1$, then $Z \in S(f_1) \subset J(f_1)$ and $A - Z \in S(f_2) \subset J(f_2)$ and thus the ideals $J(f_1)$ and $J(f_2)$ are distinct, because otherwise $A = Z \cup (A - Z) \in J(f_1)$ in contradiction to the definition of prime ideal. Thus the set of all prime ideals is at least of the same power as the set of all functions $f \in \{0, 1\}^S$, that is, at least of power $2^{2^{\mathfrak{m}}}$. On the other hand, every prime ideal is contained in the family $P(A)$ and thus the set of all prime ideals has at most power $2^{2^{\mathfrak{m}}}$.

It remains to prove the lemma. For this purpose we shall carry out a series of reductions.

Let X be an arbitrary set and let R be a family of subsets of X. We shall say that R is

(a) *independent*,

(b) *weakly independent*,

(c) *very weakly independent*,

if respectively

(a) every finite subfamily $\boldsymbol{R}_1 \subset \boldsymbol{R}$ is independent,

(b) $M - \bigcup_{i<k} M_i \neq \emptyset$ for arbitrary distinct $M, M_0, \ldots, M_{k-1} \in \boldsymbol{R}$, $(k = 0, 1, 2, \ldots)$.

(c) $M_1 - M_2 \neq \emptyset$ for arbitrary distinct $M_1, M_2 \in \boldsymbol{R}$.

We say that the *cardinal* $\mathfrak{m}$ *satisfies condition* (A), (B), or (C) provided there exists a set X of power $\mathfrak{m}$ and a family $\boldsymbol{R} \subset \boldsymbol{P}(X)$ of power $2^{\mathfrak{m}}$ satisfying the condition (a), (b), or (c), respectively.

It is clear that if $\mathfrak{m}$ satisfies condition (A), then for every set X of power $\mathfrak{m}$ there exists an independent family $\boldsymbol{R} \subset \boldsymbol{P}(X)$ of power $2^{\mathfrak{m}}$. Similar theorems hold for conditions (B) and (C).

The lemma above is equivalent to the statement that every infinite cardinal $\mathfrak{m}$ satisfies condition (A).

We now show that conditions (A), (B), (C) are mutually equivalent. The implications (A) → (B) → (C) are obvious. Thus it suffices to show that

(1) $$(\mathrm{B}) \to (\mathrm{A}),$$

(2) $$(\mathrm{C}) \to (\mathrm{B}).$$

The scheme of proof of (1) and (2) is as follows: we assume that there is a set X of power $\mathfrak{m}$ and a family $\boldsymbol{R} \subset \boldsymbol{P}(X)$ of power $2^{\mathfrak{m}}$ satisfying condition (b) (respectively (c)) and we then construct a set K of the same power $\mathfrak{m}$ as well as a family $\boldsymbol{S} \subset \boldsymbol{P}(X)$ of power $2^{\mathfrak{m}}$ satisfying condition (a) (respectively (b)).

Proof of (1). Let X be a set of power $\mathfrak{m}$ and $\boldsymbol{R}$ a weakly independent family of power $2^{\mathfrak{m}}$ consisting of subsets of X. Let $\boldsymbol{K}$ denote the family of all finite subsets of X. Clearly, $\boldsymbol{K} = \bigcup_n \boldsymbol{K}_n$, where $\boldsymbol{K}_n$ is the family of all subsets of X which contain exactly n elements. Moreover, $\overline{\overline{\boldsymbol{K}_n}} = \mathfrak{m}^n = \mathfrak{m}$ (see p. 284), and thus $\overline{\overline{\boldsymbol{K}}} = \mathfrak{m} \cdot \mathfrak{a} = \mathfrak{m}$.

For $T \in \boldsymbol{R}$, let

$$X(T) = \{Z \in \boldsymbol{K} \colon Z \cap T \neq \emptyset\},$$

and denote by $\boldsymbol{S}$ the family of all subsets of $\boldsymbol{K}$ having the form $X(T)$.

We shall prove that every finite family $\{X(T_1), \ldots, X(T_k)\}$, where

$T_1, \ldots, T_k$ are distinct elements of $\boldsymbol{R}$, is independent. We shall show, for example, that the set

$$X(T_1) \cap \ldots \cap X(T_p) \cap [K - X(T_{p+1})] \cap \ldots \cap [K - X(T_k)]$$

is non-empty.

Indeed, from the assumption that $\boldsymbol{R}$ is weakly independent it follows that $T_j - \bigcup_{i=p+1}^{k} T_i \neq \emptyset$ for $j = 1, 2, \ldots, p$. If x_j is an element of this difference, then the set $\{x_1, \ldots, x_p\}$ clearly belongs to $X(T_j)$ and to $K - X(T_i)$ for $j = 1, \ldots, p$, $i = p+1, \ldots, k$.

From the independence proved above it follows in particular that $X(T) \neq X(T')$ for $T \neq T'$. Therefore $\boldsymbol{S}$ is a family of power $2^{\mathfrak{m}}$ consisting of subsets of K and every finite subfamily $\boldsymbol{S}_1 \subset \boldsymbol{S}$ is independent; that is, the family $\boldsymbol{S}$ satisfies condition (a).

Proof of (2). Let X be a set of power $\mathfrak{m}$ and let $\boldsymbol{R}$ be a very weakly independent family of power $2^{\mathfrak{m}}$ consisting of subsets of X. For every set Z let $Z' = \bigcup_{n \in N} Z^n$; thus Z' is the family of all finite sequences whose terms belong to Z.

Let $K = X'$; clearly, $K = \bigcup_{n \in N} X^n$ and as $\overline{\overline{X^n}} = \mathfrak{m}^n = \mathfrak{m}$ it follows that $\overline{\overline{K}} = \mathfrak{m} \cdot \mathfrak{a} = \mathfrak{m}$. Let $\boldsymbol{S} = \{M' \colon M \in \boldsymbol{R}\}$; clearly $\boldsymbol{S}$ is a family of subsets of K and $\overline{\overline{\boldsymbol{S}}} = \overline{\overline{\boldsymbol{R}}} = 2^{\mathfrak{m}}$ since the mapping $M \to M'$ is one-to-one.

We shall now show that the family $\boldsymbol{S}$ is weakly independent. Let $M, M_0, \ldots, M_{k-1}$ be distinct elements of $\boldsymbol{R}$ and assume that $M' \subset \bigcup_{i<k} M_i'$. From the assumption it follows that $M - M_i \neq \emptyset$ for $i < k$; let $x_i \in M - M_i$ for $i < k$. Thus the k-term sequence $\langle x_0, \ldots, x_{k-1} \rangle$ belongs to M' but does not belong to M_i' because its ith term does not belong to M_i. Consequently, we have a contradiction which proves that the family $\boldsymbol{S}$ is weakly independent.

Proof of the lemma. Let X and X' be two disjoint sets of power $\mathfrak{m}$ and let a one-to-one function f map X onto X'. For $M \subset X$ let $M^* = M \cup [X' - f^1(M)]$. Let $\boldsymbol{S}$ be the family of all sets M^*. The family $\boldsymbol{S}$ consists of subsets of the union $K = X \cup X'$, where $\overline{\overline{K}} = \mathfrak{m} + \mathfrak{m} = \mathfrak{m}$ and $\overline{\overline{\boldsymbol{S}}} = 2^{\mathfrak{m}}$ because the mapping $M \to M^*$ is one-to-one.

If $M_1 \neq M_2$ and $M_1, M_2 \subset X$, then $M_1^* \neq M_2^*$. For $x \in M_1 - M_2$ implies that $x \in M_1^* - M_2^*$ whereas $x \in M_2 - M_1$ implies that $f(x) \in M_1^* - M_2^*$. Thus S is very weakly independent, which proves that the cardinal $\mathfrak{m}$ has property (C); hence, by (1) and (2), $\mathfrak{m}$ also has property (A). This completes the proof of the lemma.

Exercises

1. Prove that if a Boolean algebra is finite and has 2_4^n elements, then it contains exactly n prime ideals.

Hint: Every prime ideal has the form $\{x: x \wedge a = o\}$, where a is an atom.

2. The Boolean algebra K with unit of infinite power $\mathfrak{m}$ contains at least 1 and at most $2^{\mathfrak{m}}$ prime non-principal ideals.

Hint: Let a_t denote any element of K such that the ideal $\{x: x \leqslant a_t\}$ is prime; then there exists a proper ideal I containing all elements $-a_t$.

3. We say that a subset X of the Boolean algebra K is a *base* for the ideal I if (1) $X \subset I$, (2) for every $a \in I$ there exist finitely many $x_1, \ldots, x_n$ elements of X such that $a \leqslant x_1 \vee x_2 \vee \ldots \vee x_n$, and (3) no proper subset of X satisfies conditions (1) and (2).

Prove that an ideal is principal if and only if it has a one element base.

4. If $K = P(N)$ and I is a non-principal prime ideal, then no base for I is countable.

§ 9. $\mathfrak{m}$-disjoint sets

From the axiom of choice it follows that each family of mutually disjoint non-void subsets of a set A has a cardinal number $\leqslant \overline{\overline{A}}$. To prove this we merely remark that such a family S has the same power as any subset of A containing exactly one element in common with each set in S.

Sierpiński discovered in 1928 that the inequality $\overline{\overline{S}} \leqslant \overline{\overline{A}}$ need not be valid if we replace the assumption that elements of S are mutually disjoint by a weaker assumption that each pair of different elements of S has a finite intersection. Such a family is called a *family of almost disjoint sets*. We indicated the proof of Sierpiński's result in Exercise I.5.4.[1])

We now generalize the notion of almost disjoint sets:

[1]) For early results on almost disjoint sets see Sierpiński [11], Tarski [3]. Interesting applications of these sets were given by Rabin [1] and Jensen and Solovay [1].

DEFINITION 1: For an infinite cardinal $\mathfrak{m}$ we say that two sets X, Y are $\mathfrak{m}$-*disjoint* if $\overline{\overline{X \cap Y}} < \mathfrak{m}$.

Sierpiński's result can be generalized to higher cardinalities:

THEOREM 1: *If* $\mathfrak{m} \geqslant \aleph_0$ *and* $\overline{\overline{A}}_0 = \mathfrak{m}$ *then there exists a family* $\boldsymbol{S}_0 \subset \boldsymbol{P}(A_0)$ *of power* $\aleph(\mathfrak{m})$ *consisting of* $\mathfrak{m}$-*disjoint sets and such that* $A_0 = \bigcup (\boldsymbol{S}_0)$.

Briefly we may say that each set of power $\mathfrak{m}$ *can be decomposed into* $\aleph(\mathfrak{m})$ $\mathfrak{m}$-*disjoint subsets.*

PROOF. It will evidently be sufficient to find just one set A which can be decomposed in the way indicated in the theorem and has the power $\mathfrak{m}$. A one-one mapping of A onto A_0 will then enable us to obtain the desired decomposition of A_0.

Let $\mathfrak{m} = \aleph_\alpha$, $W = W(\omega_\alpha)$ and $F = W^W$. Two functions f, g in F (conceived as sets of ordered pairs) are $\aleph_\alpha$-disjoint if and only if the power of the set $\{\xi < \omega_\alpha: f(\xi) = g(\xi)\}$ is $< \aleph_\alpha$.

We shall construct a sequence of type $\omega_{\alpha+1}$ of $\aleph_\alpha$ disjoint functions. To carry out this construction we need the

LEMMA: *If* φ *is a transfinite sequence of type* $\gamma < \omega_{\alpha+1}$ *with range contained in* F, *then there exists a function* $f \in F$ *which is* $\aleph_\alpha$-*disjoint from each term* φ_ξ *of the sequence.*

To see this we notice that the range of φ has the power $\leqslant \aleph_\alpha$ and thus can be represented as the range of a function $\Phi: W \to F$. If $\xi < \omega_\alpha$ then the set $\{\Phi_\eta(\xi): \eta < \xi\}$ is a proper subset of W because its power is $< \aleph_\alpha$. Denoting by $f(\xi)$ the first ordinal not in this set, we obtain a function $f \in F$ which is $\aleph_\alpha$-disjoint from every Φ_η, because the equation $f(\xi) = \Phi_\eta(\xi)$ can hold only if $\eta \geqslant \xi$.

The lemma being proved, we denote by Γ a choice function for the family of non-void subsets of F and agree additionally that $\Gamma(\emptyset) = p$ where p is an element not belonging to F. For each transfinite sequence φ denote by $B(\varphi)$ the set of all functions in F which are $\aleph_\alpha$-disjoint from all the terms of φ which belong to F. Using the theorem on transfinite induction, we obtain a function $g: W(\omega_{\alpha+1}) \to F \cup \{p\}$ satisfying the equation

$$g_\xi = \Gamma\big(B(g|\xi)\big)$$

for each $\xi < \omega_{\alpha+1}$.

Thus for each $\xi < \omega_{\alpha+1}$ the term g_ξ is a function which belongs to F and is $\aleph_\alpha$-disjoint from all the functions g_η, $\eta < \xi$, which belong to F, provided that such functions exist; otherwise $g_\xi = p$.

We claim that $g_\xi \neq p$ for each $\xi < \omega_{\alpha+1}$. Otherwise there would exist a smallest ordinal $\gamma < \omega_{\alpha+1}$ such that $g_\gamma = p$ and $g|\gamma$ would be a sequence satisfying the assumptions of the lemma stated above. Hence the set $B(g|\gamma)$ would not be void and g_γ would be an element of this set. This contradicts the assumption that p is not an element of F.

If $\xi \neq \eta$ and $\xi, \eta < \omega_{\alpha+1}$ then g_ξ and g_η are $\aleph_\alpha$-disjoint and hence different. Thus the set $S = \mathrm{Rg}(g)$ has the power $\aleph_{\alpha+1} = \aleph(\mathfrak{m})$. Put $A = \bigcup(S)$. Since each element of A is an ordered pair of ordinals $< \omega_\alpha$, we infer that $\overline{\overline{A}} \leqslant \overline{\overline{W \times W}} = \aleph_\alpha$. On the other hand, $\overline{\overline{A}} \geqslant \overline{\overline{g_0}} = \aleph_\alpha$. Thus the cardinal number of A is $\aleph_\alpha$ and the decomposition $A = \bigcup(S)$ has the required properties.

Theorem 1 expressed in the language of Boolean algebras takes on the following form:

THEOREM 2: *If $\overline{\overline{A}} = \aleph_\alpha$ and $\boldsymbol{I}$ is the ideal of $\boldsymbol{P}(A)$ consisting of all sets whose powers are $< \mathfrak{m}$, then the quotient algebra $\boldsymbol{P}(A)/\boldsymbol{I}$ has power at least $\aleph_{\alpha+1}$.*

§ 10. Families of disjoint open sets

Let $\mathscr{X}$ be a topological space. A cardinal $\mathfrak{w}$ is called the *weight of $\mathscr{X}$* if it is the smallest cardinal number such that $\mathscr{X}$ has an open base of power $\mathfrak{w}$.[1]) The *degree of disjointness* (or briefly the *degree* of $\mathscr{X}$) is the least cardinal $\mathfrak{d}$ such that every family of non-void open sets contains at least one pair of sets whose intersection is non-void.

For instance for the Cantor space the weight is $\aleph_0$ and the degree is $\aleph_1$. For the generalized Cantor space $\{0, 1\}^X$ with an infinite X

[1]) The notion of weight of a topological space is of some importance in general topology. The importance of the degree of disjointness for general set theory is due to the fact that proofs of independence of several set-theoretical hypotheses, for instance GCH, rest heavily on the evaluations of these degrees. See Cohen [1].

the weight is $\overline{\overline{X}}$. The degree of this space is, surprisingly, again $\aleph_1$. This will follow from Theorem 5 which we shall prove below.

Not every cardinal is the degree of a topological space. We shall show that only regular cardinals have this property.

In order to obtain this result we need the relative notion of a degree.

For an open set $X \neq \emptyset$ we call the *degree of* X the least cardinal $\mathfrak{d}(X)$ such that in each family $\boldsymbol{F}$ of power $\mathfrak{d}(X)$ consisting of non-empty open subsets of X there are at least two elements which have a non-empty intersection.

An open set $X \neq \emptyset$ will be called *homogeneous* if $\mathfrak{d}(Y) = \mathfrak{d}(X)$ for each non-void open subset Y of X.

LEMMA 1: *Each open set $X \neq \emptyset$ contains an homogeneous subset Y.*

PROOF. Let $\mathfrak{m}$ be the least cardinal of the family $\{\mathfrak{d}(Y): (\emptyset \neq Y \subset X) \wedge \wedge (Y \text{ is open})\}$. Then $\mathfrak{m} = \mathfrak{d}(Y)$ for some Y and if $\emptyset \neq Z \subset Y$ then $\mathfrak{d}(Z) = \mathfrak{m}$. Hence Y is homogeneous.

The crucial lemma is the following:

LEMMA 2: *If the degree of $\mathscr{X}$ is $\aleph_\delta$ where δ is a positive limit number then there is a homogeneous set X such that $\mathfrak{d}(X) = \aleph_\delta$.*

PROOF. We shall derive a contradiction from the assumption that such a set does not exist. Thus we assume that

$$\mathfrak{d}(X) < \aleph_\delta \tag{1}$$

for each homogeneous set X.

Let us consider a set whose elements are all families of mutually disjoint homogeneous sets and let $\boldsymbol{H}$ be a maximal such family. Obviously $\overline{\overline{\boldsymbol{H}}} < \aleph_\delta$ because $\boldsymbol{H}$ consists of mutually disjoint open sets. For each X in $\boldsymbol{H}$ we have, by (1), $\mathfrak{d}(X) < \aleph_\delta$, whence

$$\sum_{X \in \boldsymbol{H}} \mathfrak{d}(X) \leqslant \aleph_\delta \cdot \aleph_\delta = \aleph_\delta.$$

We shall show that the sign $\leqslant$ can be replaced by identity. Since δ is a positive limit number, it is sufficient to show that for each $\xi < \delta$

$$\aleph_\xi \leqslant \sum_{X \in \boldsymbol{H}} \mathfrak{d}(X). \tag{2}$$

Thus let us assume that $\xi < \delta$ and let $\boldsymbol{G}$ be a family of power $\aleph_\xi$ consisting of mutually disjoint non-empty sets. From the maximality of $\boldsymbol{H}$ it follows that for each $X \in \boldsymbol{G}-\boldsymbol{H}$ the family $\boldsymbol{H} \cup \{X\}$ contains at least two sets which are not disjoint, i.e. that $X \cap Y \neq \emptyset$ for some Y in $\boldsymbol{H}$. The same is also true if $X \in \boldsymbol{H}$ because then we can take X as the set Y. Using the axiom of choice, we find thus a function f: $\boldsymbol{G} \to \boldsymbol{H}$ such that $X \cap f(X) \neq \emptyset$ for each $X \in \boldsymbol{G}$.

The power of the set $\{X \in \boldsymbol{G}: f(X) = Z\} = f^{-1}(Z)$ is at most $\mathfrak{d}(Z)$ for each Z in $\boldsymbol{H}$. This follows from the observation that correlating $X \cap Z$ with a set $X \in f^{-1}(Z)$ we obtain a one-to-one mapping of $f^{-1}(Z)$ into the family of mutually disjoint open subsets of Z and this family has at most $\mathfrak{d}(Z)$ elements. It follows that $\boldsymbol{G} = \bigcup \{f^{-1}(Z): Z \in \boldsymbol{H}\}$ has the power at most $\sum_{Z \in \boldsymbol{H}} \mathfrak{d}(Z)$ and hence (2) is proved. We proved thus

$$\sum_{X \in \boldsymbol{H}} \mathfrak{d}(X) = \aleph_\delta. \tag{3}$$

From this equation we see at once that if $\eta < \delta$ then there is an X in $\boldsymbol{H}$ for which $\mathfrak{d}(X) > \aleph_\eta$; otherwise the left-hand side of (3) would be $\leqslant \aleph_\eta \cdot \overline{\overline{\boldsymbol{H}}} = \max(\aleph_\eta, \overline{\overline{\boldsymbol{H}}}) < \aleph_\delta$. Using the axiom of choice, we can therefore establish the existence of an ordinal $\gamma \leqslant \delta$ and a sequence $\{X_\xi\}_{\xi<\gamma}$ where each $X_\xi \in \boldsymbol{H}$ such that, for each $\xi < \gamma$,

$$\mathfrak{d}(X_\xi) < \mathfrak{d}(X_{\xi+1}) \quad \text{and} \quad \sum_{\xi<\gamma} \mathfrak{d}(X_\xi) = \aleph_\delta.$$

In view of the above inequality there exists for each $\xi < \gamma$ a family of power $\mathfrak{d}(X_\xi)$ consisting of mutually disjoint non-empty subsets of $X_{\xi+1}$.

Using again the axiom of choice we select one such family and call it $\boldsymbol{W}_\xi$. Hence the union $\bigcup_{\xi<\gamma} \boldsymbol{W}_\xi$ has power $\sum_{\xi<\gamma} \mathfrak{d}(X_\xi) = \aleph_\delta$ and consists of mutually disjoint non-empty open sets. This contradicts the assumption that $\aleph_\delta$ is the degree of the space. Lemma 2 is thus proved.

From Lemma 2 we obtain easily

°THEOREM 3: *If X is homogeneous and* $\mathfrak{d}(X) = \aleph_\delta$ *then* $\aleph_\delta$ *is regular.*

PROOF. If δ is a successor ordinal, the theorem is obvious. If δ is a limit number and $\aleph_\delta$ is singular, then

$$\aleph_\delta = \sum_{\xi<\gamma} \aleph_{\varphi(\xi)}$$

where $\bar{\gamma} < \aleph_\delta$ and φ is an increasing mapping of $W(\gamma)$ into $W(\delta)$. From $\bar{\gamma} < \aleph_\delta$ it follows that there exists a family $\boldsymbol{H} = \{X_\xi\colon \xi < \gamma\}$ of mutually disjoint non-void open subsets X_ξ of X.

Since X is homogeneous, we obtain

$$\mathfrak{d}(X_\xi) = \aleph_\delta > \aleph_{\varphi(\xi)},$$

whence we infer that there is a family of power $\aleph_{\varphi(\xi)}$ consisting of mutually disjoint non-empty open subsets of X_ξ. Using the axiom of choice we select for each ξ a family $\boldsymbol{S}_\xi$ of this kind. Now the union $\boldsymbol{S} = \bigcup\{\boldsymbol{S}_\xi\colon \xi < \gamma\}$ has power $\sum_{\xi<\gamma} \aleph_{\varphi(\xi)} = \aleph_\delta$ which is a contradiction because $\boldsymbol{S}$ is a family of mutually disjoint non-empty open sets. This proves Theorem 3.

An immediate corollary from this theorem is[1])

°THEOREM 4. *If $\mathscr{X}$ is a topological space with an infinite degree then this degree is a regular cardinal.*

We pass now to the second result mentioned in the introductory remarks to the present section and prove

°THEOREM 5: (Marczewski) *If $\mathfrak{m}$ is an infinite cardinal and $\{\mathscr{X}_i\}_{i\in I}$ an arbitrary sequence of topological spaces each of which has weight $\leqslant \mathfrak{m}$ and degree $\leqslant \mathfrak{m}$, then the Cartesian product $\boldsymbol{P} = \prod_{i\in I}(\mathscr{X}_i)$ with Tychonoff topology has degree $\leqslant \mathfrak{m}$.*

PROOF. For each i in $\boldsymbol{I}$ let Γ_i be an open basis of $\mathscr{X}_i$ not containing $\emptyset$ and such that $\overline{\overline{\Gamma_i}} \leqslant \mathfrak{m}$. If S is a finite subset of $\boldsymbol{I}$ and $G \in \prod_{s\in S}(\Gamma_s)$ then we call the set

$$B(S, G) = \Big\{f \in P\colon \bigwedge_{s\in S} [f(s) \in G_s]\Big\}$$

a *box*. All boxes form an open basis of the space $\boldsymbol{P}$. The number of elements of $\boldsymbol{S}$ is called the *order* of the box.

[1]) Theorem 4 is due to Erdös and Tarski [1]. Theorem 5 was first proved by Marczewski [2]. Other related results can be found in Engelking and Karłowicz [1].

We shall proceed by contradiction and assume that there is a family of power $> \mathfrak{m} = \aleph_\alpha$ of disjoint open subsets of $\boldsymbol{P}$. Since each open set $\neq \emptyset$ contains a box, we infer that there is a sequence

$$\{B(S^\xi, G^\xi) \colon \xi < \omega_{\alpha+1}\} \tag{4}$$

consisting of mutually disjoint boxes. We abbreviate $B(S^\xi, G^\xi)$ as B_ξ.

The order of each box in this sequence being finite, we can partition the set of its terms into ω classes collecting in the nth class all boxes of order n. Hence one at least of these classes has power $\aleph_{\alpha+1}$ and we can assume from the start that all the terms of the sequence (4) have order n.

We shall reach a contradiction by showing that for some ξ the set S^ξ has more than n elements. In order to obtain this we need a characterization of disjoint boxes:

LEMMA 6: *Two boxes $B(S, G)$, $B(S', G')$ are disjoint if and only if there is an s in $S \cap S'$ such that $G_s \cap G'_s = \emptyset$.*

PROOF. If the condition is satisfied then the boxes are disjoint because an f belonging to these boxes would satisfy the relations $f(s) \in G_s$ and $f(s) \in G'_s$. If the condition is not satisfied then $G_s \cap G'_s \neq \emptyset$ for each $s \in S \cap S'$. Therefore there exists a function f such that $f(s) \in G_s \cap G'_s$ if $s \in S \cap S'$ and $f(s) \in G_s$ if $s \in S - S'$, $f(s) \in G'_s$ if $s \in S' - S$, $f(s) \in \mathscr{X}_s$ if $s \in I - (S \cup S')$. Such a function (whose existence may easily be established using the axiom of choice) belongs to both given boxes. The lemma is thus proved.

Returning to the proof of the theorem we select an arbitrary $\xi_0 < \omega_{\alpha+1}$ and notice that $B_{\xi_0} \cap B_\xi = \emptyset$ for each $\xi > \xi_0$. Using the lemma we infer that for each $\xi > \xi_0$ there exists an element $s_\xi \in S^{\xi_0}$ such that $G^{\xi_0}_{s_\xi} \cap \cap G^\xi_{s_\xi} = \emptyset$. There are just n possible values for s_ξ because the order of the box B_{ξ_0} is n. Hence the set of ordinals $\xi > \xi_0$ can be partitioned into n sets according to the values of s_ξ. Since one at least of these sets must have the power $\aleph_{\alpha+1}$ we obtain an element s_0 and a set $X_0 \subset W(\omega_{\alpha+1})$ of power $\aleph_{\alpha+1}$ such that

$$\xi_0 \notin X_0, \quad s_0 \in S^{\xi_0}, \quad \xi \in X_0 \to (s_0 \in S^\xi) \wedge (G^{\xi_0}_{s_0} \cap G^\xi_{s_0} = \emptyset). \tag{5}$$

We shall replace X_0 by a smaller set such that $G^{\xi}_{s_0}$ be constant for ξ in the smaller set. To obtain this we divide X_0 into disjoint subsets, putting two indices ξ and ξ' into the same subset if and only if $G^{\xi}_{s_0} = G^{\xi'}_{s_0}$. Since for each ξ the set $G^{\xi}_{s_0}$ is an element of the basis Γ_{s_0} of the space $\mathscr{X}_{s_0}$, we see that the number of subsets into which we partitioned X_0 is at most equal to $\overline{\overline{\Gamma}}_{s_0}$ and hence is $\leqslant \aleph_\alpha$. It follows that there is a subset Y_0 of X_0 which has the power $\aleph_{\alpha+1}$ and is such that for each ξ in Y_0 the set $G^{\xi}_{s_0}$ is equal to one and the same set G_{s_0}. Thus we obtain from (5)

$$\xi_0 \notin Y_0, \qquad s_0 \in S^{\xi_0},$$

$$\xi \in Y_0 \to (s_0 \in S^{\xi}) \wedge (G^{\xi}_{s_0} = G_{s_0}),$$

$$G^{\xi_0}_{s_0} \cap G_{s_0} = \emptyset.$$

We shall now iterate the above construction. Proceeding by induction we assume that we already defined sets $Y_0 \supset Y_1 \supset \ldots \supset Y_{k-1}$, indices $\xi_0 < \xi_1 < \ldots < \xi_{k-1}$, elements of I, $s_0, s_1, \ldots, s_{k-1}$, and open sets $G_{s_0}, G_{s_1}, \ldots, G_{s_{k-1}}$ so that for all i, k the following formulas hold:

$$\overline{\overline{Y}}_i = \aleph_{\alpha+1},$$

$$\xi_i \in Y_{i-1} - Y_i, \qquad s_i \in S^{\xi_i},$$

$$\xi \in Y_i \to (s_i \in S^{\xi}) \wedge (G^{\xi}_{s_i} = G_{s_i} \in \Gamma_{s_i}),$$

$$G^{\xi_i}_{s_i} \cap G_{s_i} = \emptyset.$$

Let ξ_k be an element of Y_{k-1} greater than ξ_{k-1}. Since $B_{\xi_k} \cap B_{\xi} = \emptyset$ for each ξ in $Y_{k-1} - \{\xi_k\}$, there is, for each such ξ, an element $s_{k,\xi}$ in S^{ξ_k} such that $s_{k,\xi} \in S^{\xi}$ and $G^{\xi_k}_{s_{k,\xi}} \cap G^{\xi}_{s_{k,\xi}} = \emptyset$. Since there are but n possible values for $s_{k,\xi}$, there is a subset X_k of $Y_{k-1} - \{\xi_k\}$ such that $\overline{\overline{X}}_k = \aleph_{\alpha+1}$ and all the $s_{k,\xi}$ where $\xi \in X_k$ have one and the same value s_k. Using the assumption that the weight of $\mathscr{X}_{s_k}$ is $\leqslant \aleph_\alpha$ and $\overline{\overline{\Gamma}}_{s_k} \leqslant \aleph_\alpha$, we obtain in the same way as above a subset Y_k of X_k such that $\overline{\overline{Y}}_k = \aleph_{\alpha+1}$ and for all ξ in Y_k all the sets $G^{\xi}_{s_k}$ are equal to one and the same set $G_{s_k} \in \Gamma_{s_k}$. Our inductive construction is thus finished.

We claim now that $s_i \neq s_k$ for $i < k$. To prove this we notice that $s_k \in S^{\xi_k}$ and $\xi_k \in Y_{k-1} - Y_k \subset Y_i$. From this we obtain $G^{\xi_k}_{s_i} = G_{s_i}$.

On the other hand, $G_{s_k}^{\xi_k} \cap G_{s_k} = \emptyset$. Thus the equation $s_i = s_k$ would yield $G_{s_k} = \emptyset$ against our assumption that the empty set does not belong to Γ_{s_k}.

To finish the proof we notice that for ξ in Y_n we have $\xi \in Y_i$ for each $i \leqslant n$ and thus $s_i \in S^{\xi}$ for $i \leqslant n$. Thus the set S^{ξ} has at least $n+1$ different elements $s_0, s_1, \ldots, s_n$ which contradicts the fact that the order of all boxes B_{ξ} is n. Theorem 5 is thus proved.

Exercises

1. Prove that if X is a uniform open set then $\mathfrak{d}(X) \neq \aleph_0$.

Hint: If X were uniform and $\mathfrak{d}(X) = \aleph_0$ then X would contain two disjoint subsets with the same property. Use this to obtain an infinite sequence of mutually disjoint subsets of X. [Erdös and Tarski]

2. Find the weight of the product space $P(\mathscr{X}_i)$ given the weights of the spaces $\mathscr{X}_i$.

3. Give examples of spaces with degree $> \aleph_1$.

4. Generalize the theory of degrees of disjointness by replacing the family of open sets by any ordered set without the minimal element and the relation of disjointness by the relation $d(x, y) \equiv \bigwedge_z \neg [(z \leqslant x) \wedge (z \leqslant y)]$.

§ 11. Equivalence of certain statements about cardinal numbers with the axiom of choice

In the final section of this chapter we shall discuss the role of the axiom of choice in the cardinal arithmetic. We saw that several theorems proved in the previous sections required the axiom of choice. We shall see now that this axiom cannot be eliminated from most of the proofs because the theorems which we established are not only consequences of the axiom of choice but are equivalent to it (on the basis of the axioms Σ[TR]).

We shall also establish a result showing that a form of the generalized continuum hypothesis implies the axiom of choice.

From the well-ordering theorem and thus indirectly from the axiom of choice (see p. 255) it follows that every cardinal is an aleph. Thus the laws of arithmetic for cardinal numbers coincide with those for alephs and we have the following theorem.

°THEOREM 1:

(1) $$\bigwedge_{\mathfrak{m}, \mathfrak{n} \notin N} [(\mathfrak{m} \leqslant \mathfrak{n}) \vee (\mathfrak{n} \leqslant \mathfrak{m})] \quad (\textit{law of trichotomy}),$$

(2) $$\bigwedge_{\mathfrak{m} \notin N} [\mathfrak{m}^2 = \mathfrak{m}],$$

(3) $$\bigwedge_{\mathfrak{m}, \mathfrak{n} \notin N} [(\mathfrak{m} \cdot \mathfrak{n} = \mathfrak{m}+\mathfrak{n} = \mathfrak{m}) \vee (\mathfrak{m} \cdot \mathfrak{n} = \mathfrak{m}+\mathfrak{n} = \mathfrak{n})],$$

(4) $$\bigwedge_{\mathfrak{m}, \mathfrak{n} \notin N} [(\mathfrak{m}^2 = \mathfrak{n}^2) \to (\mathfrak{m} = \mathfrak{n})],$$

(5) $$\bigwedge_{\mathfrak{m}, \mathfrak{n}, \mathfrak{p}, \mathfrak{q} \notin N} [(\mathfrak{m} < \mathfrak{n}) \wedge (\mathfrak{p} < \mathfrak{q}) \to (\mathfrak{m}+\mathfrak{p} < \mathfrak{n}+\mathfrak{q})],$$

(6) $$\bigwedge_{\mathfrak{m}, \mathfrak{n}, \mathfrak{p}, \mathfrak{q} \notin N} [(\mathfrak{m} < \mathfrak{n}) \wedge (\mathfrak{p} < \mathfrak{q}) \to (\mathfrak{m} \cdot \mathfrak{p} < \mathfrak{n} \cdot \mathfrak{q})],$$

(7) $$\bigwedge_{\mathfrak{m}, \mathfrak{n}, \mathfrak{p} \notin N} [(\mathfrak{m}+\mathfrak{p} < \mathfrak{n}+\mathfrak{p}) \to (\mathfrak{m} < \mathfrak{n})],$$

(8) $$\bigwedge_{\mathfrak{m}, \mathfrak{n}, \mathfrak{p} \notin N} [(\mathfrak{m} \cdot \mathfrak{p} < \mathfrak{n} \cdot \mathfrak{p}) \to \mathfrak{m} < \mathfrak{n}].$$

Seemingly each of the laws (1)–(8) is a special consequence of the axiom of choice. We shall show, however, that each of these laws in conjunction with the axioms Σ[TR] and VIII implies the axiom of choice.

THEOREM 2:[1]) *If for every pair of infinite cardinals* $\mathfrak{m}$ *and* $\mathfrak{n}$ *either* $\mathfrak{m} \leqslant \mathfrak{n}$ *or* $\mathfrak{n} \leqslant \mathfrak{m}$, *then there exists a choice function for every family of non-empty sets.*

PROOF. Let X be an arbitrary infinite set and let $\mathfrak{m} = \overline{\overline{X}}$. By assumption either $\mathfrak{m} \leqslant \aleph(\mathfrak{m})$ or $\aleph(\mathfrak{m}) \leqslant \mathfrak{m}$. The second case is impossible by Theorem 2.3; in the first case $\mathfrak{m}$ is an aleph and thus there exists a relation well ordering X. Therefore the hypothesis of the theorem implies the well-ordering theorem, and thus the existence of the desired choice function (see p. 255).

REMARK: Theorem 2 can be expressed more concisely: formula (1) for all infinite $\mathfrak{m}$ and $\mathfrak{n}$ implies the axiom of choice. We shall employ this sort of abbreviated language; in the statements of other theorems the word "implies" will be taken to denote the existence of a proof which does not use the axiom of choice.

LEMMA 3: (3) → (2) → (4).

[1]) Theorem 2 is due to Hartogs [1].

In fact, if $\mathfrak{m} \notin N$, then by (3) $\mathfrak{m}^2 = \mathfrak{m}$. If $\mathfrak{m}, \mathfrak{n} \notin N$, then by (2) $\mathfrak{m}^2 = \mathfrak{m}$ and $\mathfrak{n}^2 = \mathfrak{n}$, whence $\mathfrak{m}^2 = \mathfrak{n}^2 \to \mathfrak{m} = \mathfrak{n}$.

LEMMA 4: *If* $\mathfrak{p} \notin N$ and $\mathfrak{p} \cdot \aleph(\mathfrak{p}) = \mathfrak{p} + \aleph(\mathfrak{p})$, *then* $\mathfrak{p}$ *is an aleph.*[1])

PROOF. Let $\overline{\overline{A}} = \mathfrak{p}$, $\overline{\overline{B}} = \aleph(\mathfrak{p})$. By hypothesis there exists a partition $A \times B = P \cup Q$ into two disjoint sets P and Q such that $\overline{\overline{P}} = \mathfrak{p}, \overline{\overline{Q}} = \aleph(\mathfrak{p})$; thus $\overline{\overline{Q}}$ is an aleph and there exists a relation well ordering Q.

There are, a priori, two possibilities.

I. There exists $a \in A$ such that $\{a\} \times B \subset P$. However, since $\overline{\overline{\{a\} \times B}} = \aleph(\mathfrak{p})$, it would then follow that $\aleph(\mathfrak{p}) \leqslant \mathfrak{p}$ contrary to Theorem 2.3. Thus case I is impossible.

II. For every $a \in A$, $(\{a\} \times B) \cap Q \neq \emptyset$. As there exists a relation well ordering Q, there exists a function which associates with every $a \in A$ the first element $q(a)$ of $(\{a\} \times B) \cap Q$. Moreover, as $q(a)$ is of the form $\langle a, b \rangle$ where $b \in B$, it follows that $q(a') \neq q(a'')$ for $a' \neq a''$. Thus the function q establishes a one-to-one mapping of A onto a subset of Q; hence $\overline{\overline{A}} \leqslant \overline{\overline{Q}}$ and $\overline{\overline{A}}$ is an aleph. Q.E.D.

THEOREM 5: (4) *implies the axiom of choice.*

PROOF. Let $\mathfrak{k}$ be an arbitrary infinite cardinal. Let $\mathfrak{p} = \mathfrak{p}^{\aleph_0}$, $\mathfrak{m} = \mathfrak{p} + \aleph(\mathfrak{p})$ and $\mathfrak{n} = \mathfrak{p} \cdot \aleph(\mathfrak{p})$. Clearly,

$$\mathfrak{p}^2 = (\mathfrak{p}^{\aleph_0})^2 = \mathfrak{p}^{\aleph_0} = \mathfrak{p},$$

which implies that $\mathfrak{p} \leqslant \mathfrak{p}+1 \leqslant \mathfrak{p} \cdot \mathfrak{p} = \mathfrak{p}^2 = \mathfrak{p}$ and thus

$$\mathfrak{p}+1 = \mathfrak{p}.$$

Since $2\aleph(\mathfrak{p}) = \aleph(\mathfrak{p}) = \aleph(\mathfrak{p})+1 = [\aleph(\mathfrak{p})]^2$, we obtain from these formulas that

$$\begin{aligned} \mathfrak{m}^2 &= [\mathfrak{p}+\aleph(\mathfrak{p})]^2 \\ &= \mathfrak{p}^2 + 2\mathfrak{p} \cdot \aleph(\mathfrak{p}) + [\aleph(\mathfrak{p})]^2 \\ &= \mathfrak{p} + \mathfrak{p}[\aleph(\mathfrak{p})2] + \aleph(\mathfrak{p}) \\ &= \mathfrak{p} + \mathfrak{p} \cdot \aleph(\mathfrak{p}) + \aleph(\mathfrak{p}) \\ &= \mathfrak{p}[1+\aleph(\mathfrak{p})] + \aleph(\mathfrak{p}) \\ &= \mathfrak{p} \cdot \aleph(\mathfrak{p}) + \aleph(\mathfrak{p}) \\ &= (\mathfrak{p}+1)\aleph(\mathfrak{p}) \end{aligned}$$

[1]) Lemma 4 and Theorems 5–8 given below are due to Tarski [2].

$$= \mathfrak{p} \cdot \aleph(\mathfrak{p})$$
$$= \mathfrak{p}^2 [\aleph(\mathfrak{p})]^2$$
$$= \mathfrak{n}^2.$$

Thus by (4) we have $\mathfrak{m} = \mathfrak{n}$, that is, $\mathfrak{p}+\aleph(\mathfrak{p}) = \mathfrak{p} \cdot \aleph(\mathfrak{p})$, and thus $\mathfrak{p}$ is an aleph by Lemma 3. Since $\mathfrak{k} \leqslant \mathfrak{p}$, $\mathfrak{k}$ is also an aleph. Q.E.D.

COROLLARY 6: *Each of the formulas* (2) *and* (3) *implies the axiom of choice.*

THEOREM 7: *Each of the formulas* (5) *and* (6) *implies the axiom of choice.*

PROOF. Let $\mathfrak{k}$ be an infinite cardinal. Clearly

$$\mathfrak{k} \leqslant \mathfrak{k}+\aleph(\mathfrak{t}) \quad \text{and} \quad \aleph(\mathfrak{k}) \leqslant \mathfrak{k}+\aleph(\mathfrak{k}).$$

If the strict inequality were to hold in both of these formulas, then by (5) we would obtain the false inequality $\mathfrak{k}+\aleph(\mathfrak{k}) < \mathfrak{k}+\aleph(\mathfrak{k})$. Thus either $\mathfrak{k} = \mathfrak{k}+\aleph(\mathfrak{k})$ or $\aleph(\mathfrak{k}) = \mathfrak{k}+\aleph(\mathfrak{k})$. But the first equation implies $\mathfrak{k} \geqslant \aleph(\mathfrak{k})$ which contradicts Theorem 2.3. Thus the second equation holds, which implies $\mathfrak{k} \leqslant \aleph(\mathfrak{k})$ and thus $\mathfrak{k}$ is an aleph. It follows that (5) implies the axiom of choice.

The proof of the second part of the theorem is similar, except that instead of $\mathfrak{k}+\aleph(\mathfrak{t})$ we examine the product $\mathfrak{k} \cdot \aleph(\mathfrak{k})$.

THEOREM 8: *Each of the formulas* (7) *and* (8) *implies the axiom of choice.*

PROOF. Let $\mathfrak{k}$ be an arbitrary infinite cardinal. Let $\mathfrak{m} = \aleph_0 \cdot \mathfrak{k}$; then $\mathfrak{m}+\mathfrak{m} = \mathfrak{m}$. Now let $\mathfrak{n} = \aleph(\mathfrak{m})$ and $\mathfrak{p} = \mathfrak{m}$. Hence $\mathfrak{m}+\mathfrak{p} = \mathfrak{m}$ and $\mathfrak{n}+\mathfrak{p} = \mathfrak{m}+\aleph(\mathfrak{m})$, which implies that $\mathfrak{m}+\mathfrak{p} \leqslant \mathfrak{n}+\mathfrak{p}$. If the equation $\mathfrak{m}+\mathfrak{p} = \mathfrak{n}+\mathfrak{p}$ were true, then the inequality $\mathfrak{m} \geqslant \aleph(\mathfrak{m})$ would also be true, in contradiction to Theorem 2.3. Hence $\mathfrak{m}+\mathfrak{p} < \mathfrak{n}+\mathfrak{p}$ and by (7) $\mathfrak{m} < \mathfrak{n}$, that is $\mathfrak{m} < \aleph(\mathfrak{m})$. Thus $\mathfrak{k} < \aleph(\mathfrak{m})$, which implies that $\mathfrak{k}$ is an aleph.

The proof of the second part of the theorem is similar; we replace $\mathfrak{m}$ by $\mathfrak{k}^{\aleph_0}$ and discuss the products $\mathfrak{m} \cdot \mathfrak{p}$ and $\mathfrak{n} \cdot \mathfrak{p}$ instead of the sums $\mathfrak{m}+\mathfrak{p}$, $\mathfrak{n} \cdot \mathfrak{p}$.

Similar to (2) is the following theorem:

$$(+) \qquad \bigwedge_{\mathfrak{m} \notin N} [\mathfrak{m}+\mathfrak{m} = \mathfrak{m}].$$

This theorem is obviously provable in the system $\Sigma^\circ[\mathrm{TR}]$. It has been proved that it is not provable in the system $\Sigma[\mathrm{TR}]$. In this respect theorem (+) is similar to Theorems (1)–(8). However the adjunction of (+) to the axioms of $\Sigma[\mathrm{TR}]$ does not yield a system equivalent to $\Sigma^\circ[\mathrm{TR}]$ and so theorem (+) is essentially weaker than Theorems (1)–(8)[1].)

From Theorem 1 (proved with the axiom of choice) it follows that for every infinite cardinal $\mathfrak{m}$ and for $n \in N$, $\mathfrak{m}^n < 2^{\mathfrak{m}}$. Specker has shown that the weaker theorem $\mathfrak{m}^n$ non $\geqslant 2^{\mathfrak{m}}$ can be proved without the axiom of choice. We shall use this fact later (see p. 313).

THEOREM 9:[2]) *If* $\mathfrak{m}$ *is infinite and* $n \in N$, *then* $\mathfrak{m}^n$ non $\geqslant 2^{\mathfrak{m}}$.

PROOF. Assume that $\overline{\overline{A}} = \mathfrak{m}$ and that there exists a one-to-one function F mapping $\boldsymbol{P}(A)$ into A^n. We shall show that there is a function f which associates with every transfinite sequence φ of distinct elements of A an element of A which is not a term of φ.

The theorem then follows from the existence of such a function f as follows: Let α be the least ordinal of power $\aleph(\mathfrak{m})$. By the theorem on inductive definitions there exists an α-sequence ψ such that $\psi_\xi = f(\psi|\xi)$ for $\xi < \alpha$. By construction, ψ_ξ is not a term of $\psi|\xi$ and thus $\psi_\xi \neq \psi_\eta$ for $\xi \neq \eta$. We conclude that $\overline{\overline{A}} \geqslant \overline{\alpha} = \aleph(\mathfrak{m})$ in contradiction to Theorem 2.3.

It remains to construct the function f. For this purpose we choose an integer $k_0 > 1$ such that $2^{k_0} > k_0^n$ and k_0 distinct elements of A: $a_0, \ldots, a_{k_0-1}$. If $\varphi = \langle \varphi_0, \ldots, \varphi_{k-1} \rangle$ with $k < k_0$, then we put $f(\varphi) = a_j$ where $j = \min_i(a_i \neq \varphi_s \text{ for } s < k)$. Assume now that $\varphi = \langle \varphi_0, \ldots, \varphi_{k-1} \rangle$ with $k \geqslant k_0$ distinct terms. Denote by $S(\varphi)$ the set $\{\varphi_0, \ldots, \varphi_{k-1}\}$. There are 2^k subsets of $S(\varphi)$; we may represent them as $\{\varphi_i : i \in Z\} = S(\varphi, Z)$ where Z is contained in $\{0, 1, \ldots, k-1\}$. By the definition of F each $F(S(\varphi, Z))$ is an n-termed sequence. Since there are

[1]) The proof that the formula (+) does not imply the axiom of choice is very difficult. It was found by Sageev [1]. The proof of independence of (+) from axioms $\Sigma[\mathrm{TR}]$ is very easy.

Many results about formulas equivalent to the axiom of choice can be found in Rubin and Rubin [1]. See also Jech [2].

[2]) Theorem 9 is really a lemma which we will need in Theorem 11 below.

only k^n sequences in A^n whose terms are elements of $S(\varphi)$ and since $2^k > k^n$, we infer that there is at least one Z such that not all terms of $F(S(\varphi, Z))$ belong to $S(\varphi)$. We order sets Z similarly to the lexicographical ordering of their characteristic functions and choose the first Z_0 with the property stated above. Now we define $f(\varphi)$ as the first term of $F(S(\varphi, Z_0))$ which does not belong to $S(\varphi)$.

Finally assume that the type of φ is an infinite ordinal α. Since $\overline{\alpha}$ is an aleph, there exists a one-to-one mapping of the set $W(\alpha)$ onto the set $W(\alpha)^n$. Let the ordinal $\xi < \alpha$ correspond to the sequence $\langle \xi^{(0)}, \ldots, \xi^{(n-1)} \rangle$ under this mapping.

Let $H_\xi = \emptyset$ if $\langle \varphi_{\xi^{(0)}}, \ldots, \varphi_{\xi^{(n-1)}} \rangle$ does not belong to the set of values of the function F, let $H_\xi = F^c\langle \varphi_{\xi^{(0)}}, \ldots, \varphi_{\xi^{(n-1)}} \rangle$ otherwise, and let $X_0 = \{\varphi_\xi \colon \varphi_\xi \notin H_\xi\}$, $\langle a_0, \ldots, a_{n-1} \rangle = F(X_0)$. If all of the elements a_j were terms of φ, then for some $\xi_0 < \alpha$ we would have $\langle a_0, \ldots, a_{n-1} \rangle = \langle \varphi_{\xi_0^{(0)}}, \ldots, \varphi_{\xi_0^{(n-1)}} \rangle$; thus this sequence would belong to the set of values of F, which implies that $H_{\xi_0} = X_0$ and $\varphi_{\xi_0} \in X_0 \equiv \varphi_{\xi_0} \in H_{\xi_0}$. On the other hand, from the definition of the set X_0 it follows that $\varphi_\xi \in X_0 \equiv \varphi_\xi \notin H_{\xi_0}$ for every ξ. Thus some a_j is not a term of φ and it suffices to set $f(\varphi) = a_j$ where j is the least index such that a_j has this property.

COROLLARY 10: *If* $\mathfrak{m}$ *is an infinite cardinal then* $\mathfrak{m}+1 < 2^{\mathfrak{m}}$.

PROOF. From the fact that $2^{\mathfrak{m}} > \mathfrak{m}$ it follows that $2^{\mathfrak{m}} \geqslant \mathfrak{m}+1$. If $\mathfrak{m}+1 = 2^{\mathfrak{m}}$ then, since $\mathfrak{m}^2 \geqslant \mathfrak{m} \cdot 2 \geqslant \mathfrak{m}+1$, we would conclude that $\mathfrak{m}^2 \geqslant 2^{\mathfrak{m}}$, contrary to Theorem 9.

We conclude this section with the proof of a theorem which implies as an immediate consequence that the axiom of choice follows from the hypothesis (C) (see p. 291).

THEOREM 11:[1]) *If* $\mathfrak{m}$ *is an infinite cardinal and if neither between* $\mathfrak{m}$ *and* $2^{\mathfrak{m}}$ *nor between* $2^{\mathfrak{m}}$ *and* $2^{2^{\mathfrak{m}}}$ *lies any cardinal, then* $\mathfrak{m}$ *is an aleph.*

PROOF. Let us abbreviate the formula $\bigwedge_{\mathfrak{x}} \neg(\mathfrak{m} < \mathfrak{x} < 2^{\mathfrak{m}})$ as $H(\mathfrak{m})$. Using Corollary 10 we obtain $\mathfrak{m} \leqslant \mathfrak{m}+1 < 2^{\mathfrak{m}}$, whence $\mathfrak{m} = \mathfrak{m}+1$ by $H(\mathfrak{m})$. From $\mathfrak{m} \leqslant \mathfrak{m}+\mathfrak{m} \leqslant 2^{\mathfrak{m}} \cdot 2 = 2^{\mathfrak{m}+1}$ it follows now $\mathfrak{m} \leqslant 2\mathfrak{m} \leqslant 2^{\mathfrak{m}}$. $H(\mathfrak{m})$ implies that either $\mathfrak{m} = 2\mathfrak{m}$ or $2\mathfrak{m} = 2^{\mathfrak{m}}$. But the second

[1]) Theorem 11 was proved by Sierpiński [18]. The proof given in the text follows Specker [1].

equation is impossible by Theorem 9 because $\mathfrak{m}^2 \geqslant 2\mathfrak{m}$.

We prove next that $H(\mathfrak{m})$ implies $\mathfrak{m}^2 = \mathfrak{m}$. For $\mathfrak{m} \leqslant \mathfrak{m}^2 \leqslant 2^{\mathfrak{m}}$, $2^{\mathfrak{m}} = 2^{\mathfrak{m}+\mathfrak{m}} = 2^{\mathfrak{m}}$. The equation $\mathfrak{m}^2 = 2^{\mathfrak{m}}$ is impossible by Theorem 9. Thus by $H(\mathfrak{m})$ we have that $\mathfrak{m}^2 = \mathfrak{m}$.

By Theorem 2.7 $2^{\aleph(\mathfrak{m})} \leqslant 2^{2^{\mathfrak{m}}}$ and

$$2^{\mathfrak{m}} \leqslant 2^{\mathfrak{m}}+\aleph(\mathfrak{m}) < 2^{(2^{\mathfrak{m}}+\aleph(\mathfrak{m}))} = 2^{2^{\mathfrak{m}}} \cdot 2^{\aleph(\mathfrak{m})}$$
$$\leqslant 2^{2^{\mathfrak{m}}} \cdot 2^{2^{\mathfrak{m}}} = 2^{2^{\mathfrak{m}} \cdot 2} = 2^{2^{\mathfrak{m}+1}} = 2^{2^{\mathfrak{m}}},$$

thus, by $H(2^{\mathfrak{m}})$, $2^{\mathfrak{m}} = 2^{\mathfrak{m}}+\aleph(\mathfrak{m})$ and $\aleph(\mathfrak{m}) \leqslant 2^{\mathfrak{m}}$. From

$$\mathfrak{m} < \mathfrak{m}+\aleph(\mathfrak{m}) \leqslant \mathfrak{m} \cdot \aleph(\mathfrak{m}) \leqslant 2^{\mathfrak{m}} \cdot \aleph(\mathfrak{m}) \leqslant 2^{\mathfrak{m}} \cdot 2^{\mathfrak{m}} = 2^{\mathfrak{m}+1} = 2^{\mathfrak{m}}$$

it follows by $H(\mathfrak{m})$ that $\mathfrak{m}+\aleph(\mathfrak{m}) = \mathfrak{m} \cdot \aleph(\mathfrak{m})$, which implies (see Lemma 4) that $\mathfrak{m}$ is an aleph.

COROLLARY 12:[1]) *The hypothesis* (C) *implies the axiom of choice.*

Exercises

1. Derive the Cantor inequality $\mathfrak{m} < 2^{\mathfrak{m}}$ and its strengthened version $\mathfrak{k} \cdot \mathfrak{m} < 2^{\mathfrak{m}}$ (for finite $\mathfrak{k}$ and infinite $\mathfrak{m}$) from Theorem 9. [Specker]

2. Prove that the following formulas are equivalent to the axiom of choice:

(a) $$\bigwedge_{\mathfrak{m},\mathfrak{n} \notin N} [(\mathfrak{m}+\mathfrak{n} = \mathfrak{m}) \vee (\mathfrak{m}+\mathfrak{n} = \mathfrak{n})],$$

(b) $$\bigwedge_{\mathfrak{m},\mathfrak{n} \notin N} [(\mathfrak{m} \cdot \mathfrak{n} = \mathfrak{m}) \vee (\mathfrak{m} \cdot \mathfrak{n} = \mathfrak{n})].$$ [Leśniewski]

3. Prove that the following formula is equivalent to the axiom of choice:

$$\bigwedge_{\mathfrak{m},\mathfrak{n} \notin N} [(\mathfrak{m} \leqslant^* \mathfrak{n}) \vee (\mathfrak{n} \leqslant^* \mathfrak{m})]$$

(the relation $\leqslant^*$ is defined on p. 182). [Lindenbaum]

Hint: Replace $\aleph(\mathfrak{m})$ by $\aleph^*(\mathfrak{m})$ (defined in Exercise 2.2) in the proof of Hartogs' theorem.

4. Prove that the following formula is equivalent to the axiom of choice:

$$\bigwedge_{\mathfrak{m},\mathfrak{p},\mathfrak{q} \notin N} [(\mathfrak{m}^{\mathfrak{p}} < \mathfrak{m}^{\mathfrak{q}}) \rightarrow (\mathfrak{p} < \mathfrak{q})].$$ [Tarski]

Hint: Let $\mathfrak{m} = 2^{(\mathfrak{p}^{\aleph_0})}$, $\mathfrak{q} = \aleph(\mathfrak{m})$ and show that $\mathfrak{m}^{\mathfrak{p}} = \mathfrak{m} \leqslant \mathfrak{m}^{\mathfrak{q}}$ and that the formula $\mathfrak{m}^{\mathfrak{p}} = \mathfrak{m}^{\mathfrak{q}}$ would imply $\mathfrak{m} \geqslant \aleph(\mathfrak{m})$.

[1]) The final corollary given on this page was discovered by Lindenbaum and Tarski [1], who never published a proof of their result. Sierpiński [18] reconstructed the proof of Lindenbaum and Tarski's result.

CHAPTER IX

TREES AND PARTITIONS

The present chapter is devoted to the concept of a tree which plays an important role in the recent works on abstract set theory. In the last section we prove some simple partition theorems in order to introduce the reader to the ideas of so-called combinatorial set theory.

§ 1. Trees

The notion of a pseudo-tree was introduced in Chapter II, Section 9. We now define a special class of pseudo-trees:

DEFINITION 1: A pseudo-tree T is called a *tree* if for every t in T the set $O(t) = \{s \in T: s \leqslant t\}$ is well ordered.[1])

Let T be a tree; we define by transfinite induction the levels of T. The 0th level L_0 is the set of the minimal elements of T. For $\alpha > 0$ we define $L_\alpha = \{t: \bigwedge_{s<t} s \in \bigcup_{\xi<\alpha} L_\xi\} - \bigcup_{\xi<\alpha} L_\xi$.

For $\alpha < \beta$ the levels L_α and L_β are disjoint; if L_α is empty, then so are all higher levels. It follows that there is a least ordinal α such that $L_\alpha = \emptyset$. This ordinal is called the *height* of the tree.

Each tree is the union of its levels. To prove this statement we show that if $x_0 \in T - \bigcup_{\xi<\alpha} L_\xi$ then x must have a predecessor $x_1 < x_0$ which also belongs to $T - \bigcup_{\xi<\alpha} L_\xi$, whence by the axiom of choice we obtain

[1]) The concept of a tree as defined in this book is widely used in various branches of mathematics (e.g. in topology), in logic and in philosophy. It would not be easy to pinpoint the origin of this concept. Some remarks about this question can be found in Beth [1], p. 196 and Weyl [1], p. 53. Jech [1] contains a survey of recent mathematical results on trees.

an infinite descending sequence $x_0 > x_1 > \dots$ However this contradicts the assumption that T is a tree.

Finally we note the two easy theorems:

Each level is an antichain.

If B is a branch of a tree, then B intersects each level in exactly one point.

An immediate successor of an element x of a tree is an element $y > x$ such that no element lies between x and y. The cardinal number of all immediate successors of x is called the *order* of x. The order of T is the supremum of orders of all its elements.

A tree T is said to be of *finite order* if it has finitely many minimal elements and its order is finite.

Example. Let $T_\alpha = \{0, 1\}^{W(\alpha)}$ be the set of all sequences $\{\varphi_\xi\}_{\xi<\alpha}$ with domain $W(\alpha)$ and terms equal to 0 or 1. The union $D_\alpha = \bigcup \{T_\xi: \xi < \alpha\}$ ordered by inclusion is called the *full binary tree* of height α. The elements of this tree are zero-one sequences with domains of the form $W(\xi)$ where $\xi < \alpha$. A sequence f is a successor of g if $g \subset f$. It is easy to see that D_α is indeed a tree: the predecessors of f form a well-ordered set. They are the partial functions $f|W(\eta)$ where η belongs to the domain of f.

Sets T_ξ with $\xi < \alpha$ are the levels of D_α. The initial level contains just one element, the void set. The nth level consists of 2^n functions $\{\langle 0, \varepsilon_0\rangle, \langle 1, \varepsilon_1\rangle, \dots, \langle n-1, \varepsilon_{n-1}\rangle\}$ where for each $j < n$ the element ε_j is either 0 or 1.

A branch B of D_α is an increasing sequence $\{f_\xi\}_{\xi<\alpha}$ where $f_\xi \in T_\xi$ for each $\xi < \alpha$. Remarking that T_ξ has the power $2^{\bar{\xi}}$, we obtain

THEOREM 1: *The full binary tree of height α has $\sum_{\xi<\alpha} 2^{\bar{\xi}}$ elements and $2^{\bar{\alpha}}$ branches.*

To prove the statement about branches it is sufficient to notice that if B is a branch then $\bigcup(B) = f$ is a function with domain $W(\alpha)$ and range $\{0, 1\}$. Conversely, if f is such a function then the set $\{f|W(\xi): \xi < \alpha\}$ is a branch. Thus there is a one-to-one correspondence between branches of D_α and elements of T_α.

Two distinct branches $B = \{f|W(\xi): \xi < \alpha\}$ and $B' = \{f'|W(\xi): \xi < \alpha\}$ have $\bar{\delta}$ common elements $f|W(\xi)$, $\xi < \delta$, where δ is the least ordinal such that $f(\delta) \neq f'(\delta)$. In particular, if $\alpha = \omega_\gamma$ then any two branches are $\aleph_\gamma$-disjoint and if $\alpha = \omega$ then any two branches are almost disjoint (see p. 301). The decomposition of D_ω into the union of its branches gives thus the decomposition of a denumerable set into $2^{\aleph_0}$ almost disjoint subsets.

Later in this chapter we shall give two lemmas on the linear ordering of a tree (in order not to interrupt proofs to be given later).

Let T be a tree of height α. For each $\xi < \alpha$ we select (using the axiom of choice) an arbitrary linear ordering $\leqslant_\xi$ of the ξth level L_ξ of T. The set of immediate successors of a given $x \in L_\xi$ is an interval. As usual we shall abbreviate the conjunction $(x \neq y) \wedge (x \leqslant_\xi y)$ as $x <_\xi y$.

We denote by $O(x)$ the set $\{z \in T\colon z \leqslant x\}$. If $x \in L_\xi$ and $\eta \leqslant \xi$ then the intersection $L_\eta \cap O(x)$ consists of exactly one element which we shall denote by $O_\eta(x)$.

If $x \in L_\xi$, $x' \in L_{\xi'}$, and neither $x \leqslant x'$ nor $x' \leqslant x$ then there is a least ordinal $\delta = \delta(x, x') \leqslant \min(\xi, \xi')$ such that $O_\delta(x) \neq O_\delta(x')$.

DEFINITION: We say that x *precedes* y in T if either $x \leqslant y$ or x and y are incomparable in T and $O_\delta(x) <_\delta O_\delta(y)$ where δ stands for $\delta(x, y)$. We write then $x \preccurlyeq y$ and $x \prec y$ if $x \neq y$.

LEMMA A: *$\preccurlyeq$ is a linear ordering of T.*

We omit the proof of this lemma because it is very similar to the proof that the lexicographical ordering of a cartesian product of linearly ordered sets taken over the well–ordered argument is a linear ordering (see Chapter VI, Section 5).

LEMMA B: *$\preccurlyeq$ is an extension of $\leqslant$.*

PROOF. If $x \leqslant y$ then $x \preccurlyeq y$ by the definition of $\preccurlyeq$.

LEMMA C: *If x and y are immediate successors* (*in T*) *of an element $z \in L_\zeta$ then $x \preccurlyeq y \equiv x \leqslant_{\zeta+1} y$.*

PROOF. From the assumptions it follows that $x, y \in L_{\xi+1}$ and $\delta(x, y) = \zeta+1$.

LEMMA D: *If $t \in T$ then the set $T_t = \{z \in T\colon t \leqslant z\}$ is an interval of T in the linear ordering $\preccurlyeq$.*

PROOF. The statement says that if $x, y \in T_t$ and $x \preccurlyeq z \preccurlyeq y$ then $z \in T_t$.

To prove this we first remark that if $x \leqslant z$, then $t \leqslant z$ in view of the transitivity of $\leqslant$. If $z \leqslant y$, then z and t are comparable in T because they both precede y. The case $z < t$ would imply $z < x$ which is incompatible with the assumption $x \preccurlyeq z$ and hence $t \leqslant z$.

It remains to consider the case where x is incomparable with z, z is incomparable with y and

$$O_\eta(x) <_\eta O_\eta(z), \qquad O_\zeta(z) <_\zeta O_\zeta(y),$$

where $\eta = \delta(x, z)$, $\zeta = \delta(z, y)$.

Let $t \in L_\tau$. If $\eta > \tau$, then $O_\tau(x) = O_\tau(z)$ and from $t \leqslant y$ it follows that $O_\tau(x) = t$. Thus we obtain $t = O_\tau(x) = O_\tau(z)$, whence $t \leqslant z$. We show similarly that $t \leqslant z$ if $\zeta > \tau$. Therefore the lemma will be proved if we show that the inequalities $\eta \leqslant \tau$, $\zeta \leqslant \tau$ cannot hold simultaneously.

Assume otherwise. From the definitions of η and ζ we obtain

$$(1) \qquad [O_\eta(x) <_\eta O_\eta(z)] \wedge \bigwedge_{\xi<\eta} [O_\xi(x) = O_\xi(z)],$$

$$(2) \qquad [O_\zeta(z) <_\zeta O_\zeta(y)] \wedge \bigwedge_{\xi<\zeta} [O_\xi(z) = O_\xi(y)].$$

If $\eta = \zeta$ we obtain from these formulas

$$(3) \qquad O_\eta(x) <_\eta O_\eta(y)$$

because $<_\eta$ is transitive. If $\eta < \zeta$ then we obtain from (1) the inequality $O_\eta(x) <_\eta O_\eta(z)$ and from (2) the equation $O_\eta(z) = O_\eta(y)$, whence we see that (3) is true also in this case. Finally, if $\eta > \zeta$ then $O_\zeta(z) <_\zeta O_\zeta(y)$ by (2) and $O_\zeta(x) = O_\zeta(z)$ by (1), whence

$$(4) \qquad O_\zeta(x) <_\zeta O_\zeta(y).$$

Thus in every case either (3) or (4) holds. On the other hand, it follows from $\eta \leqslant \tau$ and $\zeta \leqslant \tau$ that $O_\eta(x) = O_\eta(t) = O_\eta(y)$ and $O_\zeta(x) = O_\zeta(t) = O_\zeta(y)$ because $t \leqslant x$ and $t \leqslant y$. Thus neither (3) nor (4) is possible. Lemma D is thus proved.

Collecting the results established in the lemmas we obtain

°THEOREM 2: *If T is a tree ordered by the relation $\leqslant$ then $\leqslant$ can be extended to a relation $\preccurlyeq$ which orders T linearly and has the additional property that each set $T_t = \{z \in T: t \leqslant z\}$ is an interval of T with respect to the relation $\preccurlyeq$.*

This theorem follows immediately from Lemmas A, B, D.

°THEOREM 3: *If T is a tree ordered by the relation $\leqslant$ and if each element of T has denumerably many immediate successors then $\leqslant$ can be extended to a relation $\preccurlyeq$ which linearly orders T and has the property that for any t in T the set of immediate successors of t has the order type ω with respect to the relation $\preccurlyeq$.*

To prove this theorem we use Lemmas A, B, C selecting suitable linear orderings $\leqslant_\eta$ of the levels. For limit ordinals η including 0 we select $\leqslant_\eta$ arbitrarily. If $\eta = \eta_1+1$ we first divide L_η into disjoint classes collecting in one class all the elements of L_η which have a common immediate predecessor. Each class is ordered in type ω and the set of all classes is ordered arbitrarily. The ordering $\leqslant_\eta$ thus obtained has the property that immediate successors of an arbitrary element z belonging to L_{η_1} are ordered in type ω.

§ 2. The lexicographical ordering of zero-one sequences. η_ξ sets

In this section we shall continue the study of linear orderings of trees. More specifically, we shall consider the tree $T_{\omega\alpha}$ introduced in Section 1, and its lexicographical ordering. We shall find that these orderings have interesting applications in the general theory of linearly ordered sets.

In order to abbreviate the formulas we shall denote ω_α by ϱ.

Let $\preccurlyeq$ be the relation of lexicographical ordering of T_ϱ; thus if $f, g \in T_\varrho$ then $f \prec g$ if and only if $f \neq g$ and there is a $\xi < \varrho$ such that $f(\xi) = 0$, $g(\xi) = 1$ and $f(\zeta) = g(\zeta)$ for all $\zeta < \xi$. We shall denote ξ by $\delta(f, g)$. Hence

$$f \prec g \equiv (f \neq g) \wedge [f|W(\delta(f,g)) = g|W(\delta(f,g))] \wedge \\ \wedge [f(\delta(f,g)) = 0] \wedge [g(\delta(f,g)) = 1].$$

As we proved on p. 221 the set T_ϱ is linearly ordered by $\preccurlyeq$.

THEOREM 1: *Let A be a non-empty subset of T_ϱ and let f_0 be a sequence in T_ϱ defined thus: $f_0(\zeta) = 1$ if and only if there is a g in A such that $f_0|W(\zeta) = g|W(\zeta)$ and $g(\zeta) = 1$. Then f_0 is the least upper bound of A.*

PROOF. We have to show

(i) $h \in A \to h \preccurlyeq f_0$;

(ii) if $h \preccurlyeq f$ for each h in A, then $f_0 \preccurlyeq f$.

If (i) were false there would exist h in A such that $f_0 \prec h$, whence $h|W(\delta(f_0, h)) = f_0|W(\delta(f_0, h))$ and $h(\delta(f_0, h)) = 1$, $f_0(\delta(f_0, h)) = 0$. However, from the last but one equation it follows $f_0(\delta(f_0, h)) = 1$ in view of the definition of f_0. In this way we obtain a contradiction.

To prove (ii) let us assume that $h \preccurlyeq f$ for each h in A and $f \prec f_0$. Hence we obtain the equations $f|W(\delta) = f_0|W(\delta)$, $f(\delta) = 0$ and $f_0(\delta) = 1$ where we wrote δ for $\delta(f, f_0)$.

The last equation proves by the definition of f_0 that there exists a g in A such that $f_0|W(\delta) = g|W(\delta)$ and $g(\delta) = 1$. Since $g \neq f$ and, by assumption, $g \preccurlyeq f$, we obtain $g|W(\delta(g,f)) = f|W(\delta(g,f))$ and $g(\delta(g,f)) < f(\delta(g,f))$. The equations $f_0|W(\delta) = g|W(\delta)$ and $f_0(\delta) = g(\delta)$ prove that $\delta(g,f) > \delta$, whence we obtain $f(\delta) = g(\delta)$. Since $f(\delta) = 0$, we obtain $g(\delta) = 0$ which is a contradiction.

Theorem 1 is thus proved. As an immediate corollary we obtain:

THEOREM 2: *Each non-empty subset A of T_ϱ has a least upper bound.*

A similar theorem holds for lower bounds. The set T_ϱ is thus continuously ordered.

Example. Let λ be a limit ordinal and $f_0 \in T_\varrho$ be such that $f_0(\xi) = 0$ for all $\xi \geqslant \lambda$ but for each $\xi < \lambda$ there is a ζ satisfying $\xi < \zeta < \lambda$ and $f_0(\zeta) = 1$. For $0 < \zeta < \lambda$ denote by f_ζ a function which is 0 for all the arguments $\geqslant \zeta$ and otherwise coincides with f_0. Then f_0 is the least upper bound of $\{f_\zeta: 0 < \zeta < \lambda\}$.

PROOF. $f_\zeta \prec f_0$ because selecting a smallest σ such that $\zeta < \sigma < \lambda$ and $f_0(\sigma) = 1$ we obtain $\delta(f_\zeta, f_0) = \sigma$ and $f_\zeta(\sigma) = 0$, $f_0(\sigma) = 1$.

Let us now assume that $f_\zeta \preccurlyeq g$ for each $0 < \zeta < \lambda$. We shall show that $f_0 \preccurlyeq g$. Assume otherwise. Then $f_0(\delta(f_0, g)) = 1$ and $g(\delta(f_0, g)) = 0$ whereas $f_0(\xi) = g(\xi)$ for all $\xi < \delta(f_0, g)$. The first equation proves that $\delta(f_0, g) < \lambda$ because f_0 vanishes for all arguments $\geqslant \lambda$. Select σ such that $\delta(f_0, g) < \sigma < \lambda$ and $f_0(\sigma) = 1$. Hence the function f_σ satisfies the relation $f_\sigma|W(\delta(f_0, g)) = f_0|W(\delta(f_0, g))$, since $f_\sigma(\xi) = f_0(\xi)$ for all $\xi < \sigma$; for the same reason the equations $f_\sigma(\delta(f_0, g)) = f_0(\delta(f_0, g)) = 1$ are also true. Therefore, since $g \prec f_0$, $g(\delta(f_0, g)) = 0$, whence $g \prec f_\sigma$ which contradicts the assumption that $f_\sigma \preccurlyeq g$. Our assertion is thus true.

THEOREM 3: *Each non-empty subset A of T_ϱ is cofinal and coinitial with an ordinal $\leqslant \omega_\alpha$.*

PROOF. Let A be a subset of T_ϱ without a last element and let f_0 be the least upper bound of A. We denote by λ the least ordinal $< \varrho$ such that $\xi \geqslant \lambda \to f_0(\xi) = 0$ and put $\lambda = \varrho$ if such an ordinal does not exist. Hence if $\xi < \lambda$ then there is a σ such that $\xi \leqslant \sigma < \lambda$, $f_0(\sigma) = 1$ and $f_0(\xi) = 0$ for all $\xi \geqslant \lambda$.

First we show that λ is a limit number. Suppose conversely that $\lambda = \mu+1$. Hence $\lambda < \varrho$. From the definitions of λ we obtain $f_0(\mu) = 1$, whence, in view of Theorem 1, there is g in A such that $g|W(\mu) = f_0|W(\mu)$ and $g(\mu) = 1$. It follows that $\delta(f_0, g) > \mu$, i.e. $\delta(f_0, g) \geqslant \lambda$. However f_0 takes on the value 0 for all arguments $\geqslant \lambda$ and so $f_0\big(\delta(f_0, g)\big) = 0$ whereas the relation $g \preccurlyeq f_0$ resulting from the definition of f_0 proves that either $g = f_0$ or $f_0\big(\delta(f_0, g)\big) = 1$. Hence we obtain $f_0 = g \in A$, i.e. f_0 is the last element of A. Since this contradicts our assumption, we see that λ cannot be a successor ordinal.

From the example given above it follows now that f_0 is the least upper bound of the set $\{f_\zeta \colon O < \zeta < \lambda\}$ and hence A is cofinal with a set whose order type is λ.

COROLLARY 4: *If X is a subset of T_ϱ which is well ordered by $\preccurlyeq$ then the order type of X is $< \omega_{\alpha+1}$; similarly if X is well ordered by the inverse relation $\succcurlyeq$ then the order type of this set is $< \omega_{\alpha+1}$.*

PROOF. Otherwise X would contain a set ordered by $\preccurlyeq$ (or by $\succcurlyeq$) in the type $\omega_{\alpha+1}$ and could not be cofinal with a type $\lambda \leqslant \omega_\alpha = \varrho$.

We shall now apply the ordering of T_{ω_ξ} to the following problem concerning linearly ordered sets. Given a cardinal $\mathfrak{m}$, does there exist a linearly ordered set H such that every linearly ordered set of power $\leqslant \mathfrak{m}$ is similar to a subset of H? For the case $\mathfrak{m} = \aleph_0$ we have already discussed this problem and solved it affirmatively (Chapter VI, p. 214).

DEFINITION: A linearly ordered set H (of arbitrary power) is said to be an *η_ξ-set*[1]) if $H \neq \emptyset$ and if for every two subsets A, B of power $< \aleph_\xi$ such that

(1) $$(a \in A) \wedge (b \in B) \to (a < b)$$

[1]) The η_ξ sets were defined by Hausdorff; see his book Hausdorff [1], p. 181. In our exposition we follow Sierpiński [21].

Similar problems for other types of relations are discussed in model theory; see Chang and Keisler [1].

there exist $u, v, w \in H$ such that

$$(a \in A) \wedge (b \in B) \to (u < a < v < b < w).$$

The η_0-sets are simply sets which are densely ordered and have no first and no last element.

°THEOREM 1: *If H is an η_ξ-set and X is a linearly ordered set of power $\leqslant \aleph_\xi$, then X is similar to a subset of H.*

PROOF. Let $\overline{\overline{X}} = \aleph_\alpha$, $\alpha \leqslant \xi$ and let τ be a one-to-one sequence of type ω_α whose range is equal to X; moreover, let χ be a one-to-one sequence (of type ω_γ) whose range is the set H.

We shall define by transfinite recursion a sequence of elements of H such that the range of the sequence is similar to X. The construction is almost identical to the construction used in the analogous case of sets of type η (p. 214).

For $\mu < \omega_\alpha$ let

$$\varphi_\mu = \min_{\zeta} \bigwedge_{\eta<\mu} \{(\chi_\zeta \neq \chi_{\varphi(\eta)}) \wedge [(\tau_\mu \prec_X \tau_\eta) \equiv (\chi_\zeta \prec_H \chi_{\varphi(\eta)})]\},$$

where $\prec_X$ and $\prec_H$ denote the "less than" relation in the set X and in the set H, respectively. The function min is to be understood in such a way that $\min_{\zeta} A(\zeta) = 0$ in the case where $\neg A(\zeta)$ for all ζ.

We shall prove by induction that $\varphi_\mu \neq \varphi_\eta$ for $\eta < \mu < \omega_\alpha$. In fact, let

$$C = \{\mu < \omega_\alpha : \bigwedge_{\eta<\mu} (\varphi_\mu \neq \varphi_\eta)\}$$

and assume that $\mu < \omega_\alpha$ and $W(\mu) \subset C$. To show that $\mu \in C$ it suffices to show that there exists an element $a \in H$ such that for all $\eta < \mu$,

$$a \neq \chi_{\varphi(\eta)} \quad \text{and} \quad (a \prec_H \chi_{\varphi(\eta)}) \equiv (\tau_\mu \prec_X \tau_\eta).$$

For this purpose let

$$A = \{\chi_{\varphi(\eta)} : (\eta < \mu) \wedge (\tau_\eta \prec_X \tau_\mu)\},$$

$$B = \{\chi_{\varphi(\eta)} : (\eta < \mu) \wedge (\tau_\mu \prec_X \tau_\eta)\}.$$

The sets A and B are of power $\leqslant \overline{\mu} < \overline{\omega}_\alpha \leqslant \aleph_\xi$ and satisfy (1), thus there exist u, v, w satisfying (2). If $B = \emptyset$ then let $a = w$, if $A = \emptyset$ let $a = u$, and if $A \neq \emptyset \neq B$ let $a = v$.

Since $\varphi_0 = 0$, it follows from the inequality obtained above that

$$\chi_{\varphi(\mu)} \prec_H \chi_{\varphi(\eta)} \equiv \tau_\mu \prec_X \tau_\eta,$$

which completes the proof of the theorem.

We shall construct an η_ξ-set, assuming that ω_ξ is a regular ordinal (see Exercise 3).

Let H_ξ be the subset of T_{ω_ξ} consisting of those sequences $\varphi \in T_{\omega_\xi}$ for which there exists a number $\varkappa < \omega_\xi$ such that $\varphi_\varkappa = 1$ and $\varphi_\sigma = 0$ for $\sigma > \varkappa$.

THEOREM 2: *If ω_ξ is a regular ordinal then H_ξ is an η_ξ-set.*

PROOF. Let A and B be subsets of H_ξ of powers $< \aleph_\xi$ and let α and β be one-to-one sequences of the types ω_μ and ω_ν respectively ($\mu < \xi, \nu < \xi$) such that the range of the sequence α is A and the range of β is B. Assume that A and B satisfy condition (1).

For every $\varrho < \omega_\mu$ the term α_ϱ is itself a sequence of type ω_ξ whose terms are either 0 or 1 and there exists an ordinal $\varkappa = \varkappa(\varrho) < \omega_\xi$ such that $\alpha_{\varrho,\varkappa(\varrho)} = 1$ and $\alpha_{\varrho\sigma} = 0$ for $\sigma > \varkappa(\varrho)$, We shall call the ordinal $\varkappa$ "critical" for the sequence α_ϱ and we shall employ similar terminology for the sequences β_ϱ, $\varrho < \omega_\nu$, where by $\lambda(\sigma)$ we denote the critical ordinal of the sequence β_σ.

The regularity of ω_ξ implies that the sequence of critical ordinals for the sequences α_ϱ, $\varrho < \omega_\mu$ is not cofinal with ω_ξ. Thus there exists an ordinal $\zeta < \omega_\xi$ such that $\varkappa(\varrho) < \zeta$ for all $\varrho < \omega_\mu$. Let $\varphi_\gamma = 0$ for $\gamma \neq \zeta$ and let $\varphi_\zeta = 1$. Then $\varphi \in H_\xi$ and $\varphi \prec \alpha_\varrho$ for all $\varrho < \omega_\mu$. Similarly there exists an ordinal $\zeta' < \omega_\xi$ such that $\lambda(\sigma) < \zeta'$ for all $\sigma < \omega_\nu$. If $\psi_\gamma = 1$ for $\gamma \leqslant \zeta'$ and $\psi_\gamma = 0$ for $\gamma > \zeta'$, then $\psi \in H_\xi$ and $\beta_\sigma \prec \psi$ for all $\sigma < \omega_\gamma$.

To prove the theorem it remains to construct a sequence $\vartheta \in H_\xi$ such that $\alpha_\varrho \prec \vartheta \prec \beta_\sigma$ for all $\varrho < \omega_\mu$ and $\sigma < \omega_\nu$.

First we construct by transfinite induction a sequence ϑ^* of type ω_ξ which does not necessarily belong to H_ξ but is such that

$$\alpha_\varrho \preccurlyeq \vartheta^* \preccurlyeq \beta_\sigma \quad \text{for} \quad \varrho < \omega_\mu,\ \sigma < \omega_\nu.$$

Let τ be an ordinal $\leqslant \omega_\xi$ and let $\varphi \in T_\tau$. For $\gamma < \tau$ we put $F(\varphi, \gamma) = 1$ if there exists an ordinal $\varrho < \omega_\mu$ such that $\varphi|\gamma = \alpha_\varrho|\gamma$ and $\alpha_{\varrho,\gamma} = 1$ and $F(\varphi, \gamma) = 0$ if there is no such ϱ.

From the theorem on transfinite induction it follows that there exists a sequence ϑ^* of type ω_ξ such that

$$\vartheta^*_\gamma = F(\vartheta^*|W(\gamma), \gamma) \quad \text{for} \quad \gamma < \omega_\xi;$$

thus

$$(\vartheta^*_\gamma = 1) \equiv \bigvee_{\varrho<\omega_\mu} \bigwedge_{\delta<\gamma} [(\alpha_{\varrho\delta} = \vartheta^*_\delta) \wedge (\alpha_{\varrho\gamma} = 1)].$$

Assume that $\alpha_\varrho \succ \vartheta^*$; then there exists an ordinal $\gamma < \omega_\xi$ such that $\alpha_{\varrho\gamma} = 1$, $\vartheta^*_\gamma = 0$ and $\alpha_{\varrho\delta} = \vartheta^*_\delta$ for $\delta < \gamma$. By the definition of the sequence ϑ^* it follows that $\vartheta^*_\gamma = 1$, which is impossible. Thus $\alpha_\varrho \preccurlyeq \vartheta^*$ for every $\varrho < \omega_\mu$.

In turn, assume that $\vartheta^* \succ \beta_\sigma$; thus there exists an ordinal $\gamma < \omega_\xi$ such that $\vartheta^*_\gamma = 1$, $\beta_{\sigma\gamma} = 0$ and $\vartheta^*_\delta = \beta_{\sigma\delta}$ for $\delta < \gamma$. From the definition of the sequence ϑ^* it follows that for some $\varrho < \omega_\mu$

$$\alpha_\varrho|W(\gamma) = \vartheta^*|W(\gamma) \quad \text{and} \quad \alpha_{\varrho\gamma} = 1.$$

Hence $\alpha_\varrho \succ \beta_\sigma$ contrary to our assumptions.

We now modify the sequence ϑ^* so as to obtain the desired sequence ϑ. We examine two cases:

Case I: *For every $\gamma_0 < \omega_\xi$ there exist $\gamma > \gamma_0$ such that $\vartheta^*_\gamma = 1$.* In this case $\vartheta^* \notin H_\xi$ and thus the strict inequality $\alpha_\delta < \vartheta^* < \beta_\sigma$ holds for arbitrary $\varrho < \omega_\mu$ and $\sigma < \omega_\nu$. Let γ_0 be an ordinal such that $\vartheta^*_{\gamma_0} = 1$. Let $\vartheta_\gamma = \vartheta^*_\gamma$ for $\gamma \leqslant \gamma_0$ and $\vartheta_\gamma = 0$ for $\gamma > \gamma_0$. Clearly for every γ_0 the sequence ϑ belongs to H_ξ. We shall show that we can choose γ_0 in such a way that the sequence ϑ satisfies the desired conditions.

Let $\gamma_0 > \max(\zeta, \zeta')$. From $\vartheta^* < \beta_\sigma$ it follows that there exist ordinals δ such that $\vartheta^*_\delta \neq \beta_{\sigma\delta}$ and the least such ordinal δ_0 satisfies the equations $\vartheta^*_{\delta_0} = 0$ and $\beta_{\sigma,\delta_0} = 1$. Thus $\delta_0 \leqslant \zeta'$ (since $\beta_{\sigma,\delta} = 0$ for $\delta > \zeta'$); hence $\vartheta_{\delta_0} < \beta_{\sigma,\delta_0}$ and $\vartheta_\delta = \beta_{\sigma,\delta}$ for $\delta < \delta_0$. Therefore $\vartheta \prec \beta_\sigma$.

For $\varrho < \omega_\mu$ there exists an ordinal $\delta_0(\varrho)$ such that $\alpha_{\varrho,\delta_0(\varrho)} = 0$ and $\vartheta^*_{\delta_0(\varrho)} = 1$ and for all $\gamma < \delta_0(\varrho)$, $\alpha_{\varrho,\gamma} = \vartheta^*_\gamma$. Choose $\gamma_0 > \delta_0(\varrho)$ for all $\varrho < \omega_\mu$. The sequence ϑ obtained from ϑ^* by the modification just described satisfies the condition $\alpha_\varrho \prec \vartheta$ for $\varrho < \omega_\mu$.

Case II. *There exists $\gamma_0 < \omega_\xi$ such that $\vartheta^*_\gamma = 0$ for $\gamma > \gamma_0$.*

Let $\gamma_1 > \max(\gamma_0, \zeta, \zeta')$ and modify the sequence ϑ^* so that $\delta_\gamma = \vartheta^*_\gamma$ for $\gamma \neq \gamma_1$ and $\vartheta^*_{\gamma_1} = 1$. It is easy to check that the sequence obtained in this way belongs to H_ξ and satisfies the condition $\alpha_\varrho \prec \vartheta \prec \beta_\sigma$.

°THEOREM 3: $\overline{\overline{H}}_{\alpha+1} = 2^{\aleph_\alpha}$; *if* ξ *is a limit ordinal, then* $\overline{\overline{H}}_\xi = \sum_{\alpha<\xi} 2^{\aleph_\alpha}$.

PROOF. Assume that $\xi = \alpha+1$ and $\gamma_0 < \omega_\xi$. The set Z_{γ_0} of those sequences $\varphi \in T_{\omega_\xi}$ for which $\varphi_{\gamma_0} = 1$ and $\varphi_\gamma = 0$ for $\gamma > \gamma_0$ has a power $\leqslant 2^{\aleph_\alpha}$. Since $H_\xi = \bigcup_{\gamma_0<\omega_\xi} Z_{\gamma_0}$, it follows that

$$\overline{\overline{H}}_\xi \leqslant 2^{\aleph_\alpha}\overline{\omega}_\xi = 2^{\aleph_\alpha}\aleph_{\alpha+1} \leqslant 2^{\aleph_\alpha}2^{\aleph_\alpha} = 2^{\aleph_\alpha}.$$

Since Z_{ω_α} has power $2^{\aleph_\alpha}$ and $Z_{\omega_\alpha} \subset H_\xi$, we conclude that $\overline{\overline{H}}_\xi = 2^{\aleph_\alpha}$.

The proof of the second part of the theorem is similar.

COROLLARY 4: *If* $2^{\aleph_\alpha} = \aleph_{\alpha+1}$, *then there exists an* $\eta_{\alpha+1}$-*set of power* $\aleph_{\alpha+1}$.

This corollary follows immediately from Theorems 2 and 3.

It can be shown that if $2^{\aleph_\alpha} > \aleph_{\alpha+1}$ then no $\eta_{\alpha+1}$-set has power $\aleph_{\alpha+1}$. On the other hand, for ordinals ξ where ω_ξ is not regular, every η_ξ-set has a power $> \aleph_\xi$. This fact follows from the theorem (proved by Hausdorff) which states that every $\eta_{\alpha+1}$-set contains a subset of power $2^{\aleph_\alpha}$.[1])

Exercises

1. Every two η_ξ-sets of power $\aleph_\xi$ are isomorphic. [Hausdorff]

Hint: Use an argument similar to that used in the proof of Theorem VI.2.2.

2. If H is an η_ξ-set and if H' is a subset of H, dense in H, then H' is also an η_ξ-set.

3. If ω_ξ is a singular ordinal, then every η_ξ-set is also an $\eta_{\xi+1}$-set.

4. If $\overline{R} = \eta$ and I is a non-principal prime ideal in the family $P(N)$, then the reduced direct product R^N/I is of type η_1. [Kochen]

Hint: Let f_n and g_m $(n, m \in N)$ be elements of R^N such that $\{i: f_n(i) \geqslant g_m(i)\} \in I$ for arbitrary n, m. The essential step in the proof depends upon the construction of a sequence $h \in R^N$ such that the sets $\{i: f_n(i) \geqslant h(i)\}$ and $\{i: h(i) \geqslant g_m(i)\}$ belong to I. If $f_n(i) \leqslant f_{n+1}(i)$ and $g_{m+1}(i) < g_m(i)$ for all $n, m, i \in N$, then let X_j be a de-

[1]) A simple proof of Hausdorff's theorem about powers of $\eta_{\alpha+1}$-sets can be found in Sierpiński [21]. In connection with Exercise 4 see Kochen [1]. Interesting algebraic applications of the η_1-sets were found by Erdös, Gillman and Henriksen [1].

creasing sequence of sets X not belonging to I such that $X_0 = N$ and $\bigcap_j X_j = \emptyset$ and for $i \in X_j - X_{j+1}$ let

$$h(i) = \tfrac{1}{2}[f_{K(j)}(i) + g_{L(j)}(i)].$$

If the sequences f_n and g_m do not satisfy the inequalities above, then they can be modified in such a way as to obtain functions belonging to the same classes mod I and satisfying the desired inequalities. For example, monotonicity can be obtained by letting $f_0'(i) = f_0(i)$, $f_{n+1}'(i) = \max(f_n'(i), f_n(i))$ and similarly for g_m.

§ 3. König's infinity lemma

The oldest and simplest but probably the most widely known theorem on trees is the following "König's lemma".[1])

THEOREM 1: *If T is an infinite tree with finite levels which for each integer n has chains with at least n elements, then T has an infinite chain.*

PROOF. Let f be a choice function for the family of non-empty finite subsets of T. Also we put $f(\emptyset) = \emptyset$.

Consider the following property of elements of T:

$\Phi(x) \equiv$ *for each integer n there is a branch B such that B contains x and at least n successors of x.*

We claim that if x satisfies Φ then so does at least one immediate successor of x. Otherwise for each y which immediately succeedes x there is an integer $n(y)$ such that every branch of T which contains y has at most $n(y)$ successors of y. Since there are only finitely many immediate successors of x we obtain an integer $n = \max\{n(y): y$ is an immediate successor of $x\}$ such that for each immediate successor y of x and each branch B containing y, this branch contains at most n successors of y. Now consider a branch B_0 containing x and calculate the number of successors of x which belong to B_0. If $x_1 > x$ and $x_1 \in B_0$ then B_0 must contain an immediate successor y_0 of x and so the number of successors of y which belong to B_0 is at most n. The element y being an immediate successor of x, we infer that the number of successors of x which belong to B_0 is at most $n+1$. This however contradicts the assumption that x satisfies Φ.

[1]) König's lemma was first published in D. König [1].

We prove in the same way that there is a minimal element x_0 of T which has the property Φ.

Now put $\Gamma(x) = \{y\colon y \text{ is an immediate successor of } x \wedge \Phi(y)\}$, $g(x) = f(\Gamma(x))$ and define $x_{n+1} = g(x_n)$. We prove by induction that $\Phi(x_n)$ for each n. For $n = 0$ this is true because of the definition of x_0. If $\Phi(x_n)$ is true for an integer n then the set $\Gamma(x_n)$ is non-empty as we proved above and hence $g(x_n) \in \Gamma(x_n)$, whence the element $g(x_n)$, i.e. x_{n+1}, satisfies Φ.

Since $x_n < x_{n+1}$ for each n, we obtained an infinite chain. Extending this chain to a branch we obtain an infinite branch.

Example. Let I be the closed interval $[0, 1] = \{x\colon 0 \leqslant x \leqslant 1\}$ of the real numbers and suppose that $I_{j,n} = \{x\colon j/2^n \leqslant x \leqslant (j+1)/2^n\}$ for $j < 2^n$, $n \in N$. Let G be a family of open intervals of the real numbers. We say that G covers $I_{j,n}$ if there is an interval $X \in G$ such that $I_{j,n} \subset X$.

THEOREM 2: *If* $\bigcup(G) = I$ *then there is an integer* n *such that* G *covers all the intervals* $I_{j,n}$ *for* $j < 2^n$.

PROOF. Let T_0 be the family consisting of all the intervals $I_{j,n}$ where $j < 2^n$ and n is an integer. T_0 is a tree of order 2 under the relation inverse to inclusion: $X \leqslant Y \equiv X \supset Y$. Consider the sub-tree T of T_0 consisting of those intervals $X \in T_0$ which are not covered by G. Note that each element $I_{j,n}$ of T has exactly n predecessors.

The theorem will be proved if we show that there is an integer n such that all chains in T have at most n elements. For suppose that n satisfies this condition; if $I_{j,n}$ were not covered by G it would belong to T and so would all its predecessors and hence we would obtain a chain with $n+1$ elements.

Now the theorem follows easily from König's lemma. If T contained chains of each finite power, it would contain an infinite chain $\{I_{j_n,n}\colon n \in N\}$. Since the intervals $I_{j_n,n}$ are closed, their intersection $\bigcap_n I_{j_n,n}$ is non-empty. But if x belongs to this intersection then $x \notin \bigcup(G)$ and so we obtain a contradiction with the assumption of the theorem.

A slight generalization of König's lemma is the following

°THEOREM 3: *If* T *is a tree of height* ω_α *all levels of which are finite then* T *has a branch of order type* ω_α.

PROOF. Let f be a choice function for the family of non-empty subsets of T. From the assumption about the height of T it follows that for each $\beta < \omega_\alpha$ there is at least one chain of T whose order type is β. We abbreviate as $\Phi(a)$ the formula

for each $\beta < \omega_\alpha$ there is a chain of type β with the minimal element a.

From the assumptions of the theorem we infer that there are elements of L_0 with the property Φ. Let a_0 be one of them. We now define a transfinite sequence $\{a_\xi\}$ of order type ω_α. For $\xi = 0$ the element a_0 has been selected. It will therefore be sufficient to show how to define a_γ $(\gamma > 0)$ from a sequence $\{a_\xi\}_{\xi<\gamma}$. Let A be the set of all terms of this sequence.

If A is not a chain or if $\neg\Phi(a)$ for at least one a in A or if $a_\xi \notin L_\xi$ for at least one $\xi < \gamma$ we put $a_\gamma = a_0$. Now assume that A is a chain, $\Phi(a)$ for $a \in A$, and $a_\xi \in L_\xi$ for each $\xi < \gamma$.

Case I: $\gamma = \delta+1$. In this case we consider the finitely many immediate successors of a_δ and show just as in the proof of König's lemma that the set $X = \{a \in L_\gamma\colon (a_\delta < a) \wedge \Phi(a)\}$ is not empty. The element a_γ is defined as $f(X)$.

Case II: γ is a limit number. In this case we first remark that the set $X = \{a \in L_\gamma\colon \Phi(a)\}$ is not empty. Otherwise a_0 would not satisfy Φ because there would be no chains of type $\gamma+1$ with the minimal element a_0. Secondly, we remark that for each $y \in L_\gamma - X$ there is an ordinal $\beta(y) < \omega_\alpha$ such that no chain with the minimal element y has the order type $> \beta(y)$.

We claim that if $\xi < \gamma$ then a_ξ has a successor in X. Otherwise each chain with the minimal element a_ξ would have the order type at most $(\gamma-\xi)+\max\{\beta(y)\colon y \in L_\gamma - X\}$. Indeed, such a chain either ends below L_γ or its part lying above L_γ has the order type at most $\max\{\beta(y)\colon y \in L_\gamma - X\}$. Since this contradicts the assumption $\Phi(a_\xi)$, our claim is established.

For each x in X put $C_x = \{b \in T\colon b \leqslant x\}$ and claim that $A \subset C_x$ for some x in X. For if $A - C_x \neq \emptyset$ then putting $b_x = f(A - C_x)$ we see that x is not a successor of b_x and therefore also not a successor of any $z \geqslant b_x$. If $A - C_x \neq \emptyset$ for each x in X, we would obtain finitely many elements b_x corresponding to finitely many elements of X.

Selecting the largest element from among $\{b_x: x \in X\}$ we would therefore obtain an element of A such that no x in X were its successor. This contradicts the claim established above. Of course, the element x such that $A \subset C_x$ is unique. This unique element is taken as a_γ.

The sequence $\{a_\xi\}_{\xi<\omega_\alpha}$ is thus defined. We prove by induction that it is strictly increasing and this gives us the desired chain of type ω_α.

Exercises

1. Give examples showing that the assumption that the tree T has finite order is essential for the validity of the Theorem 1.

2. Derive the compactness of the Cantor set C (see p. 135) from König's lemma.

3. Derive König's lemma from Tychonoff's theorem (see p. 138).

4. Prove that if A is a set of finite sequences with terms 0 or 1 such that $\bigwedge_n [e \in A \rightarrow e|n \in A]$ and such that for every n there is a sequence e in A with n terms, then there exists an infinite sequence φ such that for every n in N its restriction $\varphi|n$ belongs to A. Does this theorem hold for sequences with arbitrary integral terms?

5. Let A be a set of binary relations between non-negative integers, $A \subset P(N \times N)$, satisfying the following conditions:

(i) for every $n \in N$ there exists in A a relation whose field contains n;

(ii) if $R \in A$ then $R \cap (n \times n) \in A$ for each n in N.

Prove that there exists a relation R_0 with field N such that $R_0 \cap (n \times n) \in A$ for every n in N.

6. If mankind is to last forever, then at least one man living now will have male descendents in every generation [König].

§ 4. Aronszajn's trees[1])

A cardinal $\mathfrak{m}$ is said to have the *tree property* if every tree T of power $\mathfrak{m}$ such that each level of T has power $< \mathfrak{m}$ has a branch of power $\mathfrak{m}$. According to König's lemma the cardinal $\aleph_0$ has the tree property. We now ask what other cardinals have the tree property.

°THEOREM 1: *No singular cardinal has the tree property.*

PROOF. Suppose that $\overline{\overline{A}} = \mathfrak{m}$ where $\mathfrak{m}$ is singular. Thus A can be decomposed into a union $\bigcup_\xi A_\xi$ of less than $\mathfrak{m}$ mutually disjoint sets such that $\overline{\overline{A_\xi}} < \mathfrak{m}$ for each ξ. We may assume that A is well ordered

[1]) The main result of this section (Theorem 2) is due to Aronszajn whose proof appeared first in Kurepa [2]. See also Kurepa [1], p. 92. For further results on Aronszajn's trees, their generalizations to higher cardinalities and other related types of trees see Jech [1].

by a relation $\leqslant$. Define a partial ordering $\prec$ on A by putting $x \prec y \equiv \bigvee_{\xi} [(x \in A_\xi) \wedge (y \in A_\xi) \wedge (x \leqslant y)]$. It is easy to check that A becomes a tree under this ordering and that the sets A_ξ are the unique branches of A. Hence $\mathfrak{m}$ does not have the tree property.

For regular cardinals the problem is much more difficult. Given a regular cardinal $\mathfrak{m} > \aleph_0$, it cannot be generally decided on the basis of axioms Σ°[TR] whether $\mathfrak{m}$ has or has not the tree property. However for $\mathfrak{m} = \aleph_1$ the problem has a solution:

°THEOREM 2: *There exists a tree A of height ω_1 such that each level of A and each chain of A are at most countable.*

Trees with the property stated in the theorem are called *Aronszajn trees.*

PROOF. Let Q be the set of rational numbers and let Γ_α be the set of transfinite sequences of type α (i.e. functions) $f\colon W(\alpha) \to Q$ which are increasing: $f(\xi) < f(\eta)$ for $\xi < \eta$ $(\alpha < \omega_1)$ and have the property that $\sigma_f = \sup\{f(\xi)\colon \xi < \alpha\} \in Q$.

We denote by Φ a choice function for the family of all sets Γ_α.

In order to carry out the construction we need some preparatory steps. For each limit ordinal $\alpha < \omega_1$ we select (using the axiom of choice) a strictly increasing sequence $\{\beta_{\alpha n}\}_{n<\omega}$ which converges to α: $\lim_{n<\omega} \beta_{\alpha n} = \alpha$. Also for each pair $v < w$ of rational numbers we select a strictly increasing sequence r_n such that $v < r_n < w$ for each n and $\lim_{n<\omega} r_n = w$.

Elements of the tree to be constructed will be transfinite sequences. The αth level of the tree will consist of certain elements of Γ_α. The ordering will be the relation of inclusion, i.e. $x \leqslant y \equiv x \subset y$.

The definition of the tree will proceed by induction. We put $L_0 = \{\emptyset\}$, assume that $\alpha > 0$ and show how to define L_α from levels L_ξ where $\xi < \alpha$.

Let us assume that the set $\bigcup \{L_\xi\colon \xi < \alpha\} = L'_\alpha$ is at most countable and that it has the following property:

(*) *Either α is not a limit number or for arbitrary $\xi < \alpha$, arbitrary f in L_ξ, arbitrary w in Q such that $\sigma_f < w$ and arbitrary η such that $\xi < \eta$*

$< \alpha$ *there is a* g *in* L_η *such that* $\sigma_g = w$ *and* $f \subset g$. (If these conditions are not satisfied, we put $L_\alpha = \emptyset$.)

We consider separately the cases in which α is a successor or a limit number.

Case 1. $\alpha = \gamma+1$. In this case let L_α consist of all functions $f \in \Gamma_\alpha$ such that $f|W(\alpha) \in L_\gamma$.

Case 2. α is a limit number. For each f in L_ξ ($\xi < \alpha$) and each rational w satisfying $\sigma_f < w$ we define a function $g \in \Gamma_\alpha$ which we call an *extension* of f and denote by $g_{f,w}$. This function is equal to $\bigcup_{n<\omega} g_n$ where $g_0 = f$ and each g_{n+1} is an extension of g_n.

In order to define the functions g_n we consider the sequence r_n correlated with the pair σ_f and w. Thus $\sigma_f < r_0 < r_1 < \ldots < r_n < \ldots < w$ and $\lim r_n = w$. Furthermore we consider the increasing sequence $\beta_{\alpha n}$ and denote by n_0 the first integer such that $\beta_{\alpha n_0} > \xi$. Let F_0 be the family of functions h which extend f, belong to $L_{\beta_{\alpha n_0}+1}$, and satisfy the equation $h(\beta_{\alpha n_0}) = r_0$. By (*), the family F_0 is non-empty. Let $g_0 = \Phi(F_0)$ be a function selected from this family. Now let us assume that g_k has already been defined and $g_k \in L_{\beta_{\alpha, n_0+k+1}}$ and $g_k(\beta_{\alpha, n_0+k}) = r_k$. By (*) the family F_{k+1} of all functions h extending g_k which satisfy the conditions $h \in L_{\beta_{\alpha, n_0+k+1}}$ and $h(\beta_{\alpha, n_0+k+1}) = r_{k+1}$ is non-empty. Let $g_{k+1} = \Phi(F_{k+1})$.

By an easy induction we prove that $g_k \subset g_{k+1}$, $g_k \in L_{\beta_{\alpha, n_0+k+1}}$ and $g_k(\beta_{\alpha, n_0+k}) = r_k$ for each h. Hence $\bigcup g_k = g$ is a function in Γ_α because $\sigma_g = \lim r_n = w$ and g extends f.

We define now L_α as the set of all functions $g_{f,w}$ where $f \in \bigcup_{\xi<\alpha} L_\xi$ and $\sigma_f < w$.

The inductive definition of L_α is thus finished.

We shall now prove that (*) is valid for each α. Thus let α be a limit number, $f \in L_\xi$ where $\xi < \alpha$, $\sigma_f < w \in Q$ and let $\xi < \eta < \alpha$. We want to show that there is a g in L_η such that $\sigma_g = w$.

We use induction on η. If $\eta = \xi+1$ then we obtain g simply by adding to f the pair $\langle \xi, w \rangle$.

If η is a limit number then we apply the same procedure as we did above but replacing α by η and r by w and obtain an extension g of f

which by definition belongs to L_η. Since $\sigma_g = w$, the proof that $(*)$ is valid for all α is finished.

Now in order to prove Theorem 2 we remark that L_α is at most denumerable. For this is true for $\alpha = 0$ and if $\alpha = \gamma+1$ then each f in L_γ gives rise to exactly $\aleph_0$ elements of $L_{\gamma+1}$. Finally if α is a limit number and L_ξ is at most denumerable for each $\xi < \alpha$ then so is L_α because each element of L_α is uniquely determined by a function $f \in \bigcup_{\xi<\alpha} L_\xi$ and a rational number $w > \sigma_f$.

Since $(*)$ is valid for each α, we infer that $L_\alpha \neq \emptyset$ for all $\alpha < \omega_1$. Hence all the levels of the tree are at most denumerable and the height of the tree is ω_1.

Exercises

1. Construct a binary Aronszajn tree.

Hint: Start from an Aronszajn tree in which the order of each element is infinite and insert between each pair of consecutive levels L_α and $L_{\alpha+1}$ suitable binary trees each with exactly one minimal point lying on L_α.

2. Prove that there are $2^{\aleph_1}$ isomorphism types of Aronszajn trees.

§ 5. Souslin trees

A tree S is a *Souslin tree* if it has the height ω_1 and each of its chains as well as each of its antichains is at most denumerable. Note that a Souslin tree is an Aronszajn tree but not necessarily conversely.

The existence of Souslin trees is unprovable on the basis of axioms Σ°; if these axioms are consistent then they remain so if we add to them a sentence stating that Souslin trees exist.

Souslin trees are important chiefly because of their connections with a famous question asked by Souslin in 1920: Let us call a linearly ordered set X a *Souslin set* if every set of disjoint non-empty intervals is at most denumerable but X is not similar to any subset of the real numbers.

Souslin's problem was whether such linear orders exist.[1]) We shall prove

[1]) Souslin's problem was stated in Fundamenta Mathematicae **1** (1920) 223. The unsolvability of the Souslin problem on the basis of set theoretical axioms was shown by Solovay and Tennenbaum [1].

°THEOREM 1: *If Souslin trees exist then so do Souslin sets and conversely.*

PROOF. First let us assume that X is a Souslin set and let $\leqslant$ be its ordering. By Theorem 6, p. 215 no denumerable subset of X is dense in X for otherwise X would be similar to a subset of the real numbers. We construct a Souslin tree T whose elements will be intervals of X ordered by inverse inclusion.

Let Φ be a function which correlates with every non-empty family of infinite intervals of X a proper infinite sub-interval of an interval belonging to the family. The existence of Φ follows from the axiom of choice. Now we construct a transfinite sequence of type ω_1 consisting of infinite intervals of X.

Let $I_0 = X$; if $0 < \alpha < \omega_1$, then let F_α be the family of infinite intervals I of X with the property that for each $\xi < \alpha$ the interval I is either contained in I_ξ or disjoint from I_ξ. If the family F_α is non-void then we put $I_\alpha = \Phi(F_\alpha)$, otherwise $I_\alpha = X$.

The sequence $\{I_\xi\}_{\xi<\omega_1}$ is thus defined and for arbitrary $\xi < \eta < \omega_1$, we have either $I_\xi \cap I_\eta = \emptyset$ or $I_\xi \supset I_\eta$.

Let us note that the condition $F_\alpha = \emptyset$ is equivalent to the statement: the set of endpoints of intervals I_ξ, $\xi < \alpha$, is dense in X.

From Theorem 6, p. 215 we know that a linearly ordered set containing an at most denumerable dense subset is similar to a set of real numbers. Hence X does not have dense denumerable subsets and therefore $F_\alpha \neq \emptyset$ for each $\alpha < \omega_1$.

We can now construct a Souslin tree. Let T be the set of all the intervals I_ξ, $\xi < \omega_1$, and let $I_\xi \leqslant I_\eta$ hold if and only if $\xi < \eta$ and $I_\xi \supset I_\eta$. Obviously T is a tree and has the power $\aleph_1$. An antichain of T consists of mutually disjoint intervals and hence is at most denumerable.

We shall now show that T contains no non-denumerable chain. For suppose that $C = \{I_\nu\}$ is such a chain. The left endpoints of the intervals I_ν form a non-decreasing sequence of elements of X and the right endpoints of the intervals I_ν form a non-increasing such sequence.

The book of Devlin and Johnsbraten [1] contains a comprehensive exposition of recent results connected with the Souslin problem.

The connection between the Souslin sets and Souslin trees which is established in Theorem 1 was discovered by Miller [1].

One at least of these sequences must have a non-denumerable set of terms. Hence if T contained a non-denumerable chain, then so would X and so X would not be a Souslin set.

We pass now to the proof of the converse implication. Thus we assume the existence of a Souslin tree T_0 and shall derive the existence of a Souslin set.

The proof will proceed in two steps. First we shall modify T_0 so as to obtain a Souslin tree T with the additional property that for each x in T there are exactly $\aleph_0$ immediate successors of x. Such a tree will be called *normal.* Secondly we shall linearly order the branches of a normal tree T so as to obtain a Souslin set.

Construction of a normal Souslin tree. It is obvious from the definitions that a non-denumerable sub-tree of a Souslin tree is again a Souslin tree. Also it is easy to see that at each level of a Souslin tree there are elements which have non-denumerably many successors. Otherwise the tree would be denumerable since every level of it is at most denumerable.

It follows from these remarks that if we remove from T_0 elements which have only $\aleph_0$ successors, we still obtain a Souslin tree. Thus we may assume that each $x \in T_0$ has $\aleph_1$ successors in T_0.

From Theorem 3, p. 327 we infer that for each x in T_0 there is an ordinal $\eta(x)$ such that x has infinitely many successors on the $\eta(x)$th level of T_0. Of course, x has then infinitely many successors on each level of T_0 higher than $\eta(x)$.

We now construct a normal subtree T of T_0. We shall denote the levels of T by L_ξ and the levels of T_0 by L_ξ^0. Let $L_0 = L_0^0$ and assume that $\alpha > 0$ and levels L_ξ are already defined for $\xi < \alpha$ so that

(1) each L_ξ is contained in a level $L^0_{\lambda(\xi)}$.

If $\alpha = \beta + 1$ then we take $\lambda(\alpha) = \sup\{\eta(x)\colon x \in L_\beta\}$ and $L_\alpha = L^0_{\lambda(\alpha)}$. Thus condition (1) is satisfied. If α is a limit number we define $\lambda(\alpha) = \sup\{\lambda(\xi)\colon \xi < \alpha\}$ and $L_\alpha = L^0_{\lambda(\alpha)}$ and see again that (1) holds. Thus the levels L_α are defined for $\alpha < \omega_1$, each $x \in L_\xi$ has infinitely many successors on the level $L_{\xi+1}$ and each L_ξ with $\xi > 0$ is denumerable. Hence $T = \bigcup\{L_\xi\colon \xi < \omega_1\}$ is a normal Souslin tree. This accomplishes the first part of the proof.

The second part will rely on Theorem 3 proved in Section 1, p. 319. According to this theorem there exists a linear ordering $\preccurlyeq$ of T which is an extension of $\leqslant$ and has the property that for any t in T the immediate successors of t are ordered by $\preccurlyeq$ in type ω.

Let now X be the set of all maximal branches of T. Each B is denumerable; let χ_B be the height of B, i.e. the least ordinal α such that $B \cap L_\alpha = \emptyset$. For $\xi < \chi_B$ let B_ξ be the unique element of $B \cap L_\xi$.

We order X by agreeing that $B' \mathrel{\lessdot} B''$ if $B' = B''$ or the least ξ such that $B'_\xi \neq B''_\xi$ satisfies $B'_\xi \prec B''_\xi$. There must be such a ξ because B' and B'' are branches, i.e. maximal chains.

We shall show that X is a Souslin set with respect to this ordering. First we have to show that $\mathrel{\lessdot}$ is a linear ordering of X. The reflexivity and connectedness of $\mathrel{\lessdot}$ are obvious, and transitivity and antisymmetry easy to prove (compare the proofs given for lexicographic orderings, p. 221).

For each t in T denote by I_t the set of branches containing t. We easily show that if $B', B'' \in I_t$ and $B' \mathrel{\lessdot} B''$ then each branch lying between B' and B'' belongs to I_t. Hence I_t is an interval of X.

In order to prove that X is a Souslin set we have to show that (1) *there is no denumerable set $S \subset X$ dense in X* and that (2) *each set of disjoint intervals is at most denumerable.*

PROOF OF (1). Let $S \subset X$ and let S be denumerable. There is $\alpha < \omega_1$ larger than heights of all branches in S. If $t \in L_\alpha$ then I_t is an interval of X and $I_t \cap S = \emptyset$.

PROOF OF (2). Let $J = [B, C]$ be an interval of X such that $B \mathrel{\lessdot} C$ and $B \neq C$. Let ξ_J be the least ordinal satisfying $B_{\xi_J} \neq C_{\xi_J}$, hence $B_{\xi_J} \mathrel{\lessdot} C_{\xi_J}$.

Since immediate successors of B_{ξ_J} are ordered by $\prec$ in type ω we can find a $z_J \in L_{\xi_J+1}$ such that $B_{\xi_J} \prec z_J$ and z_J is the immediate successor of B_{ξ_J+1} with respect to the relation $\prec$. It is clear that each branch containing z_J lies inside the interval $[B, C]$.

We claim that if $J = [B, C]$ and $J' = [B', C']$ are two non-overlapping intervals of X, i.e. $C \mathrel{\lessdot} B'$ or $C' \mathrel{\lessdot} B$ then z_J and $z_{J'}$ are incomparable in T. Otherwise there would be a branch Z containing both z_J and $z_{J'}$. But then Z would lie inside J and inside J' contradicting the fact that the intervals J, J' do not overlap.

Now we see that if F is a family of non-overlapping intervals of X then $\{z_J: J \in F\}$ is a set of the same cardinality as F consisting of mutually incomparable elements of T. Since T is a Souslin tree, the family F must be at most denumerable. Our theorem is thus proved.

§ 6. Some partition theorems

Let A be a set and n an integer. We denote by $[A]^n$ the set of all subsets of A with exactly n elements. Certain combinatorial problems led Ramsey to ask the following question: suppose that $[A]^n$ is decomposed into two disjoint sets: $[A]^n = X \cup Y$. Are there big subsets B of A such that $[B]^n \subset X$ or $[B]^n \subset Y$? If $n = 2$ we can rephrase this question in a more suggestive way: let us call elements of A "points" and pairs $\{x, y\}$ "edges". If $\{x, y\} \in X$ we say that the edge is white and if $\{x, y\} \in Y$ we say that it is black. Ramsey's question can thus be formulated as follows: if each edge is either white or black, is it true that there is a big subset B of A such that all edges joining points of B are of the same colour?

We shall generalize Ramsey's question. Let f be a mapping of $[A]^n$ into a set L. If $B \subset A$ and the function f is constant on $[B]^n$ then we say that B is a subset of A *homogeneous* for f.

Let A and L be two sets, $\overline{\overline{A}} = \mathfrak{m}$, $\overline{\overline{L}} = \mathfrak{l}$. If for every f: $[A]^n \to L$ there is a set B homogeneous for f such that $\overline{\overline{B}} = \mathfrak{n}$ then we write $\mathfrak{m} \to (\mathfrak{n})^n_{\mathfrak{l}}$. If there is an f: $[A]^n \to L$ for which no homogeneous set of power $\mathfrak{n}$ exists then we write $\mathfrak{m} \nrightarrow (\mathfrak{n})^n_{\mathfrak{l}}$. Theorems of this kind are called *partition theorems*.[1])

Below we shall prove 3 rather simple but by no means trivial partition theorems. They have found numerous applications particularly in logic.

We start with some obvious observations:

[1]) The first partition theorem was established by Ramsey [1]. The theory of partitions was developed later mainly by Erdös and his collaborators; see Erdös and Rado [1] and Erdös, Hajnal and Rado [1]. The symbols $\mathfrak{m} \to (\mathfrak{n})^n_{\mathfrak{l}}$ and $\mathfrak{m} \to (\mathfrak{n})^n_{\mathfrak{l}}$ were introduced by Erdös.

1. *If* $\mathfrak{m}' \geqslant \mathfrak{m}$ *and* $\mathfrak{m} \to (\mathfrak{n})^n_{\mathfrak{l}}$ *then* $\mathfrak{m}' \to (\mathfrak{n})^n_{\mathfrak{l}}$.
2. *If* $\mathfrak{n} \geqslant \mathfrak{n}'$ *and* $\mathfrak{m} \to (\mathfrak{n})^n_{\mathfrak{l}}$ *then* $\mathfrak{m} \to (\mathfrak{n}')^n_{\mathfrak{l}}$.
3. *If* $\mathfrak{l} \geqslant \mathfrak{l}'$ *and* $\mathfrak{m} \to (\mathfrak{n})^n_{\mathfrak{l}}$ *then* $\mathfrak{m} \to (\mathfrak{n})^n_{\mathfrak{l}'}$.

Also obvious are the following facts:

4. $\mathfrak{m} \to (\mathfrak{m})^n_1$.
5. *If* $\mathfrak{m}$ *is regular and* $\mathfrak{m} > \mathfrak{l}$ *then* $\mathfrak{m} \to (\mathfrak{m})^1_{\mathfrak{l}}$.

Remark 5 says simply that if a set of a regular power $\mathfrak{m}$ is divided in less than $\mathfrak{m}$ disjoint parts, one at least of these parts has power $\mathfrak{m}$.

We shall now prove a result which initiated the whole theory.

°THEOREM 1 (Ramsey): *If m, n are finite then* $\aleph_0 \to (\aleph_0)^n_m$.

PROOF. If $n = 1$ the theorem follows from 4. We make now the inductive assumption

(1) $$\aleph_0 \to (\aleph_0)^n_m \quad \text{for each } m \geqslant 1$$

and shall derive $\aleph_0 \to (\aleph_0)^{n+1}_m$ for an arbitrary $m \geqslant 1$. For $m = 1$ this statement is obvious in view of 4. Hence we can assume that $m > 1$ and that

(2) $$\aleph_0 \to (\aleph_0)^n_{m-1}.$$

Let A be a denumerable set and let f: $[A]^{n+1} \to \{0, 1, \ldots, m-1\}$.

Let Φ, Ψ be choice functions for the family of non-empty subsets of A and $P(A)$ respectively. For each a in A define a function f_a: $[A - \{a\}]^n \to \{0, 1, \ldots, m-1\}$ by the equation $f_a(X) = f(\{a\} \cup X)$. Call a set $B \subset A - \{a\}$ an *a−0-set* (or *a−1-set*) if it is infinite and $f_a(X) < m-1$ or $= m-1$ for each X in $[B]^n$. Each infinite set contained in $A - \{a\}$ and homogeneous for f_a is obviously either an a−0-set or a−1-set. Now we distinguish two cases.

Case I. There is an infinite set $A_1 \subset A$ such that for each a in A_1 each infinite set $B \subset A_1 - \{a\}$ homogeneous for f_a is an a−1-set.

In view of the inductive assumption, each infinite set $X \subset A_1 - \{a\}$, where a is an arbitrary element of A_1, contains an infinite subset which is homogeneous for the function f_a limited to $[X]^n$. Since we deal now with the case I, this subset must be an a−1-set. Therefore we can use the function Ψ and correlate with each infinite set $X \subset A_1 - \{a\}$ an infinite set $B(a, X) \subset X$ such that $B(a, X)$ is an a−1-set.

Now we define a sequence $\{a_n\}_{n<\omega}$ of elements of A_1 and a decreasing sequence B_n of subsets of A_1. Put $a_0 = \Phi(A_1)$, $B_0 = B(a_0, A_1 - \{a_0\})$; thus B_0 is an infinite $a_0 - 1$-set. Let us assume that a_n, B_n are defined and B_n is an infinite subset of $A_1 - \{a_n\}$. We put $a_{n+1} = \Phi(B_n)$ and $B_{n+1} = B(a_{n+1}, B_n - \{a_{n+1}\})$. Thus B_{n+1} is an infinite subset of $B_n - \{a_{n+1}\}$ and hence also of A_1.

We prove by induction that $a_n \in B_n - B_{n+1}$ and B_n is an $a_n - 1$-set. Hence the set $H = \{a_n\colon n < \omega\}$ is infinite. If $X \subset [H]^{n+1}$ and $X = \{a_{p_0}, \ldots, a_{p_n}\}$ where $p_0 < p_1 < \ldots < p_n$ then $a_{p_1}, \ldots, a_{p_n}$ are elements of B_{p_0}, whence $\{a_{p_1}, \ldots, a_{p_n}\} \in [B_{p_0}]^n$ and since B_{p_0} is an $a_{p_0} - 1$-set we obtain $f_{a_{p_0}}(\{a_{p_1}, \ldots, a_{p_n}\}) = m-1$, i.e. $f(\{a_{p_0}, \ldots, a_{p_n}\}) = m-1$. Thus H is homogeneous for f.

Case II. For each infinite set $A_1 \subset A$ there exists $a \in A$, and an infinite subset $B \subset A_1 - \{a\}$ which is homogeneous for f_a and is an $a - 0$-set.

In this case we use the functions Φ and Ψ to correlate with each infinite set $A_1 \subset A$ an element $a(A_1)$ and an infinite set $B(A_1) \subset A_1 - \{a(A_1)\}$ which is an $a(A_1) - 0$-set. Again we define by induction two sequences $\{a_n\}_{n<\omega}$, $\{B_n\}_{n<\omega}$. For $n = 0$ we put $a_0 = a(A)$, $B_0 = B(A)$. If a_n, B_n are defined and B_n is infinite we put $a_{n+1} = a(B_n)$, $B_{n+1} = B(B_n)$. We can then prove by induction that for each n, the set B_n is infinite, $B_{n+1} \subset B_n$, $a_n \in B_n - B_{n+1}$ and B_n is an $a_n - 0$-set. Thus the set $H = \{a_n\colon n < \omega\}$ is infinite. If $X = \{a_{p_0}, a_{p_1}, \ldots, a_{p_n}\} \in [H]^{n+1}$ where $p_0 < p_1 < \ldots < p_n$ then $a_{p_1}, \ldots, a_{p_n}$ belong to B_{p_0} and hence $\{a_{p_1}, \ldots, a_{p_n}\} \in [B_{p_n}]^n$. Since B_{p_n} is an $a_{p_0} - 0$-set we infer $f_{a_{p_0}}(\{a_{p_1}, \ldots, a_{p_n}\}) < m-1$, i.e. $f(\{a_{p_0}, \ldots, a_{p_n}\}) = f(X) < m-1$. Hence $f\colon [H]^{n+1} \to \{0, 1, \ldots, m-2\}$.

In view of the inductive assumption (2) we obtain an infinite subset of H homogeneous for the function f restricted to $[H]^{n+1}$. This subset is the required infinite subset of A homogeneous for f.

Ramsey's theorem is thus proved.

Interesting problems arise when we discuss the existence of non-denumerable homogeneous sets. In this connection we prove

°THEOREM 2:[1]) *If n is an integer*, $\mathfrak{m}$ *is infinite*, $\mathfrak{d} \geqslant 2^{\mathfrak{m}}$ *and* $\aleph(\mathfrak{d}) \to (\aleph(\mathfrak{m}))^n_{\mathfrak{m}}$ *then* $\aleph(2^{\mathfrak{d}}) \to (\aleph(\mathfrak{m}))^{n+1}_{\mathfrak{m}}$.

[1]) Theorem 2 is due to Simpson; see Keisler [1], p. 76.

PROOF. Let E be a set of power $\aleph(2^{\mathfrak{d}})$ and f: $[E]^{n+1} \to I$ where I is a set of power $\mathfrak{m}$. We shall assume that $n > 0$ because for $n = 0$ the theorem follows immediately from Remark 4, p. 337. For $X \in [E]^n$ and $a \in E - X$ we put $f_a(X) = f(X \cup \{a\})$; thus f_a: $[E]^n \to I$.

A subset S of E will be called *saturated with respect to* f if there is a function g: $\boldsymbol{P}(S) \times E \to S$ such that for each set $M \subset S$ of power at most $\mathfrak{d}$ and each $a \in E - M$ the following formulas hold:

$$(1) \qquad g(M, a) \notin M \quad \text{and} \quad f_a | [M]^n = f_{g(M, a)} | [M]^n.$$

LEMMA: *There is a saturated subset S of E of power* $2^{\mathfrak{d}}$.

In order to construct this set we start with an arbitrary set $S_0 \subset E$ of power $2^{\mathfrak{d}}$ and extend it successively to form a transfinite sequence $\{S_\gamma\}$ where γ ranges over ordinals such that $\overline{\gamma} \leqslant \mathfrak{d}$. For limit numbers γ we take as S_γ the union of all preceding S_ξ's. Hence if all the S_ξ's have power $2^{\mathfrak{d}}$ then so has S_γ. Now we shall define $S_{\gamma+1}$ assuming that the cardinal number of S_γ is $2^{\mathfrak{d}}$.

Let M range over subsets of S_γ of power $\leqslant \mathfrak{d}$; thus the family of all M's has power $\sum_{\mathfrak{x} \leqslant \mathfrak{d}} (2^{\mathfrak{d}})^{\mathfrak{x}} = \mathfrak{d} \cdot 2^{\mathfrak{d}} = 2^{\mathfrak{d}}$ (see p. 284).

For each M let $F(M)$ be the set of functions h: $[M]^n \to I$; the cardinal number of $F(M)$ is thus $\leqslant \mathfrak{d}^{\mathfrak{d}} = 2^{\mathfrak{d}}$ and $f_a | [M]^n \in F(M)$ for each $a \in E - M$.

We fix an arbitrary well-ordering of E and denote by $g(M, a)$ the earliest a' in $E - M$ satisfying the equation

$$f_a | [M]^n = f_{g(M, a)} | [M]^n.$$

Since the family $\{f_a | [M]^n: a \in E - M\}$ is contained in $F(M)$, its power is $\leqslant 2^{\mathfrak{d}}$.

Thus for a given M of power $\leqslant \mathfrak{d}$ there are at most $2^{\mathfrak{d}}$ elements a' having the form $g(M, a)$.

We now define $S_{\gamma+1}$ as the set consisting of all the elements of S_γ and all the elements $g(M, a)$ where $M \subset S$ and the power of M is $\leqslant \mathfrak{d}$. Hence the power of $S_{\gamma+1}$ is $\leqslant 2^{\mathfrak{d}} \cdot 2^{\mathfrak{d}} = 2^{\mathfrak{d}}$.

The inductive definition of $S_{\gamma+1}$ is thus finished. The cardinal of the set $S = \bigcup S_\gamma$ where γ ranges over ordinals satisfying $\overline{\gamma} \leqslant \mathfrak{d}$ is $\leqslant 2^{\mathfrak{d}} \cdot \aleph(\mathfrak{d}) = 2^{\mathfrak{d}}$. If M is a subset of S and the cardinal number of M is $\leqslant \mathfrak{d}$ then there is a γ such that $M \subset S_\gamma$. Hence if a is an arbitrary element of $E-M$ then $g(M, a) \in E-M$ and $g(M, a)$ satisfies (1). Thus S is saturated with respect to f. The lemma is thus proved.

We can now complete the proof of Theorem 2. Let $c \in E-S$ and let h be defined by transfinite induction:

$$h(\gamma) = g\big(\mathrm{Rg}(h|W(\gamma)), c\big)$$

where, as before, γ ranges over ordinals of power $\leqslant \mathfrak{d}$.

Since g satisfies (1), we obtain $h(\gamma) \notin \mathrm{Rg}\big(h|W(\gamma)\big)$ which proves that the function h is one-to-one. Hence the range $\mathrm{Rg}(h)$ of h is a subset of E and has power $\aleph(\mathfrak{d})$. In view of our assumptions there exists a subset H of $\mathrm{Rg}(h)$ which has power $\aleph(\mathfrak{m})$ and is homogeneous for f_c. Let $f_c(X) = i_0$ for an arbitrary X in $[H]^n$. We claim that H is homogeneous for f. To prove this let us consider arbitrary $n+1$ elements of H, e.g. $h(\gamma_0), \ldots, h(\gamma_n)$ where $\gamma_0 < \ldots < \gamma_n$.

Putting $M = \mathrm{Rg}\big(h|W(\gamma_n)\big)$ in (1), we obtain

$$f_{g(M,c)}\big(\{h(\gamma_0), \ldots, h(\gamma_{n-1})\}\big) = f_c\big(\{h(\gamma_0), \ldots, h(\gamma_{n-1})\}\big) = i_0$$

because $h(\gamma_i) \in M$ for $i < n$. The left-hand side is equal to $f_{h(\gamma_n)}\big(\{h(\gamma_0), \ldots$ $\ldots, h(\gamma_{n-1})\}\big)$ in view of the definition of h. Remembering the definition of the functions f_a we obtain finally $f\big(\{h(\gamma_0), \ldots, h(\gamma_n)\}\big) = i_0$ which proves that H is homogeneous for f.

In order to express conveniently the contents of Theorem 2 let us introduce the following inductive definition: for each cardinal $\mathfrak{m}$ we put $\mathfrak{a}_0(\mathfrak{m}) = \mathfrak{m}$, $\mathfrak{a}_{n+1}(\mathfrak{m}) = 2^{\mathfrak{a}_n(\mathfrak{m})}$. Remarking that $\aleph(\mathfrak{m}) \to \big(\aleph(\mathfrak{m})\big)^1_{\mathfrak{m}}$ for each infinite $\mathfrak{m}$ and using induction, we obtain from Theorem 2 the following corollary:

°THEOREM 3 (Erdös–Rado):[1]) *If* $\mathfrak{m} \geqslant \aleph_0$ *and* $0 \leqslant n < m$ *then*

$$\aleph\big(\mathfrak{a}_n(\mathfrak{m})\big) \to \big(\aleph(\mathfrak{m})\big)^{n+1}_{\mathfrak{m}}.$$

Examples. 1. Putting $\mathfrak{m} = \aleph_0$ we obtain for $n = 1, 2, 3$ the following non-trivial partition theorems:

(*) $$\aleph(2^{\aleph_0}) \to (\aleph_1)^2_{\aleph_0},$$

(**) $$\aleph(2^{2^{\aleph_0}}) \to (\aleph_1)^3_{\aleph_0},$$

(***) $$\aleph(2^{2^{2^{\aleph_0}}}) \to (\aleph_1)^4_{\aleph_0}.$$

[1]) Theorem 3 was proved by Erdös and Rado [1].

2. $\aleph(2^{\mathfrak{m}}) \to (\aleph(\mathfrak{m}))^2_{\mathfrak{m}}$. This results directly from the Erdös–Rado theorem for $m = 1$.

We finish this section by a theorem showing that $\aleph(2^{\mathfrak{m}})$ in the last example cannot be replaced by a smaller number.

°THEOREM 4:[1]) $2^{\mathfrak{m}} \nrightarrow (\aleph(\mathfrak{m}))^2_2$.

PROOF. Let $\mathfrak{m} = \overline{\omega}_\alpha$ and consider the set T_{ω_α} ordered lexicographically by the relation $\prec$. Now consider any well ordering $\leqslant$ of T_{ω_α} and put for $\{x, y\} \in [T_{\omega_\alpha}]^2$

$$f(\{x, y\}) = 0 \quad \text{if} \quad [x \prec y \equiv x \leqslant y],$$
$$f(\{x, y\}) = 1 \quad \text{otherwise}.$$

We shall show that no set $H_0 \subset T_{\omega_\alpha}$ of power $\aleph(\mathfrak{m})$ is homogeneous for f. For assume otherwise. Thus either $x \prec y \equiv x \leqslant y$ for all $x, y \in H_0$ or $x \prec y \equiv y \leqslant x$ for all $x, y \in H_0$. Since H_0 is well-ordered by $\leqslant$, it is well-ordered either by $\prec$ or by the relation inverse to $\prec$. However this contradicts Corollary 4, p. 321 and so the theorem is proved.

Exercises

1. If X is a set quasi-ordered by a relation $\leqslant$, then the following conditions are equivalent:

(i) *every descending sequence of elements of X is decreasing and every antichain is finite*;

(ii) *for each sequence f: $N \to X$ there are integers $i < j$ such that $f_i \leqslant f_j$.*

2. Use Ramsey's theorem to prove: *If X is an infinite set of integers then X contains an infinite subset Y such that either any two elements of Y are relatively prime or no two elements of Y are relatively prime.*

3. Let X be a linearly ordered set and let $R_0, R_1, \ldots, R_{k-1}$ be binary relations with the field X. Prove that there is an infinite set $Y \subset X$ such that if $y_1, y_2 \in Y$ and $y_1 < y_2$ then $y_1 R_j y_2 \equiv y_1 R_i y_2$ for arbitrary $i, j < k$.

Hint: For $x_1 < x_2$ put $\varepsilon_j(x_1, x_2) = 0$ or 1 according as $x_1 R_j x_2$ or x_1 non-$R_j x_2$ and define a mapping f: $[X]^2 \to \{0, 1\}^k$ by putting $f(\{x_1', x_2'\}) = (\varepsilon_0(z_1, z_2), \ldots$ $\ldots, \varepsilon_{k-1}(z_1, z_2))$ where $z_1 = x_1', z_2 = x_2'$ if $x_1 < x_2'$ and $z_1 = x_2', z_2 = x_1'$ if $x_2' < x_1'$. Define Y as a set homogeneous for f.

[1]) For Theorem 4 see Sierpiński [17].

CHAPTER X

INACCESSIBLE CARDINALS

§ 1. Normal functions and stationary sets

Before proceeding to the theory of inaccessible cardinals we collect in this section some auxiliary definitions and theorems.

In the whole chapter ω_α denotes a *regular* initial ordinal $> \omega$. We shall treat the set $W(\omega_\alpha)$ as a topological space with the order topology. Thus for each $X \subset W(\omega_\alpha)$ the closure $\overline{X}$ is the set of ordinals $\xi \in W(\omega_\alpha)$ with the property that for some $Z \subset X$ the least upper bound of Z is ξ. We leave to the reader the proof that $W(\omega_\alpha)$ is a topological space and that an increasing mapping $f\colon W(\omega_\alpha) \to W(\omega_\alpha)$ is continuous if and only if for each limit number $\lambda < \omega_\alpha$ and each increasing function $\varphi\colon W(\omega_\alpha) \to W(\omega_\alpha)$

$$f\bigl(\lim_{\xi<\lambda} \varphi(\xi)\bigr) = \lim_{\xi<\lambda} f\bigl(\varphi(\xi)\bigr).$$

A mapping $f\colon W(\omega_\alpha) \to W(\omega_\alpha)$ is called *normal* if f is increasing and continuous.[1]) Obviously, the range of f is then cofinal with $W(\omega_\alpha)$.

THEOREM 1: *If $f\colon W(\omega_\alpha) \to W(\omega_\alpha)$ is normal then for each $\gamma < \omega_\alpha$ there is a number $\xi > \gamma$ critical* (see p. 233) *for f.*

PROOF. Define a sequence $\{\gamma_n\}_{n<\omega}$ by recursion: $\gamma_0 = \gamma+1$, $\gamma_{n+1} = f(\gamma_n)$. The limit $\lim\limits_{n<\omega} \gamma_n$ is then critical for f and is $> \gamma$.

Closed sets of the space $W(\omega_\alpha)$ can be characterized as sets $X \subset W(\omega_\alpha)$ such that for each $\lambda < \omega_\alpha$ and each increasing function $\varphi\colon W(\lambda) \to X$, $\lim\limits_{\xi<\lambda} \varphi(\xi) \in X$.

In order to abbreviate the formulation of the next few theorems we introduce the

[1]) Normal functions were first investigated by Veblen [1].

DEFINITION: A set $X \subset W(\omega_\alpha)$ is *normal* in $W(\omega_\alpha)$ if it is closed and cofinal with $W(\omega_\alpha)$.

If α is fixed we shall refer briefly to "normal sets" without mentioning α.

THEOREM 2: *If f: $W(\omega_\alpha) \to W(\omega_\alpha)$ is normal then so is the set of its critical numbers.*

PROOF. The set is closed because a limit of an increasing sequence of critical numbers is itself critical. The cofinality with $W(\omega_\alpha)$ follows from Theorem 1.

The next two theorems establish a connection between normal mappings and normal sets.

THEOREM 3: *The range of a normal mapping is a normal set.*

PROOF. If f is normal then its range is obviously cofinal with $W(\omega\ )$.

In order to prove that $\mathrm{Rg}(f)$ is closed let us assume that $\xi \in \overline{\mathrm{Rg}(f)}$.

Hence ξ is either an element of $\mathrm{Rg}(f)$ or $\xi \in W(\omega_\alpha)$ and ξ is the limit of an increasing sequence of type $\lambda < \omega_\alpha$ whose terms belong to $\mathrm{Rg}(f)$.

If $\xi = \lim f(\gamma_\delta)$ where the sequence $\{f(\gamma_\delta)\}$ is increasing then the sequence $\{\gamma_\delta\}$ is also increasing and since $\lambda < \omega_\alpha$, the limit μ of this sequence belongs to $W(\omega\ )$. Since f is normal we obtain $\lim\limits_{\delta<\lambda} f(\gamma_\delta) = f(\mu)$, whence $\xi = f(\mu)$ and therefore $\xi \in \mathrm{Rg}(f)$.

THEOREM 4: *Each normal set $X \subset W(\omega_\alpha)$ is the range of a normal mapping f_X: $W(\omega_\alpha) \to W(\omega_\alpha)$.*

PROOF. We construct f_X by induction: $f_X(\xi)$ is the minimal element of $X - \{f_X(\eta)$: $\eta < \xi\}$. It is easy to see that there is exactly one function satisfying these conditions for each $\xi < \omega_\alpha$; this unique function is increasing, has the domain $W(\omega_\alpha)$ and the range X. The function f_X is continuous, for if φ is an increasing sequence and $\varphi \in W(\omega_\alpha)^{W(\lambda)}$ where $\lambda < \omega_\alpha$ then $f_X(\varphi(\xi)) \in X$ for $\xi < \lambda$ and hence $\lim\limits_{\xi<\lambda} f_X(\varphi(\xi)) \in X$ since X is closed. Putting $\gamma = \lim\limits_{\xi<\lambda} (\varphi(\xi))$ we obtain $f_X(\gamma) = \lim\limits_{\xi<\lambda} f_X(\varphi(\xi))$ from the definition of f_X.

Theorem 4 is thus proved.

The function f_X constructed above will be called an *enumeration* of X.

It is easy to prove that this is the unique normal function with the range X.

DEFINITION: For each normal set $X \subset W(\omega_\alpha)$ we denote by X' the set of critical numbers of f_X. The set X' is called the *derivative* of X.

Higher order derivatives are defined by recursion: $X^{(0)} = X$; $X^{(\xi+1)} = [X^{(\xi)}]'$; $X^{(\lambda)} = \bigcap \{X^{(\xi)}: \xi < \lambda\}$.

The normal function enumerating $X^{(\xi)}$ is called the ξth *derivative* of f_X.

THEOREM 5: *If X is a normal set and $\xi < \omega_\alpha$ then $X^{(\xi)}$ is also normal.*

PROOF. We use induction. For $\xi = 0$ the theorem is obvious. The inductive step from ξ to $\xi+1$ is immediate in view of Theorem 2.

Let $\lambda < \omega_\alpha$ be a limit ordinal and suppose that the theorem is true for $\xi < \lambda$. The set $X^{(\lambda)}$ is closed because every intersection of closed sets is closed. Thus it remains to show that $X^{(\lambda)}$ is cofinal with $W(\omega_\alpha)$.

Let $\gamma < \omega_\alpha$. Since, by assumption, each $X^{(\xi)}$ where $\xi < \lambda$ is cofinal with $W(\omega_\alpha)$, there exists an increasing sequence $\{x_\xi\}_{\xi<\lambda}$ such that $x_\xi \in X^{(\xi)}$ for each $\xi < \lambda$ and $\gamma < x_0$. The limit $\delta = \lim_{\xi<\lambda} x_\xi$ belongs to $W(\omega_\alpha)$ because ω_α is regular and $\lambda < \omega_\alpha$. Moreover, $x_\eta \in X^{(\xi)}$ for $\eta \geqslant \xi$ because the sets $X^{(\xi)}$ decrease when ξ increases.

Hence $\delta \in X^{(\xi)}$ for each $\xi < \lambda$ and therefore $\delta \in \bigcap_{\xi<\lambda} X^{(\xi)} = X^{(\lambda)}$. Since $\delta > \gamma$, the theorem is proved.

°THEOREM 6: *If $\boldsymbol{F}$ is a non-empty family consisting of less than $\aleph_\alpha$ normal sets then $\bigcap(\boldsymbol{F})$ is normal.*

PROOF. The intersection of closed sets being closed, it remains to show that for every $\gamma < \omega_\alpha$ there exists an ordinal $> \gamma$ which belongs to $\bigcap(\boldsymbol{F})$.

We can assume that $\boldsymbol{F}$ is the range of a transfinite sequence $\{X_\xi\}_{\xi<\beta}$ whose terms are normal sets.

We define now a double sequence $\{\gamma_{\xi,n}\}$ of ordinals ($\xi < \beta$, $n < \omega$) such that $\gamma < \gamma_{\xi,n} \in X^{(\xi)}$, $\gamma_{\eta,n} < \gamma_{\xi,m}$ for $n \leqslant m$ and $\eta < \xi < \beta$ and $\gamma_{\eta,n} < \gamma_{0,n+1}$ for $\eta < \beta$ and arbitrary $n < \omega$.

To obtain this sequence we select first an ordinal $\gamma_{0,0} > \gamma$ which belongs to X_0 and then an increasing sequence $\gamma_{\xi,0}$ ($\xi < \beta$) such that

$\gamma_{\xi,0} \in X_\xi$. After this sequence is defined, we select the first element $\gamma_{0,1}$ in X_0 which is greater than all the ordinals $\gamma_{\xi,0}$ and elements $\gamma_{\xi,1}$ in X_ξ such that the sequence $\{\gamma_{\xi,1}\}_{\xi<\beta}$ is increasing. Continuing in this way, we obtain the desired sequence. Putting $\delta_\xi = \lim_{n<\omega} \gamma_{\xi,n}$ we obtain $\delta_\xi \in X_\xi$ because the sets X_ξ are closed.

We claim that $\delta_\xi = \delta_\eta$ for arbitrary $\xi, \eta < \beta$. Let us assume that $\eta < \xi$. Hence $\gamma_{\eta,n} < \gamma_{\xi,n+k} < \delta_\xi$, whence $\delta_\eta \leqslant \delta_\xi$. On the other hand, $\gamma_{\xi,n} < \gamma_{0,n+1} < \gamma_{\eta,n+1}$, whence by letting n converge to ω we obtain $\delta_\xi \leqslant \delta_\eta$.

Denoting now by δ the common value of all δ_ξ, we obtain $\delta = \delta_\xi \in X_\xi$ for each $\xi < \lambda$ and hence $\delta \in \bigcap_{\xi<\lambda} X_\xi = \bigcap (F)$.

Another way of expressing Theorem 6 is this: *The family of sets $Y \subset W(\omega_\alpha)$ such that Y contains a normal set is an ω_α-complete filter.*

The intersection of a family consisting of $\aleph_\alpha$ normal sets is not necessarily normal. For instance each set of the form $W(\omega_\alpha) - W(\xi)$ is normal for $\xi < \omega_\alpha$ but the intersection of these sets is empty and thus not normal. However we can establish the following result:

THEOREM 7:[1]) *If $\{X_\xi\}_{\xi<\omega_\alpha}$ is a transfinite sequence of normal sets then the set $X = \{\zeta < \omega_\alpha\colon \zeta \in \bigcap_{\eta<\zeta} X_\eta\}$ is normal.*

Remark. The set X is sometimes called the *diagonal intersection* of the sets X_ξ.

PROOF. First we prove that X is closed. Let $\gamma = \lim_{\xi<\lambda} \varphi(\xi)$ where $\lambda < \omega_\alpha$ and φ is an increasing sequence of elements of X. If $\varrho < \gamma$ then the inequality $\varrho < \varphi(\xi) < \gamma$ holds from a certain ξ_0 on. Since $\varphi(\xi) \in \bigcap \{X_\eta\colon \eta < \varphi(\xi)\}$, we obtain $\varphi(\xi) \in X_\varrho$. Letting ξ increase to λ and taking the limit, we infer that $\gamma \in X_\varrho$ because X_ϱ is closed. Hence $\gamma \in \bigcap \{X_\varrho\colon \varrho < \gamma\}$ and $\gamma \in X$.

Next we prove that X is cofinal with ω_α. To establish this we define by transfinite induction a function $f\colon W(\omega_\alpha) \to W(\omega_\alpha + 1)$ such that

[1]) Theorem 7 is due to Fodor [1] who obtained it by generalizing a result of Alexandroff and Urysohn [1]. Other results concerning regressive functions and stationary sets can be found in Bachmann [1], Section 9.

for each ξ, $f(\xi)$ is the least element of the set

$$A_\xi = \bigcap \{[X_\eta - W(f(\eta)+1)]: \eta < \xi\}$$

or $f(\xi) = \omega_\alpha$ if A_ξ is void.

We prove by transfinite induction that $f(\xi) < \omega_\alpha$ for each $\xi < \omega_\alpha$. Since $A_0 = W(\omega_\alpha)$, this is certainly true for $\xi = 0$. Let us assume that there are ordinals ξ such that $f(\xi) = \omega_\alpha$ and let ξ_0 be the least of them; thus $\xi_0 > 0$. For $\xi < \xi_0$ the set $X_\xi - W(f(\xi)+1)$ is normal since it is an intersection of two normal sets: X_ξ and $W(\omega_\alpha) - W(f(\xi)+1)$. It follows from Theorem 6 that the intersection of sets $X_\xi - W(f(\xi)+1)$ taken over ordinals $\xi < \xi_0$, i.e., the set A_{ξ_0} is normal and hence non-empty. Thus $f(\xi_0)$ is the least element of A_{ξ_0} contrary to our assumption that it is equal to ω_α.

We have thus proved that f is a mapping of $W(\omega_\alpha)$ into $W(\omega_\alpha)$. We shall now show that it is increasing and continuous.

It follows from the definition that if $\eta < \xi$ then $f(\xi) \in X_\eta$ and $f(\xi) \notin W(f(\eta)+1)$; in particular, $f(\xi) > f(\eta)$ and hence f is increasing. In order to prove that f is continuous let us consider an increasing sequence $\varphi\colon W(\lambda) \to W(\omega_\alpha)$ where $\lambda < \omega_\alpha$ and put $\gamma = \lim_{\xi<\lambda} \varphi(\xi)$. We have to show that $f(\gamma) = \lim_{\xi<\lambda} f(\varphi(\xi))$. The inequality $f(\gamma) \geqslant \lim_{\xi<\lambda} f(\varphi(\xi))$ results immediately from the fact that f is an increasing function. Now let $\eta < \gamma$. Thus $\eta < \varphi(\xi)$ from an ordinal ξ_0 on and hence $f(\varphi(\xi)) \in X_\eta - W(f(\eta)+1) \subset X_\eta$. Taking the limit for $\xi \to \lambda$ and remembering that X_η is closed, we obtain $\lim_{\xi<\lambda} f(\varphi(\xi)) \in X_\eta$. Since, in view of the monotonicity of f, $\lim_{\xi<\lambda} f(\varphi(\xi)) \notin W(f(\eta)+1)$, we obtain $\lim_{\xi<\lambda} f(\varphi(\xi)) \in X_\eta - W(f(\eta)+1)$ for each $\eta < \gamma$, whence $\lim_{\xi<\lambda} f(\varphi(\xi)) \in A_\gamma$ and therefore $\lim_{\xi<\lambda} f(\varphi(\xi)) \geqslant f(\gamma)$.

Thus the function f is normal. If $\varkappa$ is a critical number for f then $\varkappa = f(\varkappa) \in A_\varkappa \subset \bigcap \{X_\eta\colon \eta < \varkappa\}$ which proves that $\varkappa \in X$. Since the set of critical numbers of f is cofinal with $W(\omega_\alpha)$, so is X. Thus X is a normal set. Q.E.D.

We shall now introduce the notion of a stationary set.

DEFINITION: (Bloch [1]) A set $S \subset W(\omega_\alpha)$ is called *stationary* in $W(\omega_\alpha)$ if $S \cap X \neq \emptyset$ for each normal set X.

For instance, each normal set is stationary in view of Theorem 6. An example of a stationary set which is not normal is furnished by the set $S_0 = \{\xi < \omega_\alpha : cf(\xi) = 0\}$ $(\alpha > 1)$. To see that S_0 is stationary it is sufficient to select in an arbitrary normal X an increasing sequence $\{\xi_n\}_{n<\omega}$. Since $\lim_{n<\omega} \xi_n$ belongs to X and is cofinal with ω, we obtain that $\lim_{n<\omega} \xi_n \in S_0$ and so the intersection $S_0 \cap X$ is non-empty. More generally, the set $\{\xi < \omega_\alpha : cf(\xi) = \beta\}$ is stationary whenever $\beta < \alpha$.

Each stationary set has obviously the power $\aleph_\alpha$ because it intersects every set of the form $W(\omega_\alpha) - W(\xi)$.

An interesting result concerning stationary sets is connected with the following notion:

DEFINITION: A function g is called *regressive* in $W(\omega_\alpha)$ if its domain X is a subset of $W(\omega_\alpha) - \{0\}$ and $g(\xi) < \xi$ for each ξ in X.

°THEOREM 8: *If $g: X \to W(\omega_\alpha)$ is a regressive funcion whose domain is a stationary set X then there is a stationary set $S \subset X$ such that $g|S$ is constant.*

PROOF. Let $A = \text{Rg}(g)$, let α be an increasing but not necessarily normal mapping of an ordinal $\varrho \leqslant \omega_\alpha$ onto A and put $X_\xi = g^{-1}(\{\alpha(\xi)\})$ for $\xi < \varrho$. The theorem will be proved if we show that one at least of the sets X_ξ, $\xi < \varrho$, is stationary. Let us assume that this is not the case; we shall derive a contradiction from this assumption.

Using the axiom of choice, we select for each $\xi < \varrho$ a set Z_ξ which is normal and disjoint from X_ξ. We shall distinguish two cases:

Case I: $\varrho < \omega$. In this case the family $\{Z_\xi : \xi < \varrho\}$ has power $< \aleph_\alpha$ and hence $\bigcap \{Z_\xi : \xi < \varrho\}$ is normal in view of Theorem 6. Since X is stationary, there is an ordinal $\zeta < \omega_\alpha$ such that $\zeta \in X \cap \bigcap \{Z_\xi : \xi < \varrho\}$. If $g(\zeta) = \alpha(\eta)$ then $\eta < \varrho$ and $\zeta \in g^{-1}(\{\alpha(\eta)\})$, whence $\zeta \notin Z_\eta$ and so $\zeta \notin \bigcap \{Z_\xi : \xi < \varrho\}$ which is a contradiction.

Case II: $\varrho = \omega_\alpha$. In this case we consider the set $\{\xi < \omega_\alpha : \xi \in \bigcap_{\tau<\xi} Z_\tau\}$ which is normal by Theorem 7. If ζ is an ordinal in this set which belongs to X and if $g(\zeta) = \alpha(\eta)$ then $\zeta \in g^{-1}(\{\alpha(\eta)\})$ and so $\zeta \notin Z_\eta$. On the other hand, $g(\zeta) < \zeta$ since g is regressive, whence we obtain $\alpha(\eta) < \zeta$

and therefore $\eta < \zeta$ because α is an increasing function. It follows that $\zeta \notin \bigcap \{Z_\xi : \xi < \zeta\}$ and we again obtain a contradiction. Theorem 8 is thus proved.

§ 2. Weakly and strongly inaccessible cardinals[1])

The idea of accessibility is as follows: $\mathfrak{m} = \aleph_\alpha$ being a cardinal number and Φ an operation on sequences of cardinals, we say that $\mathfrak{m}$ is *Φ-accessible* if we can reach a cardinal $\geqslant \mathfrak{m}$ by applying Φ to a sequence of less than $\mathfrak{m}$ cardinals each of which is $< \mathfrak{m}$. Thus, for instance, $\aleph_0$ is Φ-inaccessible for an arbitrary operation which yields finite cardinals when applied to finite arguments.

In this section we shall give an exact definition of inaccessible cardinals and derive their simplest properties.

DEFINITION: (a) A cardinal $\aleph_\alpha$ is *weakly inaccessible* if $\alpha > 0$, ω_α is regular and α is a limit number.

(b) $\aleph_\alpha$ is *inaccessible* if it is weakly inaccessible and satisfies the condition

$$\mathfrak{x} < \aleph_\alpha \rightarrow 2^{\mathfrak{x}} < \aleph_\alpha. \tag{1}$$

Sometimes we say *strongly inaccessible* instead of inaccessible. We also often speak of *strong* or *weak inaccessibility* of initial ordinals ω_α instead of cardinals $\aleph_\alpha$.

Cardinals satisfying (1) are said to be *strong limit cardinals.*

In order to tie up the definitions with the introductory remark about the concept of "Φ-inaccessibility" we notice the following evident

°THEOREM 1:[2]) *A cardinal* $\mathfrak{m} = \aleph_\alpha > 0$ *is weakly inaccessible if and only if it satisfies the conditions*:

[1]) The notion of weakly inaccessible numbers is due to Hausdorff who spoke about them as "*Zahlen von exorbitanten Grösse*" (see Hausdorff [1], p. 131). Strongly inaccessible cardinals were introduced by Sierpiński and Tarski [1]. An early exposition of inaccessible cardinals is in Tarski [6]; theorems proved in this section are taken from the last named paper.

[2]) Theorem 1 is stated in Tarski [6]. It is closely related to Theorem 3 of von Neumann [4], p. 227, and also with investigations of Zermelo [3].

(i) $\mathfrak{n} < \mathfrak{m} \to \aleph(\mathfrak{n}) < \mathfrak{m}$;

(ii) *if* $\lambda < \omega_\alpha$ *and* $\mathfrak{f}$ *is a sequence of cardinals such that* $\mathrm{Dom}(\mathfrak{f}) = W(\lambda)$ *and* $f_\xi < \mathfrak{m}$ *for* $\xi < \lambda$, *then* $\sum_{\xi<\lambda} \mathfrak{f}_\xi < \mathfrak{m}$.

$\mathfrak{m}$ *is inaccessible if and only if it satisfies* (ii) *and*

(iii) $\mathfrak{n} < \mathfrak{m} \to 2^{\mathfrak{n}} < \mathfrak{m}$.

PROOF. Condition (i) is equivalent to the statement that α is a limit number. Condition (ii) states that ω_α is regular. Hence the conjunction of (i) and (ii) is equivalent to the condition given in part (a) of the definition above. This proves the first part of the theorem.

To prove the second it is sufficient to notice that condition (iii) implies (ii) because $\aleph(\mathfrak{n}) \leqslant 2^{\mathfrak{n}}$ (see p. 287).

THEOREM 2: *If* $\alpha > 0$, *then the ordinal* ω_α *is weakly inaccessible if and only if* $\alpha = \omega_\alpha = cf(\alpha)$.

PROOF. Assume that $\alpha > 0$ and that ω_α is weakly inaccessible. The ordinal α is then a limit ordinal $\neq 0$; let $\gamma = cf(\alpha)$. If $\alpha < \omega_\alpha$ then, since $\omega_\alpha = \lim_{\xi<\alpha} \omega_\xi$, it follows that ω_α is not regular. Thus $\alpha = \omega_\alpha$ since $\beta \leqslant \omega_\beta$ for all β.

Let $\gamma = cf(\alpha)$. By the definition of γ there is an increasing sequence φ of type ω_γ such that $\alpha = \lim_{\xi<\omega\gamma} \varphi(\xi)$, thus $\omega_\alpha = \lim_{\xi<\omega\gamma} \omega_{\varphi(\xi)}$; if $\gamma < \alpha$ then ω_α would not be regular.

Conversely, assume that $\alpha = \omega_\alpha = cf(\alpha) > 0$. Thus α is a limit ordinal and to show that ω_α is weakly inaccessible it suffices to show that it is regular. Assume thus that β is a limit ordinal $\leqslant \omega_\alpha$ and that $\omega_\alpha = \lim_{\xi<\beta} \varphi(\xi)$. If $\gamma = cf(\beta)$ then β is the limit of an increasing sequence ψ of type ω_γ. Thus $\omega_\alpha = \lim_{\xi<\omega\gamma} \varphi\big(\psi(\xi)\big)$ and $cf(\alpha) \leqslant \gamma$. On the other hand, $\gamma \leqslant \omega_\gamma$ and $\omega_\gamma \leqslant \beta$, because $\psi(\xi) \geqslant \xi$ for $\xi > \omega_\gamma$. Thus we have the inequality $cf(\alpha) \leqslant \gamma \leqslant \omega_\gamma \leqslant \beta \leqslant \omega_\alpha$; hence from our initial assumptions it follows that $\alpha = \beta$. Thus ω_α is cofinal with no ordinal $< \omega_\alpha$. Q.E.D.

REMARK: The equation $\alpha = \omega_\alpha$ does not characterize inaccessible ordinals. In fact, if β is a sequence of type ω defined inductively by the formulas $\beta_0 = 0$, $\beta_{n+1} = \omega_{\beta_n}$ then the ordinal $\alpha = \lim_n \beta_n$ satisfies the equation $\alpha = \omega_\alpha$ (see p. 237), but ω_α is not regular since $cf(\alpha) = 0$.

°THEOREM 3: *The cardinal* $\mathfrak{m} = \aleph_\alpha$ *is inaccessible if and only if* $\alpha = \omega_\alpha = cf(\alpha) = \pi(\alpha) > 0$.

PROOF. Let $\mathfrak{m}$ be inaccessible. From Theorem 2 we obtain $\alpha = \omega_\alpha = cf(\alpha) > 0$. Since α is a limit ordinal, we obtain $\pi(\alpha) = \lim_{\beta<\alpha} \pi(\beta)$ because the function π is continuous. Since the condition (iii) is satisfied, we obtain that if $\beta < \alpha$ then $\aleph_{\pi(\beta)} = 2^{\aleph_\beta} < \aleph_\alpha$ and hence $\pi(\beta) < \alpha$. Thus $\pi(\alpha) = \lim \pi(\beta) \leqslant \alpha$. Since $\alpha \leqslant \pi(\alpha)$ as we proved on p. 288, we finally obtain $\pi(\alpha) = \alpha$.

Now let us assume $\alpha = \omega_\alpha = cf(\alpha) = \pi(\alpha) > 0$. From Theorem 2 we infer that $\aleph_\alpha$ is weakly inaccessible. If $\mathfrak{n} < \aleph_\alpha$ and $\mathfrak{n}$ is infinite, then $\mathfrak{n} = \aleph_\beta$ where $\beta < \alpha$, whence $\mathfrak{n} \leqslant \mathfrak{a}_\beta$ and $2^{\mathfrak{n}} \leqslant 2^{\mathfrak{a}_\beta} = \mathfrak{a}_{\beta+1} = \aleph_{\pi(\beta+1)} < \aleph_{\pi(\alpha)} = \aleph_\alpha$. Thus $\aleph_\alpha$ satisfies (iii) and is, therefore, an inaccessible cardinal.

Our next aim is to exhibit a large number of operations Φ such that inaccessible cardinals are Φ-inaccessible. For this end we first prove

°THEOREM 4: $\aleph_\alpha$ *is inaccessible if and only if there exists a family* $\boldsymbol{R} \neq \emptyset$ *whose elements are sets and which satisfies the following conditions*:

(i) $X \in \boldsymbol{R} \to X \subset \boldsymbol{R}$,

(ii) $X \in \boldsymbol{R} \to \boldsymbol{P}(X) \in \boldsymbol{R}$,

(iii) $X \subset \boldsymbol{R} \wedge \overline{\overline{X}} < \overline{\overline{\boldsymbol{R}}} \to X \in \boldsymbol{R}$,

(iv) $\overline{\overline{\boldsymbol{R}}} = \aleph_\alpha > \aleph_0$.

PROOF. If $\aleph_\alpha$ is inaccessible then we put $\boldsymbol{R} = \boldsymbol{R}_{\omega_\alpha}$ (see p. 285). Property (i) results from Theorem 1, p. 285. If $X \in \boldsymbol{R}$ then $X \in \boldsymbol{R}_\beta$ for some $\beta < \omega_\alpha$, whence $\boldsymbol{P}(X) \in \boldsymbol{R}_{\beta+1} \subset \boldsymbol{R}_{\omega_\alpha}$. In order to prove (iii) let us assume that $X \subset \boldsymbol{R}$ and $\overline{\overline{X}} < \overline{\overline{\boldsymbol{R}}}$ and let $\beta(x)$ where $x \in \boldsymbol{R}$ be the first ordinal $< \omega_\alpha$ such that $x \in \boldsymbol{R}_{\beta(x)}$. Since $\beta(x) < \omega_\alpha$ and $\overline{\overline{X}} < \overline{\overline{\boldsymbol{R}}}$ the set $\{\beta(x): x \in X\}$ is not cofinal with ω_α because otherwise ω_α would not be regular. Hence there is an ordinal $\sigma < \omega_\alpha$ surpassing all the ordinals $\beta(x)$ and therefore $X \subset \boldsymbol{R}_\sigma$, whence $X \in \boldsymbol{R}_{\sigma+1} \subset \boldsymbol{R}_{\omega_\alpha}$. Finally (iv) results from the equations $\overline{\overline{\boldsymbol{R}}} = \mathfrak{a}_\alpha = \aleph_{\pi(\alpha)} = \aleph_\alpha$, the last of which was established in Theorem 3.

Let now (i)–(iv) be satisfied. We first prove that $\aleph_\alpha$ satisfies condition (iii) (p. 349). Thus let $\mathfrak{n} < \aleph_\alpha$ and let X be a subset of $\boldsymbol{R}$ of power $\mathfrak{n}$. By (iii) $X \in \boldsymbol{R}$, whence by (ii) and (i) $\boldsymbol{P}(X)$ and $\boldsymbol{PP}(X)$ belong to $\boldsymbol{R}$ and $\boldsymbol{PP}(X) \subset \boldsymbol{R}$. Thus $2^{2^{\mathfrak{n}}} \leqslant \aleph_\alpha$ and therefore $2^{\mathfrak{n}} < \aleph_\alpha$.

It remains to prove that $\aleph_\alpha$ satisfies (ii), p. 349.

Let $\mathfrak{f}_\xi < \aleph_\alpha$ for $\xi < \lambda$ where $\lambda < \omega_\alpha$. Using the axiom of choice we correlate with each $\xi < \lambda$ a set $X_\xi \subset \boldsymbol{R}$ of power $\mathfrak{f}_\xi$. Let $H_\xi = \boldsymbol{P}(X_\xi)$ and let Z be a subset of $\boldsymbol{R}$ of power $\overline{\lambda}$. By (ii) $H_\xi \in \boldsymbol{R}$ and by (iii) $Z \in \boldsymbol{R}$. Let φ be a one-one mapping of Z onto $W(\lambda)$. Each function $f\colon Z \to \boldsymbol{R}$ is a subset of $\boldsymbol{R}$ because if $x, y \in \boldsymbol{R}$, then by (iii) $\{x\}$, $\{x, y\}$ and $\langle x, y\rangle$ are elements of $\boldsymbol{R}$. Since the cardinal number of f is $\overline{\overline{Z}} = \overline{\lambda}$, we infer that $f \in \boldsymbol{R}$. Thus the set of all mappings $f\colon Z \to \boldsymbol{R}$ is a subset of $\boldsymbol{R}$. Consider in particular mappings f such that $f(z) \in H_{\varphi(z)}$ for each z in Z. The set of all these mappings has power $\prod_{z\in Z} \mathfrak{h}_{\varphi(z)} = \prod_{\xi<\lambda} \mathfrak{h}_\xi$ where $\mathfrak{h}_\xi = \overline{\overline{H}}_\xi$ for each $\xi < \lambda$. Hence $\prod_{z\in Z} \mathfrak{h}_{\varphi(z)} \leqslant \aleph_\alpha$ and since, by König's theorem (p. 198), $\sum_{\xi<\lambda} \mathfrak{h}_\xi < \prod_{\xi<\lambda} \mathfrak{h}_\xi$, we infer that $\sum_{\xi<\lambda} \mathfrak{f}_\xi < \aleph_\alpha$ which proves the theorem.[1])

Remark. Noting that the cardinal number of $\bigcup \{X_\xi\colon \xi < \lambda\}$ is at most $\sum_{\xi<\lambda} \mathfrak{f}_\xi$ we can also infer from the above proof that the union $\bigcup X_\xi$ of less than $\aleph_\alpha$ sets each of which belongs to $\boldsymbol{R}$ is itself an element of $\boldsymbol{R}$.

°COROLLARY 5: *If* $\overline{\overline{A}} = \aleph_\alpha$ *where* $\aleph_\alpha$ *is inaccessible, then the family* $\boldsymbol{P}_{\aleph_\alpha}(A)$ *of subsets of* A *whose power is* $< \aleph_\alpha$ *has power* $\aleph_\alpha$.

PROOF. Condition (iii) shows that $\boldsymbol{P}_{\aleph_\alpha}(\boldsymbol{R}) \subset \boldsymbol{R}$ and so the statement is true for $A = \boldsymbol{R}$ and hence for each set of the same power as $\boldsymbol{R}$.

[1]) What we did in proving Theorem 4 amounts to an explicit definition of types of relational systems in the model $\mathfrak{R}$. This definition is due, in effect, to Scott [1] who obtained it by a slight transformation of the construction due to Russell and Whitehead [1] of integers within the type-theory.

In recent expositions of set theory, ordinals are usually identified with the von Neumann ordinals N_α and the notion of a relational type is avoided altogether. In such expositions of set theory Theorem 2 is of course superfluous and can be replaced by Theorem 1.

§ 3. A digression on models of Σ°[TR]

A model of the axioms Σ° is a relational system of characteristic $\langle 1, 2\rangle$ in which all axioms of Σ° are true. In other words, it is an ordered triple $\mathfrak{M} = \langle U, S, E\rangle$ consisting of a set U, called the *universe of the model*, a set $S \subset U$, and a binary relation $E \subset U \times U$. (See p. 86.) The elements of S are "sets of the model" and E is the "elementhood relation of the model"; it should be stressed however that the elements of S need not be sets and the relation E need not coincide with the relation $\in$.

In order to verify whether an axiom is or is not true in $\mathfrak{M}$ we replace the atomic formulas "x is a set" and "$x \in y$" (see p. 46) by statements of the form "$x \in S$" and "$\langle x, y\rangle \in E$", let the variables range over U and verify whether the resulting proposition concerning $\mathfrak{M}$ is or is not true.

Similar explanations can be given for the system Σ°[TR]. Since we have in this system one more atomic formula $x\mathrm{TR}y$ (x is the type of the relational system y, see p. 88) we define models of Σ°[TR] as relational systems of the form $\langle U, S, E, T\rangle$ where $S \subset U$, $E \subset U \times U$ and $T \subset U \times U$. Again we require that all axioms of Σ°[TR] be true in the model.

In the present section we shall consider relational systems of the form $\mathfrak{R} = \langle \boldsymbol{R}, \boldsymbol{R}, E_{\boldsymbol{R}}\rangle$ where $\boldsymbol{R}$ is a family with the properties (i)–(iv) given on p. 350 and $E_{\boldsymbol{R}} = \{\langle X, Y\rangle \in \boldsymbol{R} \times \boldsymbol{R}\colon X \in Y\}$.

Relational systems of this kind are called the *natural models of set theory*.

°THEOREM 1: *All axioms of Σ° are true in $\mathfrak{R}$.*

PROOF. Notice that in the relational system $\mathfrak{R}$ the atomic formulas "x is a set" and "$x \in y$" are interpreted as $x \in \boldsymbol{R}$ and $\langle x, y\rangle \in E_{\boldsymbol{R}}$, i.e. $(x \in y) \wedge (x \in \boldsymbol{R}) \wedge (y \in \boldsymbol{R})$. In order to prove the theorem we replace in all axioms of Σ° the atomic formulas by their interpretations and verify that the resulting propositions are theorems of set theory.

1. The axiom of extensionality is true in $\mathfrak{R}$ because if $X, Y \in \boldsymbol{R}$ and if, for each x in $\boldsymbol{R}$, the formulas $\langle x, X\rangle \in E_{\boldsymbol{R}}$ and $\langle x, Y\rangle \in E_{\boldsymbol{R}}$ are equivalent, then $\bigwedge_{x \in \boldsymbol{R}} [x \in X \equiv x \in Y]$ and hence $X \cap \boldsymbol{R} = Y \cap \boldsymbol{R}$. Since, by (i), $X \cap \boldsymbol{R} = X$ and $Y \cap \boldsymbol{R} = Y$, we obtain $X = Y$.

2. The axiom of the empty set is true in $\mathfrak{R}$ because $\emptyset \in \boldsymbol{R}$ by (iii).

3. The axiom of unions is true in view of the final remark of Section 2.

4. The axiom of power sets is true in $\mathfrak{R}$ in view of (ii).

5. The axiom of infinity is true in $\mathfrak{R}$ because the first infinite von Neumann ordinal $N_\omega = \{N_0, N_1, \ldots\}$ is contained in $\boldsymbol{R}$, as we easily prove by induction and hence, in view of (iv) and (iii), $N_\omega \in \boldsymbol{R}$.

6. The axiom of choice is true in $\mathfrak{R}$ because if X is a non-empty family of non-empty disjoint sets and $X \in \boldsymbol{R}$ then any choice set C for X is contained in $\boldsymbol{R}$ in view of (i) and has the same cardinal number as X. Hence $C \subset \boldsymbol{R}$ and $C \in \boldsymbol{R}$.

7. All instances of the axiom scheme of replacement are true in $\mathfrak{R}$.

We consider an instance of the scheme of replacement for a formula Φ with at least two free variables x, y. Assume that for every x in $\boldsymbol{R}$ there is exactly one y in $\boldsymbol{R}$ such that $\Phi(x, y)$ is true in $\mathfrak{R}$. We say that *y corresponds to x*. If there are free variables other than x, y in Φ we assume that they were given fixed values in $\boldsymbol{R}$. If $X \in \boldsymbol{R}$ then the set Y of those y which correspond to elements x of X is contained in $\boldsymbol{R}$ and its cardinal number is at most equal to that of X, hence it is $\subset \boldsymbol{R}$ and so $Y \in \boldsymbol{R}$. This proves that this instance of the scheme of replacement for formula Φ is valid in $\mathfrak{R}$.

In applications of Theorem 1 it is convenient to bear in mind the following facts:

THEOREM 2: *If $\boldsymbol{R}$ satisfies* (i)–(iv) *then each von Neumann ordinal N_ξ whose power is $< \aleph_\alpha$ belongs to $\boldsymbol{R}$; also if $\xi < \omega_\alpha$ then $\boldsymbol{R}_\xi \in \boldsymbol{R}$.*

PROOF. N_0 and R_0 are $= \emptyset$ and so belong to $\boldsymbol{R}$. Let ξ be the least ordinal $< \omega_\alpha$ such that $N_\xi \notin \boldsymbol{R}$. Hence $N_\xi \subset \boldsymbol{R}$ and since $\overline{\overline{N_\xi}} = \overline{\xi} < \aleph_\alpha$, it follows $N_\xi \in \boldsymbol{R}$.

If ξ is the smallest ordinal $< \omega_\alpha$ such that $\boldsymbol{R}_\xi \notin \boldsymbol{R}$ then $\{\boldsymbol{R}_\eta \colon \eta < \xi\} \subset \boldsymbol{R}$ and thus $\{\boldsymbol{R}_\eta \colon \eta < \xi\} \in \boldsymbol{R}$. Since axiom 3 (of unions) is satisfied in $\mathfrak{R}$, it follows that the set $S = \bigcup \{\boldsymbol{R}_\eta \colon \eta < \xi\} \in \boldsymbol{R}$, whence $\boldsymbol{R}_\xi \in \boldsymbol{R}$ because $S = \boldsymbol{R}_\xi$ if ξ is a limit number or $\boldsymbol{P}(S) = \boldsymbol{R}_\xi$ if ξ has a predecessor.

We shall now extend the relational system $\mathfrak{R}$ so as to obtain a model for Σ°[TR].

LEMMA: *Each $X \in \boldsymbol{R}$ has power $< \overline{\overline{\boldsymbol{R}}}$ and there is a ξ such that $\boldsymbol{R}_\xi \in \boldsymbol{R}$ and X has the same power as a subset of $\boldsymbol{R}_\xi$.*

PROOF. Since $\boldsymbol{P}(X) \in \boldsymbol{R}$, we have $\overline{\overline{X}} < 2^{\overline{\overline{X}}} = \overline{\overline{\boldsymbol{P}(X)}} \leqslant \overline{\overline{\boldsymbol{R}}}$. There is an ordinal ξ such that $\overline{\overline{X}} = \overline{\overline{N_\xi}}$, whence $\overline{\xi} < \overline{\overline{\boldsymbol{R}}}$ and $\boldsymbol{R}_\xi \in \boldsymbol{R}$. Since $N_\xi \subset \boldsymbol{R}_\xi$, the lemma is proved.

For each relational system $\langle A, X\rangle \in \boldsymbol{R}$ let $\beta = \beta(A, X)$ be the least ordinal $< \alpha$ such that $\boldsymbol{R}_\beta \in \boldsymbol{R}$ and there are relational systems $\langle B, Y\rangle \in \boldsymbol{R}_\beta$ isomorphic to $\langle A, X\rangle$. The existence of $\langle A, X\rangle$ follows from the lemma above. We denote by $\langle A, X\rangle^*$ the set of relational systems $\langle B, Y\rangle$ which belong to $\boldsymbol{R}_{\beta(A,X)}$ and are isomorphic to $\langle A, X\rangle$.

°THEOREM 3:[1]) *If $\aleph_\alpha$ is inaccessible then all axioms of the system Σ°[TR] are true in the relational system $\mathfrak{R}'_\alpha = \langle \boldsymbol{R}_\alpha, \boldsymbol{R}_\alpha, E_{\boldsymbol{R}_\alpha}, T\rangle$ where T is the binary relation $\{\langle U, V\rangle \in \boldsymbol{R}_\alpha \times \boldsymbol{R}_\alpha$: U is a relational system $\bigwedge (V = U^*)\}$.*

The theorem is proved by showing that axiom VII of relational types is true in $\mathfrak{R}'_\alpha$. We leave the details to the reader.

The following remarks show the importance of Theorems 1 and 3 just proved. For each formula Φ with two free variables x, y such that the formula $\bigwedge\limits_x \bigvee\limits_y ! \, \Phi(x, y)$ is provable in Σ°[TR] we can infer from Theorem 3 that if $x \in R_\alpha$ then there is exactly one y in R_α such that the formula Φ is true in $\mathfrak{R}'_\alpha$ for the elements x, y. Denoting by Φ_α the relation

$$\{\langle x, y\rangle \in R_\alpha \times R_\alpha \colon \Phi(x, y) \text{ is true in } \mathfrak{R}'_\alpha\},$$

we obtain thus the result that Φ_α is an operation and R_α is closed under Φ_α. This allows us to establish the Ψ-inaccessibility of $\aleph_\alpha$ for operations Ψ which can be represented as Φ_α.

°*Examples.* 1. Let $\aleph_\alpha$ be inaccessible and let $\mathfrak{f}$ be a function with domain $W(\lambda)$ where $\lambda < \omega_\alpha$ such that $\mathfrak{f}_\xi$ is a positive cardinal $< \aleph_\alpha$. Then $\prod\limits_{\xi<\lambda} \mathfrak{f}_\xi < \aleph_\alpha$.

PROOF. Let $\varphi(\xi)$ be an ordinal such that $N_{\varphi(\xi)}$ has the power $\mathfrak{f}_\xi$. Hence $N_{\varphi(\xi)} \in \boldsymbol{R}_\alpha$.

Since the family $\boldsymbol{F} = \{N_{\varphi(\xi)} \colon \xi < \lambda\}$ is contained in $\boldsymbol{R}_\alpha$ and has power $< \aleph_\alpha$, we obtain $\boldsymbol{F} \in \boldsymbol{R}_\alpha$. Let $\Phi(x, y)$ be the formula: "y is the

[1]) Theorem 3 is due to Sierpiński and Tarski [1].

set of all functions with the common domain x such that $f(z) \in z$ for each $z \in x$".[1]) Since the sentence $\bigwedge_x \bigvee_y ! \, \Phi(x, y)$ is provable in Σ°[TR], we obtain from Theorem 3 that for every x in $\boldsymbol{R}_\alpha$ there is exactly one y in $\boldsymbol{R}$ such that the relation Φ_χ holds between x and y. Taking $x = \boldsymbol{F}$ we obtain y in $\boldsymbol{R}_\alpha$ which is in relation Φ_α to $\boldsymbol{F}$. Hence x and y satisfy in $\mathfrak{R}'_\alpha$ the formula Φ. It is easy to verify that then $y = \prod_{z \in F} z$ and hence $\overline{\overline{y}} = \prod_{\xi<\lambda} \mathfrak{f}_\xi$. Thus $\overline{\overline{y}} < \overline{\overline{\boldsymbol{R}}}_\alpha$ which proves that

$$\prod \mathfrak{f}_\xi < \aleph_\alpha .$$

2. Let $\mathfrak{f}$ be a sequence as in the above example and let the type λ of $\mathfrak{f}$ be a limit number. We define by induction the sequence $\mathfrak{g}$: $\mathfrak{g}_0 = \mathfrak{f}_0$, $\mathfrak{g}_{\xi+1} = \mathfrak{f}_\xi^{\mathfrak{g}_\xi}$, $\mathfrak{g}_\eta = \sum_{\zeta<\eta} \mathfrak{g}_\zeta$ for limit numbers η. Then $\sum_{\xi<\lambda} \mathfrak{g}_\xi < \aleph_\alpha$.

PROOF. By the theory of inductive definitions (see pp. 233–239) there is a formula Φ such that $\bigwedge_x \bigvee_y ! \, \Phi(x, y)$ is provable in Σ°[TR] and that for each transfinite sequence $x = \{x_\xi\}_{\xi<\lambda}$ of sets the unique y satisfying $\Phi(x, y)$ has power $\sum_{\xi<\lambda} \mathfrak{g}_\xi$ where $\mathfrak{g}_0 = \overline{\overline{x}}_0$, $\mathfrak{g}_{\xi+1} = (\overline{\overline{x}}_\xi)^{\mathfrak{g}_\xi}$, $\mathfrak{g}_\eta = \sum_{\xi<\eta} \mathfrak{g}_\xi$. Taking for x the sequence $\{N_{\varphi(\xi)}\}_{\xi<\lambda}$ where $\overline{\overline{N_{\varphi(\xi)}}} = \mathfrak{f}_\xi$, we easily prove that $x \in \boldsymbol{R}_\alpha$ and the unique y such that x, y satisfy Φ in $\mathfrak{R}'_\alpha$ has power $\sum_{\xi<\lambda} \mathfrak{g}_\xi$. Thus this cardinal number is $< \aleph_\alpha$.

We end this section by a result concerning fields of sets which will be needed later.

We say that a family $\boldsymbol{F}$ of subsets of a fixed set X is $\mathfrak{m}$-*complete* if $\bigcap \boldsymbol{F}' \in \boldsymbol{F}$ and $\bigcup \boldsymbol{F}' \in \boldsymbol{F}$ for every $\boldsymbol{F}' \subset \boldsymbol{F}$ such that $\overline{\overline{\boldsymbol{F}'}} < \mathfrak{m}$.

We can express this definition more briefly when we denote by $\bigcap_\mathfrak{m}(\boldsymbol{F})$ and $\bigcup_\mathfrak{m}(\boldsymbol{F})$ the families of all sets $\bigcap(\boldsymbol{F}')$ and $\bigcup(\boldsymbol{F}')$ respectively, where $\boldsymbol{F}'$ ranges over subfamilies of $\boldsymbol{F}$ of power $< \mathfrak{m}$. Then $\boldsymbol{F}$ is $\mathfrak{m}$-complete if and only if $\bigcup_\mathfrak{m}(\boldsymbol{F}) = \bigcap_\mathfrak{m}(\boldsymbol{F}) = \boldsymbol{F}$.

If $\mathfrak{m}$ is infinite then obviously $\bigcup_\mathfrak{m}\bigcup_\mathfrak{m}(\boldsymbol{F}) = \bigcup_\mathfrak{m}(\boldsymbol{F})$ and $\bigcap_\mathfrak{m}\bigcap_\mathfrak{m}(\boldsymbol{F}) = \bigcap_\mathfrak{m}(\boldsymbol{F})$.

°THEOREM 4: *If* $\mathfrak{m}$ *is inaccessible then* $\bigcap_\mathfrak{m}\bigcup_\mathfrak{m}(\boldsymbol{F}) = \bigcup_\mathfrak{m}\bigcap_\mathfrak{m}(\boldsymbol{F})$.

[1]) Strictly speaking, the formula Φ should be written without abbreviations.

PROOF. If $A \in \bigcap_{\mathfrak{m}} \bigcup_{\mathfrak{m}} \boldsymbol{F}$ then $A = \bigcap_{i \in I} \bigcup_{j \in J_i} A_{ij}$ where the powers $\mathfrak{p}$, $\mathfrak{g}_i$ of I, J_i are $< \mathfrak{m}$ and $A_{ij} \in \boldsymbol{F}$. Using the general law of distributivity (see p. 111), we obtain $A = \bigcup_f \bigcap_{i \in I} A_{if(i)}$ where f ranges over $\prod_{i \in I} J_i$. Since the power of this set is $= \prod_{i \in I} \mathfrak{g}_i < \mathfrak{m}$, we obtain $A \in \bigcup_{\mathfrak{m}} \bigcap_{\mathfrak{m}} \boldsymbol{F}$. The converse implication is proved similarly.

THEOREM 5: *If* $\mathfrak{m}$ *is inaccessible*, $\boldsymbol{F}$ *a family of power* $\mathfrak{m}$ *consisting of subsets of a fixed set* X, *then* $\boldsymbol{F}$ *can be extended to an* $\mathfrak{m}$*-complete field of sets* $\boldsymbol{B} \subset \boldsymbol{P}(X)$ *such that the power of* $\boldsymbol{B}$ *is* $\mathfrak{m}$.

PROOF. Let $F_c = \{X - A : A \in \boldsymbol{F}\}$. The family $\boldsymbol{B} = \bigcup_{\mathfrak{m}} \bigcap_{\mathfrak{m}} (\boldsymbol{F} \cup \boldsymbol{F}_c)$ obviously satisfies the equation $\bigcup_{\mathfrak{m}} \boldsymbol{B} = \boldsymbol{B}$. In view of Theorem 4 it also satisfies the equation $\boldsymbol{B}_c = \bigcap_{\mathfrak{m}} \bigcup_{\mathfrak{m}} (\boldsymbol{F} \cup \boldsymbol{F}_c) = \bigcup_{\mathfrak{m}} \bigcap_{\mathfrak{m}} (\boldsymbol{F} \cup \boldsymbol{F}_c) = \boldsymbol{B}$ and hence $\boldsymbol{B}$ is an $\mathfrak{m}$-complete field of sets.

In order to evaluate the cardinal number of $\boldsymbol{B}$ it will be sufficient to show that if $\boldsymbol{G}$ is a family of power $\leqslant \mathfrak{m}$ then so are $\bigcup_{\mathfrak{m}} \boldsymbol{G}$ and $\bigcap_{\mathfrak{m}} \boldsymbol{G}$. But this follows at once from Corollary 2.5 because the set of subfamilies $\boldsymbol{G}'$ of $\boldsymbol{G}$ such that $\overline{\overline{\boldsymbol{G}'}} < \mathfrak{m}$ has power $\mathfrak{m}$.

Exercises

Prove directly (without using theorems of this section) that the following conditions are equivalent:

(a) $\aleph_\alpha$ is inaccessible,

(b) $\aleph_\alpha > \aleph_0$ and if $\lambda < \omega_\alpha$ and $\mathfrak{m}_\xi < \aleph_\alpha$ for $\xi < \lambda$ then $\prod_{\xi < \lambda} \mathfrak{m}_\xi < \aleph_\alpha$,

(c) $\aleph_\alpha > \aleph_0$, $\sum_{\mathfrak{p} < \aleph_\alpha} \aleph_\alpha^{\mathfrak{p}} = \aleph_\alpha$ and $\mathfrak{m} < \aleph_\alpha$ implies $2^{\mathfrak{m}} < \aleph_\alpha$. [Tarski]

§ 4. Higher types of inaccessible numbers[1])

Beginning with this section we shall discuss special types of inaccessible cardinals. Since their existence cannot be established on the

[1]) The "combinatorial" approach to inaccessible cardinals is due to Erdös and Tarski [2]. The basic paper in this field is Keisler–Tarski [1]. See also Drake [1].

Weakly compact cardinals were first defined by Tarski [9]; the name "weakly compact" was chosen because these cardinals play a role in establishing for certain infinitary languages properties similar to the compactness property of the usual first order logic. [Continued on p. 357.]

basis of the axioms Σ°[TR], we shall carry out our discussion under an additional hypothesis stating that inaccessible cardinals exist.

We first remark that for each ordinal ξ there exists at most one sequence S_ξ of type $\xi+1$ with the property (*) that for each $\eta \leqslant \xi$ the ηth term $S_\xi(\eta)$ of S_ξ is the least inaccessible cardinal which is greater than all $S_\xi(\zeta)$, $\zeta < \eta$.

The hypothesis referred to above is that for each ordinal ξ there exists a sequence S_ξ of type $\xi+1$ satisfying (*). We call this hypothesis the *axiom of inaccessible cardinals.*

We denote by θ_ξ the initial ordinal such that $\overline{\theta}_\xi$ is the last term of the sequence S_ξ. Thus $\overline{\theta}_\xi$ is an enumeration of all inaccessible cardinals.

It is obvious that θ_ξ increases with ξ:

$$\xi < \eta \rightarrow \theta_\xi < \theta_\eta .$$

The function θ is not continuous: in general, if λ is a limit ordinal, then $\lim_{\xi<\lambda} \theta_\xi$ is different from θ_λ. E.g. if $\lambda = \omega$ then $\lim_{\xi<\omega} \theta_\xi$ is cofinal with ω whereas θ_ω is not because θ_ω is not cofinal with any number $< \theta_\omega$. As a matter of fact the places where θ is continuous are extremely rare if they exist at all.

THEOREM 1: $\theta_\lambda = \lim_{\xi<\lambda} \theta_\xi$ *if and only if* $\lambda = \theta_\lambda$.

PROOF. If $\lambda < \theta_\lambda$ then $\lim_{\xi<\lambda} \theta_\xi$ is cofinal with λ and so cannot be equal to θ_λ because θ_λ is inaccessible. If $\lambda = \theta_\lambda$ then $\lim_{\xi<\lambda} \theta_\xi = \theta_\lambda$ because $\theta_\xi \geqslant \xi$ and hence $\lim_{\xi<\lambda} \theta_\xi \geqslant \lambda = \theta_\lambda$.

DEFINITION: A cardinal number $\mathfrak{m}$ is 1-*inaccessible* if there is an ordinal λ such that $\lambda = \theta_\lambda$ and $\mathfrak{m} = \overline{\theta}_\lambda$.

Other equivalent characterizations of 1-inaccessible cardinals are given in the next theorem:

THEOREM 2: *The following conditions are equivalent*:

Theorems given in this section are due to Erdös and Tarski [2]. More exact references can be found in Keisler and Tarski [2]. The importance of weakly compact cardinals in meta-mathematics was particularly stressed by Silver [1]. A detailed account of the results of all these authors can be found in Drake [1].

(i_1) $\lambda = \theta_\lambda$;

(ii_1) *There is an increasing sequence* $\{\varphi(\xi)\}_{\xi<\gamma}$ *where* $\gamma \leqslant \theta_\lambda$ *such that* $\theta_\lambda = \lim_{\xi<\gamma} \theta_{\varphi(\xi)}$;

(iii_1) *The set* $\{\theta_\xi: \xi < \lambda\}$ *has the order type* θ_λ.

Proof. The implication (i_1) → (ii_1) results easily from Theorem 1. The implication (iii_1) → (i_1) follows from the fact that the order type of the set $\{\theta_\xi: \xi < \lambda\}$ is λ.

It remains to prove the implication (ii_1) → (iii_1). Let us assume (ii_1). Since θ_λ is regular, we obtain $\gamma = \theta_\lambda$, whence $\gamma \geqslant \lambda$. Since $\theta_\lambda > \theta_{\varphi(\xi)}$ for each $\xi < \gamma$, we obtain $\lambda > \varphi(\xi) \geqslant \xi$ for each $\xi < \gamma$ and hence $\lambda \geqslant \gamma$. Thus we obtain $\lambda = \gamma = \theta_\lambda$ and the theorem is proved.

Higher classes of inaccessible numbers can be defined similarly. We shall limit ourselves to indicating how to pass from the definition of n-inaccessible cardinals to the definition of $(n+1)$-inaccessible cardinals.

First of all we assume that for each ordinal ξ there exists an n-inaccessible cardinal $\overline{\theta_\xi^{(n)}}$ such that the order type of the set $\{\mathfrak{m}$: ($\mathfrak{m}$ is n-inaccessible) $\wedge$ $(\mathfrak{m} < \overline{\theta_\xi^{(n)}})\}$ is ξ. This assumption is similar to the axiom of inaccessible cardinals.

Definition: We call the cardinal $\overline{\theta_\lambda^{(n)}}$ an $(n+1)$-*inaccessible cardinal* if $\lambda = \theta_\lambda^{(n)}$.

It follows from this definition that an $(n+1)$-inaccessible cardinal is an n-inaccessible cardinal but not necessarily conversely.

Repeating *mutatis mutandum* the proof of Theorem 2, we obtain

Theorem 3: *The following three conditions are equivalent*:

(i_{n+1}) $\lambda = \theta_\lambda^{(n)}$;

(ii_{n+1}) *There is an increasing sequence* $\{\varphi(\xi)\}_{\xi<\gamma}$ *where* $\gamma \leqslant \theta_\lambda^{(n)}$ *such that* $\theta_\lambda^{(n)} = \lim_{\xi<\gamma} \theta_{\varphi(\xi)}^{(n)}$;

(iii_{n+1}) *The set* $\{\theta_\xi^{(n)}: \xi < \lambda\}$ *has the order type* $\theta_\lambda^{(n)}$.

We leave the proof of this theorem to the reader.

The hierarchy of n-inaccessible cardinals can be extended letting n be transfinite. For instance a cardinal $\mathfrak{m}$ is called an ω-*inaccessible cardinal* if it is an n-inaccessible cardinal for each $n < \omega$. We shall not pursue this matter here.

To achieve a greater symmetry we can call all inaccessible cardinals the 0-*inaccessible cardinals.*

More interesting than the n-inaccessible cardinals are *hyper-inaccessible cardinals* also called *Mahlo cardinals.*

DEFINITION:[1]) A cardinal $\overline{\theta_\alpha}$ is called a *Mahlo cardinal* if the set $\{\theta_\xi\colon \xi < \alpha\}$ is stationary in $W(\theta_\alpha)$.

The following theorem can be proved by induction:

THEOREM 4: *If θ_α is a Mahlo cardinal then for each $n < \omega$ the set $S_n = \{\theta_\xi^{(n)}\colon \theta_\xi^{(n)} < \theta_\alpha\}$ is stationary in $W(\theta_\alpha)$.*

PROOF. For $n = 0$ the theorem follows from the definition of Mahlo cardinals. Let us assume that it is true for an integer n and let θ_α be a Mahlo cardinal.

We have to prove that for each normal set $X \subset W(\theta_\alpha)$ the intersection $X \cap S_{n+1}$ is non-empty. To establish this we define by transfinite induction a normal function f such that all the elements of $S_n \cap X$ be values of f.

Let $f(0)$ be the least element of $S_n \cap X$, $f(\xi+1) =$ the least element of $S_n \cap X$ greater than $f(\xi)$ and $f(\lambda) = \lim\limits_{\xi<\lambda} f(\xi)$ if λ is a limit number $< \theta_\alpha$. The function f is a normal mapping of $W(\theta_\alpha)$ into $W(\theta_\alpha)$; the values of $f(\lambda)$ for a limit argument need not be inaccessible but $f(\xi+1) \in S_n \cap X$ for each $\xi < \theta_\alpha$.

The set of critical numbers of f being normal in $W(\theta_\alpha)$, there is a critical number $\varkappa$ of f which belongs to the stationary set $\{\theta_\xi\colon \xi < \alpha\}$. Hence $\varkappa = \theta_\varrho$ for a $\varrho < \alpha$ and the order type of the set $\{\theta_\xi^{(n)}\colon \theta_\xi^{(n)} < \theta_\varrho\}$ is θ_ϱ because each number $f(\xi+1)$ where $\xi < \varkappa$ belongs to this set. Thus by (iii_{n+1}) $\theta_\varrho \in S_{n+1}$ and since $\theta_\varrho \in X$, we obtain the desired result.

An immediate corollary to Theorem 4 is

COROLLARY 5: *For each n, a Mahlo cardinal θ_α is an n-inaccessible cardinal.*

[1]) Ideas developed in Section 4 are due essentially to Mahlo [1] although this author considered weakly and not as we do strongly inaccessible cardinals: His cardinal numbers are sometimes called "weakly Mahlo". What we call here "Mahlo cardinals" were originally called "ϱ_0-numbers." See Mahlo [2].

An excellent exposition of the Mahlo's theory and further references can be found in Drake [1].

Corollary 5 shows that the assumption that Mahlo cardinals exist is incomparably stronger than the assumption that n-inaccessible cardinals exist for any fixed n in N. Much stronger results can be proved by similar methods. For instance, one can show that a Mahlo cardinal is ξ-inaccessible for an arbitrary $\xi < \theta_\alpha$.

The concepts introduced above can be further generalized by introducing hyper-hyper-inaccessible cardinals, i.e. such cardinals θ_ξ that the set $\{\theta_\varrho: (\varrho < \xi) \wedge (\theta_\varrho \text{ is a Mahlo cardinal})\}$ is stationary in $W(\theta_\xi)$. Furthermore we can iterate this definition transfinitely many times introducing higher and higher classes of Mahlo cardinals. It is not clear how far these constructions can be extended. The existence of cardinals belonging to these classes but not to previous ones is probably consistent with the axioms $\Sigma°$[TR]. At present it can neither be said how such consistency could be established nor whether the consideration of the high inaccessible cardinals of this type can be of any use in set theory.

§ 5. Weakly compact cardinals

The approach to the study of large cardinals via the notion of accessibility is not the only one possible. A different approach consists of the study of combinatorial properties of cardinals and the search for cardinals which possess these properties. A paradigm of very large cardinals is, of course, $\aleph_0$ which before the creation of set theory was thought to be "the infinite". As an example of "combinatorial" properties of this cardinal we may cite the theorems of Ramsey and König. The study mentioned above centers around the question whether there are cardinals other than $\aleph_0$ which possess these properties and if so how are they interrelated to inaccessible and Mahlo classes.

In the remaining part of this chapter we shall describe some more introductory results of these studies. For a mere complete presentation the reader should consult sources quoted in the notes to this chapter.

We start with a property related to the Ramsey theorem. It follows from this theorem that $\aleph_0 \rightarrow (\aleph_0)_2^2$.

DEFINITION 1: A cardinal $\mathfrak{m}$ is *weakly compact* if $\mathfrak{m} > \aleph_0$ and

$$\mathfrak{m} \rightarrow (\mathfrak{m})_2^2.$$

THEOREM 1: *If* $\mathfrak{m}$ *is weakly compact then* $\mathfrak{m}$ *is inaccessible.*

PROOF. Let $\mathfrak{m} = \overline{\overline{\omega_\alpha}}$. We have to show that ω_α is (i) regular and (ii) strong limit (i.e. $\mathfrak{n} < \mathfrak{m} \to 2^{\mathfrak{n}} < \mathfrak{m}$).

(i) Let us assume that $\omega_\alpha = \lim_{\xi<\beta} \varphi(\xi)$ where $\beta < \omega_\alpha$ and $\varphi \in W(\omega_\alpha)^{W(\beta)}$. Putting $X_\xi = \{\eta\colon \varphi(\xi) \leqslant \eta < \varphi(\xi+1)\}$ we obtain $W(\omega_\alpha) = X = \bigcup_\xi X_\xi$ where the sets X_ξ are disjoint. We define now a function $f\colon [X]^2 \to \{0, 1\}$ which has no homogeneous set of power $\mathfrak{m}$ by putting for any $U \in [X]^2$ $f(U) = 0$ if there is a $\xi < \beta$ such that $U \subset X_\xi$ and $f(U) = 1$ otherwise. Let Y be a homogeneous set for f.

Hence we either have: (1) $f(U) = 0$ for each $U \in [Y]^2$ or (2) $f(U) = 1$ for each $U \in [Y]^2$. If (1) is the case then any two elements of Y belong to one and the same set X_ξ, whence $\overline{\overline{Y}} \leqslant \overline{\overline{\varphi(\xi+1)}} < \overline{\overline{\omega_\alpha}}$. If (2) is the case then no two elements of Y can belong to the same set X_ξ, whence we infer that $\overline{\overline{Y}} \leqslant \overline{\overline{\beta}} < \overline{\overline{\omega_\alpha}}$.

(ii) Assume that $\mathfrak{n} < \mathfrak{m}$ and let $\mathfrak{n} = \overline{\overline{\omega_\beta}}$, $T_{\omega_\beta} = \{0, 1\}^{W(\omega_\beta)}$. We denote by $\prec$ the relation of lexicographical order of T_{ω_β} and assume that there exists a one-one mapping F of $W(\omega_\alpha)$ into T_{ω_β}.

For $\{\xi, \eta\} \in [W(\omega_\alpha)]^2$ we put $f(\{\xi, \eta\}) = 0$ if $\xi \leqslant \eta \equiv F(\xi) \prec F(\eta)$ and $f(\{\xi, \eta\}) = 1$ otherwise. By weak compactness of $\overline{\overline{\omega_\alpha}}$ there exists a homogeneous set for f of power $\aleph_\alpha$, whence it follows that T_{ω_β} contains a subset of power $\overline{\overline{\omega_\alpha}}$ which is either well-ordered by $\prec$ or well-ordered by the relation inverse to $\prec$. This contradicts Corollary 4, p. 321 and so the assumption that $W(\omega_\alpha)$ can be mapped in a one-one way into T_{ω_β} is impossible. Hence $2^{\mathfrak{n}} < \mathfrak{m}$.

It can be shown that weakly compact cardinals are Mahlo cardinals. Even much stronger results can be proved but we shall not give the proofs here.

In the following theorems we shall establish various properties of weakly compact cardinals.

°THEOREM 2: *Each weakly compact cardinal* $\mathfrak{m}$ *has the property*

(A) *If* X *is a linearly ordered set of power* $\mathfrak{m}$ *then* X *contains a subset of power* $\mathfrak{m}$ *which is well-ordered or inversely well-ordered.*

The proof is the same as that of Theorem 4, p. 341; see also part (ii) of Theorem 1.

°THEOREM 3: *If* $\mathfrak{m}$ *is a cardinal with property* (A), *then it has the property*

(B) *Each tree of power* $\mathfrak{m}$ *all of whose levels are of power* $< \mathfrak{m}$ *has a branch of power* $\mathfrak{m}$.

PROOF. Let T be a tree which satisfies the assumptions of (B). According to Theorem 3, p. 319 we extend the ordering relation $\leqslant$ of T to a linear ordering $\prec$ such that each sub-tree $T_x = \{y \in T\colon x \leqslant y\}$ is an interval with respect to $\prec$. By (A) there is a subset A of T of power $\mathfrak{m}$ such that A is either well-ordered by $\prec$ or inversely well-ordered. We shall consider only the former case; the latter can be treated analogously by replacing $\prec$ by the converse ordering $\succ$. Let A° be an initial segment of A of power $\mathfrak{m}$ such that no proper initial segment of A° is of power $\mathfrak{m}$ and put $A_y = \{z \in A^\circ\colon y \prec z\}$; hence the power of A_y is $\mathfrak{m}$ for each y in A°. We claim that the set

$$C = \{x \in T\colon \bigvee_{y \in A} A_y \subset T_x\}$$

is linearly ordered by $\leqslant$ and has power $\mathfrak{m}$. Once this is established the theorem will be proved because C can be extended to a branch.

Let $x', x'' \in C$ and $A_{y'} \subset T_{x'}$, $A_{y''} \subset T_{x''}$. We can assume for instance that $y' \prec y''$ or $y' = y''$, i.e. $A_{y'} \subset A_{y''}$. It follows that $A_{y'} \subset T_{x'} \cap T_{x''}$ and hence x', x'' are comparable under $\leqslant$ because for incomparable x', x'' the intersection $T_{x'} \cap T_{x''}$ is empty.

In order to evaluate the power of C it will be sufficient to show that for each ξ less than the height χ of T there is an element of C which belongs to the ξth level L_ξ of T.

First of all we remark that for each ξ such that $\xi < \chi$ the level L_ξ cannot be empty. Otherwise T would be equal to a union of its levels L_η where η ranges over ordinals $< \xi_0$ and $\xi_0 < \chi$. Hence, by our assumption concerning the cardinals of levels, the power of T would be $< \mathfrak{m}$.

Let ξ be arbitrary such that $\xi < \chi$.

There are less than $\mathfrak{m}$ elements y of A° which lie on levels L_η with $\eta < \xi$. Thus for some y_0 in A° the set A_{y_0} consists exclusively of elements of the levels L_η with $\eta > \xi$. If x ranges over L_ξ then each z in A_{y_0} belongs to one of the sub-trees T_x because if $z \in L_\eta$ with $\eta > \xi$ then z has (exactly one) predecessor $x \in L_\xi$. Thus the set A_{y_0} of power $\mathfrak{m}$ has been represented as a union of less than $\mathfrak{m}$ disjoint sets $A_{y_0} \cap T_x$ and hence there exists $x \in L_\xi$ such that $A_{y_0} \cap T_x$ has power $\mathfrak{m}$. Thus $A_{y_0} \cap T_x$ is cofinal with A_0. Since T_x is an interval (with respect to the ordering $\prec$), it follows that $A_{y_0} \subset T_x$ and hence $x \in C$. Thus C contains an element which lies on the level ξ which proves the theorem.

The next property of weakly compact cardinals is expressed in terms of $\mathfrak{m}$-complete fields of sets (see p. 355).

°THEOREM 4: *If* $\mathfrak{m}$ *is inaccessible and satisfies condition* (B) *then it also satisfies the condition*

(C) *Every* $\mathfrak{m}$-*complete proper ideal* I *in an* $\mathfrak{m}$-*complete field of sets* B *such that* $\overline{\overline{B}} = \mathfrak{m}$ *can be extended to an* $\mathfrak{m}$-*complete maximal ideal.*

PROOF. Let $\mathfrak{m} = \aleph_\alpha$ and let $\{A_\xi\}_{\xi<\omega_\alpha}$ be a transfinite sequence consisting of all the elements of B. We shall write $1 = \bigcup(B)$, $A^\circ = A$ and $A^1 = 1-A$ as in Chapter I, p. 21.

Let T be a sub-tree of the full binary tree of height ω_α (see p. 316) consisting of sequences $f\colon W(\gamma) \to \{0,1\}$ such that $\gamma < \omega_\alpha$ and $1 - \bigcup_{\xi<\gamma} A_\xi^{f(\xi)} \notin I$. The ordering of T is given by the ordinary relation of inclusion.

The ξth level of T consists of sequences $f\colon W(\xi) \to \{0,1\}$ and thus has power $2^{\bar{\xi}} < \mathfrak{m}$ because $\mathfrak{m}$ is inaccessible.

The height of T is ω_α. To see this we remark that

$$1 = \bigcap_{\xi<\gamma} (A_\xi^0 \cup A_\xi^1) = \bigcup_g \bigcap_{\xi<\gamma} A_\xi^{g(\xi)}$$

where g ranges over $\{0,1\}^{W(\gamma)}$. Since this set has power $2^{\bar{\gamma}} < \mathfrak{m}$ and I is proper and $\mathfrak{m}$-complete, it cannot be the case that all the intersections $\bigcap_{\xi<\gamma} A_\xi^{g(\xi)}$ are in I. Thus for at least one g the set $1 - \bigcup_\xi A_\xi^{1-g(\xi)}$ is not in I and we obtain a function $f(\xi) = 1-g(\xi)$ which belongs to the level L_γ.

We apply now the assumption (B) and obtain a branch B of T of power $\mathfrak{m}$. The union $\bigcup(B)$ is a mapping $f\colon W(\omega_\alpha) \to \{0,1\}$ such that each restriction $f|W(\gamma)$, $\gamma < \omega_\alpha$ can be extended to a function which belongs to B.

We shall prove that the family $J = \{A_\xi\colon f(\xi) = 0\}$ is the required extension of I.

First we show that $I \subset J$. From $f(\gamma) = 1$ it follows $1 - A_\gamma \subset \bigcup_{\xi<\gamma} A_\xi^{f(\xi)}$

and hence $1-\bigcup_{\xi<\gamma} A_\xi^{f(\xi)} \subset A_\gamma$. Since the left-hand side is not an element of I, we obtain $A_\gamma \notin I$. Hence $A_\gamma \in I$ implies $f(\gamma) = 0$, i.e. $A_\gamma \in \boldsymbol{J}$.

Next we show that $\boldsymbol{J}$ is an $\mathfrak{m}$-complete ideal.

Let $A_\eta \subset A_\xi$ and $A_\xi \in \boldsymbol{J}$. We have to show $A_\eta \in \boldsymbol{J}$. Since $f(\xi) = 0$, we obtain $A_\eta \subset A_\xi \subset \bigcup_{\zeta<\gamma} A_\zeta^{f(\zeta)}$ for any $\gamma > \max(\xi, \eta)$. Also $A_\eta^{f(\eta)}$ is a subset of $\bigcup_{\zeta<\gamma} A_\zeta^{f(\zeta)}$. If $f(\eta)$ were equal to 1 we would obtain $\bigcup_{\zeta<\gamma} A_\zeta^{f(\zeta)} = 1$ and hence $-\bigcup_{\zeta<\gamma} A_\zeta^{f(\zeta)} \in I$. Since this is impossible, we obtain $f(\eta) = 0$, i.e. $A_\eta \in \boldsymbol{J}$.

Let now $\boldsymbol{F} \subset \boldsymbol{J}$ be a family of power $< \mathfrak{m}$ and $A_\sigma = \bigcup(\boldsymbol{F})$. There exists an ordinal $\gamma > \sigma$ such that each set in $\boldsymbol{F}$ occurs among the first γ terms of the sequence A_ξ. Since $\boldsymbol{F} \subset \boldsymbol{J}$, the union $A_\sigma^{f(\sigma)} \cup \bigcup(\boldsymbol{F})$ is contained in $\bigcup_{\xi<\gamma} A_\xi^{f(\xi)}$. If $f(\sigma)$ were equal to 1 we would obtain $\bigcup_{\xi<\gamma} A_\xi^{f(\xi)} = 1$ which is impossible. This proves $f(\sigma) = 0$ and $A_\sigma \in \boldsymbol{J}$.

Finally, in order to show that $\boldsymbol{J}$ is maximal it is sufficient to notice that if $A_\xi \notin \boldsymbol{J}$ then $f(\xi) = 1$, whence if $1-A_\xi = A_\eta$ we obtain $f(\eta) = 0$ and so $1-A_\xi \in \boldsymbol{J}$. Thus the ideal is prime and hence maximal.

As the final result in this section we prove

°THEOREM 5: *If* $\mathfrak{m}$ *is inaccessible and satisfies condition* (C), *it is weakly compact.*

PROOF. Let A be a set of power $\mathfrak{m}$ and $f\colon [A]^2 \to \{0, 1\}$. For $\varepsilon = 0, 1$ and each a in A denote by $K_\varepsilon(a)$ the set of all x in $A-\{a\}$ for which $f(\{x, a\}) = \varepsilon$. Obviously, $K_0(a) \cup K_1(a) = A-\{a\}$.

Let B be an $\mathfrak{m}$-complete field of subsets of A which has power $\mathfrak{m}$ and contains all singletons $\{a\}$, $a \in A$ and all the sets $K_\varepsilon(a)$ for $a \in A$, $\varepsilon = 0, 1$. The existence of such a field was proved in Section 3 (Theorem 5). The family of sets $X \subset A$ of power $< \mathfrak{m}$ is an $\mathfrak{m}$-complete ideal in B and thus by (C) can be extended to a maximal $\mathfrak{m}$-complete ideal $\boldsymbol{J}$. Note that if $X \notin \boldsymbol{J}$ then $\overline{\overline{X}} = \mathfrak{m}$ because all sets of a smaller cardinality belong to I.

We now repeat with minor changes the proof of Ramsey's theorem (p. 337).

Case I. There is a set $A^\circ \notin \boldsymbol{J}$ such that for each a in A° and each

$B \subset A° - \{a\}$ if $B \notin \boldsymbol{J}$ then $B \cap K_0(a) \in \boldsymbol{J}$ (since $\boldsymbol{J}$ is prime, we can also write this as $B \cap K_1(a) \notin \boldsymbol{J}$).

Let Φ be a choice function for non-empty subsets of A. We define transfinite sequences $\{a_\xi\}$ and $\{B_\xi\}$ by taking $a_0 = \Phi(A°)$, $B_0 = (A° - \{a_0\}) \cap K_1(a)$ and, inductively, $a_\xi = \Phi\,(\bigcap_{\eta<\xi} B_\eta)$, $B_\xi = (\bigcap_{\eta<\xi} B_\eta - \{a_\xi\}) \cap K_1(a)$. We can then show by induction that the sequence B_ξ is decreasing, $a_\xi \in \bigcap_{\eta<\xi} B_\eta - B_\xi$ and, since $\boldsymbol{J}$ is $\mathfrak{m}$-complete, $B_\xi \notin \boldsymbol{J}$ for each ξ. If $\zeta \neq \xi$, for instance $\zeta > \xi$, then $a_\zeta \in B_\xi$ but $a_\xi \notin B_\xi$ and hence $a_\zeta \neq a_\xi$. Moreover, $a_\zeta \in K_1(a_\xi)$, i.e., $f(\{a_\zeta, a_\xi\}) = 1$. Thus the set $\{a_\xi: \xi < \omega_\alpha\}$ is homogeneous for f and has power $\mathfrak{m}$.

Case II. For each set $A° \notin \boldsymbol{J}$ there are a in $A°$, $B \subset A° - \{a\}$ such that $B \cap K_0(a) \notin \boldsymbol{J}$.

Using the axiom of choice, we correlate with each $A° \notin \boldsymbol{J}$ a pair $a = \Phi(A°)$, $B = \Psi(A°)$ with these properties. Starting with $A° = A$ we define sequences $\{B_\xi\}$, $\{a_\xi\}$ as follows:

$$a_0 = \Phi(A), \qquad B_0 = (A - \{a_0\}) \cap K_0(a),$$

$$a_\xi = \Phi(\bigcap_{\eta<\xi} B_\eta - \{a_\eta: \eta < \xi\}),$$

$$B_\xi = \Psi(\bigcap_{\eta<\xi} B_\eta - \{a_\eta: \eta \leqslant \xi\}) \cap K_0(a).$$

We show by induction that $B_\xi \notin \boldsymbol{J}$ for each ξ. For if this is true for all $\eta < \xi$ then $\bigcap_{\eta<\xi} B_\eta \notin \boldsymbol{J}$ because $\boldsymbol{J}$ is maximal and $\mathfrak{m}$-complete and hence the difference $\bigcap_{\eta<\xi} B_\eta - \{a_\eta: \eta < \xi\}$ does not belong to $\boldsymbol{J}$ either because the set $\{a_\eta: \eta < \xi\}$ having the power $< \mathfrak{m}$ belongs to $\boldsymbol{J}$. Since $a_\xi \in \bigcap_{\eta<\xi} B_\eta - \{a_\eta: \eta < \xi\}$, we obtain $a_\eta \neq a_\xi$ for $\eta < \xi$. Now we can show that the set $\{a_\xi: \xi < \omega_\alpha\}$ is homogeneous. For let $\xi < \zeta$; then $a_\zeta \in B_\xi$ because $a_\zeta \in \bigcap_{\eta<\zeta} B_\eta - \{a_\eta: \eta < \zeta) \subset B_\xi$ and hence $a_\zeta \in K_0(a_\xi)$, i.e. $f(\{a_\xi, a_\zeta\}) = 0$.

Theorem 5 is thus proved.

As a corollary from Theorems 2–5 we obtain

°COROLLARY 6: *If* $\mathfrak{m}$ *is inaccessible then each of the properties* (A), (B), (C) *is equivalent to weak compactness of* $\mathfrak{m}$.

§ 6. Measurable cardinals

The notion of a measure was introduced to mathematics early in the present century in connection with problems of the theory of real functions. The theory of measure has lead to some purely set-theoretical problems. In the present and next section we shall present an introduction to these problems, limiting ourselves only to the simplest "classical" properties.[1])

Before defining measures we shall introduce a notation. If $\mathscr{X}$ is a set and $f\colon \mathscr{X} \to \mathscr{E}^+$ a mapping of $\mathscr{X}$ into non-negative real numbers then we denote by $\sum_{x\in\mathscr{X}} f(x)$ the least upper bound of the set consisting of real numbers of the form $\sum_{x\in\mathscr{F}} f(x)$ where $\mathscr{F}$ is a finite subset of $\mathscr{X}$.

In the case where $\mathscr{X} = N$ the sum $\sum_{n\in N} f(n)$ is equal to $\lim_{n\to\infty} s_n$ where $s_n = \sum_{i<n} f(i)$.

DEFINITION 1: Whenever $\mathfrak{q}$ is an infinite cardinal and A a set we call a function $m\colon \boldsymbol{P}(A) \to [0, 1]$ a $\mathfrak{q}$-*additive real-valued measure* on A if the following conditions are satisfied:

(1) $$m(\emptyset) = 0, \quad m(A) = 1,$$

(2) $$m(\{x\}) = 0 \quad \text{for each } x \text{ in } A,$$

(3) $m(\bigcup \mathscr{X}) = \sum_{X\in\mathscr{X}} m(X)$ for each family $\mathscr{X} \subset \boldsymbol{P}(A)$ consisting of mutually disjoint sets and such that $\overline{\overline{\mathscr{X}}} < \mathfrak{q}$.

[1]) The measure problem originated with Lebesgue in 1904. Lebesgue dealt only with σ-additive real valued measures defined on sets of real numbers and required that congruent sets have the same measure. Banach and Kuratowski [1] showed that if the continuum hypothesis is true, then there exists no σ-additive real-valued measure defined for all sets of real numbers even if one does not require that the measure be invariant under isometric transformations. Further results were obtained by Ulam [1] who proved most of the theorems given in the present section. The problem whether there exists a σ-additive two-valued measure defined for all subsets of a set whose power is the first inaccessible cardinal was solved by Tarski [9]. Tarski's paper was based on a new method which we shall present in Section 7. After 1962 the theory of measurable cardinals made a quick progress and led to important and unexpected results. More information about this subject can be found in Drake [1].

We call (3) the *condition of* $\mathfrak{q}$*-additivity of m.*

For $\mathfrak{q} = \aleph_0$ and $\mathfrak{q} = \aleph_1$ we use the terms "finite additivity" and "σ-additivity."

In the case in which the range of m consists of but two numbers 0, 1, we call m a *two-valued* $\mathfrak{q}$*-additive measure.* A two-valued measure is real valued but not necessarily conversely.

In order to simplify the terminology we shall use the expression "$\mathfrak{q}$-$\mathscr{E}$ measure" instead of "real-valued $\mathfrak{q}$-additive measure" and "$\mathfrak{q}$-2 measure" instead of "two-valued $\mathfrak{q}$-additive measure."

It is obvious that if a $\mathfrak{q}$-$\mathscr{E}$ measure ($\mathfrak{q}$-2 measure) exists on a set A of power $\mathfrak{m}$ then the same is true for every set of power $\mathfrak{m}$. This justifies the following

DEFINITION 2: A cardinal $\mathfrak{m}$ is $\mathfrak{q}$-$\mathscr{E}$ *measurable* (or $\mathfrak{q}$-2 *measurable*) if a $\mathfrak{q}$-$\mathscr{E}$ measure (or a $\mathfrak{q}$-2 measure) exists on a set of power $\mathfrak{m}$.

We also use terms "σ-$\mathscr{E}$ measurable" and "σ-2 measurable" with an obvious meaning.

It is also evident that if $\mathfrak{m}$ is $\mathfrak{q}$-$\mathscr{E}$ measurable then so is each cardinal $\mathfrak{n} > \mathfrak{m}$. For if m is a $\mathfrak{q}$-$\mathscr{E}$ measure on a set A and $B \supset A$ then the function m' defined by $m'(Y) = m(A \cap Y)$ is a $\mathfrak{q}$-$\mathscr{E}$ measure on B. A similar remark applies to $\mathfrak{q}$-2 measure.

In the real function theory the problems of measure were concentrated on finding a sufficiently large class of subsets of space for which a measure could be defined so as to satisfy some invariance properties. E.g. in Euclidean space one requires that two congruent measurable sets should have the same measure. In abstract set theory we drop the requirements of invariance but assume from the start that measures be defined for all subsets of the set under consideration. The main problem is concerned with the cardinal number of a set on which a measure exists.

For further reference we notice some obvious properties of measures. Let A be a set and m a $\mathfrak{q}$-$\mathscr{E}$ measure on A.

LEMMA 1: (a) *If* $X \subset Y \subset A$ *then* $m(X) \leqslant m(Y)$;

(b) *If* $X \subset A$ *then* $m(X) + m(A - X) = 1$;

(c) *If* $X, Y \subset A$ *then* $m(X \cup Y) = m(X - Y) + m(Y - X) + m(X \cap Y)$;

(d) *If* $F \subset \boldsymbol{P}(A)$, $\overline{\overline{F}} < \mathfrak{q}$ *and* $m(X) = 0$ *for each* X *in* F *then* $m\left(\bigcup(F)\right) = 0$.

We omit the obvious proof of this lemma.

Another obvious result is given in

LEMMA 2: *If m is a $\mathfrak{q}$-$\mathscr{E}$ measure on A and $B \subset A$ a set of positive measure then the function $m'(X) = m(X)/m(B)$ is a $\mathfrak{q}$-$\mathscr{E}$ measure on B.*

In order to establish a connection of the measures and Boolean algebras, we introduce the notion of an ideal of a measure:

DEFINITION 3: For each m: $\boldsymbol{P}(A) \to [0, 1]$ we put $I_m(A) = \{X \subset A: m(X) = 0\}$.

LEMMA 3: *If m is a $\mathfrak{q}$-$\mathscr{E}$ measure on A then $I_m(A)$ is a $\mathfrak{q}$-additive $\mathfrak{m}$-ideal in $\boldsymbol{P}(A)$. If in addition m is two-valued then $I_m(A)$ is prime.*

PROOF. The first part follows immediately from Lemma 1 (a) and (d). Now let m be two-valued; in order to prove that $I_m(A)$ is prime we have to show that if $X \cap Y \in I_m(A)$ then at least one of the sets X, Y is in $I_m(A)$. Let us assume therefore that $X \notin I_m(A)$. Hence $m(X) \neq 0$, whence $m(A-X) = 0$ and, since $Y-X \subset A-X$, we obtain $m(Y-X) = 0$. From $X \subset X \cup Y$ it follows that $X \cup Y \notin I_m(A)$ and hence $m(X \cup Y) = 1$. Using Lemma 1 (c) and the assumption $m(X \cap Y) = 0$, we obtain $m(X-Y) = 1$, whence $m(A-Y) = 1$ which, by Lemma 1(b), entails $m(Y) = 0$, i.e. $Y \in I_m(A)$.

Lemma 3 allows us to solve completely the problem of finitely additive two-valued measures. To formulate this result we shall denote by Fin(A) the ideal of $\boldsymbol{P}(A)$ consisting of finite subsets of A.

THEOREM 4: *There is a one-one correspondence between finitely additive two-valued measures on an infinite set A and prime ideals of $\boldsymbol{P}(A)$ containing* Fin(A).

PROOF. We saw in Lemma 2 that to each finitely additive two-valued measure m there corresponds a prime ideal $I_m(A) \subset \boldsymbol{P}(A)$ containing Fin(A). Two different measures give rise to different ideals. If I is a prime ideal containing Fin(A) then putting $m(X) = 0$ if $X \in I$ and $m(X) = 1$ if $X \notin I$ we obtain a finitely additive two-valued measure and $I = I_m(A)$.

°COROLLARY: *If $\overline{\overline{A}} = \mathfrak{m}$ then there are $2^{2^{\mathfrak{m}}}$ finitely additive two-valued measures on A* (see p. 297).

The problem of measures with a higher degree of additivity is much more difficult.

We shall evaluate the sizes of the least σ-$\mathscr{E}$ and σ-2 measurable cardinals. We shall denote them by $\mathfrak{m}_{\mathscr{E}}$ and $\mathfrak{m}_2$. Obviously, $\mathfrak{m}_{\mathscr{E}} \leqslant \mathfrak{m}_2$.

THEOREM 5:[1]) *$\mathfrak{m}_{\mathscr{E}}$ and $\mathfrak{m}_2$ are regular and* $> \aleph_0$.

PROOF. If m is a σ-$\mathscr{E}$ measure (or a σ-2 measure) then $m(X) = 0$ for every denumerable set X. Hence both $\mathfrak{m}_{\mathscr{E}}$ and $\mathfrak{m}_2$ are $> \aleph_0$.

Let us assume that $\mathfrak{m}_{\mathscr{E}}$ is singular, i.e. representable as a sum $\sum_{i \in I} \mathfrak{n}_i$ where $\mathfrak{n}_i < \mathfrak{m}_{\mathscr{E}}$ and $\overline{\overline{I}} < \mathfrak{m}_{\mathscr{E}}$.

Let $\overline{\overline{A}} = \mathfrak{m}_{\mathscr{E}}$. We can represent A as $\bigcup \{A_i \colon i \in I\}$ where $\overline{\overline{A_i}} = \mathfrak{n}_i$ for $i \in I$ and the sets A_i, A_j are disjoint for $i \neq j$. Denote by m a σ-$\mathscr{E}$ measure on A and put for $Y \subset I$

$$\overline{m}(Y) = m\Big(\bigcup_{i \in Y} A_i\Big).$$

We easily verify that $\overline{m}$ is a σ-$\mathscr{E}$ measure on I. The verification of conditions (1) and (3) is obvious; verifying (2) we use the assumption $\overline{\overline{A_i}} < \mathfrak{m}_{\mathscr{E}}$ which gives $\overline{m}(\{i\}) = m(A_i) = 0$. Since $\overline{\overline{I}} < \mathfrak{m}_{\mathscr{E}}$, we obtain a contradiction.

For $\mathfrak{m}_2$ the proof is similar.

°THEOREM 6: *$\mathfrak{m}_{\mathscr{E}}$ is weakly inaccessible.*

PROOF. Let $\mathfrak{m}_{\mathscr{E}} = \aleph_\alpha$; it is sufficient to show that α is a limit number. We shall derive a contradiction from the assumption $\alpha = \beta + 1$.

We denote by A a set of power $\aleph_{\beta+1}$ and by m a σ-$\mathscr{E}$ measure on A. In order to simplify the subsequent formulas we shall agree that the letter ξ with or without indices is a variable ranging over ordinals $< \omega_{\beta+1}$ and η, η' are variables ranging over $W(\omega_\beta)$.

°LEMMA: *There is a function*

$$F\colon\ W(\omega_\beta) \times W(\omega_{\beta+1}) \to \boldsymbol{P}(A)$$

with the properties

(i) $F(\eta, \xi_1) \cap F(\eta, \xi_2) = \emptyset$ *if* $\xi_1 \neq \xi_2$,

(ii) $\overline{\overline{A - \bigcup_{\eta} F(\eta, \xi)}} \leqslant \aleph_\beta$.

Remark. We can think of F as a matrix whose elements are sets and which has $\aleph_\beta$ rows each consisting of mutually disjoint sets and

[1]) Theorems 5–10 are due to Ulam [1]. The lemma given here is also due to Ulam; it has numerous applications in general topology. See Kuratowski [1].

$\aleph_{\beta+1}$ columns such that the union of sets in any column differs from A in at most $\aleph_\beta$ elements.

PROOF OF THE LEMMA. For each $\xi \geqslant \omega_\beta$ let Φ_ξ be the family of all one-one mappings of $W(\xi)$ onto $W(\omega_\beta)$. It is obvious that $\Phi_\xi \neq \emptyset$. Let Γ be a choice function for the family consisting of all the Φ_ξ and put $f_\xi = \Gamma(\Phi_\xi)$. Thus if $\xi \geqslant \omega_\beta$ then f_ξ is a one-one mapping of $W(\xi)$ onto $W(\omega_\beta)$.

Put $F(\eta, \xi) = \{\xi' : (\xi < \omega_\beta + \xi') \wedge (f_{\omega_\beta+\xi'}(\xi) = \eta)\}$.

Condition (i) is satisfied because $\xi' \in F(\eta, \xi_1) \cap F(\eta, \xi_2)$ implies $\xi_1 < \omega_\beta + \xi'$, $\xi_2 < \omega_\beta + \xi'$ and $f_{\omega_\beta+\xi'}(\xi_1) = f_{\omega_\beta+\xi'}(\xi_2)$ from which it follows that $\xi_1 = \xi_2$ because $f_{\omega_\beta+\xi'}$ is a one-one mapping.

Condition (ii) is also satisfied because $\xi' \notin \bigcup_\eta F(\eta, \xi)$ implies that for each η either $\xi \geqslant \omega_\beta + \xi'$ or $f_{\omega_\beta+\xi'}(\xi) \neq \eta$. If the inequality $\xi < \omega_\beta + \xi'$ were true, the value of $f_{\omega_\beta+\xi'}(\xi)$ would be an ordinal $< \omega_\beta$ and hence the condition $\xi' \notin \bigcup_\eta F(\eta, \xi)$ would not be satisfied. Hence $\xi \geqslant \omega_\beta + \xi'$ and $\xi' < \xi$. Thus $A - \bigcup_\eta F(\eta, \xi) \subset W(\xi)$.

Having proved the lemma we now distinguish two cases:

Case 1. In each representation of A as a union of $\aleph_\beta$ sets one at least of these sets has a positive measure.

In this case each column of the matrix F has at least one element with a positive measure. To see this we notice that $A = \bigcup_\eta F(\eta, \xi) \cup [A - \bigcup_\eta F(\eta, \xi)]$ and the difference $A - \bigcup_\eta F(\eta, \xi)$ has the measure 0 because of Lemma 2 and our assumption that there is no σ additive real valued measure on a set of power $< \aleph_{\beta+1}$.

Hence for each ξ there is a least $\eta(\xi)$ such that $m(F(\eta(\xi), \xi)) > 0$. Since $\eta(\xi) < \omega_\beta$ and ξ ranges over $W(\omega_{\beta+1})$, there exists an ordinal $\eta_0 < \omega_\beta$ such that for a non-denumerable set $\Xi \subset W(\omega_{\beta+1})$ the inequality $m(F(\eta_0, \xi)) > 0$ holds. The sequence $\{m(F(\eta_0, \xi))\}_{\xi\in\Xi}$ is thus an increasing non-denumerable set of real numbers which is impossible.

Case 2. There is a representation $A = \bigcup_\eta A_\eta$ where $m(A_\eta) = 0$ for each η. Putting $A'_\eta = A_\eta - \bigcup_{\zeta<\eta} A_\zeta$, we obtain

$$A = \bigcup_{\eta} A'_\eta$$

where $A'_\eta \cap A'_{\eta'} = \emptyset$ for $\eta \neq \eta'$ and $m(A'_\eta) = 0$ for each η.

Putting $\bar{m}(Y) = m(\bigcup_{\eta \in Y} A'_\eta)$ we convince ourselves easily that $\bar{m}$ is a σ-$\mathscr{E}$ measure on $W(\omega_\beta)$. Since this contradicts the definition of $\aleph_\alpha$ as the least σ-$\mathscr{E}$ measurable cardinal, Theorem 6 is proved.

For the cardinal $\mathfrak{m}_2$ we have a stronger result which we shall give in Theorem 8 below. Before we formulate it we must first prove an auxiliary but important

THEOREM 7: *If* $\mathfrak{n}$ *is a* $\mathfrak{q}$-2 *measurable cardinal and* $\mathfrak{p}$ *is the least* $\mathfrak{q}$-2 *measurable cardinal, then* $\mathfrak{n}$ *is* $\mathfrak{p}$-2 *measurable.*

PROOF. Let m be a $\mathfrak{q}$-2 measure on a set A of power $\mathfrak{n}$ and let $\mathfrak{h}$ be the least cardinal such that m is not $\mathfrak{h}$ additive. Obviously, $\mathfrak{h} \leqslant \mathfrak{n}$ because A is the union of $\mathfrak{n}$ sets of the form $\{x\}$ and $m(A) = 1$, $m(\{x\}) = 0$. On the other hand, $\mathfrak{q} < \mathfrak{h}$.

From the definition of $\mathfrak{h}$ it follows that there is a family $F \subset P(A)$ of power $< \mathfrak{h}$ consisting of mutually disjoint sets such that $m(X) = 0$ for each $X \in F$ and satisfying the equation $m[\bigcup(F)] = 1$. We define now a measure $\bar{m}$ on F putting for $Y \subset F$

$$\bar{m}(Y) = m\big(\bigcup Y\big).$$

We easily verify that $\bar{m}$ is a $\mathfrak{q}$-2 measure on F and hence $\mathfrak{p} \leqslant \bar{\bar{F}} < \mathfrak{h}$ which proves the theorem because m is $\mathfrak{w}$ additive for any $\mathfrak{w} < \mathfrak{h}$.

COROLLARY: $\mathfrak{m}_2$ *is* $\mathfrak{m}_2$-2 *measurable.*

°THEOREM 8: $\mathfrak{m}_2$ *is strongly inaccessible.*

In view of Theorem 6 it is sufficient to show that $\mathfrak{m} < \mathfrak{m}_2$ implies $2^{\mathfrak{m}} < \mathfrak{m}_2$. We assume the contrary and put $\mathfrak{m} = \aleph_\alpha$. Hence there is a σ-additive and therefore, by Theorem 7, an $\mathfrak{m}_2$-additive two-valued measure on every set of power $2^{\aleph_\alpha}$. In particular we can take the set T_{ω_α}. Let m be an $\mathfrak{m}_2$-2 measure on T_{ω_α}.

If $\varphi \in T_\xi$ where $\xi < \omega_\alpha$ then we put $T(\varphi) = \{f \in T_{\omega_\alpha} \colon \varphi \subset f\}$ and $\varphi\varepsilon = \varphi \cup \{\langle \xi, \varepsilon \rangle\}$ for $\varepsilon = 0, 1$. It is obvious that $T(\varphi) = T(\varphi 0) \cup \cup T(\varphi 1)$ and $T(\varphi 0) \cap T(\varphi 1) = \emptyset$. Hence if $m\big(T(\varphi)\big) = 1$ then exactly one of the sets $T(\varphi 0)$, $T(\varphi 1)$ has measure 1. Let $\varepsilon(\varphi)$ be 0 or 1 according as whether $m(T(\varphi 0)) = 1$ or $m(T(\varphi 1)) = 1$. If $m(T(\varphi)) = 0$ let $\varepsilon(\varphi) = 0$.

Using transfinite induction we define now a function $g \in T_{\omega_\alpha}$ which satisfies the recursive equation

$$g(\xi) = \varepsilon(g|W(\xi))$$

for each $\xi < \omega_\alpha$. We can then prove inductively that $m(T(g|W(\xi))) = 1$ for each $\xi < \omega_\alpha$. For $\xi = 0$ the equation is true because $T(\emptyset) = T_{\omega_\alpha}$ and if the equation holds for all $\xi < \xi_0$ then it holds for ξ_0. If ξ_0 is a limit number then $T(g|W(\xi_0)) = \bigcap_{\xi<\xi_0} T(g|W(\xi))$ and since the measure m is $\aleph_\alpha$ additive, we obtain $m(T(g|W(\xi_0))) = 1$. If $\xi_0 = \xi_1 + 1$ then by assumption $m(T(g|W(\xi_1))) = 1$, whence $\varepsilon(g|W(\xi_1)) = 0$ or 1 according as $m(T(g|W(\xi_1)0)) = 1$ or $m(T(g|W(\xi_1)1)) = 1$, whence we infer that $T(g|W(\xi_0)) = T(g|W(\xi_1)g(\xi_1))$ has measure 1.

Since m is $\mathfrak{m}_2$ additive and $\aleph_\alpha < \mathfrak{m}_2$, the intersection $\bigcap_{\xi<\omega_\alpha} T(g|W(\xi))$ has measure 1. It is easy to see that this intersection consists of but one element g. For

$$f \in \bigcap_{\xi<\omega_\alpha} T(g|W(\xi)) \equiv \bigwedge_{\xi<\omega_\alpha} (f|W(\xi) = g|W(\xi)) \equiv \bigwedge_{\xi<\omega_\alpha} [f(\xi) = g(\xi)] \equiv f = g.$$

We obtain thus the result that a set consisting of but one element g of T_{ω_α} has measure 1. Since this is a contradiction, the theorem is proved.

Theorems 6 and 8 are not true for arbitrary σ-$\mathscr{E}$ or σ-2 measurable cardinals because every cardinal greater than $\mathfrak{m}_{\mathscr{E}}$ is σ-$\mathscr{E}$ measurable and every cardinal greater than $\mathfrak{m}_2$ is σ-2 measurable. However it is easy to see that without modifying the proofs of Theorems 6 and 8 we can establish the following result:

°THEOREM 9: *If $\mathfrak{m}$ is $\mathfrak{m}$-$\mathscr{E}$ measurable then $\mathfrak{m}$ is weakly inaccessible; if $\mathfrak{m}$ is $\mathfrak{m}$-2 measurable then $\mathfrak{m}$ is strongly inaccessible.*

The next theorem was discovered by Ulam.

°THEOREM 10: *$\mathfrak{m}_{\mathscr{E}}$ is either $\leqslant 2^{\aleph_0}$ or $\mathfrak{m}_{\mathscr{E}} = \mathfrak{m}_2$.*

PROOF. Let A be a set of power $\mathfrak{m}_{\mathscr{E}}$ and m a σ-additive real-valued measure on A. Let us call a set $X \subset A$ an *atom* if $m(X) > 0$ and for each decomposition $X = X_1 \cup X_2$ into disjoint sets either $m(X_1) = m(X)$ or $m(X_2) = m(X)$.

Case I. There is an atom X. In this case we put $m(Y) = m(X \cap Y)$: $m(X)$ and obtain a σ-additive two-valued measure on A.

Case II. There are no atoms. In this case we shall show that there exists a σ-additive real-valued measure on T_ω. Since $\mathfrak{m}_\mathscr{E}$ is defined as the least cardinal for which there exists a σ-additive real-valued measure, it will then follow that $\mathfrak{m}_\mathscr{E} \leqslant \overline{\overline{T}}_\omega = 2^{\aleph_0}$.

°LEMMA: *If* $X \subset A$, $m(X) > 0$ *then there is a decomposition* $X = X_1 \cup X_2$ *into disjoint sets such that* $m(X_1) \geqslant \frac{1}{3}m(X)$ *and* $m(X_2) \geqslant \frac{1}{3}m(X)$.

We prove this by contradiction. We assume therefore that for each pair of disjoint sets X_1, X_2 such that $X = X_1 \cup X_2$ and $m(X_1) \leqslant m(X_2)$ the inequality $m(X_1) < \frac{1}{3}m(X)$ is true. Let us fix a function which correlates with each set $Z \subset A$ of positive measure a pair $\langle Z', Z''\rangle$ of disjoint subsets of Z such that $Z = Z' \cup Z''$ and $0 < m(Z') \leqslant m(Z'')$; if $m(Z) = 0$ let $Z' = \emptyset$, $Z'' = Z$.

We use now the theorem on definitions by transfinite induction and obtain a sequence $\{V_\xi\}_{\xi<\omega_1}$ such that $V_0 = X'$ and for each positive ξ

$$V_\xi = (A - \bigcup_{\eta<\xi} V_\eta)'.$$

It follows from the definition of the set Z' that $m(V_\xi) < \frac{1}{3}m(X)$.

Let ξ_0 be the least ordinal such that $V_{\xi_0} = \emptyset$. Such an ordinal must be $< \omega_1$ because the sets $V_\xi \neq \emptyset$ are disjoint and have positive measures and thus there can be only denumerably many of them. Rearranging the sequence $\{V_\xi\}_{\xi<\xi_0}$, we obtain a sequence $\{Y_n\}_{n\in N}$ of disjoint sets of positive measures. From the definition of ξ_0 we infer that $m(A - \bigcup_n Y_n) = 0$, whence $\sum_{n\in N} m(Y_n) = 1$. Let s be the least integer such that $\sum_{n=0}^{s} m(Y_n) > \frac{1}{3}m(X)$. Since, by assumption, $m(Y_0) < \frac{1}{3}m(X)$, we have $s > 0$ and $\sum_{n=0}^{s-1} m(Y_n) \leqslant \frac{1}{3}m(X)$.

Put $P = \bigcup_{i<s} Y_i$, $Q = Y_s$, $R = X-(P\cup Q)$ and $p = m(P)$, $q = m(Q)$, $r = m(R)$ and $a = m(X)$. Hence $p \leqslant a/3$, $q \leqslant a/3$ and $p+q+r = a$. It follows that $p+q \leqslant (2/3)a$, $r \geqslant a/3$ and hence $p+q \geqslant a/3$ and $r \geqslant a/3$. Thus the decomposition $X = (P\cup Q)\cup R$ has the properties $(P\cup Q)\cap R = \emptyset$, $m(P\cup Q) \geqslant a/3$, $m(R) \geqslant a/3$ contradicting the assumption that no such decomposition exists.

Using the lemma and the axiom of choice, we infer that there exists a function $F\colon \boldsymbol{P}(A) \to \boldsymbol{P}(A)\times \boldsymbol{P}(A)$ such that if $X \subset A$ and X has a positive measure then F correlates with X a pair $\langle X^*, X^{**}\rangle$ satisfying the conditions

$$X = X^* \cup X^{**}, \quad X^*\cap X^{**} = \emptyset, \quad \tfrac{1}{3}m(X) \leqslant m(X^*) \leqslant m(X^{**}).$$

We shall now define a real-valued σ-additive measure on T_ω. To reach this result we shall correlate a set $X(s)$ with every finite sequence $s \in \bigcup_{n\in N} \{0, 1\}^n = D_\omega$.

If $s = \emptyset$, then we put $X(s) = A$. If $s \in \{0, 1\}^{n+1}$ then we put $X(s) = X(s|n)^*$ if $s_n = 0$ and $X(s) = X(s|n)^{**}$ if $s_n = 1$. By induction we prove that $X(s)$ has the measure $\leqslant\left(\frac{2}{3}\right)^n m(X)$ for each s of length n.

We can now prove that the function

$$\overline{m}(Y) = m\Big(\bigcup_{s\in Y}\bigcap_{n\in N} X(s|n)\Big)$$

is a real-valued σ-additive measure on T_ω.

Since any union $\bigcup_{s\in\Phi}$ taken over the dense set is dense, we obtain $\overline{m}(\emptyset) = 0$. If $Y = T_\omega$ then $\bigcup_{s\in Y}\bigcap_{n\in N} X(s|n) = A$. To see this we notice that for every $a \in A$ either $a \in X(\langle 0\rangle)$ or $a \in X(\langle 1\rangle)$ where $\langle 0\rangle$ and $\langle 1\rangle$ are sequences of length 1 whose unique terms are respectively 0 and 1. If $a \in X(\langle \varepsilon_0\rangle)$ then either $a \in X(\langle \varepsilon_0, 0\rangle)$ or $a \in X(\langle \varepsilon_0, 1\rangle)$ where $\langle \varepsilon_0, \varepsilon_1\rangle$ is a two-term sequence. Continuing in this way we obtain $a \in X(\langle \varepsilon_0, \varepsilon_1, \ldots, \varepsilon_{n-1}\rangle)$ and hence there exists a sequence $s \in T_\omega$ such that $a \in X(s|n)$ for each n. It follows now that $\overline{m}(T_\omega) = m(A) = 1$.

If Y has just one element s and k is any integer then $\overline{m}(Y)$

$= m(\bigcap X(s|n)) \leqslant m(\bigcap_{n<k} X(s|n)) \leqslant (\frac{2}{3})^k$. Since this inequality holds for every k, we obtain $\overline{m}(Y) = 0$.

Finally, let $Y = \bigcup_{j \in N} Y_j$, where the sets Y_j are disjoint. We easily see that

$$\bigcup_{s \in Y} \bigcap_n X(s|n) = \bigcup_{j \in N} \Big[\bigcup_{s \in Y_j} \bigcap_n X(s|n)\Big].$$

If $j \neq k$, then $\bigcup_{s \in Y_j} \bigcap_n X(s|n) \cap \bigcup_{s \in Y_k} \bigcap_n X(s|n) = \emptyset$ for the intersection on the left-hand side is equal to $\bigcup_{s' \in Y_j} \bigcup_{s'' \in Y_k} \bigcap_n [X(s'|n) \cap X(s''|n)]$ and for $s' \neq s''$ the intersection $\bigcap_n [X(s'|n) \cap X(s''|n)]$ is empty. Hence we obtain $\overline{m}(Y) = m(\bigcup_{j \in N} [\bigcup_{s \in Y_j} \bigcap_n X(s|n)]) = \sum_{j \in N} m(\bigcup_{s \in Y_j} \bigcap_n X(s|n)) = \sum_{j \in N} \overline{m}(Y_j)$. Theorem 10 is thus proved.

°COROLLARY:[1]) *If there is no weakly inaccessible cardinal* $\leqslant 2^{\aleph_0}$, *then* $2^{\aleph_0}$ *is not* σ-$\mathscr{E}$ *measurable.*

This corollary can be considerably strengthened. For instance, if we assume that the set of weakly inaccessible cardinals $\leqslant 2^{\aleph_0}$ has power less than $2^{\aleph_0}$, then $2^{\aleph_0}$ is not σ-$\mathscr{E}$ measurable and still stronger results are known. However the simple question whether $2^{\aleph_0}$ is or is not σ-$\mathscr{E}$ measurable cannot be decided on the basis of the axioms Σ°[TR] alone.

§ 7. Measurable cardinals and reduced products[2])

In this section we shall deal with cardinals $\mathfrak{m}$ which are $\mathfrak{m}$-2-measurable. We shall call them *measurable.*

[1]) A result much stronger than the corollary given on this page has been proved by Solovay [2].

[2]) The present section contains a brief sketch of the method invented by Tarski in [9] and later reformulated by Keisler and Tarski in [1] which has permitted these authors to establish several strong results about measurable cardinals and also about other types of large cardinals. The use of reduced products is not essential: in [9] Tarski used instead a kind of compactness property for certain infinitary languages.

For a detailed presentation of the methods and results see Drake [1].

As we proved in Section 6, $\mathfrak{m}$ is measurable if and only if for each set A of power $\mathfrak{m}$ the algebra $\boldsymbol{P}(A)$ contains an $\mathfrak{m}$-additive prime ideal I containing all sets $\{a\}$, $a \in A$.

We shall sketch a method of obtaining results about measurable cardinals by using the reduced products which we introduced in Chapter IV, Section 9.

We shall need two auxiliary facts:

°THEOREM 1: *Each measurable cardinal is inaccessible.*

°THEOREM 2: *Each measurable cardinal is weakly compact.*

Theorem 1 follows from Theorem 9, p. 372. In order to prove Theorem 2 we repeat the proof of Theorem 5, Section 5 (p. 364) replacing the ideal J by the ideal of sets of measure 0.

Let us now describe the construction of reduced products in a way adapted to the present situation.

Until the end of this chapter let $\mathfrak{m} = \aleph_\alpha$ be a measurable cardinal, and $U = U_\alpha$ the set of all transfinite sequences $\varphi\colon W(\alpha) \to W(\alpha)$. We shall simplify formulas by omitting the index α whenever possible.

As we proved in Chapter IV, p. 141, the relation

$$\varphi \sim \psi \equiv \{\xi\colon \varphi(\xi) = \psi(\xi)\} \notin I$$

is an equivalence relation on U. We denote by $\tilde{\varphi}$ the equivalence class containing φ (or, as we shall say, generated by φ) and denote by $\tilde{U}$ the set of these equivalence classes. On $\tilde{U}$ we define a relation $\tilde{\leqslant}$ by the formula

$$(*) \qquad \tilde{\varphi} \mathrel{\tilde{\leqslant}} \tilde{\psi} \equiv \{\xi\colon \varphi(\xi) \leqslant \psi(\xi)\} \notin I$$

If we agree to say that $\Phi(\xi)$ is valid almost everywhere (or for almost all ξ) if $\{\xi\colon \Phi(\xi)\} \notin I$, then the relation $(*)$ can be expressed thus: φ is almost everywhere smaller than or equal to ψ.

It can be verified immediately that if $\varphi \sim \varphi'$ and $\psi \sim \psi'$ then the sets $\{\xi\colon \varphi(\xi) \leqslant \psi(\xi)\}$ and $\{\xi\colon \varphi'(\xi) \leqslant \psi'(\xi)\}$ either both belong to I or none of them belongs to I. Thus the previous definition is correct.

The relational system $(\tilde{U}, \tilde{\leqslant})$ is the reduced product of the system $(W(\alpha), \leqslant_\alpha)$, where $\leqslant_\alpha$ is the relation $\leqslant$ restricted to $W(\alpha)$.

In some of the subsequent proofs we shall use a more general construction in which the simple relational system $(W(\alpha), \leqslant_\alpha)$ will be replaced by a family of relational structures $\mathfrak{B}_\xi = (W(\varphi_0(\xi)), \leqslant_{\varphi_0(\xi)}, A_\xi, B_\xi, \ldots)$ with the same characteristic (see p. 86) where ξ ranges over $W(\alpha)$ and φ_0 is a mapping φ_0: $W(\alpha) \to W(\alpha)$.

The reduced product of the systems $\mathfrak{B}_\xi$ is the relational system $(\tilde{V}, \tilde{\leqslant}, \tilde{A}, \tilde{B}, \ldots)$, where $\tilde{V}$ consists of elements $\tilde{\varphi} \in \tilde{U}$ generated by functions φ: $W(\alpha) \to W(\alpha)$ satisfying the inequalities $\varphi(\xi) < \varphi_0(\xi)$ for each $\xi < \alpha$. The relation $\tilde{A}$ is defined by the equivalence

$$(**) \qquad \tilde{A}(\tilde{\varphi}, \tilde{\psi}, \ldots) \equiv \{\xi\colon A_\xi(\varphi(\xi), \psi(\xi), \ldots)\} \notin I;$$

the definitions of $\tilde{B}$ and the other relations are similar. We prove as in Chapter IV, p. 141 that these definitions are correct in the sense that the right-hand side of (**) depends on $\tilde{\varphi}, \tilde{\psi}, \ldots$ and not on the particular functions $\varphi, \psi, \ldots$ selected from the equivalence classes $\tilde{\varphi}, \tilde{\psi}, \ldots$

It should be borne in mind that all the notions defined above depend on α and I. In cases where it is important to stress this dependence we shall add suitable indices to symbols $\tilde{\varphi}$, $\sim$, etc.

The notation which we have just explained will be used throughout the rest of the chapter.

It follows from Łoś's theorem (p. 143) that the set $\tilde{U}$ is linearly ordered by the relation $\tilde{\leqslant}$. Moreover the following theorem holds.

°THEOREM 3: *The relation* $\tilde{\leqslant}$ *well orders* $\tilde{U}$.

In the proof we use only the fact that the union of countably many elements of I belongs to I.

Let us assume that $\{\tilde{\varphi}_n\}$ is a decreasing sequence of elements of $\tilde{U}$. Hence $\tilde{\varphi}_{n+1} \tilde{\gtrless} \varphi_n$ for each n and therefore the set $\{\xi\colon \varphi_{n+1}(\xi) < \varphi_n(\xi)\}$ does not belong to I. The intersection of these sets for $n = 0, 1, 2, \ldots$ does not belong to I either and hence is not void. For any ξ in this intersection we have $\varphi_{n+1}(\xi) < \varphi_n(\xi)$ for each n. This however is impossible because $W(\alpha)$ is well ordered by $\leqslant$.

For $\gamma < \alpha$ we shall denote by c_γ the constant function $c_\gamma(\xi) = \gamma$ and by ι the diagonal function $\iota(\xi) = \xi$. Obviously, c_γ and ι belong to U.

LEMMA 1: *For each* $\varphi \in U$ *and* $\gamma < \alpha$

$$\tilde{\varphi} \tilde{\leqslant} \tilde{c}_\gamma \text{ if and only if } \tilde{\varphi} = \tilde{c}_\delta \text{ for some } \delta \leqslant \gamma.$$

PROOF. The verification that $\tilde{c}_\delta \lessapprox \tilde{c}_\gamma$ for $\delta \leqslant \gamma$ is left to the reader. If $\tilde{\varphi} \gtrsim \tilde{c}_\gamma$ then the set $\{\xi: \varphi(\xi) < \gamma\} = \bigcup_{\delta<\gamma} \{\varphi(\xi) = \delta\} = \bigcup_{\delta<\gamma} \varphi^{-1}(\delta)$ does not belong to I. Since I is $\mathfrak{m}$-complete, one at least (and, in fact, exactly one) set $\varphi^{-1}(\delta)$ does not belong to I, whence we obtain $\varphi(\xi) = \delta$ for almost all ξ, i.e. $\tilde{\varphi} = \tilde{c}_\delta$.

We shall call the elements $\tilde{c}_\gamma$ the *constants* of $\tilde{U}$ and denote by $\tilde{C}$ (or $\tilde{C}_\alpha$) their set.

LEMMA 2: *$\tilde{C}$ is ordered by $\lessapprox$ in type α.*

This follows immediately from Lemma 1.

LEMMA 3: (i) *There exists a function φ_0: $W(\alpha) \to W(\alpha)$ such that the predecessors of $\tilde{\varphi}_0$ in $\tilde{U}$ coincide with the constants; thus the order type of $\{\tilde{\varphi} \in \tilde{U}: \tilde{\varphi} \lesssim \tilde{\varphi}_0\}$ is α.*

(ii) *Moreover, given $\beta < \alpha$, we can select φ_0 such that $\varphi_0(\xi) > \beta$ for each $\xi < \alpha$.*

PROOF. The diagonal function ι has the property that $\iota(\xi) > \gamma$ for $\xi \notin W(\gamma+1)$. Since $W(\gamma+1)$ has $\bar{\gamma}$ elements, it belongs to I because I is $\mathfrak{m}$-complete. Hence $\{\xi: c_\gamma(\xi) < \iota(\xi)\}$ is not in I, i.e. $\tilde{c}_\gamma < \tilde{\iota}$ for each $\gamma < \alpha$. Thus there are functions ψ such that $\tilde{\psi}$ succeeds in $\tilde{U}$ all the elements $\tilde{c}_\gamma$ $(\gamma < \alpha)$ and in order to prove (i) it is sufficient to take as φ_0 the first ψ with this property.

PROOF OF (ii). Let φ_0 be a function satisfying (i) and $\beta < \alpha$ any ordinal. Since $\tilde{\varphi}_0$ is not a constant, all the sets $\varphi^{-1}(\xi)$ where $\xi \leqslant \beta$ belong to I. Therefore if φ_0' differs from φ_0 at most at points belonging to $\bigcup_{\xi\leqslant\beta} \varphi^{-1}(\xi)$, then $\varphi_0' \sim \varphi_0$ and thus $\tilde{\varphi}_0' = \tilde{\varphi}$. Thus we can change the values of φ_0 without affecting $\tilde{\varphi}_0$ and so that $\varphi_0(\xi)$ is never $\leqslant \beta$.

DEFINITION: We shall say that a function φ_0 *represents* α if it satisfies Lemma 3(i).

LEMMA 4: *If φ_0 represents α then $\varphi_0(\xi)$ is an initial ordinal for almost all ξ.*

PROOF. Put $\overline{\varphi_0(\xi)} = \aleph_{\varrho(\xi)}$. Obviously, $\varrho(\xi) \leqslant \omega_{\varrho(\xi)} \leqslant \varphi_0(\xi) < \omega_{\varrho(\xi)+1}$ for all $\xi < \alpha$. If the set $\{\xi: \varrho(\xi) < \varphi_0(\xi)\}$ were not in I, we would infer $\tilde{\varrho} \lesssim \tilde{\varphi}_0$ and hence $\tilde{\varrho} = \tilde{c}_\gamma$ for some $\gamma < \alpha$. Thus we would infer $\varrho(\xi) = \gamma$ for almost all ξ, whence $\overline{\varphi_0(\xi)} = \aleph_\gamma$ and $\varphi_0(\xi) < \omega_{\gamma+1}$. Since $\omega_{g+1} < \alpha$, it would follow that φ_0 is a constant. Since this contradicts the definition of φ_0 we infer that $\varphi_0(\xi) = \varrho(\xi)$ almost everywhere and hence $\varphi_0(\xi)$ is an initial ordinal almost everywhere.

Using these lemmas, we can now prove

°THEOREM 4: *If* $\mathfrak{m} = \aleph_\alpha$ *is measurable, then there are* $\aleph_\alpha$ *weakly compact cardinals* $< \mathfrak{m}$.

PROOF. We assume the contrary and derive a contradiction. Let β be an ordinal $< \alpha$ such that for no initial ordinal γ between β and α the cardinal $\bar{\gamma}$ is weakly compact. We select a function $\varphi_0 \in U$ representing α such that $\varphi_0(\xi) > \beta$ for each ξ.

The main line of the argument will be as follows. We shall consider relational systems $\mathfrak{B}_\xi = \big(W(\varphi_0(\xi)), \leqslant_{\varphi_0(\xi)}, A_\xi^0, A_\xi^1, S_\xi\big)$ where $\leqslant_{\varphi_0(\xi)}$ is the "less than or equal to" relation restricted to $W(\varphi_0(\xi))$ and A_ξ^0, A_ξ^1, S_ξ are relations to be defined later. A_ξ^0 and A_ξ^1 will have two arguments and S_ξ one, i.e. S_ξ is a subset of $W(\varphi_0(\xi))$. The precise definition of these relations will be given later; here we remark only that A_ξ^0 and A_ξ^1 will determine a function f_ξ: $[W(\varphi_0(\xi))]^2 \to \{0, 1\}$ such that $f_\xi(\{\eta, \zeta\}) = 0$ if $\eta A_\xi^0 \zeta$ and $f_\xi(\{\eta, \zeta\}) = 1$ if $\eta A_\xi^1 \zeta$. We shall select A_ξ^0, A_ξ^1 so that each set homogeneous for f_ξ must have power less than $\overline{\varphi_0(\xi)}$. The set S_ξ will be a set homogeneous for f_ξ.

The above facts concerning A_ξ^0, A_ξ^1, S_ξ are expressible in the form that some formulas are true in $\mathfrak{B}_\xi$. Hence, by Łoś's theorem (p. 143), these formulas are also true in the reduced product $\mathfrak{P} = \langle \tilde{V}, \tilde{\leqslant}, \tilde{A}^0, \tilde{A}^1, \tilde{S} \rangle$ of the $\mathfrak{B}_\xi$'s. Hence $\tilde{A}^0$, $\tilde{A}^1$ will determine a function $\tilde{f}$: $[\tilde{V}]^2 \to \{0, 1\}$ and $\tilde{S}$ will be homogeneous for $\tilde{f}$. We shall arrange the construction so that $\tilde{S}$ be cofinal with $\tilde{V}$. This is possible because $\tilde{V}$ has power $\mathfrak{m}$ and $\mathfrak{m}$ is weakly compact in view of Theorem 2. Now we again invoke Łoś's theorem. The formula that $\tilde{S}$ is cofinal with $\tilde{V}$ being true in $\mathfrak{P}$ must be true for almost all $\mathfrak{B}_\xi$, again in view of Łoś's theorem. But this would imply that the homogeneous set S_ξ is cofinal with $W(\varphi_0(\xi))$ and thus has the same power as $W(\varphi_0(\xi))$ which is a contradiction with the choice of A_ξ^0, A_ξ^1.

We proceed now to the details of the proof.[1])

[1]) We gave the proof of Theorem 4 as an illustration of the methods used in recent papers on measurable cardinals. Much stronger results than Theorem 4 are known. See Keisler and Tarski [1] and Drake [1].

Let f_ξ: $[W(\varphi_0(\xi))]^2 \to \{0, 1\}$ be a function such that there is no set $X \subset W(\varphi_0(\xi))$ homogeneous for f_ξ of power $\overline{\varphi_0(\xi)}$. We define two binary relations $A_\xi^0, A_\xi^1 \subset W(\varphi_0(\xi))^2$:

$$\langle \eta, \zeta \rangle \in A_\xi^\varepsilon \equiv f(\{\eta, \zeta\}) = \varepsilon.$$

Now we consider the relational systems

$$\mathfrak{B}_\xi(S_\xi) = \big(W(\varphi_0(\xi)), \leqslant_{\varphi_0(\xi)}, A_\xi^0, A_\xi^1, S_\xi\big)$$

where S_ξ is a subset of $W(\varphi_0(\xi))$ which at present we leave arbitrary. Whatever the final choice of S_ξ will be, we can state that if $\varphi_0(\xi)$ is an initial ordinal and S_ξ is homogeneous with respect to f_ξ, then S_ξ is not cofinal with $W(\varphi_0(\xi))$. Remembering Lemma 4 we see that (independently of the final choice of S_ξ) the following formula (Φ) is true in $\mathfrak{B}_\xi(S_\xi)$ for almost all $\xi < \alpha$ (the variables η, ζ below range over $W(\varphi_0(\xi))$):

$$(\Phi)\quad \Big\{\bigwedge_{\eta \neq \zeta} [S_\xi(\eta) \wedge S_\xi(\zeta) \to A_\xi^0(\eta, \zeta)] \vee \bigwedge_{\eta \neq \zeta} [S_\xi(\eta) \wedge S_\xi(\zeta) \to A_\xi^1(\eta, \zeta)]\Big\} \to \bigvee_\eta \bigwedge_\zeta [S_\xi(\zeta) \to \zeta < \eta].$$

The antecedent of this implication states that S_ξ is homogeneous for f_ξ and the consequent states that S_ξ is not cofinal with $W(\varphi_0(\xi))$.

Consider now the relations $\tilde{A}^0$ and $\tilde{A}^1$ in the reduced power $\mathfrak{P} = (\tilde{V}, \tilde{\leqslant}, \tilde{A}^0, \tilde{A}^1, \tilde{S})$. The universe $\tilde{V}$ of $\mathfrak{P}$ consists of elements $\tilde{\psi}$ in $\tilde{U}$ such that $\psi(\xi) < \varphi_0(\xi)$ for each ξ. Hence in view of Lemma 3(i) $\tilde{\psi} \in \tilde{C}$ and each element of $\tilde{C}$ belongs to $\tilde{V}$. Thus $\tilde{V} = \tilde{C}$.

The formulas stating that A_ξ^0 and A_ξ^1 are symmetrical, disjoint with each other and that for each pair $\{\eta, \zeta\}$ with $\eta \neq \zeta$ either $A_\xi^0(\eta, \zeta)$ or $A_\xi^1(\eta, \zeta)$ holds are true in $\mathfrak{B}_\xi(S_\xi)$. Hence by Łoś's theorem they are also true in $\mathfrak{P}$. It follows that if we put

$$\tilde{f}(c_\eta, \tilde{c}_\zeta) = \varepsilon \equiv \tilde{A}^\varepsilon(c_\eta, c_\zeta)$$

we obtain a mapping $\tilde{f}$: $[\tilde{V}]^2 \to \{0, 1\}$.

The cardinal number of $\tilde{V}$ is $\mathfrak{m}$ and thus is weakly compact (see Theorem 2). Thus there exists a set $\tilde{S} \subset \tilde{V}$ homogeneous for $\tilde{f}$ cofinal with $\tilde{V}$. The key fact is the following

LEMMA: *There are sets $S_\xi \subset W(\varphi_0(\xi))$ such that for each $\gamma < \alpha$*

$$(*)\qquad \tilde{c}_\gamma \in \tilde{S} \equiv \{\xi : \gamma \in S_\xi\} \notin I.$$

It is sufficient to put $S_\xi = \{\gamma \in W(\alpha): \gamma < \varphi_0(\xi) \wedge \tilde{c}_\gamma \in \tilde{S}\}$. We have then $S_\xi \subset W(\varphi_0(\xi))$ for each $\xi < \alpha$. If $\tilde{c}_\gamma \in \tilde{S}$ then $\varphi_0(\xi) > \gamma$ implies $\gamma \in S_\xi$ and, since for almost all ξ the inequality $\varphi_0(\xi) > \gamma$ is true, we infer the right-hand side of (*). Conversely, the right-hand side of (*) obviously implies the left. This proves the lemma.

We select now as the S_ξ in the relational systems $\mathfrak{B}_\xi(S_\xi)$ the sets from the lemma. Łoś's theorem then proves that the formula (Φ) is true in the reduced product of the relational systems $\mathfrak{B}_\xi(S_\xi)$. Remembering that the universe of the reduced product of the $\mathfrak{B}_\xi(S_\xi)$ is $\tilde{C} = \tilde{V}$, we infer that

$$\Big\{\bigwedge_{\zeta \neq \eta} [\tilde{S}(\tilde{c}_\eta) \wedge \tilde{S}(\tilde{c}_\zeta) \to \tilde{A}^0(\tilde{c}_\eta, \tilde{c}_\zeta)] \vee \bigwedge_{\eta \neq \zeta} [\tilde{S}(\tilde{c}_\eta) \wedge \tilde{S}(\tilde{c}_\zeta) \to \tilde{A}^1(\tilde{c}_\eta, \tilde{c}_\zeta)]\Big\}$$
$$\to \bigvee_{\eta} \bigwedge_{\zeta} [\tilde{S}(\tilde{c}_\zeta) \to \tilde{c}_\zeta \precsim \tilde{c}_\eta]$$

where the variables η, ζ range over $W(\alpha)$. This however is a contradiction because $\tilde{S}$ is homogeneous for $\tilde{f}$ (and thus the antecedent of the above formula is true) and cofinal with $\tilde{V}$ (and thus the consequent is false).

Theorem 4 is thus proved.

From the proof of Theorem 4 we can still draw the following important observation.

°THEOREM 5: *If* $\mathfrak{m} = \aleph_\alpha$ *is measurable and* φ_0: $W(\alpha) \to W(\alpha)$ *represents* α *then* $\mathrm{Rg}(\varphi_0)$ *contains* $\mathfrak{m}$ *ordinals* γ *such that* $\bar{\gamma}$ *is weakly compact (and thus inaccessible).*

We give one more application of reduced products to the theory of measurable cardinals.

For each $T \subset W(\alpha)$ let $M(T)$ be the set of initial ordinals γ such that there is a function f normal in $W(\alpha)$ such that $\mathrm{Rg}(f) \subset T$ and $\gamma = \lim_{\xi < \gamma} f(\xi)$.

°THEOREM 6: *If* $\mathfrak{m} = \aleph_\alpha$ *is a measurable cardinal and* T_0 *is the set of those initial ordinals* $\gamma < \alpha$ *for which* $\bar{\gamma}$ *is not weakly compact, then* $\alpha \notin M(T_0)$.

PROOF. Let us assume that $\alpha \in M(T_0)$, i.e., that there is a normal function f such that $\mathrm{Rg}(f) \subset T_0$ and $\alpha = \lim_{\xi < \alpha} f(\xi)$.

We shall obtain a contradiction with Theorem 5 by showing that there exists a function φ_0 representing α such that $\mathrm{Rg}(\varphi_0) \subset T_0$.

To construct φ_0 we consider an arbitrary function $\psi\colon W(\alpha) \to W(\alpha)$ which represents α (see Lemma 3) and put

$$\varphi_0(\xi) = \sup\{f(\eta)\colon \eta < \psi(\xi)\}.$$

Since f is a normal function and $\psi(\xi) < \alpha$, the supremum exists and belongs to the range of f, i.e. to T_0.

It remains to prove that the predecessors of $\tilde{\varphi}_0$ are exactly the constants $\tilde{c}_\gamma$.

If $\gamma < \alpha$ then $\tilde{c}_\gamma < \tilde{\psi}$ and hence $\gamma < \psi(\xi)$ for almost all ξ, whence $f(\gamma) < \varphi_0(\xi)$ for almost all ξ. Since $\gamma \leqslant f(\gamma)$ in view of the normality of f, we obtain $\gamma < \varphi_0(\xi)$, whence $\tilde{c}_\gamma \precsim \tilde{\varphi}_0$. Now let ϑ be such that $\tilde{\vartheta} \precsim \tilde{\varphi}_0$. For almost all ξ we obtain $\vartheta(\xi) < \varphi_0(\xi)$ and hence there exists a smallest $\eta_\xi < \psi(\xi)$ such that $\vartheta(\xi) < f(\eta_\xi)$. The ordinal η_ξ need not be defined everywhere but only for ξ in a set $X \notin I$. Putting $\eta_\xi = 0$ for $\xi \notin X$ we obtain a function on $W(\alpha)$ which satisfies the inequality $\eta_\xi < \psi(\xi)$ almost everywhere. Thus there is a $\delta < \alpha$ such that $\eta_\xi = \delta$ almost everywhere and it follows that $\vartheta(\xi) < f(\delta)$ for almost all ξ. In view of the $\mathfrak{m}$-completeness of I the function $\vartheta(\xi)$ is almost everywhere equal to an ordinal $< f(\delta) < \alpha$ and hence $\tilde{\vartheta} \in \tilde{C}$.

Theorem 6 is thus proved.

We shall now sum up the information about the size of measurable cardinals supplied by the above theorems.

By Theorem 2 each measurable cardinal $\mathfrak{m}$ is inaccessible. By Theorem 5 there are $\mathfrak{m}$ inaccessible cardinals $< \mathfrak{m}$ and hence $\mathfrak{m}$ is 1-inaccessible. By Theorem 6 $\mathfrak{m}$ is a Mahlo cardinal. Replacing in the proof of Theorem 6 the set T_0 by $M(T_0)$, $MM(T_0)$, ..., we may obtain the result that $\mathfrak{m}$ is not only a Mahlo cardinal, i.e. hyper-inaccessible but hyper-hyper-inaccessible, hyper-hyper-hyper-inaccessible, etc.

All these results could be derived directly from Theorem 2 because one can show that weakly inaccessible cardinals are also hyper-inaccessible, hyper-hyper-inaccessible, etc. At any rate Theorem 2 shows that each measurable cardinal $\mathfrak{m}$ is preceded by $\mathfrak{m}$ weakly compact cardinals and so these two notions do not coincide.

A very unexpected phenomenon of fundamental importance was discovered in connection with measurable cardinals. Assuming their existence, we are able to solve some problems regarding sets of real numbers which cannot be decided on the basis of axioms Σ°[TR] alone. Thus, for instance, it has been proved that the question whether non-denumerable sets of the second projective class (PCA-sets) contain a perfect subset is undecidable on the basis of Σ°[TR] but becomes decidable (in the positive sense) if one assumes that there exist measurable cardinals.[1])

In view of these applications it would be highly desirable to establish the consistency of additional axioms expressing the existence of large cardinals, such as weakly compact or measurable ones. No solution of this problem is in sight. There exist on the contrary several results showing that meta-mathematical methods known at present are insufficient to cope with this problem.

[1]) Drake [1] contains several results illustrating connections which exist between large cardinals and properties of projective sets. Numerous research papers dealing with these problems have appeared recently; the interested reader should consult current issues of Annals of Mathematical Logic and other journals dealing with foundations of mathematics.

INTRODUCTION TO DESCRIPTIVE SET THEORY

Historically the origin of descriptive set theory is connected with the development of the so called "Theory of a real variable." This theory flourished at the end of the XIXth century and in the early years of the present century. The most prominent authors at that time were: René Baire [1], Emile Borel [1] and Henri Lebesgue [1]. Later, the theory of a real variable was developed and essentially extended by Felix Hausdorff [1], Hans Hahn [1], N. Lusin [5], W. Sierpiński [3] and their schools—to become a vast mathematical discipline known as descriptive set theory (see also Liapunov [2], Saks [1] and Keldysh [1]).

This discipline is closely connected to topology, measure theory, probability theory. It also has interesting applications to game theory and to optimal control theory and thus—to some extent—to applied mathematics (see Filippov [1], Olech [1], Wagner [1] where a further literature can be found, see also Aumann [1], Cole [1], Jacobs [1], Rockafellar [1], Sainte-Beuve [1], Valadier [1]).

CHAPTER XI

AUXILIARY NOTIONS

§ 1. The notion of a metric space. Various fundamental topological notions

In this section we shall recall some elementary topological notions and theorems which will be needed in the sequel. Most of the material will be presented in the form of exercises; their solution can be found in elementary textbooks on topology (see e.g. R. Engelking [1], J. L. Kelley [1], K. Kuratowski [3]).

Let us add that several topological notions have been already introduced in Chapters I–IV. In particular, the notion of *topological space*, of the *closure* operation, of the *interior*, etc.—in Chapter I, § 8, of *continuity* and of *homeomorphism*—on pp. 77–78, of *compactness*—on p. 137, of *open base and subbase*—on p. 116.

DEFINITION 1: A set X is said to be a *metric space* if to every pair of its elements, i.e. to every pair of *points* x, y belonging to the set X, there is assigned a real number $|x-y| \geqslant 0$, called the *distance from the point x to the point y*, which satisfies the following three conditions:

(1) $$|x-y| = 0 \text{ \textit{if and only if} } x = y,$$

(2) $$|x-y| = |y-x|,$$

(3) $$|x-y|+|y-z| \geqslant |x-z|;$$

the last condition is the so-called *triangle inequality*.

1. Show that:

1° Every set of real or complex numbers forms a metric space if the distance between two numbers x and y is understood to be the absolute value of the difference of these numbers. (This justifies the symbol we are using for the distance.)

2° The Euclidean n-space, $\mathscr{E}^n$ (whose points are sequences of n real numbers $(x_1, x_2, \ldots, x_n)$), is a metric space under the usual definition of the distance from the point $x = (x_1, x_2, \ldots, x_n)$ to the point $y = (y_1, y_2, \ldots, y_n)$ given by the Pythagorean formula

$$|x-y| = \{\sum_{i=1}^{n} |x_i - y_i|^2\}^{1/2}. \tag{4}$$

The same formula "metrizes" the *cartesian product* $X_1 \times X_2 \times \ldots \times X_n$ of any n metric spaces, $X_1, X_2, \ldots, X_n$.

3° The set $\Phi(X, \mathscr{E})$ of bounded functions $f\colon X \to \mathscr{E}$ forms a metric space if the distance between two functions f and g is defined by the formula

$$|f-g| = \sup|f(x)-g(x)|. \tag{5}$$

Remark. An arbitrary set can be considered to be a metric space if we assume that the distance between each pair of distinct points is 1.

DEFINITION 2: The least upper bound of the distances $|x-y|$ between all pairs of points x and y in the metric space X is called the *diameter* of the space X and is denoted by the symbol $\delta(X)$. If X is a circle or sphere, then its diameter $\delta(X)$ is the diameter in the usual sense.

Metric spaces with finite diameter are said to be *bounded.*

A mapping $f\colon X \to Y$ where Y is a metric space, is called *bounded* if the set $f^1(X)$ is bounded.

2. Show that if *f and g are bounded mappings of the (arbitrary) set X into the metric space Y, their distance* $|f-g|$ *given by formula* (5) *is finite.*

3. Show that each metric space X can be considered as a topological space defining the closure as follows

$$(p \in \bar{A}) \equiv \bigwedge_{\varepsilon>0} \bigvee_{a} (a \in A)(|a-p| < \varepsilon).$$

DEFINITION 3: Call open *r-ball* with centre p the set

$$K(p, r) = \{x\colon |x-p| < r\}.$$

4. Show that if the set $P = \{p_1, p_2, \ldots, p_n, \ldots\}$, where $n \in N$, is dense in the space (i.e. $\bar{P} = X$, in which case the space X is called *separable*), then the family of all sets $K(p_n, r)$, where $n \in N$ and $r \in R$ (the set of positive rationals), is a *base* of the space.

(i) Apply this theorem to the case of the Euclidean n-dimensional space $\mathscr{E}^n$.

(ii) Generalize Exercise V.2.5 to arbitrary metric separable spaces X; namely, show that *the set of isolated points of X is countable*,

(iii) Generalize the Theorem of Cantor–Bendixson (Exercise VII.1.5) to arbitrary metric separable spaces.

(iv) Show that each metric separable uncountable space contains a (non-empty) dense-in-itself set.

(v) Show that if $\boldsymbol{F}$ is a family of open subsets of a metric separable space, then $\boldsymbol{F}$ contains a *countable* subfamily $\boldsymbol{C}$ such that $\bigcup \boldsymbol{C} = \bigcup \boldsymbol{F}$ (Theorem of Lindelöf).

5. Show that *the distance is a continuous function of two variables.*

Definition 4: Call a topological space respectively *Hausdorff, regular, normal* if the conditions

(i) $x \neq y$,

(ii) $x \notin F$, where F is closed,

(iii) $F_1 \cap F_2 = \emptyset$, where F_1 and F_2 are closed imply the existence of two open disjoint sets G and H such that

(i′) $x \in G$, $y \in H$,

(ii′) $x \in G$, $F \subset H$,

(iii′) $F_1 \subset G$, $F_2 \subset H$, respectively.

6. Show that *each metric space is normal.*

Hint: Apply the notion of the *distance of a point p to a set A* ($\neq \emptyset$): $\varrho(p, A) = \inf|p-x|$, where $x \in A$; and use the equivalence

$$[\varrho(p, \bar{A}) = 0] \equiv (p \in \bar{A}), \tag{6}$$

and the continuity of the function ϱ (for constant A).

7. Express normality of the space in terms of the lattice of closed subsets of the space.

Definition 5: The union of a sequence of closed sets $F_1 \cup F_2 \cup \ldots$ is called an $\boldsymbol{F}_\sigma$*-set*.

Symmetrically, the intersection of a sequence of open sets $G_1 \cap \cap G_2 \cap \ldots$ is called a $\boldsymbol{G}_\delta$*-set*.

8. Show that in each metric space each closed set F is a $\boldsymbol{G}_\delta$-set (in other terms: each metric space is *perfectly normal*).

Hint. Put $K(F, r) = \{x\colon \varrho(x, F) < r\}$ and show that

$$F = \bigcap_{n=1}^{\infty} K(F, 1/n). \tag{7}$$

Definition 6. The *limit* of a sequence of points $p_1, p_2, \ldots$ in a metric space is defined by the formula

$$(p = \lim_{n=\infty} p_n) \equiv \bigwedge_{\varepsilon>0} \bigvee_{n} \bigwedge_{k} |p_{n+k} - p| < \varepsilon. \tag{8}$$

9. Show that, in a metric space, $p \in \overline{A}$ iff p is of the form

$$p = \lim_{n=\infty} p_n \quad \text{where } p_n \in A \text{ for each } n \in N. \tag{9}$$

10. Show that if X and Y are metric spaces and $f\colon X \to Y$, then f is *continuous* iff

$$\lim_{n=\infty} x_n = x \quad \text{implies} \quad \lim_{n=\infty} f(x_n) = f(x) \tag{10}$$

for each sequence $x, x_1, x_2, \ldots$ in X.

11. Let X and Y be two topological spaces and let S be an open subbase of Y. Let $f\colon X \to Y$. Suppose $f^{-1}(G)$ is open for each $G \in S$. Then f is *continuous*.

12. Show that if $p = \lim_{n=\infty} p_n$, $F = \overline{F}$ and B_n is the ball $\{y\colon \varrho(y, F) < 1/n\}$, then

$$p \in F \equiv \bigwedge_{n} \bigvee_{k} p_{n+k} \in B_n. \tag{11}$$

13. Deduce from this that if $f(x) = \lim_{n=\infty} f_n(x)$, then

$$[f(x) \in F] \equiv \bigwedge_{n} \bigvee_{k} [f_{n+k}(x) \in B_n] \tag{12}$$

and consequently

$$f^{-1}(F) = \bigcap_{n} \bigcup_{k} f_{n+k}^{-1}(B_n). \tag{13}$$

By definition (compare p. 51), the sequence of functions $f_n\colon X \to Y$, $n = 1, 2, \ldots$, converges *uniformly* to $f\colon X \to Y$ if

$$\bigwedge_{\varepsilon>0} \bigvee_{n} \bigwedge_{x} \bigwedge_{k} |f_{n+k}(x) - f(x)| < \varepsilon. \tag{14}$$

14. Show that the uniform convergence means that there is an (increasing) sequence $m_1, m_2, \ldots$ such that

$$|f(x)-f_{m_n+k}(x)| < 1/n \quad \text{for each } x \in X \text{ and } k \geqslant 0. \tag{15}$$

15. Show that if the sequence $f_1, f_2, \ldots$ converges uniformly to f and $F = \bar{F}$ and $B_n = \{y\colon \varrho(y, F) < 1/n\}$, then

$$[f(x) \in F] \equiv \bigwedge_n [f_{m_n}(x) \in \bar{B}_n] \tag{16}$$

and consequently

$$f^{-1}(F) = \bigcap_n f_{m_n}^{-1}(\bar{B}_n). \tag{17}$$

16. Deduce from (17) that *the limit of a uniformly convergent sequence of continuous functions is continuous.*

17. Show that the distance of two points

$$x = (x_1, x_2, \ldots) \quad \text{and} \quad y = (y_1, y_2, \ldots)$$

in the Hilbert cube $\mathcal{I}^N$ can be defined by the formula

$$|x-y| = \sum_{n=1}^{\infty} (1/2)^n |x_n - y_n| \tag{18}$$

(i.e. the topology induced by this formula is consistent with the definition of product-topology given in § 7 of Chapter IV).

18. $\mathcal{C}$ denoting the Cantor discontinuum, show that the spaces $\mathcal{C}$, $\mathcal{C} \times \mathcal{C}$ and $\mathcal{C}^N$ are homeomorphic. Similarly, N^N and $(N^N)^N$ are homeomorphic.

Represent each $t \in \mathcal{C}$ in the form

$$t = t_1/3 + t_2/9 + \ldots + t_n/3^n + \ldots, \tag{19}$$

where t_n is either 0 or 2.

Put

$$g(t) = \tfrac{1}{2}(t_1/2 + t_2/4 + \ldots + t_n/2^n + \ldots). \tag{20}$$

19. Show that *g is a continuous function of $\mathcal{C}$ onto $\mathcal{I}$.*

Deduce from this that *the Hilbert cube $\mathcal{I}^N$ is a continuous image of the Cantor discontinuum $\mathcal{C}$.*

20. Let $f_t\colon X \to Y_t$ be continuous for each $t \in T$. Then the *complex function* $h\colon X \to \prod_{t \in T} Y_t$ *determined by the functions* f_t *is continuous.*

Hint: Show that $h^{-1}(G)$ is open if G belongs to the subbase of $\prod Y_t$ composed of open subsets of Y_t multiplied by $\prod_{t' \neq t} Y_{t'}$.

21. *Urysohn embedding Theorem.* Show that for every metric separable space X there exists a homeomorphism $h\colon X \to \mathscr{I}^{N}$.

Hint: Let $\delta(X) \leqslant 1$ and let $(p_1, p_2, \ldots)$ be dense in X. Put $f_n(x) = |x - p_n|$ and let h be the complex function defined by the functions f_n.

22. Let $f_t\colon X_t \to Y$ be continuous for $t \in T$. Show that the set $E = \bigcap_{t \in T} f_t^{-1}(X_t)$ *is a continuous image of a closed subset of the space* $\prod_{t \in T} X_t$. *Y is supposed to be Hausdorff.*

Hint: Consider the set $\mathfrak{Z} = \{\mathfrak{z}\colon \bigwedge_{tt'} f_t(\mathfrak{z}^t) = f_{t'}(\mathfrak{z}^{t'})\}$, compare Exercise IV.6.7.

23. Under the above assumptions, show that if the mappings f_t are *one-to-one* (or *homeomorphisms*), then E is a one-to-one continuous (or homeomorphic) image of a closed subset of $\prod_{t \in T} X_t$.

24. Let $A \subset X$ and $B \subset Y$. Show that $A \times B$ is closed (open) in $X \times Y$ iff A and B are closed (open).

25. Let $A \subset X$. Show that if A is G_δ (F_σ), so is $A \times Y$; if A is a boundary set (a nowhere dense set), then so is $A \times Y$.

Theorems on compactness (see p. 137).

26. Show that each compact subset of a Hausdorff space is closed.

27. Show that the image under a continuous function of a compact space is compact.

28. Show that each compact metric space contains a countable open base.

Hint: Consider for each n the family $\boldsymbol{C}_n$ of all $(1/n)$ balls and denote by $\boldsymbol{D}_n$ a finite subcover of $\boldsymbol{C}_n$. Show that $\boldsymbol{D}_1 \cup \boldsymbol{D}_2 \cup \ldots$ is the required base.

29. Show that *each compact metric space is the continuous image of a closed subset of the Cantor discontinuum* $\mathscr{C}$.

Hint: Use the Urysohn embedding theorem and the function g defined by (20).

Remark. One can show that each closed non-empty subset of $\mathscr{C}$ is a continuous image of $\mathscr{C}$. Consequently each compact metric space is a continuous image of $\mathscr{C}$.

30. Show that each compact Hausdorff space is *normal.*

31. Show that each one-to-one continuous mapping of a compact space into a Hausdorff space is a homeomorphism.

32. Show that each sequence of points in a compact metric space contains a convergent subsequence.

33. Show that in a compact metric space each Cauchy sequence is convergent (a sequence $p_1, p_2, \ldots$ is Cauchy if for each $\varepsilon > 0$ there is k such that $|p_n - p_k| < \varepsilon$ for $n > k$).

In other terms (see § 3) *each compact metric space is complete.*

§ 2. Exponential topology. Compact-open topology

We denoted by $\boldsymbol{P}(X)$ the *power set* of X (see p. 53), that is, the family of all subsets of X. If we assume that X is a topological space, it seems reasonable—instead of considering all possible subsets of X—to restrict ourselves to *closed* subsets. This allows us to consider this restricted family as a topological space.

Thus, let us denote by 2^X the *family of all closed non-empty subsets of* X. The topology in 2^X, called *exponential topology* (or Vietoris topology), is defined by declaring that the family of sets which are either of the form

(i) $\{F\colon F \cap U \neq 0\}$

or

(ii) $\{F\colon F \subset U\}$,

where U is open and F closed, is a *subbase* for that topology (see Vietoris [1]).

Remarks. 1. One can show that *if X is compact, then so is* 2^X; the converse is also true (if X is T_1) (see Michael [1]).

2. One can show also that *if 2^X is metrizable, then X is compact* (under the continuum hypothesis a stronger result has been shown,

namely that normality of 2^X implies compactness of X) (see Michael [1] and Keesling [1]).

3. It has been shown recently that $2^{\mathscr{I}}$ is homeomorphic to the Hilbert cube $\mathscr{I}^N$ (see Schori and West [1]).

In view of the Remark 2 one cannot expect the space 2^X to be metric unless we impose very restrictive assumptions on X. Thus a further limitation of subsets of X may be desirable. This leads to the notion of the space $\boldsymbol{K}(X)$ of all *compact* non-empty subsets of X; the topology of $\boldsymbol{K}(X)$ is defined as in the case of 2^X. Moreover, this topology is *metrizable* if X is metric; namely, one can define the (Hausdorff) distance of two elements A and B of $\boldsymbol{K}(X)$ as the maximum of the two numbers

$$\sup \varrho(x, B) \text{ for } x \in A \quad \text{and} \quad \sup \varrho(y, A) \text{ for } y \in B.$$

Closed-set-valued functions

In topology we usually restrict the general notion of set-valued functions $F\colon Y \to \boldsymbol{P}(X)$ to *closed* set-valued functions, i.e. to functions $F\colon Y \to 2^X$.

Assuming that X and Y are topological spaces, it is meaningful to consider continuous closed-set-valued functions. Namely, according to the general definition of continuity, the function $F\colon Y \to 2^X$ is continuous if $F^{-1}(\boldsymbol{G})$, for every $\boldsymbol{G}$ open in 2^X, is open in Y. According to 1.11, one can restrict the range of variability of the sets $\boldsymbol{G}$ to members of an open subbase of 2^X. This leads to the conclusion that F is continuous iff both conditions (i′) and (ii′) are fulfilled:

(i′) *$\{y\colon F(y) \cap U \neq \emptyset\}$ is open in Y if U is open in X*

and

(ii′) *$\{y\colon F(y) \subset U\}$ is open in Y if U is open in X.*

If just one of these conditions is satisfied, we shall say that F is *semi-continuous*, namely *lower* semi-continuous if condition (i′) is fulfilled and *upper* semi-continuous if (ii′) is true.

Exercises

1. Show that if X is regular, then 2^X is Hausdorff (the converse is also true if X is a T_1-space).

Hint: Let K and L be closed and $p \in K-L$. Then there is an open G such that $p \in G$ and $\overline{G} \cap L = \emptyset$.

2. Let F_i: $Y \to 2^X$, $i = 0, 1$. If the functions F_0 and F_1 are continuous, then so is their union.

Hence, the union $K \cup L$, considered as a mapping of $2^X \times 2^X$ into 2^X, is continuous.

3. Show that if $A_t \subset X$ is closed and X is a T_1-space, then

$$2^{A_0 \cap A_1} = 2^{A_0} \cap 2^{A_1} \quad \text{and generally} \quad 2^{\bigcap_t A_t} = \bigcap_t 2^{A_t}.$$

4. The family of all finite subsets of a T_1-space X is dense in 2^X.

5. Let F_i: $Y \to 2^X$ and A closed $\subset X$. Show that

$$(F_0 \cup F_1)^{-1}(2^A) = F_0^{-1}(2^A) \cap F_1^{-1}(2^A)$$

and

$$\overline{(\bigcup_t F_t)}^{-1}(2^A) = \bigcap_t F_t^{-1}(2^A).$$

6. Let E be a fixed closed-open subset of X. Put $F(K) = K \cap E$. Show that F is continuous on 2^X.

7. Let $X = A_0 \cup A_1$ where A_0 and A_1 are closed and $A_0 \cap A_1 = \emptyset$. Show that $2^{A_0 \cup A_1}$ is homeomorphic to $2^{A_0} \times 2^{A_1}$.

8. Show that if X is normal, then the set $\{\langle K, L \rangle: K \subset L\}$ is closed and the set $\{\langle K, L \rangle: K \cap L = \emptyset\}$ is open in $2^X \times 2^X$.

9. Show that condition (ii′) is equivalent to

(ii″) $\{y$: $F(y) \cap K \neq \emptyset\}$ *is closed in* Y *if* K *is closed in* X.

10. Show that the union of two upper semi-continuous functions is upper semi-continuous.

The same for lower semi-continuous functions.

In Exercises 11–14 we assume that X is *compact metric* and Y *Hausdorff*.

11. Let $D \subset X \times Y$ be closed. Put $F(y) = \{x$: $(x, y) \in D\}$ (= the "horizontal" section). Show that F is upper semi-continuous.

12. Let F: $Y \to 2^X$. Show that the set

$$G(F) = \{\langle x, y \rangle x \in F(y)\},$$

the *graph of the relation* $x \in F(y)$, is closed iff F is upper semi-continuous.

13. Let f: $X \to Y$ be continuous onto; then the *inverse image* of f, i.e. f^{-1}: $Y \to 2^X$, is *upper semi-continuous*.

14. Let F be upper semi-continuous. Then the inverse images under F of open sets are F_σ-sets (in other terms: F is of the class 1 in the Baire classification).

Hint: Use the fact that each open subset of 2^X is the countable union of some members of the base of 2^X. Then represent each open subset of X as the union of a sequence of closed sets. (See Kuratowski [2], p. 70.)

15. Show that there is a *continuous choice-function* on the space $2^{\mathscr{I}}$ (namely the function assigning to each closed non-empty subset A of $\mathscr{I}$ the first point of A).

16. Show that there exists no continuous choice-function on the circle S (even for the family of two-elements subsets of S).

17. Show that there is a *choice-function of the first class* for the space $2^{\mathcal{Q}}$ where $\mathcal{Q}$ is the Hilbert cube ($\mathcal{Q} = \mathcal{I}^N$).

Hint: Let $g: \mathcal{I} \to \mathcal{Q}$ be continuous onto and let $h(A)$ denote the first point of $A \in 2^{\mathcal{I}}$. Put $f = g \circ h \circ g^{-1}$. (See Kuratowski [2], p. 425.)

18. Generalize the preceding statement to the case where H is an arbitrary compact metric space.

Remark. The following statement motivates the denomination of semi-continuity for set-valued mappings.

Let f be a real-valued function of a real variable.

According to the *classical* definition (of Baire) f is upper or lower semi-continuous if for each real c

$$\{x: f(x) < c\} \text{ or } \{x: f(x) > c\}, \text{ respectively, is open.}$$

19. Write $F(x) = \{y: y \leqslant f(x)\}$ and show that F is upper (lower) semi-continuous iff f is such.

The compact-open topology

If X and Y are topological spaces, we consider—instead of the space Y^X of *all* functions $f: X \to Y$—the (more restricted) space of all *continuous* functions f, denoted by $(Y^X)_{\text{top}}$. It becomes topologically meaningful when the following topology, called *compact-open*, is introduced. Given a compact $C \subset X$ and an open $H \subset Y$, write

$$\Gamma(C, H) = \{f: f^1(C) \subset H\}.$$

The compact-open topology of the space $(Y^X)_{\text{top}}$ is defined by considering the totality of all sets $\Gamma(C, H)$ as its open subbase. (See Fox [1], also Kuratowski [2], § 44.)

Exercises

1. Let X be Hausdorff compact and Y arbitrary topological. Given $f \in (Y^X)_{\text{top}}$, put $w(f, x) = f(x)$. Show that the function $w: (Y^X)_{\text{top}} \times X \to Y$ is continuous (this property of f is called *joint continuity*).

2. Let X be Hausdorff compact and Y metric. Show that the space $(Y^X)_{\text{top}}$ is metrizable. Namely, the distance of two elements f and g of this space can be defined as follows (see also XI.1(5)):

$$|f-g| = \sup|f(x)-g(x)| \quad \text{for } x \in X.$$

3. Show that the formulas V.1 (8)–(10) remain valid for the compact-open topology, the relation $\sim$ meaning homeomorphism, all spaces being assumed metric and moreover, A, B, X and T compact.

Hint: The proof is quite similar to the proof of the mentioned formulas (8)–(10).

§ 3. Complete and Polish spaces

Let us recall that a sequence of points $p_1, p_2, \ldots$ of a metric space is a *Cauchy sequence* if for every $\varepsilon > 0$ there is n such that $|p_n - p_k| < \varepsilon$ for every $k > n$.

The space is *complete* if every Cauchy sequence is convergent. The space is *topologically complete* if it is homeomorphic to a complete space (e.g. an open interval is not complete, but is topologically complete). See Fréchet [1].

Topologically complete and separable spaces are called *Polish spaces.*

The following theorem can easily be proved.

THEOREM 1: *The product* $X \times Y$ *of two complete spaces is complete when metrized by the formula*

$$|z - z_1| = \sqrt{|x - x_1|^2 + |y - y_1|^2}$$

where $z = \langle x, y \rangle$ *and* $z_1 = \langle x_1, y_1 \rangle$.

The product $X_1 \times X_2 \times \ldots$ of complete spaces is complete when metrized by the formula

$$|\mathfrak{z} - \mathfrak{y}| = \sum_{i=1}^{\infty} 2^{-i} \frac{|\mathfrak{z}^i - \mathfrak{y}^i|}{1 + |\mathfrak{z}^i - \mathfrak{y}^i|},$$

where $\mathfrak{z} = (\mathfrak{z}^1, \mathfrak{z}^2, \ldots)$, $\mathfrak{y} = (\mathfrak{y}^1, \mathfrak{y}^2, \ldots)$ and $\mathfrak{z}^n \in X_n$, $\mathfrak{y}^n \in X_n$.

In particular, the Euclidean space $\mathscr{E}^n$, the space $\mathscr{E}^N$, the Hilbert cube $\mathscr{I}^N$—are Polish.

THEOREM 2 (OF ALEXANDROV, see [3]): *Every* G_δ*-subset* A *of a complete space* X *is topologically complete; more precisely:* A *is homeomorphic to a closed subset of* $X \times \mathscr{E}^N$.

PROOF. Let X be a metric space and A a G_δ-subset of X. Hence $A = G_1 \cap G_2 \cap \ldots$ where G_n is open for $n = 1, 2, \ldots$ Put

$$f_n(x) = \frac{1}{\varrho(x, X - G_n)} \quad \text{for} \quad x \in G_n,$$

and for $x \in A$ put

$$f(x) = [f_1(x), f_2(x), \ldots], \quad \text{thus } f\colon A \to \mathscr{E}^N.$$

Since f_n is continuous on G_n, hence on A, it follows that f is continuous on A. Furthermore—as easily seen—the graph of f, i.e. the set

$\mathrm{Gr}(f) = \{\langle x, y\rangle : y = f(x)\}$, is closed in the complete space $X \times \mathscr{E}^N$. Consequently, $\mathrm{Gr}(f)$ is complete. Finally, A is homeomorphic to $\mathrm{Gr}(f)$.

THEOREM 3: *Every topologically complete subset E of a metric space is a* $\boldsymbol{G_\delta}$*-set (in this space).*

PROOF. (Comp. Sierpiński [10].) Let X be metric, $E \subset X$, Y complete and h a homeomorphism of E onto Y.

Denote by $B_n(y_0)$ the open ball with center y_0 and radius $< 1/n$. Since $h^{-1}[B_n(h(p))]$ is open for each $p \in E$, and contains p, there is in the space X an open set $K_n(p)$ such that

(1) $\quad p \in K_n(p), \quad K_n(p) \cap E \subset h^{-1}[B_n(h(p))], \quad \delta(K_n(p)) < 1/n.$

Put $G_n = \bigcup_{p \in E} K_n(p)$. We shall prove that

$$E = \bigcap_n G_n,$$

and this will complete the proof.

Obviously, $E \subset \bigcap_n G_n$. So let $x \in \bigcap_n G_n$. We have to show that $x \in E$.

Since $x \in G_n$ for $n = 1, 2, \ldots$, there exists in E a sequence of points $p_1, p_2, \ldots$ such that

(2) $\qquad x \in K_n(p_n), \quad \text{and hence} \quad x = \lim_{n=\infty} p_n,$

by the last part of (1).

We shall show that $h(p_1), h(p_2), \ldots$ is a Cauchy sequence in Y.

So let $\varepsilon > 0$ and $1/n_0 < \varepsilon$. By (2) we have for sufficiently large m (say $m \geqslant n_1$), $p_m \in K_{n_0}(p_{n_0})$, and consequently, by (1),

$$p_m \in h^{-1}[B_{n_0}(h(p_{n_0}))], \quad \text{i.e.} \quad |h(p_m) - h(p_{n_0})| < 1/n_0 < \varepsilon.$$

Thus, for $m \geqslant n_1$, we have $|h(p_m) - h(p_{n_1})| < 2\varepsilon$, which means that the sequence $h(p_1), h(p_2), \ldots$ is Cauchy. Since Y is complete, there is an $y = \lim_{n=\infty} h(p_n)$. Put $y = h(q)$, $q \in E$. Since h is a homeomorphism, we have

$$q = \lim_{n=\infty} p_n, \text{ hence } q = x, \text{ and therefore } x \in E.$$

Remark. By virtue of Theorems 2 and 3, the Polish spaces can be identified—from the topological point of view—with the $\boldsymbol{G_\delta}$*-subsets of the Hilbert cube* $\mathscr{I}^N$.

THEOREM 4: *The hyperspace, $\boldsymbol{K}(X)$, of compact subsets of a Polish space X is Polish.*

PROOF. According to the above remark, we may consider X as a $\boldsymbol{G}_\delta$-subset of the Hilbert cube $\mathcal{Q} = \mathcal{I}^N$. Then

$$\boldsymbol{K}(X) = \{F \in 2^{\mathcal{Q}}: F \subset X\}.$$

This set is a $\boldsymbol{G}_\delta$ in $2^{\mathcal{Q}}$; for put $X = G_1 \cap G_2 \cap \ldots$, where G_n is open; then

$$\{F: F \subset X\} = \bigcap_n \{F: F \subset G_n\},$$

and the set $\{F: F \subset G_n\}$ is open by definition of the exponential topology.

Since $2^{\mathcal{Q}}$ is compact (see Remark 2.1), hence complete, and $\boldsymbol{K}(X)$ is $\boldsymbol{G}_\delta$, it follows that $\boldsymbol{K}(X)$ is Polish (by Theorem 2).

THEOREM 5 (see Kuratowski [2], p. 543): *Every metric space X can be embedded in a complete space.*

More precisely: X is isometric to a subset of the (normed) space $\mathcal{E}^X$ of all continuous bounded functions $f: X \to \mathcal{E}$.

The proof is based on two lemmas.

LEMMA 1: *Denote as in* 1, 3°, *by $\Phi(X, Y)$ the space of all bounded functions $f: X \to Y$ metrized by the formula*

$$|f_1 - f_2| = \sup |f_1(x) - f_2(x)|. \tag{3}$$

If Y is complete, then so is $\Phi(X, Y)$.

The proof is quite elementary.

LEMMA 2: *$\mathcal{E}^X$ is closed in $\Phi(X, \mathcal{E})$, hence is complete (when metrized by formula* (3)).

This follows from the fact that the limit of a uniformly convergent sequence of continuous functions is continuous (see 1, 16).

PROOF OF THEOREM 5. We must define $f: X \to \mathcal{E}^X$ such that

$$|f_a - f_b| = |a - b|. \tag{4}$$

Let p be a fixed point of X. Put

$$f_a(x) = |x - a| - |x - p|. \tag{5}$$

Obviously, $|f_a(x)| \leqslant |a-p|$ and hence f_a is bounded and continuous. Furthermore,

$$|f_a(x)-f_b(x)| = ||x-a|-|x-b|| \leqslant |a-b|$$

and hence $|f_a-f_b| \leqslant |a-b|$. On the other hand,

$$f_a(a)-f_b(a) = -|a-p|-|a-b|+|a-p|,$$

and hence $|f_a-f_b| \geqslant |a-b|$. Formula (4) follows.

Remark. If X is bounded, (5) may be replaced by $f_a(x) = |x-a|$.

Properties of complete spaces

THEOREM 6 (OF CANTOR): *If* $A_1 \supset A_2 \supset \ldots$ *is a sequence of non-empty closed sets such that* $\lim_{n=\infty} \delta(A_n) = 0$, *then the set* $A_1 \cap A_2 \cap \ldots$ *consists of a single point.*

PROOF. Let $p_n \in A_n$. The sequence $p_1, p_2, \ldots$ is obviously a Cauchy sequence. Hence it converges to a point p. For each n, we have $p \in A_n$. The point p is the unique point belonging to the intersection of all A_n, because the diameter of this intersection is 0.

THEOREM 7 (OF BAIRE): *Every set of the first category* (*i.e. a countable union of nowhere dense sets*) *is a boundary set.*

PROOF. Let $E = N_1 \cup N_2 \cup \ldots$, where N_n is nowhere dense. Let B_0 be a closed ball. We have to show that $B_0 - E \neq \emptyset$.

Define by induction the sequence of balls $B_0 \supset B_1 \supset \ldots$ as follows. Let B_n be a ball such that

$$B_n \subset B_{n-1}-N_n \quad \text{and} \quad \delta(B_n) < 1/n.$$

By Theorem 6, there is a point $p \in B_0 \cap B_1 \cap \ldots$ It follows that $p \in B_0 - E$, since

$$\bigcap_n B_n \subset \bigcap_n (X-N_n) = X - \bigcup_n N_n = X-E.$$

COROLLARY 1: *The complement of a first category set is not of the first category* (*unless the space is void*).

COROLLARY 2: *The countable intersection of dense* $\mathbf{G}_\delta$*-sets is dense.*

THEOREM 8 (see also Chapter XII, § 6): *Every Polish space is a one-to-one continuous image of a closed subset of the space* N^N.

PROOF. First, let us note that the interval $\mathscr{I}$ (as well as the real line $\mathscr{E}$) satisfies the conclusion of the theorem.

For let A be the set of irrationals $0 < t < 1/2$ and let B be composed of a sequence $t_1, t_2, \ldots$ of irrationals between $1/2$ and 1 and converging to 1. Obviously, the set $F = A \cup B$ is closed in N^N and there is a one-to-one continuous onto mapping $f\colon F \to \mathscr{I}$ (A is mapped onto N^N and B onto $\mathscr{I} - N^N$).

It follows that the Hilbert cube $\mathscr{Q} = \mathscr{I}^N$ satisfies our conclusion. For let

$$g = f \times f \times \ldots, \qquad \text{hence } g\colon F^N \to \mathscr{Q},$$

and g is a one-to-one continuous onto mapping.

Since F^N is closed in $(N^N)^N$ and $(N^N)^N$ is homeomorphic to N^N, it follows that F^N is homeomorphic to a closed subset C of N^N. Thus there is a one-to-one continuous onto mapping $h\colon C \to \mathscr{Q}$.

Consider now the general case of an arbitrary Polish space X. According to the remark to Theorem 3, X can be considered as a $\boldsymbol{G}_\delta$ subset of $\mathscr{Q}$. Put $D = h^{-1}(X)$. Hence D is $\boldsymbol{G}_\delta$ in C, hence in N^N. Consequently (by Theorem 2), D is homeomorphic to a closed subset H of $N^N \times \mathscr{E}^N$.

But $\mathscr{E}^N$ is a one-to-one continuous image of a closed subset of N^N, and hence so is $N^N \times \mathscr{E}^N$ and finally—so is H. Our conclusion follows.

§ 4. *L*-measurable mappings

DEFINITION 1 (see Kuratowski [8] and also Hausdorff [2], p. 267): Let X and Y be metric, Y contains a countable open base and $\boldsymbol{L}$ a family of subsets of X (i.e. $\boldsymbol{L} \subset \boldsymbol{P}(X)$) containing all open sets.

A mapping $f\colon X \to Y$ is called an ***L*-*mapping*** if

(0) $\qquad f^{-1}(G) \in \boldsymbol{L}$ *for each G open in Y.*

If $\boldsymbol{L}$ is a σ-lattice, then f is obviously an $\boldsymbol{L}$-mapping if (0) is fulfilled by those G which belong to a countable open subbase of Y.

Remark. If $\boldsymbol{L}$ is a σ-algebra, the $\boldsymbol{L}$-mappings are usually called ***L*-*measurable***.

Our terminology of $\boldsymbol{L}$-mappings has been originated by the case where $\boldsymbol{L}$ is a σ-lattice (for δ-lattices it would seem to be preferable to replace G open by K closed).

Examples. If $\boldsymbol{L}$ is the family of all open subsets of X, then f is an $\boldsymbol{L}$-mapping iff f is continuous.

If $\boldsymbol{L}$ is the family of all Lebesgue measurable subsets of the interval, then $\boldsymbol{L}$-mappings coincide with the Lebesgue measurable functions.

THEOREM 1: *Let $A \subset X$ and let f be the characteristic function of A. Then f is an $\boldsymbol{L}$-mapping iff $A \in \boldsymbol{L}$* and $(X-A) \in \boldsymbol{L}$.

Because $f^{-1}(1) = A$ and $f^{-1}(0) = X-A$.

THEOREM 2: *Let $f\colon X \to Y$ and $g\colon Y \to Z$. If g is continuous and f is an $\boldsymbol{L}$-mapping then $h = g \circ f\colon X \to Z$ is an $\boldsymbol{L}$-mapping.*

Because $h^{-1}(G) = f^{-1}[g^{-1}(G)]$.

THEOREM 3: *Let $\boldsymbol{L}$ be a σ-lattice. Then the limit of a uniform convergent sequence of $\boldsymbol{L}$-mappings $f_n\colon X \to Y$ is an $\boldsymbol{L}$-mapping.*

PROOF. Let $F \subset Y$ be closed. Put $f(x) = \lim\limits_{n=\infty} f_n(x)$ and $B_n = \{y\colon \varrho(y, F) < 1/n\}$. By the uniform convergence (see 1(17)), there exists an increasing sequence m_n such that

$$f^{-1}(F) = \bigcap_n f_{m_n}^{-1}(\bar{B}_n).$$

This completes the proof. Because, for $G = Y-F$, we have

$$f^{-1}(G) = \bigcup_n f_{m_n}^{-1}(Y-\bar{B}_n)$$

and $f_{m_n}^{-1}(Y-\bar{B}_n) \in \boldsymbol{L}$ by assumption.

THEOREM 4 (on complex mappings): *Let $\boldsymbol{L}$ be a σ-lattice in the space T, let $f\colon T \to X$ and $g\colon T \to Y$, where X and Y are separable. Then the complex mapping*

$$h = \langle f, g\rangle\colon T \to X \times Y$$

is an $\boldsymbol{L}$-mapping iff the mappings f and g are $\boldsymbol{L}$-mappings.

More generally, if $f_n\colon T \to X_n$ for $n = 1, 2, \ldots$ and if X_n is separable, then the complex mapping

$$h = \prod_n f_n\colon T \to \prod_n X_n$$

is an $\boldsymbol{L}$-mapping iff f_n is an $\boldsymbol{L}$-mapping for each $n = 1, 2, \ldots$

PROOF. The condition is necessary. Let G be open in X_1. We have

(1) $$\{f_1(t) \in G\} \equiv \{h(t) \in (G \times X_2 \times X_3 \times \ldots)\}$$

and hence $f_1^{-1}(G) = h^{-1}(G \times X_2 \times X_3 \times \ldots)$.

Since $G \times X_2 \times X_3 \times \ldots$ is open and h is an $\boldsymbol{L}$-mapping by assumption, it follows that $f_1^{-1}(G) \in \boldsymbol{L}$, and hence f_1 is an $\boldsymbol{L}$-mapping.

The condition is sufficient. The proof is easily reduced to show that $h^{-1}(G \times X_2 \times X_3 \times \ldots) \in \boldsymbol{L}$, G being an open subset of X_1 and f_1 being supposed an $\boldsymbol{L}$-mapping. But this follows at once from (1).

DEFINITION 2: Let $\boldsymbol{L} \subset \boldsymbol{P}(X)$ and $\boldsymbol{M} \subset \boldsymbol{P}(Y)$. We denote by $\boldsymbol{L} \otimes \boldsymbol{M}$ the family of all sets

(2) $$E \times F \text{ such that } E \in \boldsymbol{L} \text{ and } F \in \boldsymbol{M}.$$

DEFINITION 3: For each family $\boldsymbol{L} \subset \boldsymbol{P}(X)$ we denote by $\overline{\boldsymbol{L}}$ the *σ-algebra generated by* $\boldsymbol{L}$.

LEMMA 1: *Let*

$$X \in \boldsymbol{L} \subset \boldsymbol{P}(X), \quad Y \in \boldsymbol{M} \subset \boldsymbol{P}(Y) \quad \textit{and} \quad \boldsymbol{L} \otimes \boldsymbol{M} \subset \boldsymbol{N}.$$

Then the family

$$\boldsymbol{R} = \{E \subset X\colon (E \times Y) \in \overline{\boldsymbol{N}}\}$$

is a σ-algebra and $\boldsymbol{L} \subset \boldsymbol{R}$. *Consequently,* $\overline{\boldsymbol{L}} \subset \boldsymbol{R}$.

PROOF. Let $E_n \subset X$ and $E_n \times Y \in \overline{\boldsymbol{N}}$ for $n = 1, 2, \ldots$ Put $E = E_1 \cup \cup E_2 \cup \ldots$ Then $E \subset X$ and $E \times Y \in \overline{\boldsymbol{N}}$, because

$$E \times Y = (E_1 \times Y) \cup (E_2 \times Y) \cup \ldots$$

Now let $H = X - E$, $E \subset X$ and $(E \times Y) \in \overline{\boldsymbol{N}}$. Then $H \subset X$ and $(H \times Y) \in \overline{\boldsymbol{N}}$, because

$$H \times Y = X \times Y - E \times Y.$$

Finally, $\boldsymbol{L} \subset \boldsymbol{R}$. For let $E \in \boldsymbol{L}$. Then $E \subset X$ and $(E \times Y) \in \boldsymbol{L} \otimes \boldsymbol{M}$, since $Y \in \boldsymbol{M}$.

Moreover, since $\overline{\boldsymbol{L}}$ is the smallest σ-algebra containing $\boldsymbol{L}$, we have $\overline{\boldsymbol{L}} \subset \boldsymbol{R}$.

LEMMA 2: *Under the above assumptions,* $\overline{\boldsymbol{L}} \otimes \overline{\boldsymbol{M}} \subset \overline{\boldsymbol{N}}$.

PROOF. Let $E \in \overline{\boldsymbol{L}}$. Then, by Lemma 1, $E \in \boldsymbol{R}$, which means that $(E \times Y) \in \overline{\boldsymbol{N}}$.

Symmetrically, $F \in \overline{M}$ implies that $(X \times F) \in \overline{N}$. Since $(E \times F) = (E \times Y) \cap (X \times F)$, we have $(E \times F) \in \overline{N}$.

THEOREM 5: *Let, for* $j = 0, 1$, $\boldsymbol{L}_j \subset \boldsymbol{P}(X_j)$ *and let* $f_j \colon X_j \to Y_j$ *be a* $\boldsymbol{L}_j$*-mapping. Suppose that* $\boldsymbol{L}_0 \otimes \boldsymbol{L}_1 \subset N$ *and that* N *is a σ-lattice in* $X_0 \times X_1$. *Then the product function*

$$(f_0 \times f_1) \colon X_0 \times X_1 \to Y_0 \times Y_1$$

is an N-mapping.

Since N is a σ-lattice, it suffices to show that

$$(f_0 \times f_1)^{-1}(G_0 \times G_1) \in N \quad \text{for } G_j \text{ open in } Y_j.$$

Now (by Exercise II.7.10) we have

$$(f_0 \times f_1)^{-1}(G_0 \times G_1) = f_0^{-1}(G_0) \times f_1^{-1}(G_1) \subset N,$$

because $f_j^{-1}(G_j) \in \boldsymbol{L}_j$ (by assumption).

COROLLARY: *If* $f_j \colon X_j \to Y$ *is an* $\boldsymbol{L}_j$*-mapping, and if* $h(x_0, x_1) = |f_0(x_0) - f_1(x_1)|$, *then* h *is an N-mapping.*

Consequently,

(3) $$\{\langle x_0, x_1 \rangle \colon f_0(x_0) \neq f_1(x_1)\} \in N.$$

Because this set equals $h^{-1}(\mathscr{E} - (0))$, which belongs to N.

Theorem 5 can be generalized as follows.

THEOREM 5a: *Let for* $j = 1, 2, \ldots$, $\boldsymbol{L}_j \subset \boldsymbol{P}(X_j)$ *and let* $f_j \colon X_j \to Y_j$ *be an* $\boldsymbol{L}_j$*-mapping. Suppose that each set of the form* $A_1 \times A_2 \times \ldots$, *where* $A_n \in \boldsymbol{L}_n$ *and for sufficiently high* n, $A_n = X_n$, *belongs to a σ-lattice* $N \subset \boldsymbol{P}(X_0 \times X_1 \times \ldots)$; *then the product-function*

$$\prod_{j=1}^{\infty} f_j \colon \prod_{j=1}^{\infty} X_j \to \prod_{j=1}^{\infty} Y_j$$

is an N-mapping.

It suffices to show that

$$\Big[\big(\prod_j f_j\big)^{-1}(G \times Y_2 \times Y_3 \times \ldots)\Big] \in N$$

for each G open in Y_1. Now this set equals

$$f_1^{-1}(G) \times X_2 \times X_3 \ldots$$

and hence belongs to N.

THEOREM 6: *Under the above assumptions put* $Y = Y_1 = Y_2 = \dots$ *and*

$$\mathfrak{Z} = \{\mathfrak{z}\colon f_1(\mathfrak{z}^1) = f_2(\mathfrak{z}^2) = \dots\} \quad \textit{where} \quad \mathfrak{z} \in X_1 \times X_2 \times \dots$$

Suppose that $f_j\colon X_j \to Y$ *is an* $\boldsymbol{L}_j$*-mapping. Then*

$$(\prod_{j=1}^{\infty} X_j) - \mathfrak{Z} \in \boldsymbol{N}.$$

PROOF. Obviously,

$$\mathfrak{z} \in \prod_j (X_j) - \mathfrak{Z} \equiv \bigvee_{jj'} [f_j(\mathfrak{z}^j) \neq f_{j'}(\mathfrak{z}^{j'})].$$

This completes the proof, because by (3)

$$\{\mathfrak{z}\colon f_j(\mathfrak{z}^j) \neq f_{j'}(\mathfrak{z}^{j'})\} \in \boldsymbol{N}$$

for j and j' fixed.

THEOREM 7: *Under the same assumptions, put*

$$g(\mathfrak{z}) = f_1(\mathfrak{z}^1) \quad \textit{for} \quad \mathfrak{z} \in \mathfrak{Z}.$$

Then $g\colon \mathfrak{Z} \to Y$ *is an* $\boldsymbol{N}$*-mapping and*

$$g^1(\mathfrak{Z}) = \bigcap_{j=1}^{\infty} f_j^1(X_j). \tag{4}$$

Formula (4) follows from Exercise IV.6.7.

To show that g is an $\boldsymbol{N}$-mapping, let G be open in Y. Then denoting by π the projection of $X_1 \times X_2 \times \dots$ onto X_1, we have

$$\begin{aligned} g^{-1}(G) &= (f_1 \circ \pi)^{-1}(G) \\ &= \pi^{-1}[f_1^{-1}(G)] = [f_1^{-1}(G) \times X_2 \times X_3 \times \dots] \in \boldsymbol{N}, \end{aligned}$$

since $f_1^{-1}(G) \in \boldsymbol{L}_1$ by assumption.

THEOREM 8: (FIRST GRAPH THEOREM) *Let* $\boldsymbol{L} \subset \boldsymbol{P}(X)$ *and let* $f\colon X \to Y$ *be an* $\boldsymbol{L}$*-mapping. Denote the graph of* f, *as usual, by*

$$\mathrm{Gr}(f) = \{\langle x, y\rangle\colon y = f(x)\}.$$

Let $\boldsymbol{G}$ *be the family of all open subsets of* Y *and let* $\boldsymbol{N}$ *be a* σ*-lattice* $\subset \boldsymbol{P}(X \times Y)$ *such that* $\boldsymbol{L} \otimes \boldsymbol{G} \subset \boldsymbol{N}$. *Then*

$$[X \times Y - \mathrm{Gr}(f)] \in \boldsymbol{N}. \tag{5}$$

PROOF. (5) follows from (3) by putting $f_0 = f$, f_1 = identity on Y, and $\boldsymbol{L}_1$ = the family of open subsets of Y.

COROLLARY: *Let* $\boldsymbol{L}$ *be a* σ*-lattice containing all open subsets of* X, *and let* $f\colon X \to X$ *be an* $\boldsymbol{L}$*-mapping. Put*

$$I = \{x\colon f(x) = x\}.$$

Then $(X - I) \in \boldsymbol{L}$.

PROOF. Put $\boldsymbol{N} = \boldsymbol{L} \times \boldsymbol{L}$ and $Y = X$.

Remark. Let us recall that there is in 2^Y an open subbase composed of two kinds of families of open sets

$$\{E\colon E \cap U \neq \emptyset\} \quad \text{and} \quad \{E\colon E \subset U\}$$

where U is open. (See § 2 (i) and (ii).)

Therefore, if $\boldsymbol{G}$ belongs to this subbase and if $F^{-1}(\boldsymbol{G}) \in \boldsymbol{L}$, then we have

$$F^{-1}\{E\colon E \cap U \neq \emptyset\} \in \boldsymbol{L} \quad \text{or} \quad F^{-1}\{E\colon E \subset U\} \in \boldsymbol{L}.$$

In other terms, we have either

$$\{x\colon F(x) \cap U \neq \emptyset\} \in \boldsymbol{L} \quad \text{or} \quad \{x\colon F(x) \subset U\} \in \boldsymbol{L}.$$

This leads to the notion of lower and upper-$\boldsymbol{L}$ mappings.

DEFINITION 4: Let $\boldsymbol{L} \subset \boldsymbol{P}(X)$. We call $F\colon X \to \boldsymbol{P}(Y)$ *lower-*$\boldsymbol{L}$, respectively *upper-*$\boldsymbol{L}$, if

(6) $$\{x\colon F(x) \cap U \neq \emptyset\} \in \boldsymbol{L} \quad \text{if } U \text{ is open in } Y,$$

equivalently

(6′) $$\{x\colon F(x) \not\subset K\} \in \boldsymbol{L} \quad \text{if } K \text{ is closed in } Y;$$

respectively

(7) $$\{x\colon F(x) \subset U\} \in \boldsymbol{L} \quad \text{if } U \text{ is open in } Y,$$

equivalently

(7′) $$\{x\colon F(x) \cap K = \emptyset\} \in \boldsymbol{L} \quad \text{if } K \text{ is closed in } Y.$$

Obviously if $\boldsymbol{L}$ is the family of all open subsets of X, then $F\colon X \to 2^Y$ is lower(upper)-$\boldsymbol{L}$ if F is lower (upper) *semi-continuous*.

DEFINITION 5: We denote by $G(F)$ the *graph of the formula* $y \in F(x)$, i.e.

(8) $$G(F) = \{\langle x, y\rangle\colon y \in F(x)\}.$$

THEOREM 9: (SECOND GRAPH THEOREM) *Let* $\boldsymbol{G}$ *be the lattice of all open subsets of* Y, *let* $\boldsymbol{L} \subset \boldsymbol{P}(X)$ *and let* $\boldsymbol{N}$ *be a* σ*-lattice such that* $\boldsymbol{L} \otimes \boldsymbol{G} \subset \boldsymbol{N}$.

If $F: X \to 2^Y$ *is upper-*$\boldsymbol{L}$, *then*

$$[X \times Y - G(F)] \in \boldsymbol{N}. \tag{9}$$

PROOF. Let $G_1, G_2, \ldots$ be an open base of Y. Obviously,

$$y \notin F(x) \equiv \bigvee_n (y \in G_n) \wedge \big(F(x) \cap \overline{G}_n = \emptyset\big),$$

and hence

$$X \times Y - G(F) = \bigcup_n [\{x: F(x) \cap \overline{G}_n = \emptyset\} \times G_n].$$

This completes the proof, since $G_n \in \boldsymbol{G}$ and by (7′)

$$\{x: F(x) \cap \overline{G}_n = \emptyset\} \in \boldsymbol{L}.$$

Remark: Theorem 8 can be deduced, of course, from Theorem 9.

THEOREM 10: (THIRD GRAPH THEOREM) *Let* $\boldsymbol{C}$ *be the family of all closed subsets of* Y, $X \in \boldsymbol{L} \subset \boldsymbol{P}(X)$, *and* $\boldsymbol{N}$ *a* δ*-lattice in* $X \times Y$ *such that* $\boldsymbol{L} \otimes \boldsymbol{C} \subset \boldsymbol{N}$.

If $F: X \to 2^Y$ *is lower-*$\boldsymbol{L}$, *then* $G(F) \in \boldsymbol{N}$.

PROOF. Let $G_1, G_2, \ldots$ be an open base of Y. Obviously,

$$[y \notin F(x)] \equiv \bigvee_n [(y \in G_n) \wedge \big(G_n \cap F(x)\big) = \emptyset],$$

and hence

$$G(F) = \bigcap_n [X \times (Y - G_n)] \cup [\{x: F(x) \cap G_n \neq \emptyset\} \times Y].$$

Our conclusion follows, because $\{x: F(x) \cap G_n \neq \emptyset\} \in \boldsymbol{L}$ (by (6)) and $\boldsymbol{L} \otimes \boldsymbol{C} \subset \boldsymbol{N}$.

THEOREM 11 (see Kuratowski [17]): *Let* $\boldsymbol{L}$ *be a* σ *and* δ*-lattice of subsets of* X. *Then for each mapping* $F: X \to \boldsymbol{K}(Y)$, *condition* (6) *is equivalent to*

$$\{x: F(x) \cap K \neq \emptyset\} \in \boldsymbol{L} \quad \textit{for each } K \textit{ closed in } Y. \tag{10}$$

PROOF. Put briefly $F^-(E) = \{x: F(x) \cap E \neq \emptyset\}$ for $E \subset Y$, and note that

$$F^-(\bigcup_t E_t) = \bigcup_t F^-(E_t). \tag{11}$$

1. (10) ⇒ (6). Let $U = K_1 \cup K_2 \cup \ldots$, where K_n is closed. By (11) we have $F^-(U) = \bigcup_n F^-(K_n)$ and $F^-(K_n) \in \boldsymbol{L}$ by (10).

Hence $F^-(U) \in \boldsymbol{L}$ (since $\boldsymbol{L}$ is countably additive).

2. (6) ⇒ (10). Let $U_1, U_2, \ldots$ be a countable open base of Y, closed under finite unions. By assumption, $F(x)$ is compact, and hence

$$\{F(x) \cap K = \emptyset\} \equiv \bigvee_n (F(x) \subset \overline{U}_n) \wedge (K \cap \overline{U}_n = \emptyset).$$

Hence $\{x\colon F(x) \cap K = \emptyset\} = \bigcup_n \{x\colon [F(x) \cap (Y - \overline{U}_n) = \emptyset]\}$, where n ranges over indices satisfying condition $K \cap \overline{U}_n = \emptyset$. By (6) we have

$$\{x\colon [F(x) \cap (Y - \overline{U}_n) = \emptyset\} \in \boldsymbol{L}^c$$

and this completes the proof (since $\boldsymbol{L}^c$ is countably additive).

COROLLARY 1: *If $\boldsymbol{R}$ is a σ-algebra, then under the above assumptions, the conditions of being upper and lower $\boldsymbol{R}$ are equivalent.*

Because, in this case, $\boldsymbol{R}^c = \boldsymbol{R}$. (It is thus justified to use the term "$\boldsymbol{R}$-measurable", e.g. Borel-measurable, $\overline{\boldsymbol{S}}$-measurable, etc., see also Chapter XIII, § 3.)

COROLLARY 2: *If $F\colon X \to \boldsymbol{K}(Y)$ is upper (or lower) $\boldsymbol{B}$-measurable, then F is $\boldsymbol{B}$-measurable.*

Because $\boldsymbol{K}(Y)$ is metric separable (see § 2, Remarks).

Remark (due to J. Kaniewski): *If we do not assume $F(x)$ of being compact* (assuming only that $F(x)$ is closed, i.e. $F\colon X \to 2^Y$), *then Theorem* 11 *is no more true* (even if $\boldsymbol{L}$ is supposed to be a σ-algebra and Y—Polish).

Namely, let us put $X = \mathscr{I}$ (the interval 01), $Y = \mathscr{I} \times N^N$, $\boldsymbol{L} = \boldsymbol{B}(\mathscr{I})$ (the σ-algebra of Borel subsets of the interval $\mathscr{I}$), p = the orthogonal projection of Y onto X and $F(x) = p^{-1}(x)$ for each $x \in X$. Finally, let K be a closed subset of Y such that $p^1(K)$ is not Borel (for the existence of such a set, see Chapter XIII, § 1(3) and Theorem 12).

Condition (10) is not fulfilled, since obviously

$$\{x\colon F(x) \cap K \neq \emptyset\} = \{x\colon p^{-1}(x) \cap K \neq \emptyset\} = p^1(K),$$

while (6) is true, because p is an open mapping, which means that $p^1(U)$ is open for each open $U \subset Y$.

The above remark answers in the negative a question raised by C. J. Himmelberg [1], namely whether weak "$\boldsymbol{L}$-measurability" (i.e. condition (6)) implies "$\boldsymbol{L}$-measurability".

Sets open modulo an ideal

Let $\boldsymbol{I}$ be an ideal of subsets of the space X. Denote by $\boldsymbol{L}$ the family of sets open mod $\boldsymbol{I}$. That means (see Exercise I.5.6) that $E \in \boldsymbol{L}$ iff E is of the form

(12) $$E = G - A \cup B \text{ where } G \text{ is open and } A, B \in \boldsymbol{I}.$$

THEOREM 12: *Suppose that f: $X \to Y$ and that $\boldsymbol{I}$ is a σ-ideal. Then is an $\boldsymbol{L}$-mapping iff there exists $A \in \boldsymbol{I}$ such that the partial mapping $f|(X-A)$ is continuous.*

PROOF. 1° The condition is necessary. Let $U_1, U_2, \ldots$ be an open base of Y. By assumption, $f^{-1}(U_n) \in \boldsymbol{L}$. Therefore,

$$f^{-1}(U_n) = G_n - A_n \cup B_n,$$

where G_n is open and $A_n \in \boldsymbol{I}$ and $B_n \in \boldsymbol{I}$.

Put $A = \bigcup\limits_n A_n \cup \bigcup\limits_n B_n$. Hence $A \in \boldsymbol{I}$. We have to show that the mapping $g = f|(X-A)$ is continuous. So let H be open in Y; we must show that the set $g^{-1}(H) = f^{-1}(H) - A$ is open in $X - A$.

Now $H = U_{k_1} \cup U_{k_2} \cup \ldots$ and hence

$$g^{-1}(H) = \bigcup_n f^{-1}(U_{k_n}) - A = \bigcup_n (G_{k_n} - A_{k_n} \cup B_{k_n}) - A.$$

Since $A_{k_n} \cup B_{k_n} \subset A$, it follows that

$$g^{-1}(H) = \Big(\bigcup_n G_{k_n}\Big) - A,$$

which completes the proof since $\bigcup\limits_n G_{k_n}$ is open.

2° The condition is sufficient. Let $A \in \boldsymbol{I}$ and let $g = f|(X-A)$ be continuous. Therefore, if H is open in Y, the set $g^{-1}(H) = f^{-1}(H) - A$ is open in $X - A$, that is, there is G open in X such that $f^{-1}(H) - A = G - A$. Therefore,

(13) $$\begin{aligned} f^{-1}(H) &= [f^{-1}(H) - A] \cup [f^{-1}(H) \cap A] \\ &= (G - A) \cup [f^{-1}(H) \cap A]. \end{aligned}$$

Since $A \in \boldsymbol{I}$, we have $[f^{-1}(H) \cap A] \in \boldsymbol{I}$ and it follows by (13) that $f^{-1}(H) \in \boldsymbol{L}$ (comp. (12)). Hence f is an $\boldsymbol{L}$-mapping.

§ 5. The operation $\mathscr{A}$ (see Lusin and Souslin [1])

DEFINITION: Let $\mathscr{S}$ denote the set of all finite sequences of natural numbers and let A be a set-valued function defined on $\mathscr{S}$. Thus $A(k_1, \ldots, k_n)$, denoted also by $A_{k_1 \ldots k_n}$, is a subset of the given space X, and $A\colon \mathscr{S} \to \boldsymbol{P}(X)$. Put

(1) $$\mathscr{A}(A) = \bigcup_{\mathfrak{z}} \bigcap_{n=1}^{\infty} A_{\mathfrak{z}|n},$$

where $\mathfrak{z} = (\mathfrak{z}^1, \mathfrak{z}^2, \ldots)$ ranges over N^N and $\mathfrak{z}|n$ denotes $(\mathfrak{z}^1 \ldots \mathfrak{z}^n)$.

The set $\mathscr{A}(A)$ is called "the result of the operation $\mathscr{A}$ applied to A". A is called *regular* if

(2) $$A_{\mathfrak{z}|n+1} \subset A_{\mathfrak{z}|n}.$$

Define A^* as follows

(3) $$A^*_{\mathfrak{z}|n} = A_{\mathfrak{z}^1} \cap A_{\mathfrak{z}^1\mathfrak{z}^2} \cap \ldots \cap A_{\mathfrak{z}|n}.$$

One may easily show the following formulas

(4) $$\mathscr{A}(A^*) = \mathscr{A}(A)$$

(thus every function A can be "regularized"), and for A regular

(5) $$\bigcup_m \bigcup_{\mathfrak{z}} \bigcap_k A_{(\mathfrak{y}|i)m(\mathfrak{z}|k)} = \bigcup_{\mathfrak{z}} \bigcap_k A_{(\mathfrak{y}|i)(\mathfrak{z}|k)},$$

where $(\mathfrak{y}|i)(\mathfrak{z}|k)$ means $\mathfrak{y}^1 \ldots \mathfrak{y}^i \mathfrak{z}^1 \ldots \mathfrak{z}^k$,

(6) $$A_{\mathfrak{z}|k} \subset B_{\mathfrak{z}|k} \quad \text{implies} \quad \bigcup_{\mathfrak{z}} \bigcap_k A_{\mathfrak{z}|k} \subset \bigcup_{\mathfrak{z}} \bigcap_k B_{\mathfrak{z}|k},$$

which means that the function $\mathscr{A}$ is *monotone*,

(7) $$X - \bigcup_{\mathfrak{z}} \bigcap_k A_{\mathfrak{z}|k} \subset \bigcup_{\mathfrak{z}} \bigcup_k \Big(A_{\mathfrak{z}|k} - \bigcup_m A_{(\mathfrak{z}|k)m}\Big)$$

where we assume that $A_{\mathfrak{z}|k} = X$ for $k = 0$.

THEOREM 1: *If A is regular and*

(8) $$[(\mathfrak{z}|n) \neq (\mathfrak{y}|n)] \Rightarrow (A_{\mathfrak{z}|n} \cap A_{\mathfrak{y}|n} = \emptyset),$$

then

(9) $$\bigcup_{\mathfrak{z}} \bigcap_{n} A_{\mathfrak{z}|n} = \bigcap_{n} \bigcup_{\mathfrak{z}} A_{\mathfrak{z}|n}.$$

PROOF. We have to show that the right-hand side of (9) is contained in its left-hand side.

So let p belong to the right-hand side. It follows that there is an index m_1 (and only one) such that $p \in A_{m_1}$. Similarly, there is a pair $q_1 m_2$ such that $p \in A_{q_1 m_2}$, and since $A_{q_1 m_2} \subset A_{q_1}$, we have $q_1 = m_1$. Proceeding in this way, we define a sequence $\mathfrak{z} = (m_1, m_2, m_3, \ldots)$. Thus $p \in A_{\mathfrak{z}|n}$ for each n. This completes the proof.

Applications of the operation $\mathcal{A}$ to complete spaces

Let X be a complete space. We assume that

$$A\colon \mathcal{S} \to 2^X \cup \{\emptyset\},$$

i.e. that the sets $A_{k_1 \ldots k_n}$ are *closed* (empty or not) subsets of X. Moreover, we assume that A is *regular*, i.e. that condition (2) is fulfilled, and finally that

(10) $$\lim_{k=\infty} \delta(A_{\mathfrak{z}|k}) = 0.$$

Put

(11) $$\mathfrak{Z} = \{\mathfrak{z}\colon \bigwedge_k (A_{\mathfrak{z}|k} \neq \emptyset)\}.$$

Then, for $\mathfrak{z} \in \mathfrak{Z}$, the set $A_{\mathfrak{z}|1} \cap A_{\mathfrak{z}|2} \cap \ldots$ reduces to a single point (by Theorem 2.7); let us denote this point by $f(\mathfrak{z})$. Thus

(12) $$f^1(\mathfrak{Z}) = \bigcup_{\mathfrak{z}} \bigcap_{k} A_{\mathfrak{z}|k}.$$

Put $\mathcal{N}_{n_1 \ldots n_k} = \{\mathfrak{z} \in N^N\colon (\mathfrak{z}^1 = n_1) \ldots (\mathfrak{z}^k = n_k)\}$.

Obviously the sets $\mathcal{N}_{n_1 \ldots n_k}$ are closed-open in N^N and form a base of this space.

It is easy to show that

(13) $$f^1(\mathfrak{Z} \cap \mathcal{N}_{n_1 \ldots n_k}) \subset A_{n_1 \ldots n_k},$$

and that, consequently,

(14) $\mathfrak{Z}$ *is closed in* N^N.

We shall show that

(15) *the function* $f\colon \mathfrak{Z} \to f^1(\mathfrak{Z})$ *is continuous*.

PROOF. Let $\mathfrak{z} \in \mathfrak{Z}$, $\varepsilon > 0$ and let k be an index such that $\delta(A_{\mathfrak{z}|k}) < \varepsilon$. Hence, by (13),

$$\delta[f^1(\mathfrak{Z} \cap \mathscr{N}_{\mathfrak{z}|k})] < \varepsilon .$$

This shows that f is continuous at $\mathfrak{z}$, since $\mathscr{N}_{\mathfrak{z}|k}$ is a neighborhood of $\mathfrak{z}$.

(16) *If* $X = \bigcup_i A_i$ *and* $A_{\mathfrak{z}|k} = \bigcup_i A_{(\mathfrak{z}|k)i}$ *for each* $\mathfrak{z}$ *and* k, *then* $f^1(\mathfrak{Z}) = X$.

PROOF. Let $x \in X$. We shall define by induction a sequence $n_1, n_2, \ldots$ Namely, let $x \in A_{n_1}$ and, assuming that $x \in A_{n_1 \ldots n_k}$, let n_{k+1} be such that $x \in A_{n_1 \ldots n_k n_{k+1}}$. We denote by $\mathfrak{z}$ the sequence $n_1, n_2, \ldots$ Then $x = f(\mathfrak{z})$.

(17) *If condition* (8) *is fulfilled, then* f *is one-to-one, and we have* (by Theorem 3.6)

(17′) $$f^1(\mathfrak{Z}) = \bigcap_k \bigcup_{\mathfrak{z}} A_{\mathfrak{z}|k}$$

and

(17″) $$f^1(\mathfrak{Z} \cap \mathscr{N}_{n_1 \ldots n_k}) = f^1(\mathfrak{Z}) \cap A_{n_1 \ldots n_k} .$$

This follows easily from Theorem 1.

The next statement is obvious.

(18) *If* $\mathfrak{z} \in \mathfrak{Z}$ *implies that* $\mathfrak{z}^k = 0$ *or* 2 *for each* k, *then* $\mathfrak{Z}$ *is a Cantor set* (i.e. homeomorphic to the Cantor discontinuum).

Let us note also that

(19) $$N^N = \bigcup_{\mathfrak{z}} \bigcap_k \mathscr{N}_{\mathfrak{z}|k}$$

and

(20) $$\{\mathfrak{z}\} = \bigcap_k \mathscr{N}_{\mathfrak{z}|k}, \qquad \lim_{k=\infty} \delta(\mathscr{N}_{\mathfrak{z}|k}) = 0 .$$

THEOREM 2: *Suppose that conditions* (2), (8) *and* (10) *are satisfied. Then under the assumption of* (18), *the set* $f^1(\mathfrak{Z})$ *is a Cantor set.*

Because $\mathfrak{Z}$ is (by (18)) a Cantor set, f is continuous (by (15)) and one-to-one (by (17)), hence a homeomorphism (see Exercise 1.3).

Finally, let us show that *if* f *is a continuous map of* N^N, *then, for each*

$\mathfrak{z} \in N^N$, *we have*

(21) $$\{f(\mathfrak{z})\} = \bigcap_k f^1(\mathcal{N}_{\mathfrak{z}|k}) = \bigcap_k \overline{f^1(\mathcal{N}_{\mathfrak{z}|k})}.$$

Because, by (20),

(22) $$f^1(\mathfrak{z}) = f^1(\bigcap_k \mathcal{N}_{\mathfrak{z}|k}) \subset \bigcap_k f^1(\mathcal{N}_{\mathfrak{z}|k}) \subset \bigcap_k \overline{f^1(\mathcal{N}_{\mathfrak{z}|k})},$$

and, by the continuity of f, we have

$$\lim_{k=\infty} \delta[f^1(\mathcal{N}_{\mathfrak{z}|k})] = 0, \quad \text{hence} \quad \lim_{k=\infty} \delta\overline{[f^1(\mathcal{N}_{\mathfrak{z}|k})]} = 0,$$

and finally, $\delta\left[\bigcap_k \overline{f^1(\mathcal{N}_{\mathfrak{z}|k})}\right] = 0$. Thus $\bigcap_k \overline{f^1(\mathcal{N}_{\mathfrak{z}|k})}$ reduces to a single point, namely to the point $f(\mathfrak{z})$. Thus (21) follows from (22).

Remarks on Hausdorff operations

The $\mathscr{A}$ operation can be generalized as follows. Let $A_1, A_2, \ldots$ be a sequence of sets and $B \subset N^N$. Put

$$x \in H \equiv \bigvee_{\mathfrak{z} \in B} \bigwedge_n x \in A_{\mathfrak{z}^n}.$$

We say that H is obtained from the sequence $A_1, A_2, \ldots$ by means of a *Hausdorff operation with basis B*.

See Hausdorff [2], p. 93, and independently Kolmogorov [1]. For the development and applications of Hausdorff operations see Kantorovitch and Livensohn [1], Katětov [1], Liapounov [1] and [2], where numerous references can be found. See also Oczan [1].

§ 6. The Lusin sieve (see Lusin [4])

DEFINITION 1: Let $\mathscr{R}_0$ denote the set of binary fractions

(1) $$r = \frac{1}{2^{k_1}} + \ldots + \frac{1}{2^{k_n}} \quad \text{where} \quad 1 \leqslant k_1 < \ldots < k_n.$$

A set-valued mapping $W: \mathscr{R}_0 \to \boldsymbol{P}(X)$, where X is the given space, is called a *sieve*. We denote by $L(W)$ the set of all points x such that there is an infinite sequence $r_1, r_2, \ldots$ satisfying conditions

(2) $$r_1 < r_2 < \ldots \quad \text{and} \quad x \in \bigcap_n W(r_n)$$

($L(W)$ is said to be *sifted* by W).

In other words, if

(3) $M_x = \{r\colon x \in W(r)\}$ or equivalently $W(r) = \{x\colon r \in M_x\}$,

then

(4) $L(W) = \{x\colon M_x$ is not well ordered by the relation $r \geqslant s\}$.

Remark. There is a one-to-one correspondence between the sets $\mathscr{R}_0$ and $\mathscr{S}$, namely the correspondence

$$C(r) = \{k_1, k_2 - k_1, \ldots, k_n - k_{n-1}\} \tag{5}$$

where r is given by formula (1).

THEOREM 1 (see Lusin and Sierpiński [1] and Sierpiński [9]): *Let $H\colon \mathscr{S} \to \mathbf{P}(X)$ be regular and let $W(r) = H_{C(r)}$. Then $\mathscr{A}(H) = L(W)$.*

PROOF. 1° Let $x \in \mathscr{A}(H)$. Then $x \in H_{k_1} \cap H_{k_1 k_2} \cap \ldots$ Put

$$r_n = \frac{1}{2^{k_1}} + \frac{1}{2^{k_1+k_2}} + \ldots + \frac{1}{2^{k_1+\ldots+k_n}}.$$

Then (2) is fulfilled and hence $x \in L(W)$.

2° Let $x \in L(W)$. Hence (2) is fulfilled. Put

$$\lim_{n=\infty} r_n = \sum_{n=1}^{\infty} \frac{1}{2^{m_1}} \quad \text{where} \quad 1 \leqslant m_1 < m_2 < \ldots$$

Put $k_1 = m_1$ and $k_n = m_n - m_{n-1}$ for $n > 1$. Let j_n be such that

$$\sum_{i=1}^{n} \frac{1}{2^{m_i}} < r_{j_n} < \sum_{i=1}^{\infty} \frac{1}{2^{m_i}}$$

and let

$$r_{j_n} = \frac{1}{2^{q_1}} + \ldots + \frac{1}{2^{q_s}} \quad \text{where} \quad 1 \leqslant q_1 < \ldots < q_s.$$

Hence $q_1 = m_1, \ldots, q_n = m_n$. It follows that the first n terms of $C(r_{j_n})$ are identical to $k_1, \ldots, k_n$ respectively. Since H is regular, we have $H_{C(r_{j_n})} \subset H_{k_1 \ldots k_n}$. Hence

$$x \in H_{k_1 \ldots k_n} \text{ for } n = 1, 2, \ldots, \quad \text{i.e.} \quad x \in \mathscr{A}(H).$$

Remark. If $X = \mathscr{E}$, the sieve has the following geometrical interpretation. Put $W(r)$ on the line $y = r$. Then M_{x_0} is the intersection of $\bigcup_r W(r)$ with the vertical line $x = x_0$.

DEFINITION 2: Let $A = L(W)$. We call *constituents* of $X-A$ (relative to W) the sets

$$C_\alpha = \{x\colon\ M_x \text{ has order-type } \alpha\}, \qquad \alpha < \omega_1.$$

Thus

$$X-A = \bigcup_{\alpha<\omega_1} C_\alpha.$$

Example. Let $X = \mathscr{C}$ (the Cantor discontinuum). Let $\mathfrak{z} \in \mathscr{C}$. Then

$$\mathfrak{z} = \frac{\mathfrak{z}^1}{3} + \frac{\mathfrak{z}^2}{9} + \ldots \quad \text{where} \quad \mathfrak{z}^n = 0 \text{ or } 2.$$

Denote by $\mathscr{R}_0 = \{r_1, r_2, \ldots\}$ the sequence of binary fractions. Let

$$R_\mathfrak{z} = \{r_n\colon\ \mathfrak{z}^n = 2\} \quad \text{and } \bar{\mathfrak{z}} \text{ the order-type of } R_\mathfrak{z},$$

and

$$C_\tau = \{\mathfrak{z}\colon\ \bar{\mathfrak{z}} = \tau\} \quad \text{and} \quad W(r) = \{\mathfrak{z}\colon\ r \in R_\mathfrak{z}\}.$$

Then the sets C_α are the constituents of $\bigcup_{\alpha<\omega_1} C_\alpha$ relative to the sieve W.

Note that $C_\tau \neq \varnothing$ for any countable order-type τ (because $\mathscr{R}_0$ has a dense order type).

CHAPTER XII

BOREL SETS. *B*-MEASURABLE FUNCTIONS BAIRE PROPERTY

§ 1. Elementary properties of Borel subsets of a metric space

Let us recall that the family $\boldsymbol{B}(X)$ of *Borel* subsets of the space X is, by definition, the σ-algebra (see p. 126) generated by all closed subsets of X.

In this definition, the term "closed" can be replaced by "open" since every open subset of a metric space is an $\boldsymbol{F}_\sigma$-set (countable union of closed sets, see Chapter XI, § 1,8).

We have seen (comp. p. 241) that the family of Borel subsets of X admits two natural classifications into $\aleph_1$ classes:

$$\boldsymbol{B}(X) = \bigcup_{\alpha<\omega_1} \boldsymbol{F}_\alpha \quad \text{and} \quad \boldsymbol{B}(X) = \bigcup_{\alpha<\omega_1} \boldsymbol{G}_\alpha$$

where $\boldsymbol{F}_0$ is the family of closed sets, $\boldsymbol{F}_1$ is the family $\boldsymbol{F}_\sigma$, ..., symmetrically $\boldsymbol{G}_0$ is the family of open sets, $\boldsymbol{G}_1$ is the family $\boldsymbol{G}_\delta$, and so on. The families $\boldsymbol{F}_\alpha$ with even indices as well as the families $\boldsymbol{G}_\alpha$ with odd indices are countably multiplicative. The sets belonging to such a family will be said to be of *multiplicative class* α. Similarly, the families $\boldsymbol{F}_\alpha$ with odd indices as well as the families $\boldsymbol{G}_\alpha$ with even indices are countable additive and form the *additive class* α. Thus the multiplicative class α (additive class α) consists of the countable intersections (unions) of sets of classes $< \alpha$.

The classes with finite indices are denoted as follows:

$$\boldsymbol{F}_\sigma, \boldsymbol{F}_{\sigma\delta}, \boldsymbol{F}_{\sigma\delta\sigma}, \ldots, \qquad \boldsymbol{G}_\delta, \boldsymbol{G}_{\delta\sigma}, \boldsymbol{G}_{\delta\sigma\delta}, \ldots$$

The following properties of $\boldsymbol{F}_\alpha$ and $\boldsymbol{G}_\alpha$-sets extend by induction to classes of an arbitrary index.

The *complement* of a set of class $\boldsymbol{F}_\alpha$ is of class $\boldsymbol{G}_\alpha$. The *finite union* and *finite intersection* of sets of the same class belong to that class. Thus $\boldsymbol{F}_\alpha$ with an odd index (and $\boldsymbol{G}_\alpha$ with an even index) is a *σ-lattice*. Similarly, $\boldsymbol{F}_\alpha$ with an even index (and $\boldsymbol{G}_\alpha$ with an odd index) is a *δ-lattice*.

Every set of an additive class α is the union of an *increasing* sequence of sets with indices $\xi < \alpha$. Every set of a multiplicative class α is the intersection of a *decreasing* sequence of sets with indices $\xi < \alpha$. A set is of class $\boldsymbol{F}_\alpha$ (of class $\boldsymbol{G}_\alpha$) relative to a set E if and only if it is the intersection of E with a set of class $\boldsymbol{F}_\alpha$ (of class $\boldsymbol{G}_\alpha$).

Observe that a Borel set of class α belongs to every ($\boldsymbol{F}$ or $\boldsymbol{G}$) class of a *greater index*. This follows by induction from the fact that every open set is an $\boldsymbol{F}_\sigma$-set and every closed set is a $\boldsymbol{G}_\delta$-set.

The *Cartesian product* of two sets of class $\boldsymbol{F}_\alpha$ (of class $\boldsymbol{G}_\alpha$) belongs to the same class. For, the product of two open sets is open and the product of two closed sets is closed and

$$\Big(\bigcup_n A_n\Big)\times\Big(\bigcup_m B_m\Big) = \bigcup_{nm}(A_n\times B_m),$$

$$\Big(\bigcap_n A_n\Big)\times\Big(\bigcap_m B_m\Big) = \bigcap_{nm}(A_n\times B_m).$$

In particular, the product of a set by an *axis* does not change its class. From this and from the formula

$$\prod_{i\in N} A_i = \bigcap_i (X_1\times \ldots \times X_{i-1}\times A_i\times X_{i+1}\times \ldots)$$

we derive that the *countable* Cartesian product of sets of multiplicative class α is of the multiplicative class α (the analogous theorem for additive classes is not true, even for open sets).

A countable Cartesian product of Borel sets is a Borel set.

THEOREM: *If a subset Z of the Cartesian product $X\times Y$ is of class $\boldsymbol{F}_\alpha$ or of class $\boldsymbol{G}_\alpha$, then the sets*

(1) $\qquad A_y = \{x\colon [\langle x, y\rangle \in Z]\}$ *and* $\{x\colon [\langle x, x\rangle \in Z]\}$

are of the same class (the second conclusion concerns the case where $X = Y$).

PROOF. The sets $Z\cap\{\langle x, y\rangle\colon (y = y_0)\}$ and $Z\cap\{\langle x, y\rangle\colon (x = y)\}$ are of class $\boldsymbol{F}_\alpha(\boldsymbol{G}_\alpha)$ relative to $\{\langle x, y\rangle\colon (y = y_0)\}$ and $\{\langle x, y\rangle\colon (x = y)\}$, respectively; the same is true of the sets (1).

In a *separable space*, the family of open sets (as well as the family of closed sets) has power $\leqslant \mathfrak{c}$. The same is true of each Borel class. Since the family of Borel sets is the union of $\aleph_1$ classes, it follows that its power is $\leqslant \mathfrak{c}\,\aleph_1 = \mathfrak{c}$. Therefore *in every separable space of power* $\mathfrak{c}$ *there exist non-borelian sets.*

The problem of an effective definition of a non-borelian set in the space of real numbers will be treated in § 5.

The problem of proving without the continuum hypothesis the existence of a non-borelian set in every uncountable separable space is open.

§ 2. Ambiguous Borel sets

A set E which belongs to both $\boldsymbol{F}_\alpha$ and $\boldsymbol{G}_\alpha$ is said to be *ambiguous of class* α; in symbols $E \in \boldsymbol{A}_\alpha$; for instance, a set both closed and open is ambiguous of class 0; it is ambiguous of class 1, if it is both $\boldsymbol{F}_\sigma$ and $\boldsymbol{G}_\delta$.

Every Borel set of class α *is ambiguous of the class* $\alpha+1$.

Clearly the complement of an ambiguous set is ambiguous (of the same class). Therefore $\boldsymbol{A}_\alpha$ *is an algebra*; thus, the union, intersection and difference of two ambiguous sets of class α are ambiguous of class α.

THEOREM 1: *Every set of additive class* $\alpha > 0$ *is a countable union of disjoint ambiguous sets of class* α; *hence* $\boldsymbol{G}_\alpha = \boldsymbol{A}_{\alpha\sigma}$.

PROOF. Let $A = A_1 \cup A_2 \cup \ldots \cup A_n \cup \ldots$ Then

(1) $$A = A_1 \cup [A_2 - A_1] \cup \ldots \cup [A_n - (A_1 \cup \ldots \cup A_{n-1})] \cup \ldots$$

If each A_n is of multiplicative class $< \alpha$, then it is ambiguous of class α. Hence the terms of (1) are disjoint ambiguous sets of class α.

THEOREM 2 (of Lusin, see Sierpiński [8]): *Every set of the additive class* $\alpha > 1$ *is a countable union of disjoint sets of multiplicative classes* $< \alpha$.

PROOF. Consider the decomposition (1). Since each A_n is of a multiplicative class $< \alpha$, then so is the union $A_1 \cup \ldots \cup A_{n-1}$. Therefore the set $X - (A_1 \cup \ldots \cup A_{n-1})$ is of additive class $< \alpha$. By Theorem 1 this set is of the form $\bigcup_{i=1}^{\infty} B_i^n$, where B_i^n are disjoint ambiguous sets of classes $< \alpha$ (for $\alpha > 1$). Thus

$$A = \bigcup_{n=1}^{\infty} \bigcup_{i=1}^{\infty} A_n \cap B_i^n .$$

is the desired decomposition, because $A_n \cap B_i^n$ is of a multiplicative class $< \alpha$.

The following theorem can easily be shown (see e.g. Kuratowski [1], p. 348).

THEOREM 3: *The family of Borel sets is the smallest family that contains:*

(i) *all open sets;*

(ii) *the countable intersections of its elements;*

(iii) *disjoint countable unions of its elements.*

Since, by Theorem 1, $G_\alpha = A_{\alpha\sigma}$, where A_α ($\alpha > 0$) is a field, it follows by Theorem IV.5.1 that G_α has the countable reduction property:

THEOREM 4: (REDUCTION THEOREM) *For every (finite or infinite) sequence $G_1, G_2, \ldots$ of sets of additive class $\alpha > 0$ there exists a sequence $H_1, H_2, \ldots$ of disjoint sets of additive class $\alpha > 0$ such that*

$$H_i \subset G_i \quad \text{and} \quad H_1 \cup H_2 \cup \ldots = G_1 \cup G_2 \cup \ldots$$

Consequently, if $X = G_1 \cup G_2 \cup \ldots$, then the sets H_i are ambiguous of class α.

Similarly, we deduce from Corollary IV.5.2 the following theorems.

THEOREM 5: (SEPARATION THEOREM) *If $F_1, F_2, \ldots$ is a sequence of sets of multiplicative class $\alpha > 0$ such that $F_1 \cap F_2 \cap \ldots = \emptyset$, then there exists a sequence $E_1, E_2, \ldots$ of ambiguous sets of class α such that*

$$F_i \subset E_i \quad \text{and} \quad E_1 \cap E_2 \cap \ldots = \emptyset .$$

In particular, if A and B are two disjoint sets of multiplicative class $\alpha > 0$, then there exists an ambiguous set E of class α such that

$$A \subset E \quad \text{and} \quad E \cap B = \emptyset . \tag{3}$$

In other words, if $A \subset C$ are two sets of class $\alpha > 0$ such that A is of the multiplicative class and C is of the additive class, then there exists an ambiguous set E of class α such that

$$A \subset E \subset C . \tag{4}$$

(See Sierpiński [6] and [16] where many applications are given.)

THEOREM 6. *If $F_1, F_2, \ldots$ is a sequence of sets of the multiplicative class $\alpha > 0$, then there exists a sequence $B_1, B_2, \ldots$ of the additive class α such that*

$$F_i - \bigcap_{m=1}^{\infty} F_m \subset B_i \quad \textit{and} \quad \bigcap_{i=1}^{\infty} B_i = 0.$$

Remark. If the space X is 0-dimensional (separable), i.e. contains a countable base of class A_0 (of closed-open sets), then the above theorems are true also for the case $\alpha = 0$. In particular, in that case, each F_σ-set is a countable union of disjoint closed sets.

§ 3. Borel-measurable functions

Let X and Y be metric.

DEFINITIONS: A function $f: X \to Y$ is called *Borel-measurable* (briefly *B-measurable*) if $f^{-1}(G)$ is Borel in X whenever G is open in Y.

More precisely: f is *B-measurable of class* α (briefly: is of class α) if $f^{-1}(G)$ is of additive class α.

Equivalently: f is of class α if $f^{-1}(F)$ is of multiplicative class α in X whenever F is closed in Y. (Comp. p. 400.)

Obviously, the class 0 is composed of all continuous functions.

If f is one-to-one it is said to be a (*generalized*) *homeomorphism of class* α,β if f is of class α and its inverse f^{-1} is of class β.

Clearly, homeomorphisms of class 0,0 coincide with the homeomorphisms in the usual sense.

THEOREM 1: *If $R_1, R_2, \ldots$ is an open base of Y, then f is of class α if each set $f^{-1}(R_n)$ is of additive class α.*

PROOF. Since $G = R_{k_1} \cup R_{k_2} \cup \ldots$, we have

$$f^{-1}(G) = f^{-1}(R_{k_1}) \cup f^{-1}(R_{k_2}) \cup \ldots$$

It follows also that if the sets $f^{-1}(R_n)$ *are Borel*, for $n = 1, 2, \ldots$, then the function *f is B-measurable*. In fact, its class is α, where $\alpha > \alpha_n$ and where $f^{-1}(R_n)$ is of class α_n.

In particular, if Y is the space of *real numbers*, the functions of class α can be defined by assuming that the sets $\{x: [a < f(x) < b]\}$ are of additive class α, for every a and b (which can be assumed to be rational).

COROLLARY: *If the space Y is discrete and the function f is of class α, then the set $f^{-1}(B)$ is ambiguous of class α for every $B \subset Y$.*

Because B *is both closed and open in* Y.

THEOREM 2: *If f is a function of class α and B is a set of class β, then the set $f^{-1}(B)$ is of class $\alpha+\beta$ (which is multiplicative or additive according to the class of B).*

The proof is by transfinite induction (on β). We use the identities

$$f^{-1}(\bigcup_n B_n) = \bigcup_n f^{-1}(B_n), \quad f^{-1}(\bigcap_n B_n) = \bigcap_n f^{-1}(B_n),$$

and the implication $\beta_n < \beta$ implies $\alpha+\beta_n < \alpha+\beta$.

In particular if f is *continuous*, i.e. $\alpha = 0$, then $f^{-1}(B)$ is of class β.

THEOREM 3: *If $f: X \to Y$ is of class α and $g: Y \to Z$ is of class β, then the function $h = g \circ f$ is of class $\alpha+\beta$.*

PROOF. We have

$$h^{-1}(F) = f^{-1}[g^{-1}(F)];$$

and hence, if the set F is closed, then $g^{-1}(F)$ is of the multiplicative class β, and $f^{-1}[g^{-1}(F)]$ is, by Theorem 2, of class $\alpha+\beta$.

In particular, *if g is continuous, then the functions $g \circ f$ and $f \circ g$ are of the same class as f.*

THEOREM 4: *If f is the characteristic function of the set $E \subset X$, then f is of class α iff $E \in \boldsymbol{A}_\alpha$.*

THEOREM 5: *Let Y be separable and $f: X \to Y$ of class α. Then the graph of f, i.e. the set*

$$\mathrm{Gr}(f) = \{\langle x, y\rangle: y = f(x)\}$$

is of the multiplicative class α.

This theorem is a particular case of Theorem XI.4.8, letting $\boldsymbol{L}$ = the additive class α in X and $\boldsymbol{N}$ = the additive class α in $X \times Y$.

To give a more direct proof based on a different idea (see Hausdorff [2], p. 269), let us consider the function $h(x, y) = |y-f(x)|$. Then

$$\mathrm{Gr}(f) = \{\langle x, y\rangle: [h(x, y) = 0]\} = h^{-1}(0).$$

Since the function h is of class α, the set $h^{-1}(0)$ is of the multiplicative class α.

Remark. The condition of separability of Y can be removed using a method of D. Montgomery [1], p. 532 (see also R. Engelking [2]).

THEOREM 6 (See Hausdorff [2], p. 267): *The limit of a convergent sequence of functions of class α is of class $\alpha+1$.*

PROOF. Let $f(x) = \lim_{n=\infty} f_n(x)$. Let $F = \bar{F} \subset Y$ and write $B_n = \{y: \varrho(y, F) < 1/n\}$. Then by XI.1 (13)

$$f^{-1}(F) = \bigcap_n \bigcup_k f_{n+k}^{-1}(B_n).$$

Since the functions f_n are of class α, the set $f_{n+k}^{-1}(B_n)$ is of the additive class α. Consequently the set $f^{-1}(F)$ is of the multiplicative class $\alpha+1$.

Thus, in particular, the limit of a sequence of continuous functions is of class 1. The limit of a sequence of functions of finite classes is of class $\omega+1$.

THEOREM 7: *The limit of a uniformly convergent sequence of functions of class α is of class α.*

This is a particular case of Theorem XI.4.3.

Remark. The converse of Theorem 6 is not generally true for $\alpha = 1$ (as shown by taking for Y a space composed of two points). However it is true for $Y = \mathscr{I}$.

More generally, by the Lebesgue–Hausdorff theorem, for $Y = \mathscr{I}$, the class Φ_α of analytically representable functions (see p. 236) coincides with the family of **B**-measurable functions of class α or $\alpha+1$, (according to as α is finite or infinite) see Lebesgue [1], p. 168 and Hausdorff [2], Chapter 9.

For further generalizations, see de la Vallée Poussin [1], p. 118 and S. Banach [2], p. 287.

§ 4. *B*-measurable complex and product functions

A pair of functions $f: T \to X$, $g: T \to Y$ define a "*complex*" function $h: T \to X \times Y$, namely $h(t) = \langle f(t), g(t) \rangle$.

THEOREM 1: *Let the spaces X and Y be separable. The function h is of class α iff the functions f and g (the "coordinates") are of class α.*

More generally, let $X_1, X_2, \ldots$ be a sequence of separable spaces and $\mathfrak{z}: T \to X_1 \times X_2 \times \ldots$, i.e. the function $\mathfrak{z}$ represents a sequence of func-

tions $\mathfrak{z}_1, \mathfrak{z}_2, \ldots$ *The function* $\mathfrak{z}$ *is of class* α *iff each function* $\mathfrak{z}_i$ *is of this class.*

Theorem 1 follows directly from Theorem XI.4.4 substituting for L the additive class α in T.

Let f_i: $X_i \to Y_i$ and let $\mathfrak{z} = [\mathfrak{z}^1, \mathfrak{z}^2, \ldots]$ be a variable point of $X_1 \times X_2 \times \ldots$; then putting $\mathfrak{y}(\mathfrak{z}) = [f_1(\mathfrak{z}^1), f_2(\mathfrak{z}^2), \ldots]$ we get the *product function*

$$\mathfrak{y}\colon X_1 \times X_2 \times \ldots \to Y_1 \times Y_2 \times \ldots$$

THEOREM 2: *If each function* f_i: $X_i \to Y_i$ *is of class* α, *then so is* $\mathfrak{y}$, *provided that the spaces* Y_i *are separable.*

If a separable space Y_i *is an image of a space* X_i *under a mapping of class* α, *then the space* $Y_1 \times Y_2 \times \ldots$ *is also an image of* $X_1 \times X_2 \times \ldots$ *under a mapping of class* α.

This follows from Theorem XI.4.5a substituting for L_i the family of subsets of additive class α of X_i, and for N—the family of subsets of class α of $X_1 \times X_2 \times \ldots$

THEOREM 3: *If* f_i, $i = 1, 2, \ldots$, *is a mapping of class* α *of a space* X_i *onto a subset of a separable space* Y, *then the set*

$$\mathfrak{Z} = \{\mathfrak{z}\colon [f_1(\mathfrak{z}^1) = f_2(\mathfrak{z}^2) = \ldots]\}, \quad \textit{where} \quad \mathfrak{z} \in (X_1 \times X_2 \times \ldots),$$

is of the multiplicative class α *in* $X_1 \times X_2 \times \ldots$

Moreover, the function f^* *defined by the condition*

$$f^*(\mathfrak{z}) = f_1(\mathfrak{z}^1) \quad \textit{for} \quad \mathfrak{z} \in \mathfrak{Z}$$

is a mapping of class α *of* $\mathfrak{Z}$ *onto* $f^{*1}(\mathfrak{Z}) = f_1^1(X_1) \cap f_2^1(X_2) \cap \ldots$

Finally, if the functions f_i *are homeomorphisms of class* α, β *then so is the function* f^* (*provided that* X_i *are separable*).

For the proof see Theorems XI.4.6–7.

THEOREM 4: *In order that the characteristic function of a sequence of sets* $A_1, A_2, \ldots$ *should be of class* α *it is necessary and sufficient that each of the sets be ambiguous of class* α.

This is a direct consequence of Theorem 1.

Examples. (i) If Y is separable and the functions f_1: $X_1 \to Y$, f_2: $X_2 \to Y$ are of class α, and $h(x_1, x_2) = |f_1(x_1) - f_2(x_2)|$, then h is of class α.

We apply Theorem 2 to $g(y_1, y_2) = |y_1 - y_2|$, the distance being a continuous function.

(ii) If f_1 and f_2 are two real-valued functions of class α, then $f_1(x_1) \pm \pm f_2(x_2)$, $f_1(x_1) \cdot f_2(x_2)$, $f_1(x_1) : f_2(x_2)$ define functions of the same class.

§ 5. Universal functions for Borel classes

Given a family $\boldsymbol{F}$ of sets and a set T, a *universal function* for $\boldsymbol{F}$ is a set-valued function F: $T \to \boldsymbol{F}$ onto, i.e.

$$\{A \in \boldsymbol{F}\} \equiv \bigvee_t [A = F(t)]$$

(compare Sierpiński [12], p. 82).

We shall assume subsequently that the space X (whose subsets are elements of $\boldsymbol{F}$) is separable metric. Let $T = \mathcal{N} = N^N$. If $\boldsymbol{F}$ is of power $\leqslant \mathfrak{c}$, there obviously exists a universal function for $\boldsymbol{F}$. Hence any Borel class $\boldsymbol{F}_\alpha$ or $\boldsymbol{G}_\alpha$ can be substituted for $\boldsymbol{F}$.

THEOREM: *For each $\alpha < \omega_1$ there exists a universal function G_α with respect to $\boldsymbol{G}_\alpha$ such that the set $\{\langle \mathfrak{z}, x\rangle : [x \in G_\alpha(\mathfrak{z})]\}$ is a $\boldsymbol{G}_\alpha$-set in the product $X \times N^N$* (this set is called the *graph* of the relation $x \in G_\alpha(\mathfrak{z})$. See XI, § 4 (8).

PROOF (compare Lebesgue [1], p. 209). We shall use the following notation. As usual, $\mathfrak{z} \in N^N$ is regarded as a sequence $\mathfrak{z}^{(1)}, \mathfrak{z}^{(2)}, \ldots$ of natural numbers. Since the space N^N is homeomorphic to $(N^N)^N$ (comp. Chapter XI, § 1, 18), there is a one-to-one correspondence between irrational numbers $\mathfrak{z}$ and all sequences of irrational numbers $\mathfrak{z}_{(1)}, \mathfrak{z}_{(2)}, \ldots$ such that, for a fixed n, $\mathfrak{z}_{(n)}$ is a continuous function of $\mathfrak{z}$. For example, we can put

$$\mathfrak{z}_{(n)} = [\mathfrak{z}^{(2^{n-1})}, \ldots, \mathfrak{z}^{(2^{n-1}+k \cdot 2^n)}, \ldots].$$

To every transfinite limit ordinal $\lambda < \omega_1$ let us assign a sequence $\lambda_1 < \lambda_2 < \ldots$ converging to λ (its existence follows from the axiom of choice).

Let $R_1, R_2, \ldots$ be an open base of the space (containing the empty set). We define the function G_α as follows.

1) $G_0(\mathfrak{z}) = \bigcup_n R_{\mathfrak{z}^{(n)}}$;

2) $G_{\alpha+1}(\mathfrak{z}) = \bigcap_n G_\alpha(\mathfrak{z}_{(n)})$ or $\bigcup_n G_\alpha(\mathfrak{z}_{(n)})$, depending whether α is even or odd;

3) $G_\lambda(\mathfrak{z}) = \bigcup_n G_{\lambda_n}(\mathfrak{z}_{(n)})$ if λ is a limit ordinal.

We have to prove that:

(i) the set $G_\alpha(\mathfrak{z})$ is of class $\boldsymbol{G}_\alpha$;

(ii) if $\boldsymbol{A}$ is of class $\boldsymbol{G}_\alpha$, there exists $\mathfrak{z} \in N^N$ such that $A = G_\alpha(\mathfrak{z})$;

(iii) the set $\{\langle \mathfrak{z}, x\rangle: [x \in G_\alpha(\mathfrak{z})]\}$ is of class $\boldsymbol{G}_\alpha$.

Property (i) is a direct consequence of (iii).

PROOF OF (ii). First assume that A is of class $\boldsymbol{G}_0$, i.e., an open set. By definition of the base, A has the form $A = \bigcup_n R_{k_n}$. Let $\mathfrak{z}$ be an element of N^N such that $\mathfrak{z}^{(1)} = k_1$, $\mathfrak{z}^{(2)} = k_2$, ... Then

$$G_0(\mathfrak{z}) = \bigcup_n R_{\mathfrak{z}^{(n)}} = \bigcup_n R_{k_n} = A.$$

Thus condition (ii) is satisfied for $\alpha = 0$.

Assuming it for α, we shall prove it for $\alpha+1$. Let A be a set of class $\boldsymbol{G}_{\alpha+1}$. Then

$$A = \bigcap_n A_n \quad \text{or} \quad A = \bigcup_n A_n$$

(according to as α is even or odd), where A_n is of class $\boldsymbol{G}_\alpha$. By hypothesis there exists a sequence $\{\mathfrak{z}_n\}$ such that $A_n = G_\alpha(\mathfrak{z}_n)$. By definition of the function $\mathfrak{z}_{(n)}$, there exists $\mathfrak{z}$ such that $\mathfrak{z}_n = \mathfrak{z}_{(n)}$ for every n. It follows that

$$G_{\alpha+1}(\mathfrak{z}) = \bigcap_n G_\alpha(\mathfrak{z}_{(n)}) = A \quad \text{or} \quad G_{\alpha+1}(\mathfrak{z}) = \bigcup_n G_\alpha(\mathfrak{z}_{(n)}) = A$$

according to as α is even or odd.

Now suppose that $\lambda = \lim \lambda_n$ and that (ii) holds for each λ_n. If A is a set of class $\boldsymbol{G}_\lambda$, then $A = \bigcup_n A_n$ where A_n is of class $\boldsymbol{G}_{\alpha_n}$ with $\alpha_n < \lambda$. Since the sequence $\{\lambda_n\}$ converges to λ, there exists a k_n for each n such that $\alpha_n \leqslant \lambda_{k_n}$. Therefore, A_n is of class $\boldsymbol{G}_{\lambda_{k_n}}$. Hence there exists $\mathfrak{z}_{k_n}$ with $A_n = G_{\lambda_{k_n}}(\mathfrak{z}_{k_n})$. If i is different from all k_n, let $\mathfrak{z}_i$ be such that

$G_{\lambda_i}(\mathfrak{z}_i) = \emptyset$. Hence $A = \bigcup_n G_{\lambda_n}(\mathfrak{z}_n)$. As before, let $\mathfrak{z}$ be such that $\mathfrak{z}_n = \mathfrak{z}_{(n)}$. Then

$$A = \bigcup_n G_{\lambda_n}(\mathfrak{z}_{(n)}) = G_\lambda(\mathfrak{z}).$$

PROOF OF (iii). First observe that $\{\langle x, n\rangle: (x \in R_n)\}$ is open in $X \times N$. In other words, the formula $x \in R_n$ (of two variables) is of class $\boldsymbol{G}_0$. As the function $\mathfrak{z}^n$, for a fixed n, is continuous, then the formula $x \in R_{\mathfrak{z}^{(n)}}$ is also of class $\boldsymbol{G}_0$. The same is true of the formula $\bigvee_n (x \in R_{\mathfrak{z}^{(n)}})$ which is equivalent to $x \in \bigcup_n R_{\mathfrak{z}^{(n)}}$. Consequently, the formula $x \in G_0(\mathfrak{z})$ is of class $\boldsymbol{G}_0$ and the set $\{\langle \mathfrak{z}, x\rangle: [x \in G_0(\mathfrak{z})]\}$ is open. Similarly, if the formula $x \in G_0(\mathfrak{z})$ is of class α, then so is $x \in G_\alpha(\mathfrak{z}_{(n)})$, for a fixed n, since $\mathfrak{z}_{(n)}$ is a continuous function of $\mathfrak{z}$. It follows that the formula $\bigwedge_n [x \in G_\alpha(\mathfrak{z}_{(n)})]$ (for an even α) is of class $\boldsymbol{G}_{\alpha+1}$ (the case of α odd is similar). Since

$$\bigwedge_n [x \in G_\alpha(\mathfrak{z}_{(n)})] \equiv \{x \in \bigcap_n G_\alpha(\mathfrak{z}_{(n)})\} \equiv \{x \in G_{\alpha+1}(\mathfrak{z})\},$$

it follows that condition (iii) is satisfied for $\alpha+1$. Finally, if the formula $x \in G_{\lambda_n}(\mathfrak{z})$, for every n, is of class $\boldsymbol{G}_{\lambda_n}$, then the formula $\bigvee_n [x \in G_{\lambda_n}(\mathfrak{z}_{(n)})]$ is of class $\boldsymbol{G}_\lambda$. As before we infer that $\{\langle \mathfrak{z}, x\rangle: [x \in G_\lambda(\mathfrak{z})]\}$ is of class G_λ.

Remark. An analogous theorem on the classes $\boldsymbol{F}_\alpha$ is true: For every α there exists a universal function F_α such that the set $\{\langle \mathfrak{z}, x\rangle: [x \in F_\alpha(\mathfrak{z})]\}$ is of class $\boldsymbol{F}_\alpha$. Namely, $F_\alpha(\mathfrak{z}) = X - G_\alpha(\mathfrak{z})$.

Existence of sets of class $\boldsymbol{G}_\alpha$ *which are not of class* $\boldsymbol{F}_\alpha$ (see Lusin [4], p. 290). We shall prove the existence of such sets in the space N^N. Let $X = N^N$ and consider the set

$$Z_\alpha = \{\mathfrak{z}: [\mathfrak{z} \in G_\alpha(\mathfrak{z})]\}$$

which is the projection, on the N^N axis, of the intersection of $\{\langle \mathfrak{z}, \mathfrak{z}'\rangle: [\mathfrak{z} \in G_\alpha(\mathfrak{z}')]\}$ with the *diagonal* of $N^N \times N^N$, i.e. with the set $\{\langle \mathfrak{z}, \mathfrak{z}'\rangle: [\mathfrak{z} = \mathfrak{z}']\}$. Since the set $\{\langle \mathfrak{z}, \mathfrak{z}'\rangle: [\mathfrak{z} \in G_\alpha(\mathfrak{z}')]\}$ is of class $\boldsymbol{G}_\alpha$, so is Z_α.

It remains to prove that Z_α is not of class $\boldsymbol{F}_\alpha$. Suppose it is; then $N^N - Z_\alpha$ is a $\boldsymbol{G}_\alpha$-set. Since the function G_α is universal, there exists a $\mathfrak{z}_0$

such that $N^N - Z_\alpha = G_\alpha(\mathfrak{z}_0)$. By the definition of Z_α, we have

$$\{\mathfrak{z}_0 \in G_\alpha(\mathfrak{z}_0)\} \equiv \{\mathfrak{z}_0 \in Z_\alpha\}.$$

This is a contradiction.

Remark 1: For the second part of the argument, compare the proof of the "diagonalization" theorem of Chapter V, § 3. For a different proof, see also Engelking, Holsztyński, and Sikorski [1].

Remark 2: *Problem of effectiveness.* Our proof of existence of a set of class $\boldsymbol{G}_\alpha$ which is not of class $\boldsymbol{F}_\alpha$ is *not effective*. That is, we have not specified a function which assigns, to each α, a set with the required property. By inspecting our argument, we see that the non-effectiveness is due to the fact that we have not defined a function which assigns, to each limit ordinal λ, a convergent sequence of ordinals $< \lambda$. We have, in fact, affirmed the existence without determining any specified sequence of this kind. Actually, no effective definition of a sequence convergent to λ is known. However, the problem of existence of $\boldsymbol{G}_\alpha$-sets which are not $\boldsymbol{F}_\alpha$-sets has been solved effectively using a different idea (see Kuratowski [1], p. 372).

§ 6. Borel subsets of Polish spaces

THEOREM (see Kuratowski [1], § 37): *Every Borel subset of a Polish space X is a one-to-one continuous image of a closed subset of the space N^N.*

PROOF. First let us note that every open subset of X has the property under consideration; because every Polish space has this property (by Theorem XI.3.8).

Next, this property is closed under countable disjoint unions. For let F_n be a closed subset of the set of irrationals in the interval $(n-1, n)$ and let f_n be continuous and one-to-one. Assume that the sets $f_n^1(F_n)$, $n = 1, 2, \ldots$, are disjoint; then the mapping $f = f_1 \cup f_2 \cup \ldots$ is obviously a continuous and one-to-one mapping of $\bigcup_n F_n$ onto $\bigcup_n f_n^1(F_n)$.

Finally, the considered property is countably multiplicative. For let F_n be closed in N^N and f_n be continuous and one-to-one. Then the set $F = \prod_n F_n$ is closed in $(N^N)^N$ hence is homeomorphic to a closed subset of N^N. Moreover, $\bigcap_n f_n^1(F_n)$ is a continuous and one-to-one image of a closed subset of N^N (see Chapter XI, § 1, 23).

This completes the proof, because the family $\boldsymbol{B}$ of Borel subsets is the least family containing all open sets and closed under countable disjoint unions and countable intersections (see Theorem 3.3).

COROLLARY (THEOREM OF ALEXANDROV–HAUSDORFF, see [1] and [4]): *Every uncountable Borel subset E of a Polish space X has the cardinality* $\mathfrak{c}$.

More precisely, E contains a Cantor set.

Because every uncountable closed subset of N^N contains a Cantor set.

Remark. It will be shown later (Theorem XIII.1.9) that *every one-to-one continuous image of a Borel set is Borel.* Thus the Borel subsets of a Polish space can be characterized as the one-to-one continuous images of closed subsets of N^N (they are called also *Lusin spaces*, see Bourbaki [2], p. 128).

We shall also see that *any two Borel sets of the same cardinality are always Borel-isomorphic* (in Polish spaces).

§ 7. Further properties of Borel sets

Let us note here several important properties of Borel sets (which will not be used in the next sections).

1. *Extension of continuous functions.* Let X be metric and Y complete. Let $A \subset X$ and $f\colon A \to Y$ continuous. Then there is a continuous extension of f on a $\boldsymbol{G}_\delta$-subset of X.

2. *Extension of homeomorphisms. Lavrentiev Theorem* (see [1]). Any homeomorphism between two subsets of complete spaces can be extended to $\boldsymbol{G}_\delta$-subsets of these spaces.

3. *Extension of $\boldsymbol{B}$-measurable functions* (see Kuratowski [1], p. 434, and Maitra and Rao [2]). Let X be metric and Y Polish. Let $A \subset X$ and $f\colon A \to Y$ be of class α. Then there is an extension of f of class α on a set of multiplicative class $\alpha+1$. (See also Hansell [1].)

4. *Topological invariance of Borel classes.* The Borel additive classes with $\alpha > 1$ ($\boldsymbol{G}_{\delta\sigma}$, etc.) and the multiplicative classes with $\alpha > 0$ ($\boldsymbol{G}_\delta$, etc.) are topological invariants.

A simple proof is furnished by the Lavrentiev theorem (see [1] and [2]). For historical references see Kuratowski [1], p. 432.

For the invariance under more general mappings (closed or open), see in particular Frolik [1] and [3], Hausdorff [5], Taimanov [1] and [2], and Jayne [1], where further references can be found.

§ 8. Baire property

Let us recall that a *meager* set (called also a set of the *first category*) is a countable union of nowhere dense sets. It follows at once that the *family* $\boldsymbol{I}$ *of all meager subsets of a given (metric) space* X *is a* σ*-ideal.*

DEFINITION 1: We say that a set $A \subset X$ has the *Baire property* if A is open modulo $\boldsymbol{I}$; which means that A is of the form

$$A = (G-P) \cup R \tag{1}$$

where G is open and P and R are meager.

We denote by $\boldsymbol{C}(X)$, or briefly $\boldsymbol{C}$, the family of all subsets of X which have the Baire property.

THEOREM 1: *The family* $\boldsymbol{C}$ *is a* σ*-algebra.*

PROOF. The σ-additivity of $\boldsymbol{C}$ follows from Exercise IV.4.6.

To show that $A \in \boldsymbol{C}$ implies $(X-A) \in \boldsymbol{C}$, note that $F \in \boldsymbol{C}$ for F closed. This follows from the identity $F = \mathrm{Int}(F) \cup \mathrm{Fr}(F)$, because $\mathrm{Fr}(F)$ is nowhere dense, hence a member of $\boldsymbol{I}$.

Now, by (1), $X-A = (X-G)-R \cup (P-R)$, hence $(X-A) \in \boldsymbol{C}$.

The following theorem follows from Theorem 1.

THEOREM 2: *Each Borel set has the Baire property.*

Because every open set has the Baire property and the family $\boldsymbol{B}$ of Borel sets (contained in X) is the least σ-algebra containing all open sets. Thus $\boldsymbol{B} \subset \boldsymbol{C}$.

In order to prove the next theorem, we will have to establish some properties of meager sets.

DEFINITION 2: The set $A \subset X$ is said to be *meager at the point* p if there is a neighborhood G of p such that $A \cap G$ is meager.

The set of points of X where A is not meager is denoted by $D(A)$.

The following properties of the operator D can easily be established (for metric separable spaces; however, they are valid in general for T_1 topological spaces, see Kuratowski [1], § 10).

1. $(D(A) = \emptyset) \equiv$ (*A is meager*).
2. $D(A-D(A)) = \emptyset$, i.e. *$A-D(A)$ is meager.*
3. $D(A) \subset \bar{A}$.
4. *$D(A)$ is closed.*
5. $(D(E) = \emptyset) \Rightarrow (D(A) = D(A-E) = D(A \cup E))$.

THEOREM 3: *If $A \in \mathbf{C}$, then $D(A) \cap D(X-A)$ is nowhere dense.*

In other terms, every open non-empty set contains a point where either A or $X-A$ is meager.[1])

PROOF. Assume that (1) is true. Then

$$X-A = (X-G)-R \cup (P-R).$$

By 5 we have $D(A) = D(G)$ and $D(X-A) = D(X-G)$. Therefore, by 3,

$$D(A) \cap D(X-A) = D(G) \cap D(X-G) \subset \bar{G}-G,$$

which completes the proof because $\bar{G}-G$ is nowhere dense.

COROLLARY: *If $A \in \mathbf{C}$, then $D(A)-A$ is meager.*

Since

$$\begin{aligned} D(A)-A &= (D(A)-A) \cap D(X-A) \cup (D(A)-A-D(X-A)) \\ &\subset [D(A) \cap D(X-A)] \cup [(X-A)-D(X-A)], \end{aligned}$$

the conclusion follows from Theorem 3 and Proposition 2.

THEOREM 4: *$A \in \mathbf{C}$ iff A is the union of a $\mathbf{G}_\delta$-set and of a meager set.*

PROOF. 1° The sufficiency follows from Theorem 2.

2° Let (1) be fulfilled. Since $P \cup R$ is meager, we have

$$P \cup R = N_1 \cup N_2 \cup \ldots, \qquad \text{where } N_n \text{ is nowhere dense.}$$

Put $W = \bar{N}_1 \cup \bar{N}_2 \cup \ldots$ Thus W is a meager $\mathbf{F}_\sigma$-set. Now since $(P \cup R) \subset W$, we have $A-W = G-W$. Thus $A-W$ is a $\mathbf{G}_\delta$-set and $A = (A-W) \cup (A \cap W)$ is the required decomposition of A.

THEOREM 5: *If $A \in \mathbf{C}(X)$ and $B \in \mathbf{C}(Y)$, then $A \times B \in \mathbf{C}(X \times Y)$.*

[1]) This condition was originally taken as the definition of the Baire property.

PROOF. It is sufficient to show that $A \times Y \in \boldsymbol{C}(X \times Y)$. In view of Theorem 4, this is reduced to the well-known propositions (see Exercise XI.1.25): if Q is a $\boldsymbol{G}_\delta$ in X, then $Q \times Y$ is $\boldsymbol{G}_\delta$ in $X \times Y$, and if P is meager in X, then $P \times Y$ is meager in $X \times Y$.

THEOREM 6: *Every set $E \subset X$ is contained in an $\boldsymbol{F}_\sigma$-set Z such that*

$$(E \subset A \in \boldsymbol{C}) \Rightarrow (Z - A \textit{ is meager}).$$

PROOF. By 2, the set $E - D(E)$ is meager. Hence

$$E - D(E) = N_1 \cup N_2 \cup \ldots, \quad \text{where } N_n \text{ is nowhere dense}.$$

Put $W = \overline{N}_1 \cup \overline{N}_2 \cup \ldots$ and $Z = W \cup D(E)$. Hence $E \subset Z$. Since $D(E)$ is closed (by 4), the set Z is $\boldsymbol{F}_\sigma$.

Now let $E \subset A \in \boldsymbol{C}$. Then

$$Z - A = (W - A) \cup \big(D(E) - A\big) \subset W \cup \big(D(A) - A\big),$$

because $E \subset A \Rightarrow D(E) \subset D(A)$.

Finally, since W and $D(A) - A$ are meager (by the Corollary), so is $Z - A$.

THEOREM 7: *The Baire property is invariant under the $\mathcal{A}$-operation* (see Nikodym [1], p. 149).

PROOF. Let

(2) $$E = \bigcup_{\mathfrak{z}} \bigcap_n A_{\mathfrak{z}|n}, \quad \text{where} \quad A_{\mathfrak{z}|n} \in \boldsymbol{C}.$$

We can assume that $A_{\mathfrak{z}|n+1} \subset A_{\mathfrak{z}|n}$, because $A_{\mathfrak{z}|n}$ can be replaced by $A_{\mathfrak{z}|1} \cap A_{\mathfrak{z}|2} \cap \ldots \cap A_{\mathfrak{z}|n}$ (since $\boldsymbol{C}$ is multiplicative).

Consider the set Z from Theorem 6. More generally, for a given k, let $Z_{\mathfrak{y}|k}$ be such that

(i) $$\bigcup_{\mathfrak{z}} \bigcap_n A_{(\mathfrak{y}|k)(\mathfrak{z}|n)} \subset Z_{\mathfrak{y}|k} \in \boldsymbol{C},$$

(ii) $$\Big(\bigcup_{\mathfrak{z}} \bigcap_n A_{(\mathfrak{y}|k)(\mathfrak{z}|n)} \subset A \in \boldsymbol{C}\Big) \Rightarrow (Z_{\mathfrak{y}|k} - A \text{ is meager}),$$

where $(\mathfrak{y}|k)(\mathfrak{z}|n)$ means $\mathfrak{y}^1 \ldots \mathfrak{y}^k \mathfrak{z}^1 \ldots \mathfrak{z}^n$.

Moreover, we can assume that

(3) $$Z_{\mathfrak{y}|k} \subset A_{\mathfrak{y}|k},$$

because the set $Z_{\mathfrak{y}|k} \cap A_{\mathfrak{y}|k}$ obviously fulfils the conditions imposed on $Z_{\mathfrak{y}|k}$.

Since $E = Z-(Z-E)$ and since Z is $\boldsymbol{F}_\sigma$, it is sufficient to prove that $Z-E$ is meager.

Now, by formula XI.5(7) and by virtue of (2) and (3), we have

$$Z-E = Z-\bigcup_{\mathfrak{y}}\bigcap_{k} A_{\mathfrak{y}|k} \subset Z-\bigcup_{\mathfrak{y}}\bigcap_{k} Z_{\mathfrak{y}|k} \subset \bigcup_{\mathfrak{y}}\bigcup_{k}\Big[Z_{\mathfrak{y}|k}-\bigcup_{m} Z_{(\mathfrak{y}|k)m}\Big].$$

Since the operator $\bigcup\limits_{\mathfrak{y}}\bigcup\limits_{k}$ is countable, it remains to show that the set in brackets [ ] is meager. But this follows from (ii), where we put

$$A = \bigcup_{m} Z_{(\mathfrak{y}|k)m}.$$

That can be done, because (see formula XI.5(5))

$$\bigcup_{\mathfrak{z}}\bigcap_{n} A_{(\mathfrak{y}|k)(\mathfrak{z}|n)} = \bigcup_{m}\bigcup_{\mathfrak{z}}\bigcap_{n} A_{(\mathfrak{y}|k)m(\mathfrak{z}|n)}$$

and by (i)

$$\bigcup_{\mathfrak{z}}\bigcap_{n} A_{(\mathfrak{y}|k)m(\mathfrak{z}|n)} \subset Z_{(\mathfrak{y}|k)m},$$

hence $\bigcup\limits_{\mathfrak{z}}\bigcap\limits_{n} A_{(\mathfrak{y}|k)(\mathfrak{z}|n)} \subset \bigcup\limits_{m} Z_{(\mathfrak{y}|k)m} = A.$

Remark 1: A σ-algebra $\boldsymbol{R}$ of subsets of the space X will be called an *$\mathscr{M}$-algebra* (a Marczewski algebra, see Marczewski [1], p. 209), if each set $E \subset X$ is contained in a set $Z \in \boldsymbol{R}$ such that

$$(E \subset A \in \boldsymbol{R}) \Rightarrow (Z-A \text{ is hereditarily an } \boldsymbol{R}\text{-set}),$$

i.e. such that each subset of $Z-A$ belongs to $\boldsymbol{R}$.

A slight modification of the proof of Theorem 7 leads to the following generalization of that theorem (which belongs to the general set theory)

THEOREM 7a: *Each $\mathscr{M}$-algebra is closed under the $\mathscr{A}$-operation.*

Besides the family $\boldsymbol{C}$, an important example of an $\mathscr{M}$-algebra constitutes the family of *Lebesgue-measurable* subsets of the interval. Because in this case one can take for Z a $\boldsymbol{G}_\delta$-set whose measure equals the exterior measure of E. Hence, by Theorem 7a, *the Lebesgue-measurability is invariant under the $\mathscr{A}$-operation* (see also Lusin and Sierpiński [1]).

Remark 2 (*on the duality between measure and Baire property*): The preceding remarks are connected with the duality between measure and Baire property consisting in substituting sets having the Baire property to measurable sets, and meager sets to sets of measure zero, and vice versa.[1])

Let us cite the following "duality principle" (due to P. Erdös and W. Sierpiński, see Oxtoby [1], p. 75) shown under the continuum hypothesis.

Let α be a formula involving—besides the terms of set theory—solely the terms "set of measure zero" and "meager set". Let α^* be the formula obtained from α by interchanging these two terms. Then α and α^* are equivalent.

Remark 3: *There exists in the space $\mathscr{E}$ of reals a set which does not have the Baire property.*

Such is the set of *Vitali* (which is not Lebesgue measurable either, see [1]). This set is obtained in the following way.

We decompose $\mathscr{E}$ into disjoint subsets by letting two numbers belong to the same subset iff their difference is rational. By the axiom of choice there exists a set containing a single element from each of these sets. Such a set does not have the Baire property.

***C*-measurable mappings**

According to the general definition (see Definition XI.4.1) a mapping $f\colon X \to Y$ is called ***C*-measurable** (or having the *Baire property*) if

(4) $$f^{-1}(G) \in \boldsymbol{C}(X) \quad \text{whenever } G \text{ is open in } Y.$$

Since $\boldsymbol{C}$ is a σ-lattice (even a σ-algebra), most of the theorems of Chapter XI, § 4 can be applied to $\boldsymbol{C}$. Let us mention, in particular, the following statements.

THEOREM 8: *Y being supposed metric separable, the mapping f is **C**-measurable iff there is a meager set $P \subset X$ such that the partial function $f|(X-P)$ is continuous.*

This follows from Theorem XI.4.11.

[1]) This duality has been extensively studied by J. C. Oxtoby in [1] and by various authors (see *ibidem*).

THEOREM 9: *Every* ***B****-measurable mapping is* ***C****-measurable.*

This follows at once from the fact that every Borel set has the Baire property (Theorem 2).

THEOREM 10: *The limit of a convergent sequence of* ***C****-mappings is a* ***C****-mapping.*

This follows from Theorem XI.1(13).

Theorem 11: (GRAPH THEOREM) *Assume that Y is separable. The graph*

$$\mathrm{Gr}(f) = \{\langle x, y\rangle : y = f(x)\}$$

of a ***C****-mapping has the Baire property (in $X \times Y$).*

This follows from Theorem XI.4.8, in virtue of Theorem 5.

Baire property in the restricted sense

DEFINITION: A has the *Baire property in the restricted sense* ($A \in \boldsymbol{C}_r$) if for every $E \subset X$ the set $A \cap E$ has the Baire property relative to E, i.e. $A \cap E \in \boldsymbol{C}(E)$ for each E (see Lebesgue [1], p. 187, and e.g. Kuratowski [1], p. 92).

The range of variability of E can be restricted in the above definition to *closed* sets.

Of course, *every Borel set belongs to* $\boldsymbol{C}_r$.

One shows (see Sierpiński [5], p. 319) that $A \in \boldsymbol{C}_r$ iff for every closed $F \subset X$, A is the union of a $\boldsymbol{G}_\delta$-set and a set meager in F.

It follows that in Polish spaces *the family* $\boldsymbol{C}_r$ *is a topological invariant.*

However, one can show, under the continuum hypothesis, that $\boldsymbol{C}_r$ is not an invariant of continuous one-to-one mappings, nor of the Cartesian multiplication by an axis (see Sierpiński [14], p. 54).

It is also remarkable that, under the continuum hypothesis, the graph of a $\boldsymbol{C}_r$-mapping does not need to be $\boldsymbol{C}_r$ (f is a $\boldsymbol{C}_r$-mapping iff $f^{-1}(G)$ is $\boldsymbol{C}_r$ for each open G).

For the relation of $\boldsymbol{C}_r$-sets to Lebesgue measurable sets see Saks [2], p. 277, and Lusin [3], p. 147.

CHAPTER XIII

SOUSLIN SPACES. PROJECTIVE SETS

§ 1. Souslin spaces. Fundamental properties

DEFINITION: A *Souslin space* is a metric space which is a continuous image of the space N^N (of irrationals).

The family of all Souslin subsets of a space X, including $\emptyset$, will be denoted by $\boldsymbol{S}(X)$ (or—briefly $\boldsymbol{S}$ if no confusion can occur).

If X is Polish, the members of $\boldsymbol{S}(X)$ are also called *analytic sets*. Their complements are called *coanalytic* or $\boldsymbol{CA}$-sets.

Remark: The idea of the analytic set is due to M. Souslin and N. Lusin (see [4]). The actual definition of a Souslin space is essentially due to F. Hausdorff [2], p. 177.

For a more general approach to Souslin spaces, which do not need to be metric or separable, see e.g. G. Choquet [1] and [2], Z. Frolik [2] and [3], C. A. Rogers [1], M. Sion [2].

Obviously,

(1) *Any continuous image of a Souslin space is Souslin.*

(2) *Any Souslin space is separable* (*contains a countable open base*).

(3) $\boldsymbol{S}(X)$ *coincides with the family of projections on the X axis of closed subsets of $N^N \times X$* (*for X Souslin*).

For, if $E \in \boldsymbol{S}(X)$, then E is the projection on X of the graph of a continuous function f: $N^N \to X$ such that $E = f^1(N^N)$.

THEOREM 1: *Every Polish space is Souslin.*

PROOF. Let $G_1, G_2, \ldots$ be an open base of the Polish space X. We may, of course, assume that $\delta(G_n) < 1$ and $G_n \neq \emptyset$. Put $A_i = \overline{G_i}$ and define $A_{k_1 \ldots k_n}$ by induction, assuming that

$$A_{k_1 \ldots k_n} = \bigcup_{i=1}^{\infty} A_{k_1 \ldots k_n i} \quad \text{and} \quad \delta(A_{k_1 \ldots k_n i}) < \frac{1}{n+1}.$$

Consider the set Z and the function f of formula XI.5(12). Since $A_{k_1 \ldots k_n} \neq \emptyset$, we have $Z = N^N$ and, by (16), $f^1(Z) = X$. Moreover, f is continuous (by XI.5 (15)). This completes the proof.

Remark. Theorem 1, as well as its generalization, Theorem 6, can also be deduced from Theorem XII.6.1.

THEOREM 2. *Countable unions, countable intersections and countable products of Souslin sets are Souslin.*

PROOF. Let $A_n \in S(X)$, for $n = 1, 2, \ldots$, and let $f_n \colon N^N \to A_n$ be continuous onto.

Put $B = A_1 \cup A_2 \cup \ldots$ To show that $B \in S(X)$, denote by N_n the set of irrationals of the interval $(n-1, n)$ and put

$$f(x) = f_n(x-n+1) \quad \text{for } x \in N_n,\ n = 1, 2, \ldots$$

Then f is a continuous mapping of $N^* = N_1 \cup N_2 \cup \ldots$ onto B. This completes the proof, since N^* is homeomorphic to N^N.

Put $C = A_1 \cap A_2 \cap \ldots$ To show that $C \in S(X)$, consider the space $D = (N^N)^N$ which is homeomorphic to N^N (see Chapter XI.1.18). Now C is a continuous image of a closed subset Z of D (see Chapter XI.1.22) and since D is Polish, so is Z. Therefore, by Theorem 1, C is Souslin.

Put $P = A_1 \times A_2 \times \ldots$ Then the product-mapping $f_1 \times f_2 \times \ldots$ is a continuous mapping of $(N^N)^N$ onto P (see Chapter XI.1.20). Since $(N^N)^N$ is homeomorphic to N^N, the proof is completed.

THEOREM 3: *The family $S(X)$ of Souslin subsets of X is invariant under the operation $\mathcal{A}$.*

PROOF. Let $E = \bigcup\limits_{\mathfrak{z}} \bigcap\limits_{n=1}^{\infty} A_{\mathfrak{z}|n}$ where $A_{\mathfrak{z}|n} \in S(X)$. Thus

$$(x \in E) \equiv \bigvee_{\mathfrak{z}} \bigwedge_{n} (x \in A_{\mathfrak{z}|n}).$$

Hence to show that $E \in S(X)$, it is sufficient to prove that

$$\{\langle x, \mathfrak{z}\rangle \colon x \in A_{\mathfrak{z}|n}\} \in S(X \times N^N) \quad \text{for fixed } n$$

(because the continuous image—in this case the projection—of a Souslin set is Souslin).

But this follows from the equivalence

$$(x \in A_{\mathfrak{z}|n}) = \bigvee_{k_1} \ldots \bigvee_{k_n} (\mathfrak{z}^1 = k_1) \wedge \ldots \wedge (\mathfrak{z}^n = k_n) \wedge (x \in A_{k_1 \ldots k_n}).$$

THEOREM 4: *Let X be Polish and $E \subset X$. Then $E \in S(X)$ iff E is the result of the operation $\mathscr{A}$ performed on closed subsets of X.*

PROOF. The sufficiency follows at once from Theorem 3. To prove the necessity, suppose that $E \in S(X)$. We have to define closed sets $A_{k_1 \ldots k_n}$ such that

(4) $$E = \bigcup_{\mathfrak{z}} \bigcap_{n=1}^{\infty} A_{\mathfrak{z}|n}.$$

By assumption, there is $f\colon N^N \to E$ continuous onto.

Referring to Chapter XI, § 5,

(5) $$A_{\mathfrak{z}|k} = \overline{f(\mathscr{N}_{\mathfrak{z}|k})}.$$

Then, by XI.5 (21), we have

(6) $$f(z) = \bigcap_k A_{\mathfrak{z}|k}$$

and (4) follows.

Remark: As shown each Souslin set E can be represented in the form (4) where the sets $A_{k_1 \ldots k_n}$ are closed and the system of these sets is regular (see Kaniewski [1], Lemma 1).

It is worth noticing that the term "*closed*" can be replaced by "*open*". Put namely

$$G_{k_1 \ldots k_n} = \{x\colon \varrho(x, A_{k_1 \ldots k_n} < 1/k)\}.$$

Then, as easily seen,

$$E = \bigcup_{\mathfrak{z}} \bigcap_{n=1}^{\infty} G_{\mathfrak{z}|n}.$$

It follows also that *E can be sifted by a sieve composed of open sets.*

COROLLARY: *Each $E \in S(X)$ is sifted by a sieve $W\colon \mathscr{R}_0 \to 2^X$.*

This follows from Theorem XI.6.1.

THEOREM 5: *Every Souslin space has the Baire property* (*even in the restricted sense*, Chapter XII, § 8).

PROOF. This is a consequence of the invariance of the Baire property under the operation $\mathscr{A}$ (see Theorem XII.8.7) and of Theorem 4; by this theorem a Souslin space is the result of the operation $\mathscr{A}$ performed on closed sets, hence on sets having the Baire property.

The following theorem is a generalization of Theorem 1.

THEOREM 6: *Every Borel subset of a Polish space is Souslin. Thus* $\boldsymbol{B}(X) \subset \boldsymbol{S}(X)$. *Hence* $\boldsymbol{B}(X) \subset \boldsymbol{CA}(X)$.

PROOF. Let us recall (see p. 124) that $\boldsymbol{B}(X)$ is the least σ-additive and δ-multiplicative family of sets containing all closed subsets of X. By Theorem 2 the family $\boldsymbol{S}(X)$ is also σ-additive and δ-multiplicative and by Theorem 1, $\boldsymbol{S}(X)$ contains all closed subsets of X (because a closed subset of a Polish space is Polish, hence Souslin). Therefore $\boldsymbol{B}(X) \subset \boldsymbol{S}(X)$.

THEOREM 7: *Every uncountable Souslin space* Y *has cardinality* $\mathfrak{c}$. *More precisely,* Y *contains a Cantor set* (see Souslin [1]).

PROOF. Let $f\colon N^N \to Y$ be continuous onto. We are going to define a Cantor set $A \subset N^N$ such that the mapping $f|A$ is one-to-one. This will complete the proof.

For this purpose, we shall attach to each finite system $k_1, \ldots, k_n$, where k_i is either 0 or 2, a closed set $A_{k_1 \ldots k_n} \subset N^N$ so that (assuming that the metric of N^N is complete):

(7) $$\delta(A_{k_1 \ldots k_n}) < 1/n, \qquad A_{k_1 \ldots k_n} \neq \emptyset,$$

(8) $$A_{k_1 \ldots k_n} \subset A_{k_1 \ldots k_{n-1}} \quad \text{and} \quad A_{k_1 \ldots k_n} \cap A_{m_1 \ldots m_n} = \emptyset$$

if $(k_1 \ldots k_n) \neq (m_1 \ldots m_n)$.

According to Theorem XI.5.2 the set

$$A = \bigcup_{\mathfrak{z}} \bigcap_{n} A_{\mathfrak{z}|n}, \quad \text{where} \quad \mathfrak{z} \in (0, 2)^N,$$

is a Cantor set.

Now let $g\colon Y \to N^N$ be a selector for the inverse map:

$$g(y) \in f^{-1}(y), \quad \text{thus} \quad f(g(y)) = y.$$

Put $B = g^1(Y)$. Since g is one-to-one, B is uncountable, Hence B contains a (non-empty) dense-in-itself set D (see Exercise XI.1.4 (iv)). The sets $A_{k_1 \ldots k_n}$ will be defined so that—in addition to the conditions (7) and (8)—the two following conditions be fulfilled:

(9) $$A_{k_1 \ldots k_n} = \overline{G}_{k_1 \ldots k_n}, \quad \text{where } G_{k_1 \ldots k_n} \text{ is open and } D \cap G_{k_1 \ldots k_n} \neq \emptyset,$$

(10) $$f^1(A_{k_1 \ldots k_n}) \cap f^1(A_{m_1 \ldots m_n}) = \emptyset$$

if $(k_1 \ldots k_n) \neq (m_1 \ldots m_n)$.

We proceed by induction. Let, for $j = 0, 2$, $x_j \in D$ and $x_0 \neq x_1$.

Hence $f(x_0) \neq f(x_1)$ since $f|D$ is one-to-one, and there are open $U_j \subset Y$ such that $f(x_j) \in U_j$ and $U_0 \cap U_1 = \emptyset$. Therefore

$$x_j \in f^{-1}(U_j) \quad \text{and} \quad f^{-1}(U_0) \cap f^{-1}(U_1) = \emptyset.$$

Since $f^{-1}(U_j)$ is open, there exists an open G_j such that

$$\overline{G}_j \subset f^{-1}(U_j), \quad x_j \in G_j \quad \text{and} \quad \delta(G_j) < 1.$$

Put $A_j = \overline{G}_j$. Obviously, conditions (7), (9), (10) and the second part of (8) are fulfilled for $n = 1$.

Now suppose (7)–(10) are fulfilled for a given n. We have to define $A_{k_1 \ldots k_n j}$ for $j = 0, 2$.

Let $x_{k_1 \ldots k_n j} \in D \cap G_{k_1 \ldots k_n}$ and $x_{k_1 \ldots k_n 0} \neq x_{k_1 \ldots k_n 2}$. This can be assumed by the second part of (9) since D is dense-in-itself.

It follows that $f(x_{k_1 \ldots k_n 0}) \neq f(x_{k_1 \ldots k_n 2})$. Let $U_{k_1 \ldots k_n j}$ be open and such that

$$f(x_{k_1 \ldots k_n j}) \in U_{k_1 \ldots k_n j} \quad \text{and} \quad U_{k_1 \ldots k_n 0} \cap U_{k_1 \ldots k_n 2} = \emptyset$$

and—as before—let $G_{k_1 \ldots k_n j}$ be open and such that

$$x_{k_1 \ldots k_n j} \in G_{k_1 \ldots k_n j}, \quad \overline{G}_{k_1 \ldots k_n j} \subset G_{k_1 \ldots k_n} \cap f^{-1}(U_{k_1 \ldots k_n j})$$

and $\delta(G_{k_1 \ldots k_n j}) < 1/(n+1)$.

It is easy to check that the conditions (7)–(10) are satisfied for $n+1$. Thus A is a Cantor set.

Moreover, $f|A$ is one-to-one. For let p and q be two distinct points of A. Therefore there are $(k_1 \ldots k_n) \neq (m_1 \ldots m_n)$ such that $p \in A_{k_1 \ldots k_n}$ and $q \in A_{m_1 \ldots m_n}$, and it follows by (10) that $f(p) \neq f(q)$.

Remark. The following extension of Theorem 7 will be mentioned in § 5. Namely, *every* ***PCA****-set of cardinality* $> \aleph_1$ contains a Cantor set.

Another generalization—an extension to non-complete spaces—was given by W. Hurewicz [1]. Namely, the result of the operation $\mathscr{A}$ performed on closed subsets of a metric separable space (not necessarily complete) contains a Cantor set (provided that it is uncountable).

THEOREM 8 (FIRST LUSIN'S SEPARATION THEOREM, see [4]): *Let A and B be two disjoint Souslin subsets of the Polish space X. Then there exists a Borel set E such that*

(11) $$A \subset E \quad \textit{and} \quad E \cap B = \emptyset.$$

LEMMA: *Let us say that two sets A and B are "B-separable" if the assertion of the theorem is fulfilled. Let $P_1, P_2, \ldots$ and $Q_1, Q_2, \ldots$ be sets such that each pair P_n, Q_m is B-separable. Then the sets*

$$P = P_1 \cup P_2 \cup \ldots \quad \textit{and} \quad Q = Q_1 \cup Q_2 \cup \ldots$$

are B-separable.

PROOF. Let Z_{nm} be Borel such that

$$P_n \subset Z_{nm} \subset X - Q_m .$$

It follows that

$$P_n \subset \bigcap_{m=1}^{\infty} Z_{nm} \subset \bigcap_{m=1}^{\infty} (X - Q_m) = X - \bigcup_{m=1}^{\infty} Q_m = X - Q$$

and hence

$$P = \bigcup_{n=1}^{\infty} P_n \subset \bigcup_{n=1}^{\infty} \bigcap_{m=1}^{\infty} Z_{nm} \subset X - Q .$$

That means that the Borel set $\bigcup_{n=1}^{\infty} \bigcap_{m=1}^{\infty} Z_{nm}$ separates P and Q.

PROOF OF THE THEOREM. Suppose the contrary is true. Let A and B be two disjoint Souslin sets which are not B-separable.

Let $f: N^N \to A$ and $g: N^N \to B$, both continuous onto. Using the notation of Chapter XI, § 5, we have (writing $f(E)$ instead of $f^1(E)$)

$$A = f(N_1) \cup f(N_2) \cup \ldots \quad \text{and} \quad B = g(N_1) \cup g(N_2) \cup \ldots$$

By the lemma there are two indices n_1 and m_1 such that

$$f(N_{n_1}) \text{ and } g(N_{m_1}) \text{ are not } B\text{-separable.}$$

Since

$$f(N_{n_1 \ldots n_k}) = \bigcup_{i=1}^{\infty} f(N_{n_1 \ldots n_k i}),$$

$$g(N_{m_1 \ldots m_k}) = \bigcup_{j=1}^{\infty} g(N_{m_1 \ldots m_k j}),$$

we can define by induction two sequences $\mathfrak{z} \in N^N$ and $\mathfrak{y} \in N^N$ such that

$$f(N_{\mathfrak{z}|k}) \text{ and } g(N_{\mathfrak{y}|k}) \text{ are not } B\text{-separable.} \tag{12}$$

Since $f(\mathfrak{z}) \in A$, $g(\mathfrak{y}) \in B$ and $A \cap B = \emptyset$, we have $|f(\mathfrak{z}) - g(\mathfrak{y})| > 0$.

By the continuity of f and g, we have

$$\delta[f(N_{\mathfrak{z}|k})]+\delta[g(N_{\mathfrak{y}|k})] < |f(\mathfrak{z})-g(\mathfrak{y})|$$

for sufficiently large k.

Since $f(\mathfrak{z}) \in f(N_{\mathfrak{z}|k})$ and $g(\mathfrak{y}) \in g(N_{\mathfrak{z}|k})$, it follows that

$$\overline{f(N_{\mathfrak{z}|k})} \cap g(N_{\mathfrak{y}|k}) = \emptyset$$

and hence the sets $f(N_{\mathfrak{z}|k})$ and $g(N_{\mathfrak{y}|k})$ are B-separable, contrary to (12).

COROLLARY 1 (SOUSLIN THEOREM, see [1]): *If the sets A and $X-A$ are Souslin, then they are Borel.*

For, if we let $B = X-A$, we get $A = E$.

COROLLARY 2 (SIMULTANEOUS SEPARATION): *If $A_1, A_2, \ldots$ are disjoint Souslin, then there are disjoint Borel sets $B_1, B_2, \ldots$ such that $A_n \subset B_n$ for $n = 1, 2, \ldots$*

PROOF. By Theorem 8, there are Borel sets E_{nm} such that

$$A_n \subset E_{nm} \quad \text{and} \quad E_{nm} \cap A_m = \emptyset \quad \text{for} \quad n \neq m.$$

Let

$$B_1 = \bigcap_{m=2}^{\infty} E_{1m} \quad \text{and} \quad B_n = \bigcap_{m\neq n} E_{nm} - (B_1 \cup \ldots \cup B_{n-1}).$$

Therefore $A_n \subset B_n$. Because we have $B_m \subset E_{mn}$ and hence

$$(A_n \cap B_m) \subset (A_n \cap E_{mn}) = \emptyset \quad \text{for} \quad m < n.$$

Applications of the First Separation Theorem to Borel sets

THEOREM 9. *Every one-to-one continuous image of a Borel subset of a Polish space is a Borel set* (see Souslin [1]).

PROOF. Since every Borel subset B of a Polish space is a one-to-one continuous image of a closed subset F of the space N^N (see the Theorem of Chapter XII, § 6), the proof reduces to show that if f is a one-to-one continuous mapping of F, then $f^1(F)$ is Borel.

Consider, as in § 5 of Chapter XI, the sets

$$\mathcal{N}_{n_1 \ldots n_k} = \{\mathfrak{z} \in N^N\colon (\mathfrak{z}^1 = n_1) \ldots (\mathfrak{z}^k = n_k)\}$$

and put $F_{n_1 \ldots n_k} = F \cap \mathcal{N}_{n_1 \ldots n_k}$.

Since the sets $\mathcal{N}_{n_1 \ldots n_k}$ are disjoint for fixed k, so are the sets $f^1(F_{n_1 \ldots n_k})$. Now, by Corollary 2, for each k, there exists a system of disjoint Borel sets $B_{n_1\ldots n_k}$ such that

(13) $$f^1(F_{n_1\ldots n_k}) \subset B_{n_1\ldots n_k}.$$

We are now going to define a mapping C: $\mathscr{S} \to B(X)$ (an operation $\mathscr{A}$) such that

(14) $$f^1(F) = \bigcup_{\mathfrak{z}} \bigcap_k C_{\mathfrak{z}|k}$$

Put $C_{n_1} = B_{n_1} \cap \overline{f^1(F_{n_1})}$ and generally

(15) $$C_{n_1\ldots n_k} = B_{n_1\ldots n_k} \cap \overline{f^1(F_{n_1\ldots n_k})} \cap C_{n_1\ldots n_{k-1}}.$$

We shall prove by induction that

(16_k) $$f^1(F_{n_1\ldots n_k}) \subset C_{n_1\ldots n_k}.$$

For $k = 1$, this is obvious. Suppose (16_{k-1}) is true.

Since $\mathscr{N}_{n_1\ldots n_k} \subset \mathscr{N}_{n_1\ldots n_{k-1}}$, we have

$$f^1(F_{n_1\ldots n_k}) \subset f^1(F_{n_1\ldots n_{k-1}}),$$

hence (16_k) follows by (16_{k-1}) and (13).

Thus we have by (16_k), (14) and (15)

$$f^1(F_{n_1\ldots n_k}) \subset C_{n_1\ldots n_k} \subset \overline{f^1(F_{n_1\ldots n_k})}.$$

Hence, for each $\mathfrak{z} \in F$,

$$\bigcap_k f^1(F_{\mathfrak{z}|k}) \subset \bigcap_k C_{\mathfrak{z}|k} \subset \bigcap_{\mathfrak{z}} \overline{f^1(F_{\mathfrak{z}|k})}$$

and since (see XI.2 (21))

$$f^1(F) = \bigcup_{\mathfrak{z}} \bigcap_k f^1(F_{\mathfrak{z}|k}) = \bigcup_{\mathfrak{z}} \bigcap_k \overline{f^1(F_{\mathfrak{z}|k})},$$

we have (14).

Now since the mapping C: $F \to \boldsymbol{P}(X)$ is regular and the sets $C_{n_1\ldots n_k}$ are disjoint for fixed k, we have by Theorem XI.5.1,

$$\bigcup_{\mathfrak{z}} \bigcap_k C_{\mathfrak{z}|k} = \bigcap_k \bigcup_{\mathfrak{z}} C_{\mathfrak{z}|k}, \quad \text{hence} \quad f^1(F) = \bigcap_k \bigcup_{\mathfrak{z}} C_{\mathfrak{z}|k}.$$

Since the sets $C_{\mathfrak{z}|k}$ are Borel, and, for a fixed k, the set $\bigcup_{\mathfrak{z}} C_{\mathfrak{z}|k}$ is a countable union of Borel sets, we conclude finally that $f^1(F)$ is Borel.

Theorem 9 can be generalized as follows.

COROLLARY 1: *The term "continuous" in Theorem 9 can be replaced by "**B**-measurable".*

PROOF. Let f: $F \to Y$ be **B**-measurable and one-to-one. By Theorem XII.3.5 the graph of f, i.e. the set $\mathrm{Gr}(f) = \{\langle x, y\rangle : y = f(x)\}$ is Borel in the Polish space $X \times Y$, and the projection of $\mathrm{Gr}(f)$ into the Y-axis is one-to-one. Then, according to Theorem 9, the set $f^1(F)$ is Borel.

COROLLARY 2: *Let, as before, X and Y be Polish, $E \subset X$ Borel and f: $E \to Y$ be **B**-measurable and one-to-one. Then f is a Borel isomorphism between E and $f^1(E)$,* i.e. *f^{-1} is **B**-measurable.*

PROOF. Put $g = f^{-1}$ and let G be an open set relative to E. We have to show that $g^{-1}(G)$ is Borel. But $g^{-1}(G) = f^1(G)$ and, since G is Borel (in X), $f^1(G)$ is Borel by Corollary 1.

Remarks: *Any two Borel sets of the same cardinality (contained in Polish spaces) are Borel-isomorphic* (see Kuratowski [18], and [1], p. 450, and for applications Mackey [1]).

Since the case of A and B countable is trivial, we may assume that these sets are uncountable (hence of cardinality $\mathfrak{c}$). Let, according to Theorem of Chapter XII, § 6, F and H be two closed (uncountable) subsets of the space N^N and f: $F \to A$ and g: $H \to B$ two one-to-one **B**-measurable onto functions. It can be shown that the sets F and H are Borel isomorphic. So let h: $F \to H$ be a Borel isomorphism onto. Then since f^{-1} is **B**-measurable (by Corollary 2),

$$g \circ h \circ f^{-1}$$

is the required Borel isomorphism.

More precisely, one can show that *if A and B are of multiplicative classes $\alpha+1 > 2$ and $\beta+1 > 2$ respectively, then there exists a generalized homeomorphism of A onto B of class α, β.*

Let us note also that for each pair of uncountable Polish spaces there exists a generalized homeomorphism of class 1,1 (see Kuratowski [1], p. 451).

THEOREM 10: *Let X be Polish. There exists a universal function F for the Souslin subsets of X whose graph, $G(F)$, is Souslin.*

PROOF. Put $\mathcal{N} = N^N$. We have to define F: $\mathcal{N} \to \boldsymbol{S}(X)$ onto and such that the set $G(F) = \{\langle t, x\rangle : x \in F(t)\}$ is Souslin (in $\mathcal{N} \times X$).

By the Theorem of Chapter XII, § 5, there is a universal function for closed sets, i.e. a function $H\colon \mathcal{N} \to 2^{X\times\mathcal{N}}$ onto, such that the set

$$G(H) = \{\langle t, x, s\rangle : \langle x, s\rangle \in H(t)\}$$

is closed in $\mathcal{N}\times X\times \mathcal{N}$.

Put $F(t)$ = projection of $H(t)$ on X, i.e.

$$F(t) = \{x\colon \bigvee_s \langle x, s\rangle \in H(t)\}.$$

F is universal because every Souslin subset of X is a projection of a closed subset of $X\times\mathcal{N}$ (see 1 (3)). Furthermore, $G(F)$ is Souslin, because $G(F)$ is obviously the projection of $G(H)$ on $\mathcal{N}\times X$.

THEOREM 11: (EXISTENCE THEOREM OF SOUSLIN) *There exist, in the space $\mathcal{N} = N^N$, Souslin sets which are not Borel.*

PROOF. Let F be a universal function for $S(\mathcal{N})$ such that $G(F)$ is Souslin (see Theorem 10). Then the set

$$Z = \{t\colon t \in F(t)\} \text{ is analytic and not coanalytic,}$$

and vice versa

$$Z' = \{t\colon t \notin F(t)\} \text{ is coanalytic and not analytic.}$$

This follows from the Theorem V.2.1 on diagonalization. More precisely, let Δ be the diagonal of $\mathcal{N}\times\mathcal{N}$; then Z is the projection of $\Delta \cap G(F)$ on the axis of "ordinates" and Z' is the projection of $\Delta - G(F)$. Since $\Delta \cap G(F)$ is Souslin, so is Z, and since $Z' = \mathcal{N} - Z$, Z' is $\boldsymbol{CA}$.

Now, by the theorem on diagonalization, Z' is not Souslin, and hence Z is not $\boldsymbol{CA}$. Thus Z is a Souslin not Borel set (because by Theorem 6, every Borel set is $\boldsymbol{CA}$).

Remark (see Hurewicz [2]): In the space $2^{\mathcal{I}}$ there are simple examples of Souslin non-$\boldsymbol{CA}$-sets (hence non-Borel sets). Such is *the set of all uncountable closed subsets of $\mathcal{I}$*. Such is also the set of all closed subsets containing at least one irrational number.

Another interesting example of a non-Borel $\boldsymbol{CA}$-set is the set of differentiable functions in the space $\mathcal{E}^{\mathcal{I}}$ (see Mazurkiewicz [2]. See also Christensen [1], p. 12).

§ 2. Applications of countable order types to Souslin spaces

Let us order the binary fractions $r \in \mathscr{R}_0$ in an infinite sequence $r_1, r_2, \ldots$ For a point $t = (t^1, t^2, \ldots)$ of the Cantor discontinuum $\mathscr{C}$, define C_{r_n} by the condition

(1) $$(t \in C_{r_n}) \equiv (t^n = 2).$$

Let

(2) $$R_t = \{r_n\colon t \in C_{r_n}\},$$

$\bar{t}$ = the order type of R_t ordered by the relation $r \geqslant s$,

(3) $$C_\tau = \{t\colon \bar{t} = \tau\}.$$

The mapping $C\colon \mathscr{R}_0 \to 2^{\mathscr{C}}$ is called a *sieve*.

Let us recall that if τ is an ordinal number α, then C_α is called a *constituent* of $L = \{t\colon \bar{t} < \omega_1\}$. Moreover, $C_\tau \neq \emptyset$ for any $\tau \in \mathscr{T}$ (i.e. for any countable ordinal type).

DEFINITION 1 (Kuratowski [12]): Let us call *Souslin* (*Borel* or ***CA***) a formula $\varphi(\tau_1, \ldots, \tau_n)$, where $\tau_i \in \mathscr{T}$, if the set

$$\{\langle t_1, \ldots, t_n\rangle \in \mathscr{C}^n\colon \varphi(\bar{t}_1, \ldots, \bar{t}_n)\}$$

is Souslin (Borel or ***CA***), where $\bar{t}_1 = \tau_1, \ldots, \bar{t}_n = \tau_n$.

The above definition can of course be extended to formulas $\varphi(\tau_1, \ldots, \tau_n, x_1, \ldots, x_m)$ involving variables ranging over Polish spaces $X_1, \ldots, X_m$.

Let us note that Definition 1 is "correct", i.e. it does not depend on the choice of $t_1, \ldots, t_n$. Otherwise stated, if $s_1, \ldots, s_n$ are points of $\mathscr{C}$ such that $\bar{s}_i = \tau_i$, then

$$\{\langle s_1, \ldots, s_n\rangle\colon \varphi(\bar{s}_1, \ldots, \bar{s}_n)\} = \{\langle t_1, \ldots, t_n\rangle\colon \varphi(\bar{t}_1, \ldots, \bar{t}_n)\}.$$

DEFINITION 2: We shall introduce the following relations in the domain $\mathscr{T}$ (where $\tau = \bar{t}$ and $\sigma = \bar{s}$):

(i) $(\tau \leqslant \sigma) \equiv (R_t$ is similar to a subset of $R_s)$,

(ii) $(\tau < \sigma) \equiv \bigvee_r (r \in R_s) \wedge (R_t$ is similar to $R_s \cap$ interval $r1)$,

(iii) $(\tau = \sigma) \equiv (R_t$ is similar to $R_s)$.

In proving that the formulas $\tau \leqslant \sigma$, $\tau < \sigma$ and $\tau = \sigma$ are Souslin, we will use the following abbreviations. Let $\mathfrak{z} \in \mathscr{R}_0^N$, i.e. $\mathfrak{z} = (\mathfrak{z}^1, \mathfrak{z}^2, \ldots)$

where $\mathfrak{z}^n \in \mathscr{R}_0$, and let $A \subset \mathscr{R}_0$. Then

(4) $$\mathfrak{z} \subseteq A \text{ means } \bigwedge_n \mathfrak{z}^n \in A,$$

(5) $$A \subseteq \mathfrak{z} \text{ means } \bigwedge_r \Big[r \in A \Rightarrow \bigvee_n r = \mathfrak{z}^n\Big],$$

(6) $$\mathfrak{z} \preccurlyeq v \text{ means } \bigwedge_{nm} (\mathfrak{z}^n < \mathfrak{z}^m) \Rightarrow (v^n < v^m),$$

(7) $$z \equiv A \text{ means } \mathfrak{z} \subseteq A \subseteq \mathfrak{z},$$

(8) $$\mathfrak{z} \sim v \text{ means } \mathfrak{z} \preccurlyeq v \preccurlyeq \mathfrak{z}.$$

LEMMA: *The sets*

$$\{\langle \mathfrak{z}, t\rangle : \mathfrak{z} \subseteq R_t\}, \quad \{\langle \mathfrak{z}, t\rangle : R_t \subseteq \mathfrak{z}\}, \quad \{\langle \mathfrak{z}, v\rangle : \mathfrak{z} \preccurlyeq \mathfrak{v}\}$$
$$\{\langle \mathfrak{z}, t\rangle : \mathfrak{z} \equiv R_t\} \quad \textit{and} \quad \{\langle \mathfrak{z}, \mathfrak{v}\rangle : \mathfrak{z} \sim \mathfrak{v}\}$$

are Borel.

This follows at once from (4)–(8).

THEOREM 1: *The relation* $\tau \leqslant \sigma$ *is Souslin.*

PROOF. We have

$$(\bar{t} \leqslant \bar{s}) \equiv \bigvee_{\mathfrak{z}\mathfrak{v}} \{(R_t \subseteq \mathfrak{z})(\mathfrak{v} \subseteq R_s)(\mathfrak{z} \preccurlyeq \mathfrak{v})\}$$

and the formula in braces is Borel (by the Lemma). Hence the set $\{\langle t, s\rangle : \bar{t} \leqslant \bar{s}\}$ is Souslin.

THEOREM 2: *The relation* $\tau < \sigma$ *is Souslin.*

Because

$$(\bar{t} < \bar{s}) \equiv$$
$$\bigvee_{\mathfrak{z}\mathfrak{v}r} \Big\{(r \in R_s)(R_t \subseteq \mathfrak{z})(\mathfrak{z} \preccurlyeq \mathfrak{v} \subseteq R_s) \bigwedge_p \Big[(p \in R_s)(r \leqslant p) \Rightarrow \bigvee_k p = \mathfrak{v}^k\Big]\Big\}.$$

THEOREM 3: *The relation* $\tau = \sigma$ *is Souslin.*

Because

$$(\bar{t} = \bar{s}) \equiv \bigvee_{\mathfrak{z}\mathfrak{v}} (\mathfrak{z} \equiv R_t)(\mathfrak{v} \equiv R_s)(\mathfrak{z} \sim \mathfrak{v}).$$

THEOREM 4: *The relation* $\tau = \sigma + \pi$ *is Souslin.*

PROOF. Let $\bar{p} = \pi$ be the order type of R_p $(p \in \mathscr{C})$ and let $R_p^* = R_p - 2$ denote the set obtained from R_p by shifting 2 to the left. Then, since $R_s \cap R_p^* = \emptyset$, the set $R_s \cup R_p^*$ is of order type $\bar{s} + \bar{p}$.

Thus we have

$$(\bar{t} = \bar{s}+\bar{p}) \equiv \bigvee_{\mathfrak{z}\mathfrak{v}} (\mathfrak{z} \equiv R_t)\,(\mathfrak{v} \equiv R_s \cup R_p^*)\,(\mathfrak{z} \sim \mathfrak{v})$$

and the analycity follows at once from (7) and (8).

Remark 1: Theorem 4 can be extended to the sum of any finite number of terms.

Remark 2. In a similar way one shows that the multiplication $\tau = \sigma \cdot \pi$ is Souslin (on $\mathscr{T}^3$).

THEOREM 5: *There is a Souslin relation* $\sigma = \varrho(\tau, n)$ *such that, whenever* α *and* β *are ordinals, then*

$$(9) \qquad [\varrho(\alpha, n) = \varrho(\beta, k)] \Rightarrow [n = k \text{ and } \alpha = \beta],$$

$$(10) \qquad \varrho(\tau, n) < \omega_1 \quad \text{iff} \quad \tau < \omega_1 .$$

Such is the mapping

$$(11) \qquad \varrho(\tau, n) = (\tau \cdot 2+1) \cdot 2^n .$$

PROOF. By Theorem 4, (11) is Souslin for n fixed; hence it is a Souslin relation.

Formula (9) can easily be shown (see e.g. Sierpiński [4], p. 330).

THEOREM 6: *Let* $\mu\colon X \to \mathscr{T}$ *and* $\nu\colon \mathscr{T} \to \mathscr{T}$ *be Souslin functions. Then their composition* $\nu \circ \mu$ *is Souslin.*

More precisely, if the sets

$$A = \{\langle x, t\rangle\colon \bar{t} = \mu(x)\} \quad \text{and} \quad B = \{\langle t, s\rangle\colon \bar{s} = \nu(\bar{t})\}$$

are Souslin, then so is the set

$$C = \{\langle x, s\rangle\colon \bar{s} = \nu(\mu(x))\}.$$

Because $(\langle x, s\rangle \in C) \equiv \bigvee_t \{(\bar{t} = \mu(x)) \wedge (\bar{s} = \nu(\bar{t}))\}$ and the sets A and B are Souslin.

THEOREM 7: *Let* $E \subset X$ *be Souslin and let—according to Corollary to Theorem* 1.4—W *be the corresponding "sieve," i.e.* $W\colon \mathscr{R}_0 \to 2^X$ *and*

$$x \in (X-E) \equiv (\mu(x) < \omega_1),$$

where $\mu(x)$ *denotes the order type of* $M_x = \{r\colon x \in W(r)\}$.

Then the relation $\tau = \mu(x)$ *is Souslin.*

Because, for $t \in \mathscr{C}$ and $x \in X$, we have

$$(\bar{t} = \mu(x)) \equiv \bigvee_{\mathfrak{z}\mathfrak{v}} (\mathfrak{z} \equiv R_t)\,(\mathfrak{v} \equiv M_x)\,(\mathfrak{z} \sim \mathfrak{v}).$$

§ 3. Coanalytic sets (*CA*-sets)

All spaces in §§ 3–5 will be understood to be Polish.

Let us recall that E is called *coanalytic* (or briefly a ***CA***-set) if it can be imbedded in a Polish space X so that $X-E$ is Souslin (analytic).

By Theorems 1.6 and 1.12, *every Borel set is coanalytic, and there are coanalytic sets which are not Borel* (not even Souslin). Such is also the set $L = \{t\colon \bar{t} < \omega_1\}$ (see § 2, Lusin and Sierpiński [2]).

THEOREM 1: *Countable unions, countable intersections and countable products of* ***CA***-*sets are* ***CA***-*sets.*

PROOF. The countable additivity of ***CA***-sets follows at once from the countable multiplicativity of Souslin sets (see Theorem 1.2). The same concerns the countable multiplicativity of ***CA***-sets.

Now let $E_n \subset X_n$ be ***CA*** (X_n is Polish). Since

$$E_1 \times E_2 \times \ldots = (E_1 \times X_2 \times X_3 \times \ldots) \cap (X_1 \times E_2 \times X_3 \times \ldots) \cap \ldots,$$

our proof is reduced to show that if $E \subset X$ is ***CA*** and Y is Polish, then $E \times Y$ is ***CA*** (in $X \times Y$). But this follows from the identity

$$X \times Y - E \times Y = (X - E) \times Y,$$

since $(X-E) \times Y$ is Souslin (by Theorem 1.2).

Let us now recall that, according to Corollary to Theorem 1.4, if $E \subset X$ is ***CA***, then there exists a mapping $W\colon \mathscr{R}_0 \to 2^X$ (called a sieve) such that, writing

(1) $$M_x = \{r \in \mathscr{R}_0\colon x \in W(r)\}$$

and

(2) $\mu(x)$ = order type of M_x relative to the inequality $r \geqslant s$,

we have

$$E = \{x\colon \mu(x) \text{ is an ordinal number}\}.$$

Hence (comp. Chapter XI, § 8)

(3) $$E = \bigcup_{\alpha < \omega_1} C_\alpha, \quad \text{where} \quad C_\alpha = \{x\colon \mu(x) = \alpha\}$$

(the sets C_α with $\alpha < \omega_1$ are called the *constituents* of E).

THEOREM 2 (see Lusin [1]): *The constituents C_α of a **CA**-set are Borel. Thus any **CA**-set is an $\aleph_1$ union of Borel sets.*

PROOF. Let $\alpha < \omega_1$ and $x \in C_\alpha$. We have to show that the set

$$C_\alpha = \{y \in E\colon \mu(y) = \mu(x)\}$$

is Borel.

Since the relations $\tau = \sigma$ and $\tau = \mu(x)$ are Souslin (by Theorems 2.3 and 2.7), it follows (by Theorem 2.6) that the set C_α is Souslin.

On the other hand, since $\mu(x)$ and $\mu(y)$ are ordinals, we have

$$\big(\mu(y) = \mu(x)\big) \equiv \{\mu(x) \nless \mu(y) \nless \mu(x)\},$$

and hence, by Theorem 2.2, C_α is a CA-set.

It follows that C_α is Borel (by Corollary 1 to Theorem 1.8).

For proofs under slighter assumptions, see Rogers and Willmott [2] and Maitra [3].

Remark 1: If one does not assume the continuum hypothesis, the following statement becomes meaningful.

Any coanalytic set which has the cardinality $> \aleph_1$ contains a Cantor set.

PROOF. At least one of the constituents of E is uncountable (because otherwise the cardinality of E would be $\aleph_0 \times \aleph_1 = \aleph_1$). Since any uncountable Borel set contains a Cantor set (see Chapter XII, § 6, Coroll.), our conclusion follows.

COROLLARY: *The cardinality of any uncountable coanalytic set is either $\aleph_1$ or $\mathfrak{c}$.*

Remark 2: Let us add that the existence of uncountable CA-sets that do not contain Cantor sets is consistent with Zermelo–Fraenkel axioms of set theory (announced by Gödel [1] and proved by Solovay [1]).

Even more striking is the following result: the assumption that every set of power $\aleph_1$ is CA is consistent with ZF (see Martin and Solovay [1], also Bukovský [1] and Tall [1]).

Remark 3. The following extension of Theorem 2 has been established. The set

$$C_\tau = \{x\colon M_x \text{ has order-type } \tau\},$$

where τ is *any* countable order-type (not necessarily an ordinal number), is Borel (see Kunugui [1] and D. Scott [1] and for a partial result Hartman [1]).

THEOREM 3: (LUSIN'S SECOND SEPARATION THEOREM [2]) *Let A and B be two Souslin subsets of a Polish space X. Then there are two **CA**-sets, D and H, such that*

$$(4) \qquad A-B \subset D, \quad B-A \subset H \quad \textit{and} \quad D \cap H = \emptyset.$$

PROOF (see Kuratowski [1], p. 491, also Blackwell [1]). Let $\mu(x)$ be—as in (2)—the order type of the set M_x determined by the sieve W of $X-A$, and let $\nu(x)$ be, similarly, the order-type of N_x determined by a sieve of $X-B$. Thus $\mu(x) < \omega_1$ and $\nu(x) < \omega_1$ iff $x \in X-A$ and $x \in X-B$, respectively.

By Theorem 2.1 the relation $\tau \leqslant \sigma$ is Souslin. It follows (applying Theorem 2.6 and 2.7) that

$$(5) \qquad \text{the set } \{x\colon \mu(x) \leqslant \nu(x)\} \text{ is Souslin.}$$

Now put

$$D = B^c \cap \{x\colon \mu(x) \leqslant \nu(x)\}^c \quad \text{and} \quad H = A^c \cap \{x\colon \nu(x) \leqslant \mu(x)\}^c$$

(where E^c means $X-E$).

Formula (4) holds. For, if $x \in A-B$, $\mu(x)$ is not an ordinal while $\nu(x)$ is. Hence $\mu(x) \leqslant \nu(x)$ and therefore $x \in D$. Thus $A-B \subset D$ and symmetrically $B-A \subset H$.

Next $D \cap H = \emptyset$. For suppose that $x \in D \cap H$. Hence $x \in B^c \cap A^c$. Then both, $\mu(x)$ and $\nu(x)$, are ordinals. Consequently we have either $\mu(x) \leqslant \nu(x)$ or $\nu(x) \nleqslant \mu(x)$, and hence $x \notin D \cap H$.

Finally, D and H are *CA*-sets by (5).

COROLLARY (simultaneous separation): *If $A_1, A_2, \ldots$ is a sequence of Souslin sets, then there is a sequence $C_1, C_2, \ldots$ of disjoint **CA**-sets such that*

$$A_n - (A_1 \cup \ldots \cup A_{n-1} \cup A_{n+1} \cup \ldots) \subset C_n.$$

PROOF. By Theorem 3, there are *CA*-sets D_n and H_n such that

$$A_n - \bigcup_{k \neq n} A_k \subset D_n, \quad \bigcup_{k \neq n} A_k - A_n \subset H_n, \quad D_n \cap H_n = \emptyset.$$

Put $C_n = D_n \cap \bigcap_{k \neq n} H_k$. Thus C_n are disjoint *CA*-sets, and

$$A_n - \bigcup_{k \neq n} A_k \subset A_n - A_k \subset \bigcup_{k \neq n} A_n - A_k \subset H_k$$

for every $k \neq n$. Therefore $A_n - \bigcup_{k \neq n} A_k \subset \bigcap_{k \neq n} H_k$, which completes the proof.

For proofs based on different ideas, see H. Rogers [1] and Dellachérie and Meyer [1].

THEOREM 4 (see Kuratowski [10] and Novikov [2]): *The family of CA-sets has the countable reduction property.*

In other terms, if $E_1, E_2, \ldots$ is an infinite sequence of CA-sets, then there exists a sequence $F_1, F_2, \ldots$ of disjoint CA-sets such that

$$F_n \subset E_n \quad \textit{and} \quad \bigcup_n F_n = \bigcup_n E_n.$$

Consequently, if $X = \bigcup_n E_n$, then the F_n are Borel sets.

PROOF. Let $W_n, M_{nx}, \mu_n(x)$ and $C_{n\alpha}$ be as in (1)–(3). Thus

$$E_n = \bigcup_{\alpha < \omega_1} C_{n\alpha} \quad \text{and} \quad C_{n\alpha} = \{x \in E_n \colon \mu_n(x) = \alpha\}. \tag{6}$$

By Theorem 2.7, the relation $\tau = \mu_n(x)$ is Souslin for each n. Let $\sigma = \varrho(\tau, n)$ be a Souslin relation satisfying Theorem 2.5. Hence, by Theorem 2.6, the relation $\tau = \varrho(\mu_n(x), n)$ is Souslin.

We define F_n as follows (comp. § 2 (i)):

$$(x \in F_n) \equiv (x \in E_n) \wedge \bigwedge_{k \neq n} [\varrho(\mu_k(x), k) \nless \varrho(\mu_n(x), n)]. \tag{7}$$

By Theorem 2.1, the sets F_n are *CA*.

Obviously, $F_n \subset E_n$.

Now, suppose that $x \in F_n \cap F_k$. Hence $x \in E_n \cap E_k$ and consequently, $\mu_k(x)$ are ordinals, and so are $\varrho(\mu_k(x), k)$ and $\varrho(\mu_n(x), n)$. If $k \neq n$, we have by (7)

$$\varrho(\mu_k(x), k) \nless \varrho(\mu_n(x), n) \nless \varrho(\mu_k(x), k),$$

and this implies (for ordinals) that $\varrho(\mu_k(x), k) = \varrho(\mu_n(x), n)$ and finally $n = k$ (by 2 (9)). This contradiction shows that the sets $F_1, F_2, \ldots$ are disjoint.

It remains to prove that

(8) $$E_n \subset \bigcup_{m=1}^{\infty} F_m.$$

So let, according to (6), $x \in C_{n\alpha}$ and $\mu_n(x) = \alpha$. Hence $\varrho(\mu_n(x), n) < \omega_1$. Denote by m_0 the smallest index such that

(9) $$\varrho(\mu_m(x), m) < \omega_1.$$

Hence, for $k \neq m_0$, $\varrho(\mu_k(x), k) \nless \varrho(\mu_{m_0}(x), m_0)$. Finally, $\mu_{m_0}(x) < \omega_1$ by (9) (compare § 2 (10)).

Thus $x \in E_{m_0}$ and by (7), $x \in F_{m_0}$.

This completes the proof of (8).

To prove the second part of the theorem, observe that, since the sets $F_1, F_2, \ldots$ are disjoint, we have

$$X - F_n = \bigcup_{k \neq n} F_k.$$

Thus $X - F_n$ is *CA* and hence F_n is Borel, since F_n is simultaneously a *CA* and a Souslin set.

Theorem 4 implies the following two "*separation theorems*" (Novikov [2]).

THEOREM 5: *If $A_1, A_2, \ldots$ are Souslin and $\bigcap_n A_n = \emptyset$, then there exists a sequence $B_1, B_2, \ldots$ of Borel sets such that*

$$A_n \subset B_n \quad \textit{and} \quad \bigcap_n B_n = \emptyset.$$

THEOREM 6: *If $A_1, A_2, \ldots$ are Souslin, then there exists a sequence $B_1, B_2, \ldots$ of CA-sets such that*

$$A_n - \bigcap_m A_m \subset B_n \quad \textit{and} \quad \bigcap_n B_n = \emptyset.$$

§ 4. The σ-algebra $\bar{S}$ generated by Souslin sets and the $\bar{S}$-measurable mappings

DEFINITION 1: We denote by $\bar{S}(X)$, or briefly by $\bar{S}$, the *σ-algebra generated by* $S(X)$ (*i.e. the smallest σ-algebra containing* S *and composed of subsets of* X).

DEFINITION 2 (comp. § 4 of Chapter XI): Let $f\colon X \to Y$. The mapping f is called *$\bar{S}$-measurable* (or an *$\bar{S}$-mapping*) if

(1) $$f^{-1}(U) \in \bar{S} \quad \text{for each } U \text{ open in } Y.$$

Denoting by $\boldsymbol{C}$ the σ-algebra of sets having the Baire property and lying in the space X, and by $\boldsymbol{L}$ the σ-algebra of Lebesgue-measurable sets (here $X = \mathscr{I}$), we have

THEOREM 1: $\bar{S} \subset \boldsymbol{C}$ *and if* $X = \mathscr{I}$, $\bar{S} \subset \boldsymbol{L}$.

This follows from $S \subset \boldsymbol{C}$ (see Theorem 1.5 and Theorem XII.8.7a).

COROLLARY: *Let $f\colon X \to Y$ be an $\bar{S}$-mapping. Then f is a $\boldsymbol{C}$-mapping and consequently* (see Theorem XII.8.8), *there is a set $P \subset X$ of first category such that $f|X-P$ is continuous.*

THEOREM 2: *Let, for $j = 0, 1$, X_j be Souslin and $E_j \in \bar{S}(X_j)$. Then* $(E_0 \times E_1) \in \bar{S}(X_0 \times X_1)$.

This follows from Lemma XI.4.2.

THEOREM 3: *Let $f\colon X \to Y$ and $g\colon Y \to Z$ (where X, Y, Z are Souslin). Let g be continuous. If f is $\bar{S}$-measurable, then so is $h = g \circ f$.*

Because $h^{-1}(U) = f^{-1}[g^{-1}(U)]$ and $g^{-1}(U)$ is open if U is open.

THEOREM 4: *Let f be $\bar{S}$-measurable. Then denoting, as usual, the graph of f by*

(2) $$\mathrm{Gr}(f) = \{\langle x, y\rangle : y = f(x)\},$$

we have $\mathrm{Gr}(f) \in \bar{S}(X \times Y)$.

PROOF. By Theorem XI.4.8 and Theorem 2, the complement of $\mathrm{Gr}(f)$ belongs to $\bar{S}(X \times Y)$. Since $\bar{S}(X \times Y)$ is an algebra, this completes the proof.

THEOREM 5: *Let $f\colon X \to Y$ be $\boldsymbol{B}$-measurable. If $E \subset Y$ is Souslin or $\boldsymbol{CA}$, then so is $f^{-1}(E)$.*

PROOF. Obviously the set $f^{-1}(E)$ is the projection on the X-axis of the intersection $\mathrm{Gr}(f) \cap (X \times E)$. If E is Souslin, then both factors are Souslin, since the graph of a $\boldsymbol{B}$-measurable mapping is Borel (see Theorem XII.3.5).

Now, assume that E is $\boldsymbol{CA}$. Since

(3) $$f^{-1}(E) = X - f^{-1}(Y-E),$$

it follows that $f^{-1}(E)$ is $\boldsymbol{CA}$, because $f^{-1}(Y-E)$ is Souslin.

THEOREM 6: *If f: $X \to Y$ and* Gr(f) *is Souslin, then f is* **B**-*measurable and consequently* Gr(f) *is Borelian.*

PROOF. Let $E \subset Y$ be open. Then, as before, $f^{-1}(E)$ is the projection of $\mathrm{Gr}(f) \cap (X \times E)$, hence is Souslin, and by formula (3), $f^{-1}(E)$ is **CA**. It follows by Corollary 1 to Theorem 1.8 that $f^{-1}(E)$ is Borel and hence f is **B**-measurable.

THEOREM 7: *Let f: $X \to Y$ be one-to-one,* **B**-*measurable and onto. Let $E \subset X$ be a* **CA**-*set. Then so is $f^1(E)$.*

PROOF. Put $g = f^{-1}$. By Corollary 1 to Theorem 1.9, g is **B**-measurable. Since $f = g^{-1}$, we have $f^1(E) = g^{-1}(E)$. This completes the proof, because $g^{-1}(E)$ is **CA** by Theorem 5.

Upper and lower Souslin set-valued mappings

According to the general definition of upper and lower-**L** mapping (see Definition XI.4.4) the mapping F: $X \to \boldsymbol{P}(Y)$ is called *lower-Souslin*, or *upper-Souslin*, if

(4) $\{x\colon F(x) \cap U \neq \emptyset\} \in \boldsymbol{S}$ if U is open in Y

equivalently

(4′) $\{x\colon F(x) \not\subset K\} \in \boldsymbol{S}$ if K is closed in Y,

or

(5) $\{x\colon F(x) \subset U\} \in \boldsymbol{S}$ if U is open in Y

equivalently

(5′) $\{x\colon F(x) \cap K = \emptyset\} \in \boldsymbol{S}$ if K is closed in Y.

Examples. The following two set-valued mappings are simple examples of lower-Souslin mappings which are not upper-Souslin and vice-versa.

Let A be an analytic non-Borel subset of the interval $X = (01)$. Let $Y = X$ and

$$F_0(x) = \{y\colon 0 \leqslant y \leqslant 1\} \text{ for } x \in A \quad \text{and} \quad F_0(x) = \{0\} \text{ for } x \notin A,$$

$$F_1(x) = \{y\colon 0 \leqslant y \leqslant 1\} \text{ for } x \notin A \quad \text{and} \quad F_1(x) = \{0\} \text{ for } x \in A.$$

Let F: $X \to \boldsymbol{P}(Y)$. The *graph of the formula* $y \in F(x)$ is the set (see Definition XI.4.5)

$$G(F) = \{\langle x, y\rangle\colon y \in F(x)\}.$$

1. First Graph Theorem: *Let $F: X \to 2^Y$. $G(F)$ is Souslin iff F is lower-Souslin.*

Proof. 1° Let $G(F) \in S(X \times Y)$. Obviously,

$$[F(x) \cap U \neq \emptyset] \equiv \bigvee_y [(y \in F(x))(y \in U)].$$

Hence $\{x: F(x) \cap U \neq \emptyset\}$ is the projection on the X-axis of the set $G(F) \cap (X \times U)$. This set is Souslin (by assumption), hence so is its projection. Thus F is lower-Souslin.

2° Conversely, if F is lower-Souslin, then $G(F)$ is Souslin according to Theorem XI.4.10.

Remark: In part 1° of the proof we do not need to assume that $F(x)$ is closed.

2. Second Graph Theorem: *Let $F: X \to 2^Y$. If F is upper-Souslin, then $G(F)$ is a* ***CA****-set (in $X \times Y$).*

This is a direct consequence of Theorem XI.4.9.

Corollary: *Let $F: X \to 2^Y$. If F is both lower and upper-Souslin, then the graph $G(F)$ is Borel.*

Moreover, if $F: X \to \boldsymbol{K}(Y)$, then F is ***B****-measurable.*

Proof. The first part follows from Theorems 1 and 2 by virtue of Corollary 1 to Theorem 1.8. To show the second part of the corollary, let K be closed $\subset Y$. By assumption, both sets

$$A = \{x: F(x) \cap K \neq \emptyset\} \quad \text{and} \quad B = \{x: F(x) \subset K\}$$

are Souslin.

Since the set $U = Y - K$ is a countable union of closed sets, the set $\{x: F(x) \cap U \neq \emptyset\}$ is also Souslin. But obviously this set is identical with $X - B$. Thus B and $X - B$ are Souslin and hence Borel, which means that F is lower-Borel, hence Borel (by Corollary 1 to Theorem 1.11).

3. The Inverse-Function Theorem: *If $f: Y \to X$ is continuous onto, then $f^{-1}: X \to 2^Y$ is lower-Souslin.*

Proof. Since f is continuous, the set $\text{Gr}(f) = \{\langle x, y \rangle: x = f(y)\}$ is closed in $X \times Y$, hence is Souslin. Thus $G(f^{-1}) = \text{Gr}(f)$ is Souslin and, by the First Graph Theorem, f^{-1} is lower-Souslin.

§ 5. The *PCA*-sets and sets of higher projective classes

The ***PCA*** (called also Σ^1_2)sets are the *continuous images of* ***CA****-sets* (in Polish spaces).

THEOREM 1: *Countable unions, countable intersections and countable products of* ***PCA****-sets are* ***PCA****-sets.*

The proof is completely similar to the proof of Theorem 1.2.

COROLLARY: *Every member of the σ-algebra $\bar{S}$ is a* ***PCA****-set.*

Remark 1: As we saw, the Souslin sets and the *CA*-sets have the Baire property and are Lebesgue measurable (in the case of reals). It is remarkable that *the existence of non-measurable* ***PCA****-sets is consistent with the axioms of set theory* (see Gödel [1], and also Addison [2], Novikov [1]).

THEOREM 2: *Every* ***PCA****-set is the union of $\aleph_1$ Borel sets.*

Consequently, if this set is of power $> \aleph_1$, it contains a Cantor set.

PROOF. Let X be ***CA***, $Y-$***PCA*** and $f\colon X \to Y$ continuous onto. By formula 3(3), we have

$$X = \bigcup_{\alpha<\omega_1} A_\alpha, \quad \text{where } A_\alpha \text{ is Borel.}$$

Hence $Y = \bigcup_{\alpha<\omega_1} f(A_\alpha)$, and since $f(A_\alpha)$ is analytic, it follows by 3(3) that $f(A_\alpha)$ is a union of $\aleph_1$ Borel sets. Our conclusion follows.

COROLLARY: *The cardinality of an uncountable* ***PCA****-set is either $\aleph_1$ or $\mathfrak{c}$.*

Remark 2: An interesting family of subsets of a Polish space has been introduced by N. Lusin under the name of (*C*)*-sets*. The family of (*C*)-sets is closed under the $\mathscr{A}$ operation and the subtraction (and contains all Borel sets).

One shows that the (*C*)-sets are simultaneously ***PCA***-sets and ***CPCA***-sets. The contrary is not true.

See Alexandrov [2], p. 165, Kantorovitch–Livensohn [2], p. 215–217, and [2]. Also Sielivanovski [1] and Nikodym [2].

Remark 3: As in the case of *CA*-sets, the following reduction and separation theorems are valid (Novikov [3]):

(i) one has to replace in Theorem 3.4, *CA* by *PCA*;

(ii) one has to replace in Theorem 3.5, Souslin sets by ***CPCA***-sets and Borel sets by sets which are simultaneously ***PCA*** and ***CPCA***.

Remark 4: An important property of ***PCA***-sets is formulated in the following statement, which is consistent with the axioms of set theory.

(G) *There exists a well-ordering of class **PCA** of the set $\mathscr{E}$ of reals.*

More precisely, there exists a well-ordering relation $x \prec y$ of the reals such that $\{\langle x, y\rangle : x \prec y\}$ is ***PCA***. (See Gödel [1] and Addison [2]; see also Mansfield [1] and [2], and Kuratowski [19].)

It is worth noticing that the statement (G) implies the existence of very paradoxical projective sets (Kuratowski [15]), like:

(1) sets which are totally imperfect as well as their complements (in the space of reals, F. Bernstein sets [2]),

(2) rarefied spaces, i.e. spaces in which every countable set is a $\boldsymbol{G}_\delta$-set,

(3) σ-spaces, i.e. spaces in which every $\boldsymbol{F}_\sigma$-set is $\boldsymbol{G}_\delta$ (Sierpiński [2]),

(4) ν-spaces, i.e. spaces whose every nowhere dense set is countable (Lusin [1] and Mahlo [1]).

DEFINITION: Given a Polish space X, the family of *projective sets* is the smallest family containing Borel sets and closed under complementation and continuous mappings (see Lusin [3]).

Thus the family ***B*** of Borel sets is the class 0 of projective sets, then follows the class of Souslin subsets of X—denoted also by ***PB*** or ***A***—next the ***CA***-sets, then the ***PCA***-sets, ***CPCA***-sets, and so on.

Our knowledge of projective sets is rather poor, e.g. as concerns their cardinality, measurability, Baire property. Let us mention some theorems.

THEOREM 3: *Each projective class is closed under countable union, countable intersection and countable product.*

This has been shown for the classes ***A*** and ***CA***. The general case (***PCA***... and ***CPC***...) is completely similar.

THEOREM 4: *For each projective class, there is a universal function whose graph G is of the same class.*

PROOF. There are two cases to be considered. If the projective class is of the form ***CL***, then the universal function for ***CL*** is the complement of the universal function for ***L***.

If the projective class is of the form ***PL***, then its universal function is the projection of a universal function for the class ***L*** in the space $X \times \mathcal{N}$.

THEOREM 5: (EXISTENCE THEOREM) *Each projective class contains a set which does not belong to any preceding class.*

PROOF. We consider the universal function defined in Theorem 4 and apply the Theorem on Diagonalization (exactly like in Theorem 1.12). See also Mazurkiewicz [2] for interesting examples of projective sets of functions.

THEOREM 6: *The space N^N contains non-projective sets.*

PROOF. Let E_n be—accordingly to Theorem 5—a projective set which is not of class $n-1$ ($n = 1, 2, \ldots$). We think of E_n as placed in the interval $(n-1, n)$. Then the set $E_\infty = E_1 \cup E_2 \cup \ldots$ is not projective.

Remark 5: *The projective classes, except the class **B** and the class **CA**, are invariant under the operation $\mathcal{A}$* (see Kantorovitch and Livensohn [2] and Addison and Kleene [1]).

Remark 6: We call a formula $\varphi(x)$ *projective of class L_n* if the set $\{x: \varphi(x)\}$ is of class L_n.

It is easy to see that the operations $\neg$, $\vee$, $\wedge$, $\bigvee$, $\bigwedge$ performed on projective formulas always lead to projective formulas. Together with the rules mentioned in § 4 of Chapter IV they allow the evaluation of the Borel or projective class of the set defined by this formula (Kuratowski [7]).

This explains the fact that, *by using the logical notation, one is not lead outside the projective sets* (Kuratowski and Tarski [1]).

Also—under very general hypotheses—*transfinite induction* does not lead out of the projective sets (Kuratowski [11] and Kuratowski and v. Neumann [1]).

Remark 7: As recently shown by L. Harrington [1], the following statement is consistent with the Zermelo–Fraenkel axioms of set theory.

*Every set of reals of cardinality less than $\mathfrak{c}$ is of class **CPCA**.*

CHAPTER XIV

MEASURABLE SELECTORS

There are essentially two selector problems (both closely related to the axiom of choice in its various formulations).

The first concerns the case in which we are given a set-valued mapping $F: X \to \boldsymbol{P}(Y)$. Then we call *selector for F* any function $f: X \to Y$ such that

(1) $$f(x) \in F(x) \quad \text{for each } x \in X.$$

The second case is when a partition $\boldsymbol{Q}$ of a space X into (disjoint) subsets is given. We call *selector for* $\boldsymbol{Q}$ any set W such that $W \cap E$ is a single point for each $E \in \boldsymbol{Q}$.

Let us add that the *choice problem* is a particular case of the selector problem of the first kind. Namely, we are given a family $\boldsymbol{R}$ of non-empty (not necessarily disjoint) subsets of X and we are asking for a function $f: \boldsymbol{R} \to X$ such that $f(E) \in E$ for each $E \in \boldsymbol{R}$.

§ 1. The general selector theorem (see Kuratowski and Ryll-Nardzewski [1]) and its consequences (see Kuratowski [17], Castaing [2], Čoban [1], Dauer and Van Vleck [1], Dellachérie [1], Himmelberg [1]-[3], Leese [1], Maitra and Rao [1], Parthasarathy [1], Robertson [1], Sunyah [1], and Wagner [1], where a further extensive bibliography is given)

THEOREM 1: *Let Y be a Polish space,* $\boldsymbol{A}$ *a field of subsets of X (no assumptions are made about the topology of X) and* $F: X \to 2^Y$. *Put* $\boldsymbol{R} = \boldsymbol{A}_\sigma$ (see Chapter XI, § 4). *If*

(2) $$F \text{ is lower-}\boldsymbol{R},$$

then F admits a selector f such that

(3) $$f \text{ is an } \boldsymbol{R}\text{-mapping}.$$

PROOF. Let $H = (r_1, r_2, \ldots, r_i, \ldots)$ be a countable set dense in Y. We may suppose, of course, that Y is complete and its diameter is < 1.

We shall define f as the limit of mappings $f_n\colon X \to H$, where $n = 0, 1, \ldots$, satisfying the following conditions:

$(4)_n$ $\qquad f_n$ is an $\boldsymbol{R}$-mapping,

$(5)_n$ $$\varrho[f_n(x), F(x)] < 1/2^n,$$

$(6)_n$ $$|f_n(x) - f_{n-1}(x)| < 1/2^{n-1} \quad \text{for} \quad n > 0.$$

Let us proceed by induction. Put $f_0(x) = r_1$ for each $x \in X$. Thus $(4)_0$ and $(5)_0$ are fulfilled.

Now, let us assume, for a given $n > 0$, that f_{n-1} satisfies conditions $(4)_{n-1}$ and $(5)_{n-1}$. Put:

(7) $$C_i^n = \{x\colon \varrho[r_i, F(x)] < 1/2^n\},$$

(8) $$D_i^n = \{x\colon |r_i - f_{n-1}(x)| < 1/2^{n-1}\},$$

(9) $$A_i^n = C_i^n \cap D_i^n.$$

We have

(10) $$X = A_1^n \cup A_2^n \cup \ldots$$

For, x being a given point of X, there is by $(5)_{n-1}$, $y \in F(x)$ such that $|y - f_{n-1}(x)| < 1/2^{n-1}$. As $\bar{H} = Y$, there is i such that

$$|r_i - y| < 1/2^n \quad \text{and} \quad |r_i - y| + |y - f_{n-1}(x)| < 1/2^{n-1}.$$

Hence $x \in A_i^n$.

Denote by B_i^n the open ball $\{y\colon |y - r_i| < 1/2^n\}$. It follows that

(11) $$C_i^n = \{x\colon F(x) \cap B_i^n \neq \emptyset\},$$

(12) $$D_i^n = f_{n-1}^{-1}(B_i^{n-1}).$$

Hence, by (2) and $(4)_{n-1}$, $C_i^n \in \boldsymbol{R}$ and $D_i^n \in \boldsymbol{R}$, and therefore

(13) $$A_i^n \in \boldsymbol{R}.$$

Let us apply now Theorem IV.5.1 stating that if $\boldsymbol{A}$ is a field and $\boldsymbol{R} = \boldsymbol{A}_\sigma$, then conditions (10) and (13) imply (for a fixed n) the existence of a sequence $Q_1^n, Q_2^n, \ldots$ of *disjoint* sets such that

(14) $$Q_i^n \in \boldsymbol{R}, \quad Q_i^n \subset A_i^n$$

and

(15) $$X = Q_1^n \cup Q_2^n \cup \ldots$$

We define $f_n\colon X \to H$ as follows:

(16) $$f_n(x) = r_i \quad \text{iff} \quad x \in Q_i^n.$$

We must show that f_n satisfies conditions $(4)_n$–$(6)_n$.

By (16) and (14), we have $f_n^{-1}(r_i) = Q_i^n \in \boldsymbol{R}$ for each i. Consequently, $f_n^{-1}(Z) \in \boldsymbol{R}$ for each $Z \subset H$ (because H is countable and $\boldsymbol{R}$ countably additive). Thus $(4)_n$ is fulfilled.

Let x be given. Put $x \in Q_i^n$. Then, by (14) and (9), $x \in C_i^n$, which implies $(5)_n$ in virtue of (16). Since $A_i^n \subset D_i^n$, it follows also that $x \in D_i^n$, which implies $(6)_n$.

Thus the sequence $f_0, f_1, \ldots, f_n, \ldots$ has been defined according to conditions $(4)_n$–$(6)_n$.

By $(6)_n$ and by the completeness of the space Y, this sequence converges uniformly to a mapping $f\colon X \to Y$. Hence condition $(4)_n$ implies (3) (see Theorem XI.4.3).

Finally, (1) follows from $(5)_n$.

COROLLARY 1: *If $\boldsymbol{R}$ is a σ-algebra and F is lower-$\boldsymbol{R}$, then F admits an $\boldsymbol{R}$-measurable selector f* (for F compact-valued, see Castaing [1]).

Remark: In this case, the proof of the theorem (starting with formula (13)) can be slightly simplified. Namely, we replace formula (16) by the following

$$[f_n(x) = r_i] \equiv [x \in A_i^n - (A_1^n \cup \ldots \cup A_{i-1}^n)].$$

COROLLARY 2: *For each Polish space Y there is a choice function $f\colon 2^Y \to Y$ of the first class of Baire.*

f may be assumed to be continuous if $\dim Y = 0$ (i.e. Y contains a countable base composed of closed-open sets).

PROOF. Put in the Theorem 1: $X = 2^Y$, $\boldsymbol{A}$ = the field of subsets of 2^Y which are simultaneously $\boldsymbol{F}_\sigma$ and $\boldsymbol{G}_\delta$, and F = the identity mapping defined on 2^Y. According to the exponential topology, the sets $\{K\colon K \cap G \neq \emptyset\}$ are open and the sets $\{K\colon K \cap Q \neq \emptyset\}$ are closed (in 2^Y) provided G is open and Q is closed (in Y). Since G is $\boldsymbol{F}_\sigma$ in Y, then $\{K\colon K \cap G \neq \emptyset\}$ is $\boldsymbol{F}_\sigma$ in 2^Y, hence a member of $\boldsymbol{R}$. It follows by Theorem

1, that there is a choice function $f\colon 2^Y \to Y$ satisfying (3). Since the members of $\boldsymbol{R}$ are F_σ-sets, f is of the first class.

In the particular case where $\dim Y = 0$, we denote by $\boldsymbol{A}$ the field of closed-open subsets of 2^Y. By assumption, we have $G = Q_1 \cup Q_2 \cup \dots$, where Q_n is closed-open. Consequently, $\{K\colon K \cap Q_n \neq \emptyset\} \in \boldsymbol{A}$ and $\{K\colon K \cap G \neq \emptyset\} \in \boldsymbol{R}$. Since the members of $\boldsymbol{R}$ are open sets, it follows by (4) that $f^{-1}(G)$ is open, i.e. that f is continuous.

The following theorem concerns the case of less restrictive assumption on Y.

THEOREM 2 (see Kuratowski [17], compare also Wagner [1]): *Let $F\colon X \to 2^Y$ (where X and Y are Souslin) be lower-$\boldsymbol{S}$; then there exists an $\bar{\boldsymbol{S}}$-measurable selector f for F.*

PROOF. Put $T = N^N$. Since Y is Souslin, there is a continuous function $g\colon T \to Y$ onto. Therefore, $g^{-1}\colon 2^Y \to 2^T$. Put

$$H = g^{-1} \circ F, \quad \text{i.e.} \quad H(x) = g^{-1}\big(F(x)\big), \quad \text{hence} \quad H\colon X \to 2^T.$$

We shall prove that H is lower-Souslin.

By the First Graph Theorem of Chapter XIII, § 4, it suffices to show that

$$G(H) \in \boldsymbol{S}(X \times T), \quad \text{i.e.} \quad \{\langle x, t\rangle\colon t \in H(x)\} \in \boldsymbol{S}(X \times T).$$

Now

$$[t \in H(x)] \equiv [t \in g^{-1}\big(F(x)\big)] \equiv [g(t) \in F(x)] \equiv \bigvee_y [y = g(t)]\, [y \in F(x)].$$

Therefore, $G(H)$ is the projection on $X \times T$ of the set

$$Z = \{\langle x, y, t\rangle\colon [y = g(t)]\} \cap \{\langle x, y, t\rangle\colon [y \in F(x)]\}.$$

Since $\{\langle y, t\rangle\colon [y = g(t)]\}$ is closed and $\{\langle x, y\rangle\colon [y \in F(x)]\} = G(F)$ is Souslin (by the First Graph Theorem applied to F), it follows that Z is Souslin, and so is its projection, $G(H)$.

Consequently, the mapping $H\colon X \to 2^T$ is lower-Souslin, hence lower-$\bar{\boldsymbol{S}}$. Since T is Polish and $\bar{\boldsymbol{S}}$ is a σ-algebra, Corollary 1 can be applied to H. So let h be an $\bar{\boldsymbol{S}}$-measurable selector for H, and put $f = g \circ h$. Thus f is $\bar{\boldsymbol{S}}$-measurable and, since $h(x) \in H(x) = g^{-1}\big(F(x)\big)$, it follows that $f(x) \in F(x)$.

Another extension of Theorem 1 concerns the case where $F(x)$ is not assumed to be closed (in fact, $F(x)$ will be Souslin).

THEOREM 3: *Let $F(x) \subset Y$. If the graph $G(F)$ is Souslin, then there is an $\bar{\boldsymbol{S}}$ measurable selector for F.*

PROOF. Put $T(x) = \{x\} \times F(x)$. Hence $T(x)$ is closed in $G(F)$ and thus $T\colon X \to 2^{G(F)}$. We shall show that T is lower $\boldsymbol{S}$. By the First Graph Theorem of Chapter XIII, § 4 (in which we substitute $G(F)$ for Y), it suffices to show that $G(T)$ is Souslin. Now

$$\begin{aligned} G(T) &= \{\langle xyz\rangle\colon\ [\langle yz\rangle \in T(x)]\} = \{\langle xyz\rangle\colon\ \big(y \in F(x)\big)\,(z = x)\} \\ &= \{\langle xyz\rangle\colon\ [\langle xy\rangle \in G(F)] \wedge (z = x)\}. \end{aligned}$$

It follows that $G(T)$ is Souslin (since $G(F)$ is Souslin).

Apply now Theorem 2. It follows that there exists an $\bar{\boldsymbol{S}}$-measurable selector h for T. Hence $h(x) \in T(x)$, which means that h is of the form

$$h(x) = \langle x, f(x)\rangle \in T(x), \qquad \text{hence} \qquad f(x) \in F(x).$$

Finally, f is $\bar{\boldsymbol{S}}$-measurable. Because $f^{-1}(U) = h^{-1}(X \times U)$, and assuming that U is open, we have $h^{-1}(X \times U) \in \bar{\boldsymbol{S}}$.

Remark. There was an attempt made by V. A. Rokhlin to prove the Corollary 1 to Theorem 1 (see Uspiehi **IV** (1949) Lemma 2, p. 85; see also Amer. Math. Soc. Transl. **49** (1966) 71–240).

Unfortunately his argument is not valid. This remark was communicated to the authors by Dr R. Pol. See also Wagner [1], p. 3.

*Applications to **B**-measurable mappings*

According to the general definition, the mapping $F\colon X \to 2^Y$ is said to be of *lower **B**-measurable class* α (or briefly, of *lower class* α) if (see Chapter XI, § 4 (6))

$$\{x\colon\ F(x) \cap U \neq \emptyset\} \qquad \text{is of additive class } \alpha$$

whenever $U \subset Y$ is open.

Theorem 1 implies the following statement.

THEOREM 4 (see Kuratowski and Ryll-Nardzewski [1], p. 401): *Let X be metric and Y Polish. If $F\colon X \to 2^Y$ is of lower class $\alpha > 0$, then there is a selector f of class α.*

In particular, if F is upper semi-continuous, then f is of class 1.

If X contains a countable base of closed-open sets, then the theorem is also true for $\alpha = 0$.

To show Theorem 4, we substitute to $\boldsymbol{A}$ in Theorem 1 the family $\boldsymbol{A}_\alpha$ (of ambiguous sets of class α).

Remark: As shown by J. Kaniewski and R. Pol [1], the separability of Y can be omitted under some additional conditions.

See also Engelking [3] and Čoban [1], where the problem of omitting separability is considered for lower semi-continuous mappings.

§ 2. Selectors for measurable partitions of Polish spaces

Let us recall that by a *partition* of the space X is meant a collection $\boldsymbol{Q}$ of closed, disjoint, non-empty sets, whose union is X.

DEFINITION 1: Given a family $\boldsymbol{R}$ of subsets of X, we shall say that $\boldsymbol{Q}$ is a *lower-$\boldsymbol{R}$* or *upper-$\boldsymbol{R}$ partition of X* if

(1) $$\bigcup\{E \in \boldsymbol{Q}\colon E \cap U \neq \emptyset\} \in \boldsymbol{R} \quad \text{for each open } U \subset X,$$

or

(2) $$\bigcup\{E \in \boldsymbol{Q}\colon E \cap K = 0\} \in \boldsymbol{R} \quad \text{for each closed } K \subset X,$$

respectively.

Obviously, if $\boldsymbol{R}$ denotes the family of open sets, then $\boldsymbol{Q}$ is lower (upper)-$\boldsymbol{R}$ iff $\boldsymbol{Q}$ is a lower (upper) *semi-continuous decomposition* of X in the usual sense.

Given a partition $\boldsymbol{Q}$ of X, denote by $P(x)$ the (unique) member of $\boldsymbol{Q}$ containing x. Thus

(3) $$x \in P(x) \in \boldsymbol{Q} \quad \text{and} \quad P\colon X \to 2^X.$$

P is called the "*natural projection*" of X onto $\boldsymbol{Q}$ (the quotient-space). It is easily seen that, for each $Z \subset X$, we have

(4) $$\bigcup\{E \in \boldsymbol{Q}\colon E \cap Z \neq \emptyset\} = \{x \in X\colon P(x) \cap Z \neq \emptyset\}.$$

Consequently (comp. Definition XI.4.4),

(5) $\boldsymbol{Q}$ is a lower (upper)-$\boldsymbol{R}$ partition of X

$\equiv P$ is a lower (upper)-$\boldsymbol{R}$ mapping of X into 2^X.

DEFINITION 2: A set $W \subset X$ is called a *selector for* $\boldsymbol{Q}$ if $W \cap E$ is a singleton for each $E \in \boldsymbol{Q}$.

It follows that

(6) *If f is a selector for the mapping $P: X \to 2^X$ and if f is constant on each $E \in \boldsymbol{Q}$, then the set $f^1(X)$ (the range of f) is a selector for $\boldsymbol{Q}$.*

DEFINITION 3: We call *saturated* (with respect to the partition $\boldsymbol{Q}$) those sets which are unions of some members of $\boldsymbol{Q}$ and we denote by $\boldsymbol{U}$ the *family of all saturated subsets of* X.

Obviously, $\boldsymbol{U}$ is a complete Boolean algebra of sets (i.e. closed under the complementation and arbitrary unions and intersections). Let us note that the atoms of $\boldsymbol{U}$ are just the members E of $\boldsymbol{Q}$.

Given a subset Z of X, we say that A is the *saturation of* Z if A is the union of all $E \in \boldsymbol{Q}$ which intersect Z. With this terminology, $\boldsymbol{Q}$ is lower (upper)-$\boldsymbol{R}$ iff the saturation of each open (closed) set is a member of $\boldsymbol{R}$ (of $\boldsymbol{R}^c$).

We are going to show the following theorem (see Kuratowski and Maitra [1], see also Kuratowski [18]–[20]).

THEOREM 1: *Let X be Polish and let $\boldsymbol{A}$ be any field of subsets of X. Put $\boldsymbol{R} = \boldsymbol{A}_\sigma$ and assume that every open set of X belongs to $\boldsymbol{R}$.*

If the partition $\boldsymbol{Q}$ of X is either lower or upper $\boldsymbol{A}$, then there is a selector W for $\boldsymbol{Q}$ such that $(X-W) \in \boldsymbol{R}$.

The proof will be based on the following lemmas.

LEMMA 1: *Let $\boldsymbol{R}$ be a σ-lattice containing all open subsets of X, and let $f: X \to X$ be an $\boldsymbol{R}$-*mapping. *Put*

$$I = \{x: f(x) = x\}.$$

Then $(X-I) \in \boldsymbol{R}$.

Furthermore, if f is a selector for P and if f is constant on each $E \in \boldsymbol{Q}$, then we have $f^1(X) = I$.

The first part of Lemma 1 is the Corollary to Theorem XI.4.8.

Now let $y = f(x_0)$ and $x_0 \in E \in \boldsymbol{Q}$. Since f is a selector for P, we have $f(x_0) \in P(x_0) = E$, i.e. $y \in E$, and since f is constant on E, it follows that $f(y) = f(x_0)$, i.e. $f(y) = y$, and thus $y \in I$.

LEMMA 2: *If f: $X \to X$ is a $\boldsymbol{U}$-mapping, then f is constant on each $E \in \boldsymbol{Q}$.*

PROOF. By assumption, $f^{-1}(G) \in \boldsymbol{U}$ for each G open. It follows at once that also $f^{-1}(K) \in \boldsymbol{U}$ for K closed. In particular, $f^{-1}\{x\} \in \boldsymbol{U}$ for each $x \in X$.

Now let $x_0 \in E$ and $f(x_0) = y_0$, i.e. $x_0 \in f^{-1}\{y_0\}$. It follows that $E \subset f^{-1}\{y_0\}$. Hence $f^1(E) \subset f^1 f^{-1}\{y_0\} = \{y_0\}$.

PROOF OF THEOREM 1. 1. Suppose $\boldsymbol{Q}$ is lower-$\boldsymbol{A}$. Then so is P (by (5)), and this means that, for each open $U \subset Y$, the set $\{x\colon P(x) \cap U \neq \emptyset\}$ belongs to $\boldsymbol{A}$. But this set belongs also to $\boldsymbol{U}$. Put $\boldsymbol{A}_1 = \boldsymbol{A} \cap \boldsymbol{U}$. Thus P is lower-$\boldsymbol{A}_1$. Since $\boldsymbol{A}_1$ is a field in X, we can apply Theorem 1.1. This implies the existence of a selector f: $X \to X$ for P which is an $\boldsymbol{A}_{1\sigma}$-mapping, i.e.

$$f^{-1}(G) \in \boldsymbol{A}_{1\sigma} = (\boldsymbol{A} \cap \boldsymbol{U})_\sigma \subset (\boldsymbol{A}_\sigma \cap \boldsymbol{U}_\sigma) = \boldsymbol{R} \cap \boldsymbol{U} \tag{7}$$

for each open $G \subset X$, and this means that f is an $\boldsymbol{R}$-mapping and a $\boldsymbol{U}$-mapping as well.

By Lemmas 1 and 2, $f^1(X)$ is a selector for $\boldsymbol{Q}$ and its complement belongs to $\boldsymbol{R}$.

2. Now, let $\boldsymbol{Q}$ be upper-$\boldsymbol{A}$. Hence so is P and therefore we have, for each closed $K \subset X$,

$$\{x\colon P(x) \cap K \neq \emptyset\} \in \boldsymbol{A}^c = \boldsymbol{A} \tag{8}$$

since $\boldsymbol{A}$ is a field.

Let U be open $\subset X$. Put $U = K_1 \cup K_2 \cup \ldots$, where K_n is closed. Obviously,

$$\{x\colon P(x) \cap U \neq \emptyset\} = \bigcup_n \{x\colon P(x) \cap K_n \neq \emptyset\}.$$

It follows by (8) that

$$\{x\colon P(x) \cap U \neq \emptyset\} \in (\boldsymbol{A} \cap \boldsymbol{U})_\sigma = \boldsymbol{A}_{1\sigma}$$

which means that P is lower-$\boldsymbol{A}_{1\sigma}$.

Applying Theorem 1.1, we conclude that there is a selector f: $X \to X$ for P satisfying (7). This completes the proof.

COROLLARY 1 (see also Bourbaki [2], Ex. 9(a), p. 262): *Each semi-continuous partition (either lower or upper) of a Polish space admits a $\boldsymbol{G}_\delta$-selector.*

In this case we denote by $\boldsymbol{A}$ the family of sets which are simultaneously $\boldsymbol{F}_\sigma$ and $\boldsymbol{G}_\delta$.

COROLLARY 2: *If* $\boldsymbol{R}$ *is any* σ*-algebra containing all open sets, then every lower* $\boldsymbol{R}$*-measurable partition admits an* $\boldsymbol{R}$*-measurable selector.*

In particular we can substitute for $\boldsymbol{R}$ the σ-algebra of *Borel* sets, of sets having the *Baire property*, the algebra $\overline{\boldsymbol{S}}$, the algebra of *Lebesgue measurable* sets (in the last case for $X = \mathscr{E}$).

Remark 1 (see Maitra and Rao [2]). In the case of a Borel-measurable partition, a more precise statement is true. Namely, if $\boldsymbol{R}$ is the family of sets of additive class α, then each partition of upper class α admits a selector of multiplicative class $\alpha+1$, while a partition of lower class $\alpha > 0$ admits a selector of multiplicative class α.

Remark 2. Each lower-$\boldsymbol{S}$ partition of a Borel set (in a Polish space) admits a $\boldsymbol{CA}$-selection (see Kaniewski [1], Corollary).

§ 3. Selectors for point-inverses of continuous mappings

THEOREM 1: *Let* $f: Y \to X$ *be continuous onto. Then there is an* $\overline{\boldsymbol{S}}$*-measurable selector* g *for the inverse function* $f^{-1}: X \to 2^Y$*; i.e. a function* $g: X \to Y$ *such that* $fg(x) = x$.

PROOF. By Theorem XIII.4.3, f^{-1} is lower Souslin. According to Corollary 2.2, this completes the proof.

Remark 1. The function g (according to the Corollary of Chapter XIII, § 1) has the Baire property (Chapter XII, § 8) and hence is *continuous on the complement of a set of* I *category.*

In the case of reals, g is *Lebesgue measurable.*

The existence of a Lebesgue-measurable function g such that $f \circ g = 1$ leads to the well-known *v. Neumann Theorem* (see v. Neumann [1]).

In some important cases, the evaluation of the selector can be more precise. Consider namely the following theorem.

THEOREM 2 (OF MAZURKIEWICZ [1]): *Let* $f: N^N \to X$ *be continuous onto. Then there exists a* $\boldsymbol{CA}$*-selector for the partition of* N^N *into point-inverses. Consequently, each Souslin space is a one-to-one continuous image of a* $\boldsymbol{CA}$*-set (contained in* N^N*). Moreover, the theorem remains true when* N^N *is replaced by any closed set* $F \subset N^N$.

PROOF (see e.g. Kuratowski [1], p. 480). As usually each member $\mathfrak{z} \in N^N$ is represented in the form $\mathfrak{z} = [\mathfrak{z}^1, \mathfrak{z}^2, \ldots]$, where $\mathfrak{z}^n \in N$. Let us consider their *lexicographical* ordering:

$$(1) \qquad [\mathfrak{z} \prec \mathfrak{y}] \equiv (\mathfrak{z}^1 \leqslant \mathfrak{y}^1) \wedge \bigwedge_k [(\mathfrak{z}^k \leqslant \mathfrak{y}^k) \vee \bigvee_{i<k} (\mathfrak{z}^i < \mathfrak{y}^i)].$$

It is easily seen that the set $\{\langle \mathfrak{z}, \mathfrak{y} \rangle : \mathfrak{z} \prec \mathfrak{y}\}$ is closed. Moreover, in every non-empty closed set $A \subset N^N$ there exists the first element; such is the point $\mathfrak{p}(A) = X_1 \cap X_2 \cap \ldots$, where X_1 is the set of all $\mathfrak{z} \in A$ such that $\mathfrak{z}^1$ assumes the minimum value; and X_n, with $n > 1$, is the set of the $\mathfrak{z} \in X_{n-1}$ such that $\mathfrak{z}^n$ assumes the minimum value. By the Cantor Theorem (Theorem XI.3.6) the intersection of the sets X_n reduces to a single point.

Now let W be the set of the points $\mathfrak{p}[f^{-1}(x)]$, where x runs over $f^1(F)$; that is,

$$(2) \qquad [\mathfrak{z} \in W] \equiv [\mathfrak{z} \in F] \wedge \bigwedge_{\mathfrak{y}} \{[f(\mathfrak{z}) = f(\mathfrak{y})] \rightarrow [\mathfrak{z} \prec \mathfrak{y}]\}.$$

W is a ***CA***-set, since the formula in the braces { } is borelian. Moreover, the partial fonction $f|W$ is one-to-one, since the conditions $\mathfrak{z} \in W$, $\mathfrak{y} \in W$ and $f(\mathfrak{z}) = f(\mathfrak{y})$ imply $\mathfrak{z} \prec \mathfrak{y} \prec \mathfrak{z}$, and hence $\mathfrak{z} = \mathfrak{y}$. Finally, we have $f^1(W) = f^1(F)$, since $\mathfrak{z}$ can be replaced by $\mathfrak{p}[f^{-1}(\mathfrak{x})]$ in the left side member of (2).

Remark 2. Since every uncountable Souslin set is of power $\mathfrak{c}$ (by Theorem XIII.1.7), W is of power $\mathfrak{c}$.

COROLLARY 1: *Every Souslin space is an* $\aleph_1$ *union of Borel sets.*

PROOF. By Theorem 2, each Souslin space A is a one-to-one continuous image of a ***CA***-set W. Since W is an $\aleph_1$ union of Borel sets (by Theorem 3.2) and any one-to-one continuous image of a Borel set is Borel (by Theorem 1.9), our conclusion follows.

More generally:

COROLLARY 2: *If E is a **PCA**-subset of a Polish space, then E is an $\aleph_1$ union of Borel sets, and consequently, if the cardinality of E is $> \aleph_1$, then E contains a Cantor set (and its cardinality is $\mathfrak{c}$).*

Let H be ***CA*** in a Polish space and let $f\colon H \rightarrow E$ be continuous onto. We have

$$H = \bigcup_{\alpha<\omega_1} C_\alpha, \qquad \text{hence} \qquad E = f(H) = \bigcup_{\alpha<\omega_1} f(C_\alpha).$$

C_α being supposed Borel, $f(C_\alpha)$ is Souslin, hence, according to Corollary 1, is an $\aleph_1$ union of Borel sets, and so is E.

THEOREM 3: (GENERALIZATION OF MAZURKIEWICZ THEOREM) *Let Y be an arbitrary Borel subset of a Polish space and let $f\colon Y \to X$ be continuous onto. Then there exists a **CA** selector for the partition of Y into point-inverses.*

*In other terms, there is a set $W \subset Y$ of class **CA** such that the partial function $f|W$ is one-to-one and $f^1(W) = X$.*

PROOF. Since Y is Borel, there is a closed subset $F \subset N^N$ and a continuous and one-to-one function $g\colon F \to Y$ onto (see Theorem XII.6.1). Put $h = f \circ g$. Thus $h\colon F \to X$ is continuous onto. By Theorem 2, F contains a **CA**-set H such that the function $h|H$ is one-to-one and $h^1(H) = X$.

Put $W = g^1(H)$. It is easy to see that W is a selector for the partition of Y into point-inverses $f^{-1}(x)$, where x ranges over X.

Moreover, W is a **CA**-set, because g is a one-to-one continuous function defined on a Polish space F, and H is a **CA**-subset of F (see Theorem XIII.4.7).

The following corollary is a "uniformization" theorem (see also Shchegolkov [1]).

Remark 3. Theorem 3 can be deduced from Remark 2 of § 2.

COROLLARY: *Let $Z \subset X \times Y$ be Borel. Then there is a **CA** selector for the partition of Z into sections "parallel" to the Y-axis* (see also Theorem 5).

Substitute in the preceding theorem Z for Y and denote by f the projection on the X-axis.

In view of proving another important uniformization theorem, namely the Theorem of Kunugui–Novikov, we introduce the following notation. Let $Z \subset X \times Y$. We put

(3) $$Z^x = \{y\colon \langle x, y\rangle \in Z\} \quad \text{for} \quad x \in X,$$

(4) $$Z(\varepsilon) = [\langle x, y\rangle\colon \varrho(y, Z^x) < \varepsilon\} \quad \text{for} \quad \varepsilon > 0,$$

(5) $p(x, y) = x$, i.e. p is the projection of $X \times Y$ onto X,

hence

(6) $$x \in p(Z) \equiv \bigvee_y y \in Z^x$$

(we write $p(Z)$ instead of $p^1(Z)$).

It is easy to see that

(7) $$(\bigcap_t Z_t)^x = \bigcap_t (Z_t^x).$$

(8) *If Z is Souslin, then so is $Z(\varepsilon)$ for each $\varepsilon > 0$ (X and Y being supposed Polish).*

This follows directly from the equivalence

(9) $$[\langle x, y\rangle \in Z(\varepsilon)] \equiv \bigvee_{y'} \{(|y'-y| < \varepsilon)\,(\langle x, y'\rangle \in Z)\},$$

the formula in brackets $\{\ \}$ being Souslin.

(10) *If Z^x is closed in Y for each $x \in X$, then* $\bigcap_{n=1}^{\infty} Z(1/n) = Z.$

Obviously, $Z \subset Z(1/n)$ for each n.

On the other hand, let $\langle x, y\rangle \in Z(1/n)$ for each n. Then by (4) and (9), there is y_n such that $|y_n - y| < 1/n$ and $y_n \in Z^x$. This implies that $y \in Z^x$, i.e. $\langle x, y\rangle \in Z$.

(11) *The set $(Z(\varepsilon))^x$ is open for each $x \in X$,*

because this set is the union of balls, $\bigcup_{y'} |y'-y| < \varepsilon$, such that $\langle x, y'\rangle \in Z$.

Let $Z_1 \supset Z_2 \supset \ldots$ and let Z_n^x be compact for each x and n. Then

(12) $$p(\bigcap_n Z_n) = \bigcap_n p(Z_n).$$

By the compactness of Z_n^x, we have by (6) for each x, and

$$x \in p(\bigcap_n Z_n) \equiv \bigvee_y y \in (\bigcap_n Z_n)^x \equiv \bigvee_y \bigwedge_n y \in Z_n^x \equiv \bigwedge_n \bigvee_y y \in Z_n^x$$
$$\equiv \bigwedge_n x \in p(Z_n) \equiv x \in \bigcap_n p(Z_n)$$

(where the equivalence "$\bigvee_y \bigwedge_n \equiv \bigwedge_n \bigvee_y$" follows from the compactness of Z^x).

Lemma: *Let $E \subset X \times Y$ and let $C_1, C_2, \ldots$ be CA-sets in $X \times Y$ such that*

(i) $$\bigcap_n C_n \subset E,$$

(ii) $$E(1/2n) \subset C_n,$$

(iii) C_n^x *is compact for each* x *and* n.

Then $p(E)$ *is a* **CA**-*set.*

Proof. Put $D_n = C_1 \cap \ldots \cap C_n$. Since

$$E(1/2n) \subset E(1/2k) \subset C_k \quad \text{for} \quad k \leqslant n,$$

we have

(13) $$E(1/2n) \subset D_n \quad \text{for each } n.$$

By (ii), $E \subset \bigcap_n E(1/2n) \subset \bigcap_n C_n = \bigcap_n D_n$.

Hence by (i) we have

(14) $$\bigcap_n D_n = E.$$

By (iii) and (7),

(15) $$D_n^x \text{ is compact.}$$

Hence by (12) and (14)

(16) $$p(E) = p\Big(\bigcap_n D_n\Big) = \bigcap_n p(D_n).$$

Let $R = (r_1, r_2, \ldots, r_i, \ldots)$ be dense in Y. Put

(17) $$K_n = \bigcup_i \{x \colon \langle x, r_i \rangle \in D_n\}.$$

It follows that

(18) $$K_n \subset p(D_n),$$

because (compare (6))

$$x \in K_n \equiv \Big(\bigvee_i r_i \in D_n^x\Big) \Rightarrow x \in p(D_n).$$

We shall show that

(19) $$p(E) \subset K_n \quad \text{for each } n.$$

Let $x \in p(E)$. Hence there is y such that $\langle x, y\rangle \in E$. Choose i_n so that $|r_{i_n} - y| < 1/2n$. Therefore, by (9), $\langle x, r_{i_n}\rangle \in E(1/2n)$ and hence by (13), $\langle x, r_{i_n}\rangle \in D_n$, and $x \in K_n$.

By (19), (18) and (16) we have

$$p(E) \subset \bigcap_n K_n \subset \bigcap_n p(D_n) = p(E) \quad \text{and hence} \quad p(E) = \bigcap_n K_n.$$

This completes the proof, because D_n is CA and so is K_n (by (17)).

THEOREM 4 (OF KUNUGUI–NOVIKOV): *Let X be complete and Y compact metric, $E \subset X \times Y$ a Borel set and $p: X \times Y \to X$ the projection. Let E^x be compact*[1]) *for each $x \in X$. Then $p^1(E)$ is a Borel set.*

PROOF (communicated by A. Maitra and R. Pol, see also Čoban [2]): Put $A_n = E(1/n) - E$. According to (8), A_n is Souslin, and by (10), $\bigcap_n A_n = \emptyset$. Hence, by virtue of Theorem 4.5, there is a sequence $B_1, B_2, \ldots$ of Borel sets such that

$$A_n \subset B_n \quad \text{and} \quad \bigcap_n B_n = \emptyset. \tag{20}$$

Put

$$H_n = X \times Y - (E \cup B_n) \quad \text{and} \quad C_n = X \times Y - H_n(1/2n). \tag{21}$$

Obviously,

$$E(1/n) \cap H_n = \emptyset \quad \text{and} \quad C_n \subset E \cup B_n. \tag{22}$$

Let us check that the assumptions of Lemma are fulfilled.

First, C_n is CA, because $H_n(1/n)$ is Souslin (by (8), since H_n is Souslin).

Next, (i) is fulfilled, because (by (22) and (20)):

$$\bigcap_n C_n \subset E \cup \bigcap_n B_n = E. \tag{23}$$

Since $C_n^x = Y - [H_n(1/2n)]^x$, we conclude from (11) that C_n^x is closed (in Y), hence compact.

Finally, we shall show that (ii) is satisfied. Suppose it is not. Then there is $\langle x, y\rangle \in E(1/2n) - C_n$, i.e.

$$\langle x, y\rangle \in [E(1/2n) \cap H_n(1/2n)].$$

There exist, therefore, two points y_1 and y_2 such that (see (9)):

$$|y_1 - y| < 1/2n, \quad y_1 \in E^x, \quad |y_2 - y| < 1/2n, \quad y_2 \in H_n^x,$$

[1]) For a generalization to σ-compact sections, see Arsjenin and Liapunov [1] See also Larman [1].

and consequently $|y_2 - y_1| < 1/n$, $\langle x, y_1 \rangle \in E$ and $\langle x, y_2 \rangle \in H_n$. But also $\langle x, y_2 \rangle \in E(1/n)$. Thus $E(1/n) \cap H_n \neq 0$, which contradicts (22).

By Lemma, $p(E)$ is *CA* and simultaneously—is Souslin (as a continuous image of a Borel set). Therefore $p(E)$ is Borel.

THEOREM 5 (OF KONDÔ): *Let X and Y be Polish, $E \subset X \times Y$ a **CA**-set and, as before, p: $X \times Y \to X$ the projection. Then there exists a selector D of class **CA** for the family of point-inverses $\{p^{-1}(x) \cap E\}$, $x \in p^1(E)$.*

In other terms: $p^1(D) = p^1(E)$ and p is one-to-one on D.

PROOF.[1]) In view of Theorems XI.3.8 and XII.4.7 we can assume that Y is a closed subset of the space N^N. As usually, we write $y = \{y^1, y^2, \ldots\}$ for $y \in N^N$ (where $y^n \in N$).

Let $\mathscr{R}_0 = (0, r_2, r_3, \ldots)$ be the set of binary fractions.

According to § 6 of Chapter XI and Remark to Theorem XIII.1.4, there is a sieve W: $\mathscr{R}_0 \to P(X \times Y)$ such that all the $W(r)$ are open and

(24) $E = \{\langle x, y \rangle$: *$M(x, y)$ is well ordered by the inequality* $\geqslant\}$,
where $M(x, y) = \{r\colon \langle x, y \rangle \in W(r)\}$.

Put

(25) $M(x, y)|r = \{r' \in M(x, y)\colon r' > r\}$ (thus $M(x, y)|r_1 = M(x, y)$).

Write

$$\mu(x, y, r) = \text{order type of } M(x, y)|r.$$

Of course, $\mu(x, y, r) < \omega_1$ if $\langle x, y \rangle \in E$.

For each $x \in p^1(E)$ let us define by induction three sequences: $\alpha_n(x)$ of ordinals, $k_n(x)$ of natural numbers, and $F_n(x)$ of subsets of Y, where $n = 1, 2, \ldots$; besides we put

(26) $$F_0(x) = \{y\colon \langle x, y \rangle \in E\},$$

(26a) $$\alpha_n(x) = \min\{\mu(x, y, r_n)\colon y \in F_{n-1}(x)\},$$

(26b) $$k_n(x) = \min\{y^n\colon y \in F_{n-1}(x),\ \mu(x, y, r_n) = \alpha_n(x)\},$$

(26c) $$F_n(x) = \{y\colon y \in F_{n-1}(x),\ y^n = k_n(x),\ \mu(x, y, r_n) = \alpha_n(x)\}.$$

[1]) This proof is essentially based on Sampei [1] and on ideas kindly communicated to the authors by Dr P. Zbierski. Also the help of Mr J. Kaniewski and Dr R. Pol is to be emphasized in this connection.

It is easy to see that

(27) $$F_0(x) \supset F_1(x) \supset \ldots$$

Put

$$k(x) = \langle k_1(x), k_2(x), \ldots\rangle \quad \text{for } x \in p^1(E),$$
$$D = \{\langle x, k(x)\rangle : x \in p^1(E)\}.$$

Obviously, $p^1(D) = p^1(E)$ and the mapping $p|D$ is one-to-one.

Now, we shall show that

(28) *if* $r_i \in M\big(x, k(x)\big)$ *and* $r_j < r_i$, *then* $\alpha_i(x) < \alpha_j(x)$.

Let $y_1, y_2, \ldots$ be a sequence such that $y_n \in F_n(x)$. For $n \geqslant \max(i,j)$ we have by (27) $y_n \in [F_i(x) \cap F_j(x)]$ and hence (by (26c)):

(29) $$\alpha_i(x) = \mu(x, y_n, r_i) \quad \text{and} \quad \alpha_j(x) = \mu(x, y_n, r_j).$$

Since $r_j < r_i$, it follows that $\alpha_i(x) \leqslant \alpha_j(x)$. It is sufficient to show that there exists an $n \geqslant \max(i,j)$ such that

(30) $$r_i \in M(x, y_n).$$

We have $\langle x, k(x)\rangle \in W(r_i)$ because $r_i \in M\big(x, k(x)\big)$. Since $W(r_i)$ is open, formula (30) is fulfilled for sufficiently large n, because (by the definitions of $k(x)$ and $F_n(x)$): $y_n^i = k_i(x)$ for $i \leqslant n$ and thus

$$\lim_{n=\infty} y_n = k(x) \quad \text{and hence} \quad \lim_{n=\infty} \langle x, y_n\rangle = \langle x, k(x)\rangle.$$

Next we shall deduce from (28) that

(31) $$D \subset E.$$

It is sufficient to show that, for $x \in p^1(E)$, we have $\langle x, k(x)\rangle \in E$, i.e. $k(x) \in F_0(x)$; in other terms, that

(32) *the set* $M\big(x, k(x)\big)$ *is well ordered.*

Otherwise there would exist in $M\big(x, k(x)\big)$ a sequence $r_{i_1} < r_{i_2} < \ldots$ and we would have (by (28)):

$$\alpha_{i_1}(x) > \alpha_{i_2}(x) > \ldots,$$

which is not possible.

Thus (31) is true.

Now we are going to show that for each $x \in p^1(E)$

(33) *the set* $\bigcap_n F_n(x)$ *reduces to* $k(x)$,

or equivalently—that

(34) $k(x) \in F_n(x)$ for each n.

We proceed by induction. Since $k(x) \in F_0(x)$, we have to prove (34) under the assumption that $k(x) \in F_{n-1}(x)$ (where $n \geqslant 1$). By the definition of $k(x)$, it is sufficient to prove that

(35) $\mu(x, k(x), r_n) = \alpha_n(x)$, i.e. that $\mu(x, k(x), r_n) \leqslant \alpha_n(x)$.

Suppose the contrary is true. Hence there exists i_1 such that

$$r_{i_1} > r_n, \quad r_{i_1} \in M(x, k(x)) \quad \text{and} \quad \mu(x, k(x), r_{i_1}) \geqslant \alpha_n(x).$$

It follows by (28) that $\mu(x, k(x), r_{i_1}) > \alpha_{i_1}(x)$. Proceeding further this way, we would define a sequence $r_{i_1} < r_{i_2} < \ldots$ of elements of $M(x, k(x))$ contradicting (32).

It remains to show that D is a CA-set. By (33) we have

(36) $$\langle x, y\rangle \in D \equiv \bigwedge_{n \geqslant 0} y \in F_n(x).$$

We shall show that

(37) $$\begin{aligned}(x, y\rangle \notin D \equiv [(x, y > \notin E] \vee \bigvee_{y'} \bigvee_{n \geqslant 1} \{[(\mu(x, y', r_n) < \mu(x, y, r_n)) \vee \\ \vee (\mu(x, y', r_n) = \mu(x, y, r_n)) \wedge (y'^n < y^n)] \wedge \\ \wedge \bigwedge_{j<n} [\mu(x, y', r_j) = \mu(x, y, r_j)) \wedge (y'^j = y^j)]\},\end{aligned}$$

where the sign "<" between order types is defined as in XII. § 2 (ii) (it coincides with the usual inequality in the case of ordinal numbers).

This will complete the proof of coanalycity of D, because the formula in brackets { } is analytic (according to § 2).

In order to show the implication from left to right in (37), let us assume that $\langle x, y\rangle \notin D$ and $\langle x, y\rangle \in E$. Then there is by (36) an index $n \geqslant 1$ such that

(38) $y \notin F_n(x)$ and $y \in F_j(x)$ for $j < n$.

We shall check that the formula in brackets { } is true for $y' = k(x)$. First, it follows from (33) and from the definition of $F_j(x)$ that

(39) $\mu(x, y', r_j) = \alpha_j(x)$ and $y'^j = k_j(x)$ for each $j \geqslant 1$.

By (38) we have $y \in F_{n-1}(x) - F_n(x)$. Hence there are two possibilities:

either $[\mu(x, y, r_n) \neq \alpha_n(x)]$ or

$[\mu(x, y, r_n) = \alpha_n(x)$ and $y^n \neq k_n(x)]$.

In the first case, we have (by (39))

$$\mu(x, y, r_n) > \alpha_n(x) = \mu(x, y', r_n),$$

and in the second case

$$\mu(x, y, r_n) = \mu(x, y', r_n) \quad \text{and} \quad y^n > k_n(x) = y'^n.$$

Furthermore, we have $y \in F_j(x)$ for $j < n$ and hence

$$\mu(x, y, r_j) = \alpha_j(x) = \mu(x, y', r_j) \quad \text{and} \quad y^j = k_j(x) = y'^j.$$

This completes the proof of the implication from left to right.

In order to show the implication in the opposite direction, let us assume that $\langle x, y\rangle \in D$ and that the right-hand side of (37) is fulfilled. We have to show that this leads to a contradiction.

Since, by (31), $\langle x, y\rangle \in E$, there exist an $y' \in Y$ and an $n \geqslant 1$ such that the formula $\{\ \}$ is true; it follows also that $\mu(x, y, 0)$ is an ordinal and so is $\mu(x, y', 0)$ because, if $n = 1$, then $\mu(x, y', 0) \leqslant \mu(x, y, 0)$, otherwise $\mu(x, y'\ 0) = \mu(x, y, 0)$. Hence $\langle x, y'\rangle \in E$, i.e. $y' \in F_0(x)$.

Since for each $j \geqslant 1$ we have (by (36)) $y \in F_j(x)$, it follows that, for each $j < n$,

$$\mu(x, y', r_j) = \mu(x, y, r_j) = \alpha_j(x) \quad \text{and} \quad y'^j = y^j = k_j(x).$$

This implies by induction that $y' \in F_{n-1}(x)$. Now, according to the formula in brackets $\{\ \}$, there are two possibilities:

either $[\mu(x, y', r_n) < \mu(x, y, r_n)]$ or

$[\mu(x, y', r_n) = \mu(x, y, r_n)$ and $y'^n < y^n]$.

In both cases we are led to a contradiction: the first case contradicts the definition of $\alpha_n(x)$ and the second—the definition of $k_n(x)$.

This contradiction completes the proof of the implication from right to left in (37). Thus the proof of formula (37)—and therefore—of Theorem 5 is completed.

Remark 4. For a recent generalization of Theorem 5, see Kaniewski [1].

BIBLIOGRAPHY

(cited and related papers)

Addison, J. W.

[1] *Separation principles in the hierarchies in classical and effective descriptive set theory*, Fund. Math. **46** (1959) 123–135.

[2] *Some consequences of the axiom of constructibility*, Fund. Math. **46** (1959) 338–357.

Addison, J. W. and **Kleene, S. C.**

[1] *A note on function quantification*, Proc. Amer. Math. Soc. **8** (1957) 1002–1009.

Alexandrov, P.

[1] *Sur la puissance des ensembles mesurables B*, C. R. Paris **162** (1916) 323.

[2] *Sur les ensembles complémentaires aux ensembles (A)*, Fund. Math. **5** (1924) 160–165.

[3] *Sur les ensembles de la première classe des ensembles abstraits*, C. R. Paris **178** (1924) 185.

Alexandrov, P. and **Urysohn, P.**

[1] *Mémoire sur les espaces topologiques compacts*, Verh. Akad. Wett. Amsterdam **14** (1929) 1–96.

Arsjenin, W. and **Ljapunov, A.**

[1] *Theory of A-sets* (Russian), Uspiekhi **5** (1950) 45–108.

Aumann, R. J.

[1] *Measurable utility and the measurable choice theorem*, Coll. Int. CNRS **171** (1969) 15–28, Paris.

Baire, R.

[1] *Leçons sur les fonctions discontinues* (Gauthier-Villars, Paris 1905).

Bachmann, H.

[1] *Transfinite Zahlen*, Ergebnisse der Mathematik und ihrer Grenzgebiete, 2nd edition (Springer, Berlin 1967).

Banach, S.

[1] *Un théorème sur les transformations biunivoques*, Fund. Math. **6** (1924) 236–239.

[2] *Ueber analytisch darstellbare Operationen in abstrakten Raümen*, Fund. Math. **17** (1931) 283–295.

Banach, S. and **Kuratowski, K.**
[1] *Sur une généralisation du problème de la mesure*, Fund. Math. **14** (1920) 127–131.

Banach, S. and **Tarski, A.**
[1] *Sur la décomposition des ensembles de points en parties respectivement congruentes*, Fund. Math. **6** (1924) 244–277.

Barwise, J.
[1] *Admissible sets over models of set theory. Generalized recursion theory* (North-Holland, Amsterdam 1974) 97–122.

Bell, J. L. and **Slomson, A. B.**
[1] *Models and ultraproducts: An introduction* (North-Holland, Amsterdam 1969).

Bernays, P.
[1] *A system of axiomatic set theory*, J. Symb. Logic **2** (1937) 65–77.
[2] *A system of axiomatic set theory*, J. Symb. Logic **6** (1941) 1–17, 133–145.
[3] *A system of axiomatic set theory*, J. Symb. Logic **8** (1943) 89–106.
[4] *A system of axiomatic set theory*, J. Symb. Logic **13** (1948) 65–79.
[5] *A system of axiomatic set theory*, J. Symb. Logic **19** (1954) 81–96.

Bernays, P. and **Fraenkel, A.**
[1] *Axiomatic set theory* (North-Holland, Amsterdam 1958).

Bernstein, F.
[1] *Untersuchungen aus der Mengenlehre*, Math. Ann. **61** (1905) 117–155.
[2] *Zur Theorie der trigonometrischen Reihen*, Leipziger Berichte **60** (1908) 325–338.

Beth, E. W.
[1] *The foundations of mathematics* (North-Holland, Amsterdam 1959).

Birkhoff, G.
[1] *Lattice theory*, 2nd edition (New York 1948).

Blackwell, D.
[1] *Infinite games and analytic sets*, Proc. Nat. Ac. Sci. 58 (1967) 1836.

Bloch, G.
[1] *Sur les ensembles des nombres ordinaux et les suites distinguées de fonctions regressives*, C. R. Paris **236** (1953) 265.

Bolzano, B.
[1] *Paradoxien des Unendlichen*, 1851, English edition; 1950.

Boole, G.
[1] *An investigation of the laws of thought* (London 1854) (reprint: Dover Publ., N. York 1954).

Borel, E.
[1] *Leçons sur la théorie des fonctions* (Gauthier-Villars, Paris 1898).
[2] *Eléments de la théorie des ensembles* (Albin Michel, Paris 1949).

Borsuk, K.
[1] *Sur les rétractes*, Fund. Math. **17** (1931) 152–170.

Bourbaki, N.

[1] *Théorie des ensembles* (Hermann, Paris 1970).

[2] *Topologie générale* (Hermann, 1958), Chap. IX.

Braun, S.

[1] *Sur une propriété caractéristique des ensembles* $G_{\delta\sigma\delta}$, Fund. Math. **18** (1932) 138–147.

Brouwer, L. E. J.

[1] *On linear inner limiting sets*, Proc. Akad. Amsterdam **20** (1917) 1191.

Bukovský, L.

[1] *Borel subsets of metric separable spaces*, Proc. 2nd Prague Topological Symp. (Prague 1967) 83–86.

Burali-Forti, C.

[1] *Una questione sui numeri transfiniti*, Rend. Circ. Mat. Palermo **11** (1897) 154–164.

Cantor, G.

[1] *Ein Beitrag zur Mannigfaltigkeitslehre*, J. Reine Angew. Math. **84** (1878) 242–258.

[2] *Über unendliche lineare Punktmannigfaltigkeiten*, Math. Ann. **20** (1882) 113–121.

[3] *Über unendliche lineare Punktmannigfaltigkeiten*, Math. Ann. **21** (1883).

[4] *Mitteilungen zur Lehre vom Transfiniten*, Zeitschrift für Philosophie und philosophische Kritik **91** (1887) 81–125.

[5] *Beiträge zur Begründung der transfiniten Mengenlehre*, Math. Ann. **46** (1895) 481–512.

[6] *Beiträge zur Begründung der transfiniten Mengenlehre*, Math. Ann. **49** (1897) 207–246.

[7] *Gesammelte Abhandlungen mathematischen und philosophischen Inhalts*, edited by E. Zermelo (Berlin 1932) 488.

Castaing, C.

[1] *Quelques problèmes de mesurabilité liées à la théorie de la commande*, C. R. Paris **262** (1966) 409–411.

[2] *Sur le graphe d'une multiapplication souslinienne* (*mesurable*) (Montpellier 1969).

[3] *Intégrales convexes duales*, C. R. Paris **270** (1972) 1331.

Chang, C. C. and **Keisler, J. H.**

[1] *Model theory* (North Holland, Amsterdam 1974).

Choquet, G.

[1] *Theory of capacities*, Ann. Inst. Fourier **5** (1953) 131–294.

[2] *Ensembles K-analytiques et K-Sousliniens. Cas général et cas métriques*, Ann. Inst. Fourier **9** (1959) 75–81.

Christensen, J. P. R.

[1] *Topology and Borel structure* (North-Holland, Amsterdam 1974).

Čoban, M.

[1] *Multivalued mappings and Borel sets* (Russian), Dokl. Akad. SSSR **182** (1968) 514–516. Also Trudy Mosk Ob. **22** (1970) 229–250, and **23** (1970), 277–301.

[2] *On B-measurable mappings* (Russian), Doklady **207** (1972) 42.

Cohen, P. J.

[1] *Set theory and the continuum hypothesis* (Benjamin, New York 1966).

Cohn, P. M.

[1] *Universal algebra* (Harper and Row, N. York 1965).

Cole, J. K.

[1] *A selector theorem in Banach spaces*, J. Optimization Theory Appl. **7** (1971) 170–172 and **9** (1972) 214.

Dauer, J. P. and **Van Vleck, F. S.**

[1] *Measurable selectors of multifunctions and applications* (to appear in Math. Systems Theory).

Davis, R. L.

[1] *The number of structures of finite relations*, Proc. Amer. Math. Soc. **4** (1953) 486–495.

Dedekind, R.

[1] *Was sind und was sollen die Zahlen* (Friedrich Vieweg., Braunschweig 1888).

Dellacherie, C.

[1] *Ensembles analytiques* (Univer. Strasbourg 1973).

[2] *Un crible généralisé*, Springer Lecture Notes in Mathematics **321** (Berlin 1973).

Dellacherie, C. et **Meyer, P. A.**

[1] *Ensembles analytiques et temps d'arrêt* (Université de Strasbourg, Séminaire de Probabilités, 1974).

Devlin, K. J. and **Johnsbraten, H.**

[1] *The Souslin Problem*, Springer Lecture Notes **405** (Berlin 1974).

Drake, F.

[1] *Set theory. An introduction to large cardinals* (North-Holland, Amsterdam 1974).

Easton, W.

[1] *Power of regular cardinals*, Ann. Math. Logic **1** (1970) 139–170.

Engelking, R.

[1] *Outline of general topology* (PWN—North-Holland, Warszawa—Amsterdam 1968).

[2] *Selectors of the first class for semi-continuous set-valued functions*, Bull. Pol. Acad. Sci. **16** (1968) 272–282.

[3] *On Borel sets and B-measurable functions in metric spaces*, Prace Mat. **10** (1967) 145–149.

Engelking, R., Heath, R. W. and **Michael, E.**

[1] *Topological well-ordering and continuous selections*, Invent. Math. **6** (1968) 150–158.

Engelking, R., Holsztyński, W. and **Sikorski, R.**

[1] *Some examples of Borel sets*, Coll. Math. **15** (1966) 271–274.

Engelking, R. and **Karłowicz, M.**

[1] *Some theorems of set theory and their topological consequences*, Fund. Math. **57** (1965) 275–285.

Erdös, P., Gillman, L. and **Henriksen, M.**

[1] *An isomorphism theorem for real closed fields*, Ann. Math. **61** (1955) 542–554.

Erdös, P. and **Hajnal, A.**

[1] *On a classification of denumerable order types...*, Fund. Math. **51** (1962) 117–129.

Erdös, P., Hajnal, A. and **Rado, R.**

[1] *Partition relations for cardinal numbers*, Acta Math. Acad. Sci. Hung. **16** (1965) 93–196.

Erdös, P. and **Rado, R.**

[1] *A partition calculus in set theory*, Bull. Amer. Math. Soc. **62** (1956) 427-489.

Erdös, P. and **Tarski, A.**

[1] *On families of mutually exclusive sets*, Ann. of Math. **44** (1943) 315-329.

[2] *On some problems involving inaccessible cardinals*, Essays on the Foundations of Mathematics (Magnes Press, Jerusalem 1961).

Feferman, S. and **Vaught, R. L.**

[1] *The first order properties of products of algebraic systems*, Fund. Math. **47** (1959) 51–108.

Fichtenholz, G. and **Kantorovitch, L.**

[1] *Sur les opérations linéaires dans l'espace des fonctions bornées*, Stud. Math. **5** (1934) 69–98.

Filippov, A. F.

[1] *On certain questions in the theory of optimal control*, SIAM J. Control, Ser. A (1959) 76–84.

Fodor, G.

[1] *Generalization of a theorem of Aleksandrov and Urysohn*, Acta Sci. Math. (Szeged) **16** (1955), 204–206.

Fox, R. H.

[1] *On topologies for function spaces*, Bull. Amer. Math. Soc. **51** (1945) 429–432.

Fraenkel, A. A.

[1] *Zu den Grundlagen der Cantor–Zermeloschen Mengenlehre*, Math. Ann. **86** (1922) 230–237.

[2] *Abstract set theory* (North-Holland, Amsterdam 1953).

Fraenkel, A. A., Bar-Hillel, J. and **Levy, A.**

[1] *Foundations of set theory* (North-Holland, Amsterdam 1973).

Fraïssé, R.

[1] *Sur l'extension aux relations de quelques propriétés des ordres*, Ann. Sci. Ecole Normale Sup. **71** (1954) 361–388.

Fréchet, M.

[1] *Les espaces abstraits et leur théorie considerée comme introduction à l'analyse générale* (Gauthier-Villars, Paris 1928).

Frege, G.

[1] *Die Grundlagen der Arithmetik* (Koebner, Breslau 1884).

[2] *Grundgesetze der Arithmetik*, vol. 2 (Pohle, Jena 1903).

Friedman, H.

[1] *Minimality in the Δ^1_2 degrees*, Fund. Math. **81** (1974) 183–197.

[2] *PCA well-orderings of the line*, J. Symb. Log. **39** (1974) 79–80.

Frolik, Z.

[1] *A contribution to the descriptive theory of sets*, Proc. Symp. Gen. Topology (Praha 1961) 151–173.

[2] *On the descriptive theory of sets*, Czechosl. Math. J. **13** (1963) 335–359.

[3] *On bianalytic spaces*, ibid. 561–572.

Gödel, K.

[1] *The consistency of the axiom of choice and of the generalized continuum hypothesis*, Proc. Nat. Acad. Sci. **24** (1938), and Annals of Math. Studies **3** (Princeton 1951).

Hahn, H.

[1] *Reelle Funktionen* (Akademische Verlagsgesellschaft M.B.H., Leipzig 1932).

Halmos, P. R.

[1] *Lectures on Boolean Algebras* (Van Nostrand, Princeton 1963).

Hamel, G.

[1] *Eine Basis aller Zahlen und die unstetigen Lösungen der Funktionalgleichung $f(x+y) = f(x)+f(y)$*, Math. Ann. **60** (1908) 459–462.

Hansell, R. W.

[1] *Borel measurable mappings for nonseparable metric spaces*, Trans. Amer. Math. Soc. **161** (1971) 145–169.

[2] *On Borel mappings and Baire functions*, ibid. **194** (1974) 195–211.

[3] *On characterizing non-separable analytic and extended Borel sets as types of continuous images*, London Math. Soc. **28** (1974) 683–699.

Hardy, G.

[1] *Orders of infinity*, Cambridge Tracts in Mathematics and Mathematical Physics **12**, 2nd edition (Cambridge 1924).

Harrington, L.

[1] *Long projective wellorderings*, Doctor-Dissertation (MIT) (to appear).

Hartman, S.

[1] *Zur Geometrisierung der abzählbaren Ordnungstypen*, Fund. Math. **29** (1937) 209–214.

Hartogs, F.

[1] *Über das Problem der Wohlordnung*, Math. Ann. **76** (1915) 442.

Hausdorff, F.

[1] *Grundzüge der Mengenlehre*, (Veit u. Co., Leipzig 1914) (reprinted: Chelsea, New York 1949).

[2] *Mengenlehre*, 3rd edition (Sammlung Göschen, de Gruyter, Berlin 1937) (reprinted: Chelsea, New York 1957).

[3] *Der Potenzbegriff in der Mengenlehre*, Jahr. Deutsch. Math. Ver. **13** (1904) 570.

[4] *Die Mächtigkeit der Borelschen Mengen*, Math. Ann. **77** (1916) 430–437.

[5] *Über innere Abbildungen*, Fund. Math. **23** (1934) 279–291.

[6] *Über zwei Sätze von G. Fichtenholz und L. Kantorovitch*, Stud. Math. **6** (1936) 18–19.

Henkin, L., Monk, J. D. and **Tarski, A.**

[1] *Cylindric algebras*, part I (North-Holland, 1971).

Hermes, H.

[1] *Einführung in die Verbandstheorie*, 2nd edition (Springer Verlag, 1967).

Hessenberg, G.

[1] *Grundbegriffe der Mengenlehre* (Göttingen, 1906).

Himmelberg, C. J.

[1] *Measurable relations*, Fund. Math. **87** (1975) 53–72.

Himmelberg, C. J., Jacobs, M. Q. and **Van Vleck, F. S.**

[1] *Measurable multifunctions, selectors and Filippov's implicit function lemma*, J. Math. An. and Applications **25** (1969) 276–284.

Himmelberg, C. J. and **Van Vleck, F. S.**

[1] *Some remarks on Filippov's lemma*, Proc. U.S.–Japan Seminar (Benjamin 1967).

[2] *Some selection theorems for measurable functions*, Canadian J. Math. **21** (1969) 394–399.

[3] *Selection and implicit function theorems for multifunctions with Souslin graph*, Bull. Pol. Acad. Sci. **19** (1971) 911-916.

[4] *Multifunctions with values in a space of probability measure* (preprint).

Hintikka, J.

[1] *A closure and complement result for nested topologies*, Fund. Math. **57** (1965) 97–106.

Hoborski, A.

[1] *Une remarque sur la limite des nombres cardinaux*, Fund. Math. **2** (1921) 193–198.

Hoffman-Jörgensen, J.

[1] *The theory of analytic spaces*, Aarhus Nr **10** (1970).

Hurewicz, W.

[1] *Relativ perfekte Teile von Punktmengen und Mengen* (*A*), Fund. Math. **12** (1928) 78–109.

[2] *Theorie der analytischen Mengen*, Fund. Math. **15** (1930) 4–17.

Jacobs, M. Q.

[1] *Measurable multivalued mappings and Lusin's theorem*, Trans. Amer. Math. Soc. **134** (1968) 471–481.

Jacobsthal, L.

[1] *Über der Aufbau der transfiniten Arithmetic*, Math. Ann. **66** (1909) 145–194.

Jayne, J. E.

[1] *Characterizations and metrization of proper analytic spaces*, Inventiones Mathematicae **22** (1973) 51–62.

[2] *Continuous images of proper analytic and proper Borel spaces*, Proc. Camb. Phil. Soc. **75** (1974) 185–191.

[3] *Spaces of Baire functions I*, Ann. Inst. Fourier, Grenoble **24**, **4** (1974) 47–76.

Jech, T. J.

[1] *Trees*, J. Symb. Logic. **36** (1971) 1–14.

[2] *The axiom of choice* (North-Holland, Amsterdam 1973).

Jensen, R. B. and **Solovay, R. M.**

[1] *Some applications of almost disjoint sets*, Math. Logic and Found. of Set Theory, ed. by Y. Bar-Hillel (North-Holland, Amsterdam 1970) 84–104.

Kaniewski, J.

[1] *A generalization of Kondô Theorem*, Bull. Pol. Acad. Sci. **24,6** (1976).

Kaniewski, J. and **Pol, R.**

[1] *Borel-measurable selectors for compact-valued mappings in the non-separable case*, Bull. Pol. Acad. Sci. **23** (1975) 1043-1050.

Kantorovitch, L. and **Livensohn, E.**

[1] *Sur les ensembles projectifs de M. Lusin*, C. R. Paris, **190** (1930) 1113.

[2] *Memoir on the analytical operations and projective sets I*, Fund. Math. **18** (1932) 214–279.

[3] *Memoir on the analytical operations and projective sets II*, Fund. Math. **20** (1933) 54–97.

Katětov, M.

[1] *On descriptive classes of functions*, *Theory of sets and topology* (Berlin 1972) 265–278.

Keesling, J.

[1] *On the equivalence of normality and compactness in hyperspaces*, Pacific J. Math. **33** (1970) 657–667.

Keisler, H. J.

[1] *Model theory for infinitary logic* (North–Holland, Amsterdam 1971).

Keisler, H. J. and **Tarski, A.**

[1] *From accessible to inaccessible cardinals*, Fund. Math. **53** (1964) 225–308.

Keldysh, L.

[1] *Lusin ideas in the descriptive set theory* (Russian), Uspiehi **29** (1974) 183–196.

Kelley, J. L.

[1] *The Tychonov product theorem implies the axiom of choice*, Fund. Math. **37** (1950) 75–76.

[2] *General topology* (Van Nostrand, New York 1955).

Kochen, S.

[1] *Ultraproducts in the theory of models*, Ann. Math. **74** (1961) 221–261.

Kolmogorov, A.

[1] *Opérations sur des ensembles* (Russian), Mat. Sb. **35** (1928) 415–422.

Kondô, M.

[1] *Sur l'uniformization des complémentaires analytiques et les ensembles projectifs de la seconde classe*, Japan J. Math. **15** (1939) 197–230.

König, D.

[1] *Über eine Schlussweise aus dem Endlichen ins Unendliche*, Acta Litt. Ac. Sci. Hung. Fran. Joseph. **3** (1927) 121–130.

König, J.

[1] *Zum Kontinuumproblem*, Math. Ann. **60** (1904) 177–180.

[2] *Sur la théorie des ensembles*, Compt. Rend. Sci. **143** (1906) 110–112.

Kruse, A. H.

[1] *Souslinoid and analytic sets in a general setting*, Memoirs, Amer. Math. Soc. **86** (1969).

Kunugui, K.

[1] *Contributions à la théorie des ensembles boreliens et analytiques II and III*, Journ. Fac. Sci. Hokkaido Univ. **8** (1939–40) 79–108.

Kuratowski, K.

[1] *Topology I* (Academic Press—PWN, New York—Warszawa 1966).

[2] *Topology II* (Academic Press—PWN, New York—Warszawa 1968).

[3] *Introduction to set theory and topology* (Pergamon Press, Oxford 1972).

[4] *Sur la notion de l'ordre dans la théorie des ensembles*, Fund. Math. **2** (1921) 161–171.

[5] *Une méthode d'élimination des nombres transfinis des raisonnements mathématiques*, Fund. Math. **3** (1922) 76–108.

[6] *Sur l'opération $\bar{A}$ d'analysis situs*, Fund. Math. **3** (1922) 182–199.

[7] *Evaluation de la classe borelienne ou projective d'un ensemble de points à l'aide des symboles logiques*, Fund. Math. **17** (1931) 249–272.

[8] *Sur une généralisation de la notion d'homéomorphie*, Fund. Math. **22** (1934) 206–220.

[9] *Quelques problèmes concernant les espaces métriques non-séparables*, Fund. Math. **25** (1935) 532–545.

[10] *Sur les théorèmes de séparation dans la théorie des ensembles*, Fund. Math. **26** (1936) 183–191.

[11] *Les ensembles projectifs et l'induction transfinie*, Fund. Math. **27** (1936) 269–276.

[12] *Sur la géométrisation des types d'ordre dénombrable*, Fund. Math. **28** (1937) 167–185.

[13] *Les types d'ordre définissables et les ensembles boreliens*, Fund. Math. **29** (1937) 97–100.

[14] *Sur une caractérisation des alephs*, Fund. Math. **38** (1951) 14–17.

[15] *Ensembles projectifs et ensembles singuliers*, Fund. Math. **35** (1948) 131–140.

[16] *A general approach to the theory of set-valued mappings*, Proc. III Prague Topological Symposium 1971 (Academia, Prague 1972) 271–280.

[17] *The σ-algebra generated by Souslin sets and its applications to set-valued mappings and to selector problems*, Boll. Unione Mat. Ital. **11** (1975).

[18] *On the concept of strongly transitive systems in topology*, Annali di Mat. **98** (1974) 357–363.

[19] *On partitions of complete spaces which do not admit analytic selectors and on some consequences of a theorem of Gödel*, Symposia Mathematica, Roma 1974 (Academic Press, New York).

[20] *On the selector problems for the partitions of Polish spaces and for the compact-valued mappings*, Ann. Polon. Math. **29** (1974) 421–427.

Kuratowski, K. and **Maitra, A.**

[1] *Some theorems on selectors and their applications to semi-continuous decompositions*, Bull. Acad. Pol. Sci. **22** (1974) 877–882.

Kuratowski, K. and **von Neumann, J.**

[1] *On some analytic sets defined by transfinite induction*, Annals of Math. **38** (1937) 521–525.

Kuratowski, K. and **Ryll-Nardzewski, C.**

[1] *A general theorem on selectors*, Bull. Pol. Acad. Sci. **13** (1965) 397–403.

Kuratowski, K. and **Tarski, A.**

[1] *Les opérations logiques et les ensembles projectifs*, Fund. Math. **17** (1931) 240–248.

Kurepa, G.

[1] *Ensembles ordonnés et ramifiés*, Publ. Math. Univ. Belgrad **4** (1935) 1–138.

[2] *Ensembles linéaires et une classe de tableaux ramifiés* (tableaux ramifiés de M. Aronszajn), Publ. Math. Univ. Belgrad **6** (1938) 129–160.

Lake, J.

[1] *A note on the function $\pi(\alpha)$*, Bull. Pol. Acad. Sci. **23** (1975) (to appear).

Larman, D. G.

[1] *Projecting and uniformising Borel sets with K_σ sections* I, Mathematica **19** (1972) 231–244, II ibid. **20** (1973) 233–246.

Laver, R.

[1] *On Fraïssé order type conjecture*, Ann. of Math. **93** (1971) 89–111.

[2] *An order type decomposition theorem*, Ann. of Math. **98** (1973) 96–119.

Lavrentiev, M.

[1] *Sur la recherche des ensembles homéomorphes*, C. R. Paris **178** (1924) 187.

[2] *Contribution à la théorie des ensembles homéomorphes*, Fund. Math. **6** (1924) 149–160.

Lebesgue, H.

[1] *Sur les fonctions représentables analytiquement*, Journ. Math. **(6) 1** (1905) 29–31.

Leese, S. J.

[1] *Measurable selections in normed space*, Proc. Edinburgh Math. Soc. **19** (1974) 147–150.

[2] *Multifunctions of Souslin type*, Bull. Austral. Math. Soc. **11** (1974) 395–411.

Lévy, A.

[1] *On models of set theory with urelements*, Bull. Acad. Polon. Sci., sér. math. phys. astr. **8** (1960) 463–465.

[2] *Axiom schemata of strong infinity in axiomatic set theory*, Pacific J. Math. **10** (1960) 223–238.

[3] *The definability of cardinal numbers*, *Fundations of Mathematics*, Symposium Papers commemorating the sixtieth birthday of Kurt Gödel, edited by J. J. Buloff, Th. C. Holyoke and S. W. Hahn (Springer-Verlag, New York 1969) 15–38.

Levi, B.

[1] *Intorno alla teoria degli aggregati*, R. Instituto Lombardo di science e lettere, Rendiconti (2) **35** (1902) 863–868.

Liapunov, A.

[1] *R-sets* (Russian), Trudy Inst. Stekl. **40** (Moscow, 1953).

[2] *On the work of P. S. Novikov in descriptive set theory* (Russian), ibidem **133** (1973) 11–22.

Lindenbaum, A. and Tarski, A.

[1] *Communications sur les recherches de la théorie des ensembles*, C. R. Soc. Sci., Varsovie, ed. II, **19** (1926) 299–330.

Łoś, J.

[1] *Quelques remarques, théorèmes et problèmes sur les classes définissables d'algèbres*, "Mathematical interpretations of formal systems" (North-Holland, Amsterdam 1955) 98–113.

Łoś, J. and **Ryll-Nardzewski, C.**

[1] *On the application of Tychonov's theorem in mathematical proofs*, Fund. Math. **38** (1951) 233–237.

Lusin, N.

[1] *Sur un problème de M. Baire*, C. R. Acad. Sci. **158** (1914) 1259.

[2] *Sur les ensembles projectifs de M. Henri Lebesgue*, C. R. Paris **180** (1925) 1570.

[3] *Sur une question concernant la propriété de Baire*, Fund. Math. **9** (1927) 116–118.

[4] *Sur les ensembles analytiques*, Fund. Math. **10** (1927) 1–95.

[5] *Leçons sur les ensembles analytiques* (Gauthier-Villars, Paris 1930).

[6] *Collected papers*, vol. 2 (Russian) (Moscow 1958).

Lusin, N. and **Novikov, P.**

[1] *Choix effectif d'un point dans un complémentaire analytique arbitraire, donné par un crible*, Fund. Math. **25** (1935) 559–560.

Lusin, N. and **Sierpiński, W.**

[1] *Sur quelques propriétés des ensembles (A)*, Bull. Acad. Sc. de Cracovie (1918) 35–48.

[2] *Sur un ensemble non mesurable B*, Journ. Math., sér. 9, t. **II** (1923) 53–72.

Lusin, N. and **Souslin, M.**

[1] *Sur une définition des ensembles mesurables B sans nombres transfinis*, C. R. Paris **164** (1917) 88.

Mackey, G. W.

[1] *Les ensembles boreliens...*, Journ. Math. **36** (1957) 171.

Mc. Kinsey, J. and **Tarski, A.**

[1] *On closed elements in closure algebra*, Ann. Math. **47** (1946) 122–162.

McNeille, H.

[1] *Partially ordered sets*, Trans. Amer. Math. Soc. **42** (1937) 416–460.

Mac Shane, E. J. and **Warfield, R.**

[1] *On Filippov's implicit functions lemma*, Proc. Amer. Math. Soc. **18** (1967) 41–47.

Mahlo, P.

[1] *Über lineare transfinite Mengen*, Leipziger Berichte, math.-phys. Klasse **63** (1911) 187–225.

[2] *Zur Theorie und Anwendungen der ϱ_0 Zahlen*, Ber. Säch. Akad. Wiss. Leipzig, Math. Naturwiss. Kl. **64** (1912) 108–112, **65** (1913) 268–282.

Maitra, A.

[1] *Coanalytic sets that are not Blackwell spaces*, Stat. Inst. Calcutta, Techn. Reports in Math. and Stat. (1968).

[2] *On game-theoretic methods in the theory of Souslin sets*, ibid.

[3] *A note on bimeasurable functions*, Bull. Pol. Acad. Sc. **23** (1975) 131–134.

Maitra, A. and **Rao, B. V.**

[1] *Selection theorems and the reduction principle*, Trans. Amer. Math. Soc. **202** (1975) 57–66.

[2] *Selection theorems for partitions of Polish spaces*, ibid.

Mansfield, R.

[1] *On the possibility of a Δ_2^1 well-ordering of the Baire space*, J. Symb. Logic **38** (1973) 396–398.

[2] *The non-existence of Δ_2^1 well-ordering of the Cantor set*, Fund. Math. **86** (1975) 279–282.

Marczewski, E. (Szpilrajn, E.)

[1] *On measurability and the Baire property* (Polish), C.R.I. Congrès Math. des Pays Slaves, Warsaw 1929 (Książnica Atlas, Warszawa 1930) 297–303.

[2] *Remarque sur les produits cartésiens d'espaces topologiques*, Dokl. Akad. Nauk SSSR **31** (1941) 525–528.

[3] *Concerning the symmetric difference in the theory of sets and in Boolean algebras*, Colloq. Math. **1** (1948) 200–202.

Marczewski, E. and **Sikorski, R.**

[1] *Remarks on measure and category*, Colloq. Math. **2** (1949) 13–19.

Martin, D. A. and **Solovay, R. M.**

[1] *Internal Cohen extensions*, Ann. Math. Logic **2** (1970) 143–178.

Mazurkiewicz, S.

[1] *Sur une propriété des ensembles C(A)*, Fund. Math. **10** (1927) 172–174.

[2] *Ueber die Menge der differenzierbaren Funktionen*, Fund. Math. **27** (1936) 244–249.

[3] *Eine projektive Menge der Klasse PCA im Funktionalraum*, Fund. Math. **28** (1937) 7–10.

Michael, E.

[1] *Topologies on spaces of subsets*, Trans. Amer. Math. Soc **71** (1951) 152–182.

[2] *G_δ selections and compact-covering maps*, Duke Math. J. **36** (1969) 125–127.

[3] *A survey of continuous selections*, Lecture Notes in Mathematics **171** (1970) (Springer, Berlin) 54–58.

Miller, E. W.

[1] *A note on Souslin's problem*, Amer. J. Math. **65** (1943) 673–678.

Mirimanov, D.

[1] *Les antinomies de Russell et de Burali-Forti et le problème fondamental de la théorie des ensembles*, Enseig. Math. **19** (1917) 37–52.

Monk, D. and **Scott, D.**

[1] *Additions to some results of Erdös and Tarski*, Fund. Math. **53** (1964) 335–343.

Montgomery, D.

[1] *Non-separable metric spaces*, Fund. Math. **25** (1935) 527–533.

Mostowski, A.

[1] *Constructible sets with applications* (North-Holland, Amsterdam 1969).

V. Neumann, J.

[1] *Zur Einführung der transfiniten Zahlen*, Ac. Sci. Hung. **1** (1923) 199–208.

[2] *Eine Axiomatisierung der Mengenlehre*, Crelle J. **154** (1925) 219–240.

[3] *Die Axiomatisierung der Mengenlehre*, Math. Z. **27** (1928) 669–752.

[4] *Eine Widerspruchsfreiheitsfrage in der axiomatischen Mengenlehre*, Crelle J. **160** (1929) 227.

[5] *On rings of operators*, Ann. of Math. **50** (1949) 401–485.

Nikodym, O.

[1] *Sur une propriété de l'opération A*, Fund. Math. **7** (1925) 149–154.

[2] *Sur les diverses classes d'ensembles*, Fund. Math. **14** (1929) 145–204.

Novikov, P. S.

[1] *On the consistency of some statements of the descriptive set theory* (Russian), Trudy **38** (1951) 279–316.

[2] *Généralisation du deuxième principe de séparabilité* (Russian), Dokl. Akad. URSS **5** (1934) 8–11.

[3] *Sur la séparabilité des ensembles projectifs de seconde classe*, Fund. Math. **25** (1935) 459–466.

Oczan, Iu. S.

[1] *Theory of operations on sets* (Russian), Uspiehi **10** (1955) 71–128.

Olech, C.

[1] *A note concerning set-valued measurable functions*, Bull. Polish Acad. Sci. **13** (1965) 317–321.

[2] *Existence theorems for optimal problems with vector-valued cost function*, Trans. Amer. Math. Soc. **136** (1969) 159–180.

Ostaszewski, A. J.

[1] *On Lusin's separation principle in Hausdorff spaces*, Proc. London Math. Soc., III ser. **27** (1973) 649–666.

[2] *Families of compact sets and their universals*, Mathematika **21** (1974) 116–127.

Oxtoby, J. C.

[1] *Measure and category* (Springer, Berlin 1970).

Oxtoby, J. C. and **Ulam, S.**

[1] *On the equivalence of any set of first category to a set of measure zero*, Fund. Math. **31** (1938) 201–206.

Parthasarathy, T.

[1] *Selection theorems and their applications*, Lecture Notes in Mathematics **263** (Springer, Berlin 1972).

Peano, G.

[1] *Formulaire de Mathématiques* (Torino 1895) 144.

[2] *Sulla definizione di funzione*, Atti Real Ac. Lin. **20** (1911) 3–5.

Pliś, A.

[1] *Remark on measurable set-valued functions*, Bull. Polish Acad. Sci. **9** (1961) 857–859.

Ponomarov, S.

[1] *A new space of closed sets* (Russian), Mat. Sb. **48** (1959) 191–212.

Quine, W.

[1] *Set theory and its logic* (Harvard Univ. Press, 1963).

Rabin, M. O.

[1] *Arithmetical extensions with prescribed cardinality*, Proc. Acad. Amsterdam, ser. A., **62** (1959) 439–446.

Rado, R.

[1] *Direct composition of partitions*, J. London Math. Soc. **29** (1954) 71–83.

Ramsey, F.

[1] *On a problem of formal logic*, Proc. London Math. Soc., ser. 2, **30** (1929) 264–286.

Rao, B. V.

[1] *On discrete Borel spaces and projective sets*, Bull. Amer. Soc. **75** (1969) 614.

Rasiowa, H. and **Sikorski, R.**

[1] *The mathematics of metamathematics*, Monogr. Mat. **41** (PWN, Warszawa 1963).

Robertson, A. P.

[1] *On measurable selections*, Proc. R. Soc., Edinburgh **72** (1974) 1–7.

Rockafellar, R. T.

[1] *Measurable dependence on convex sets and functions on parameters*, J. Math. Anal. Appl. **28** (1969) 4–25.

Rogers, C. A.

[1] *Analytic sets in Hausdorff spaces*, Mathematica **11** (1964) 1–8.

[2] *Descriptive Borel sets*, Proc. Roy. Soc. **286** (1965) 455–478.

Rogers, C. A. and **Willmott, R. C.**

[1] *On the projection of Souslin sets*, Mathematika **13** (1966) 147–150.

[2] *On the uniformization of sets in topological spaces*, Acta Math. **120** (1968) 1–52.

Rogers, H.

[1] *Theory of recursive functions and effective computability* (Mc Graw Hill, 1967).

Rosser, J. B.

[1] *Simplified independence proof* (Acad. Press, New York 1969).

Rubin H., and **Rubin, J.**

[1] *Equivalents of the axiom of choice* (North-Holland, Amsterdam 1963).

Rudin, M. E.

[1] *Souslin's conjecture*, Amer. Math. Monthly **76** (1969) 1113–1119.

Russell, B. and **Whitehead, A.**

[1] *Principia Mathematica*, vol. 1 (Cambridge Univ. Press, 1910).

Ryll-Nardzewski, C.

[1] *On Borel measurability of orbits*, Fund. Math. **56** (1964) 129–130.

Sageev, G.

[1] *An independence result concerning the axiom of choice*, Ann. Math. Logic **8** (1975) 1–184.

Sainte-Beuve, F.

[1] *On the extension of v. Neumann–Aumann's theorem*, J. Funct. Anal. **17** (1974) 112–129.

Saks, S.

[1] *Theory of the integral*, Monogr. Mat. **7** (Warszawa 1938).

[2] *Sur un ensemble non mesurable jouissant de la propriété de Baire*, Fund. Math **11** (1928) 277.

Sampei, I.

[1] *On the uniformization of the complement of an analytic set*, Comm. Univ. St. Paul **10** (1960) 57–62.

Schoenflies, A.

[1] *Entwickelung der Mengenlehre und ihren Anwendungen* (Leipzig–Berlin 1913).

Schori, R. and **West, J. E.**

[1] 2^I *is homeomorphic to the Hilbert cube*, Bull. Amer. Math. Soc. **78** (1972) 402–406.

Schröder, E.

[1] *Vorlesungen über die Algebra der Logik*, vols. 1–3 (1890–1895) (reprinted by Chelsea Publ., Bronx 1966).

Scott, D.

[1] *Measurable cardinals and constructible sets*, Bull. Acad. Polon. Sc. **7** (1961) 145–149.

[2] *Invariant Borel sets*, Fund. Math. **56** (1964) 117–128.

Shchegolkov, E. A.

[1] *On the uniformization of sets of certain classes* (Russian), Trudy Inst. Stekl. **133** (1973) 251–262.

Shoenfield, J.

[1] *Mathematical logic* (Addison-Wesley, Menlo Park 1967).

Selivanovski, E. A.

[1] *On a class of effective sets* (*sets C*) (Russian), Mat. Sb. **35** (1928) 279–413.

Sierpiński, W.

[1] *Cardinal and ordinal numbers*, Monogr. Mat. **34** (PWN, Warszawa 1965).

[2] *L'hypothèse du continu*, Monogr. Mat. **4** (Warszawa 1934).

[3] *Les ensembles projectifs et analytiques*, Mémorial des Sciences Mathématiques CXII (Paris, 1950).

[4] *Algèbre des ensembles*, Monogr. Mat. **23** (Warszawa, 1951).

[5] *Sur l'invariance topologique de la propriété de Baire*, Fund. Math. **4** (1923) 319–323.

[6] *Sur une propriété des ensembles ambigus*, Fund. Math. **6** (1924) 1–5.

[7] *Sur une propriété des ensembles (A)*, Fund. Math. **8** (1926) 362.

[8] *Sur une classification des ensembles mesurables (B)*, Fund. Math. **10** (1927) 320–327.

[9] *Le crible de M. Lusin et l'opération (A) dans les espaces abstraits*, Fund. Math. **11** (1928) 16–18.

[10] *Sur les ensembles complets d'un espace (D)*, Fund. Math. **11** (1928) 203–205.

[11] *Sur une décomposition d'ensembles*, Mon. Mat. and Phys. **35** (1928) 239–242.

[12] *Sur l'existence de diverses classes d'ensembles*, Fund. Math. **14** (1929) 82–91.

[13] *Le théorème d'unicité de M. Lusin pour les espaces abstraits*, Fund. Math. **21** (1933) 250–275.

[14] *Sur un problème de la théorie des relations*, Ann. Sc. Norm. Pisa (2) **2** (1933) 285–287.

[15] *Sur un problème de M. Kuratowski concernant la propriété de Baire des ensembles*, Fund. Math. **22** (1934) 54–56.

[16] *Sur la dualité entre la première catégorie et la mesure nulle*, Fund. Math. **22** (1934) 276–280.

[17] *Sur la séparabilité multiple des ensembles mesurables B*, Fund. Math. **23** (1934) 292–303.

[18] *L'hypothèse généralisée du continu et l'axiome du choix*, Fund. Math. **34** (1947) 1–5.

[19] *Sur un espace complet qui n'admet pas le théorème de Souslin*, Fund. Math. **34** (1947) 66–68.

[20] *Sur une proposition qui entraîne l'existence des ensembles non mesurables*, Fund. Math. **34** (1947) 157–162.

[21] *Sur une propriété des ensembles ordonnés*, Fund. Math. **36** (1949) 56–67.

[22] *Sur quelques propositions concernant la puissance du continu*, Fund. Math. **38** (1951) 1–13.

[23] *Oeuvres choisies*, vol. 2 (PWN, Warszawa 1975).

Sierpiński, W. and **Tarski, A.**

[1] *Sur une propriété caractéristique des nombres inaccessibles*, Fund. Math. **15** (1930) 292–300.

Sikorski, R.

[1] *Boolean algebras*, 2nd edition (Springer, Berlin 1964).

[2] *A characterisation of alephs*, Fund. Math. **38** (1951) 18–22.

Silver, J. H.

[1] *Some applications of model theory in set theory*, Annals of Mathematical Logic **3** (1971) 45–110.

Sion, M.

[1] *On uniformization of sets in topological spaces*, Trans. Amer. Math. Soc. **96** (1960) 237–245.

[2] *On analytic sets in topological spaces*, Trans. Amer. Math. Soc. **96** (1960) 341–354.

Skolem, T.

[1] *Untersuchungen über die Axiome des Klassenkalküls...*, Skrifter utgit av Videnskapsselkapet i Kristiania, I Klasse, **3** (Oslo 1919).

[2] *Einige Bemerkungen zur axiomatischen Begründung der Mengenlehre*, Wissenschaftliche Vortrage gehalten auf dem fünften Kongress der Skandinavischen Matematiker in Helsingfors (1922) 217–232.

Solovay, R. M.

[1] *On the cardinality of Σ_2^1 sets of reals*, Symp. on Found. of Math. (Springer, Berlin 1969) 58–73.

[2] *Real valued measurable cardinals. Axiomatic set theory*, (edited by D. Scott) (Providence 1971) 397–428.

Solovay, R. and **Tennenbaum, S.**

[1] *Iterated Cohen extensions and Souslin's problem*, Ann. of Math. **94** (1971) 201–245.

Souslin, M.

[1] *Sur une définition des ensembles mesurables B sans nombres transfinis*, Compt. Rend. Acad. Sci. **164** (1917) 89.

Specker, E.

[1] *Verallgemeinerte Kontinuumhypothese und das Auswahlaxiom*, Arch. Math. **5** (1954) 332–337.

Stone, A. H.

[1] *Non-separable Borel sets*, Dissertationes Math. **28** (1962).

Stone, M. H.

[1] *The theory of representations for Boolean algebras*, Trans. Amer. Math. Soc. **40** (1936) 37–111.

Sunyah, C. and **de Wilde, M.**

[1] *Un théorème de sélection pour des applications à graphe borélien*, C. R. Paris **269** (1969) 273–274.

Takeuki, C. and **Zaring, W. M.**

[1] *Axiomatic set theory* (Springer, Berlin 1973).

Taimanov, A. D.

[1] *On closed mappings I* (Russian), Mat. Sb. **36** (1955) 349–352.

[2] *On open mappings of Borel sets* (Russian), Mat. Sb. **37** (1955) 293–300.

[3] *Certain works connected with descriptive set theory and topology* (Russian), Trudy Mat. Inst. im. Steklova, **133** (1973) 203–213.

Tall, F. D.

[1] *Some set-theory consistency results in topology*, Proc. 3rd Topological Symp., Prague 1972, 419–425.

Tarski, A.

[1] *Sur quelques théorèmes qui équivalent à l'axiome du choix*, Fund. Math. **5** (1924) 147–154.

[2] *Quelques théorèmes sur les alephs*, Fund. Math. **7** (1925) 7.

[3] *Sur la décomposition des ensembles en sous-ensembles presque disjoints*, Fund. Math. **12** (1928) 186–205, and **14** (1929) 205–216.

[4] *Sur les classes d'ensembles closes par rapport à certaines opérations élémentaires*, Fund. Math. **16** (1930) 181–304.

[5] *Zur Grundlegung der Boole'schen Algebra I*, Fund. Math. **24** (1935) 195–197.

[6] *Über unerreichbare Kardinalzahlen*, Fund. Math. **30** (1938) 68–89.

[7] *Ideale in vollständigen Mengenkörpern*, Fund. Math. **32** (1939) 45–63.

[8] *Cardinal algebras* (Oxford Univ. Press, New York 1949).

[9] *Some problems and results connected with foundations of set theory*, Logic, Methodology and Philosophy of Science (Edited by Nagel, P. Suppes and A. Tarski) (Stanford 1962) 125–135.

Teichmüller, O.

[1] *Braucht der Algebraiker das Auswahlaxiom*?, Deutsche Zeit. Math. **4** (1939) 567–577.

Tychonoff, A.

[1] *Über topologische Erweiterung von Räumen*, Math. Ann. **102** (1930) 544–561.

Ulam, S.

[1] *Zur Masstheorie in der allgemeinen Mengenlehre*, Fund. Math. **16** (1930) 140–150.

[2] *What is measure*?, Amer. Math. Monthly **50** (1943) 597–602.

Valadier, M.

[1] *Multi-applications mesurables à valeurs convexes compactes*, J. Math. **50** (1971) 265–297.

de la Vallée Poussin, Ch.

[1] *Intégrales de Lebesgue, fonctions d'ensemble, classes de Baire*, 2nd edition, (Gauthier-Villars, Paris 1950).

Vaught, R. L.

[1] *On the equivalence of the axiom of choice and a maximal principle*, Bull. Amer. Math. Soc. **58** (1952) 66.

Veblen, O.

[1] *Continuous increasing functions of finite and transfinite ordinals*, Trans. Amer. Math. Soc. **9** (1908) 280–292.

Vietoris, L.

[1] *Stetige Mengen*, Mon. f. Math. u. Ph. **31** (1921) 173–204.

Vitali, G.

[1] *Sul problema della misura dei gruppi di punti di una retta* (Bologna 1905).

Wagner, D.

[1] *Brief survey of measurable selection theorems. Operations research* (Paoli, Pennsylvania 1976).

Wesley, E.

[1] *Extensions of the measurable choice theorem by means of forcing*, Israel Journ. Math. **14** (1973) 104–114.

Weyl, H.

[1] *Philosophy of Mathematics and Natural Sciences* (Princeton Univ. Press 1949).

Wiener, N.

[1] *A simplification of the logic of relations*, Proc. Cambridge Phil. Soc. **17** (1914) 387–390.

Willmott, R. C.

[1] *On the uniformization of Souslin F sets*, Proc. Amer. Mat. Soc. **22** (1969) 148–155.

Young, W. H. and **Young, G. Ch.**

[1] *The theory of sets of points* (Cambridge, Univer. Press 1906).

Zermelo, E.

[1] *Beweis dass jede Menge wohlgeordnet werden kann*, Math. Ann. **59** (1904) 514–516.

[2] *Untersuchungen über die Grundlagen der Mengenlehre I*, Math. Ann. **65** (1908) 261–281.

[3] *Über Grenzzahlen und Mengenbereiche*, Fund. Math. **16** (1930) 29–47.

Zorn, M.

[1] *A remark on method in transfinite algebra*, Bull. Amer. Math. Soc. **41** (1935) 667–670.

LIST OF IMPORTANT SYMBOLS

$p \wedge q$	conjunction (logical product) of the sentences p and q 1
$p \vee q$	disjunction (logical sum) of p and q 1
$p \rightarrow q$	implication "if p then q" 2
$p \equiv q$	equivalence of p and q 2
$\neg p$	negation of the sentence p 2
$Z(x)$	"x is a set" 5
$x \in y$	"x is an element of y" 5
$x \notin y$ or $\neg(x \in y)$	"x is not an element of y" 5
$x\mathrm{TR}y$	"x is the relational type of y" 5
$A \cup B$	sum of the sets A and B 6
$A - B$	difference of the sets A and B 6
$A \cap B$	intersection of the sets A and B 7
$A \dot{-} B$	symmetric difference of A and B 7, 14
$\subset$	inclusion relation 7
$\varnothing$	empty set 9
$A \doteq B \pmod{I}$	"the sets A and B are congruent modulo the ideal I" 17
1	space (universe) 18
A^c, $-A$	complement of the set A 19
$\bar{A}$, A^-	closure of the set A 27
$\mathrm{Int}(A)$	interior of the set A 28
$\mathrm{Fr}(A)$	boundary of the set A 32
A'	derivative of the set A 33
$a \vee b$	sum of elements of a Boolean algebra 33, 34
$a - b$	difference of elements of a Boolean algebra 33, 34
$a \triangle b$	symmetric difference of elements of a Boolean algebra 33, 34
$a \wedge b$	product of elements of a Boolean algebra 33, 34
o	zero element of a Boolean algebra 33, 34
$f_K(a_1, a_2, \ldots)$	value of the Boolean polynomial f 35
$f \sim g$	"Boolean polynomial f is immediately transformable into the polynomial g" 35
$a \leqslant b$	order relation in a Boolean algebra 37
i	unit of a Boolean algebra 37
$a \overset{*}{-} b$	pseudo-difference of elements of a Brouwerian lattice 43

$\overset{*}{-}b$	pseudo-complement of an element of a Brouwerian lattice 44
$\bigwedge$	general quantifier 46, 47
$\bigvee$	existential quantifier 46, 47
$\bigcup(A)$, $\bigcup_{A\in X} X$	union of sets belonging to the family A 57
$P(A)$	power set of the set A 53
$\{x \in A\colon \Phi(x)\}$	the set of x which belong to A and for which $\Phi(x)$ 55
$\{\Phi\}$“A	image of the set A obtained by the transformation Φ 55
ZFCΣ°	system of axioms I–VII$_\Phi$ of set theory 56
ZFΣ	system Σ° with the axiom of choice deleted 56
$\{a, b\}$	unordered pair of the elements a and b 58
$\langle a, b\rangle$	ordered pair of the elements a and b 58
$\bigcap(A)$, $\bigcap_{A\in X} X$	intersection of sets belonging to the family A 60
$X \times Y$	cartesian product of the sets X and Y 62
$\mathscr{E}$	the set of real numbers 63
$\langle a, b\rangle \in R$, $a\,R\,b$	“a is in the relation R to b” or: “the relation R holds between a and b” 64
D_1	left domain of a relation 64
D_r	right domain (range, counter-domain) of a relation 64
$D_1(f)$	domain of the function f 69
$D_r(f)$	range of the function f 69
$F(R)$	field of the relation R 64
R^i	inverse of the relation R 64
$R \circ S$	composition of the relations R and S 64
x/R	equivalence class containing the element x 68
C/R	quotient class of the set C with respect to the relation R 68
Y^X	set of all mappings of X into Y 69
$f\colon X \to Y$, $X \underset{f}{\to} Y$	“f maps X into Y” 69
$f(x), f_x$	value of the function f at x 69
$g\vert A$	restriction f of g for which $D_1(f) = A$ 71
$\underset{x\in A}{\boldsymbol{F}}[\ldots x \ldots]$	function defined by the formula … x … 72
$R^1(X)$	image of the set X under the relation R 74
$f^1(X)$	set of values of the function f on the set X 74
$R^{-1}(Y)$	inverse image of the set Y under the relation R 74
$f^{-1}(Y)$	inverse image of the set Y under the function f 75
K/I	factor Boolean algebra 79

$\bigvee_{t\in T} f_t$	least upper bound of a set 82
$\bigwedge_{t\in T} f_t$	greatest lower bound of a set 82
0, 0_A	zero element of an ordered set 83
1, 1_A	unit element of an ordered set 83
$\langle A, R_0, R_1, \ldots, R_{k-1}\rangle$	relational system of characteristic $(p_0, p_1, \ldots, p_{k-1})$ 85
$\langle A, R\rangle$	relational system of characteristic (2) 86
$\langle A, R\rangle \approx \langle B, S\rangle$, $R \approx S$	isomorphism of the relational systems $\langle A, R\rangle$ and $\langle B, S\rangle$ 86
X'	successor of the set X 89
a^n	nth iteration of the function a 94
$\lvert X\rvert = n$	"the set X has n elements" 102
$\bigcup_t F_t$, $\bigcup(W)$	union of the sets F_t belonging to the family W being the range of F 107
$\bigcap_t F_t$, $\bigcap(W)$	intersection of the sets belonging to the family W being the range of F 107
$\operatorname{Limsup}_{n=\infty} F_n$	limit superior of the sequence $F_0, F_1, \ldots$ 118
$\operatorname{Liminf}_{n=\infty} F_n$	limit inferior of the sequence $F_0, F_1, \ldots$ 118
$\operatorname{Lim}_{n=\infty} F_n$	limit of the sequence $F_0, F_1, \ldots$ 118
$\boldsymbol{R}_s$	least family of sets containing the family $\boldsymbol{R}$ and closed under the finite union of sets 122
$\boldsymbol{R}_d$	least family containing the family $\boldsymbol{R}$ and closed under the finite intersection of sets 123
$\boldsymbol{R}_\sigma$	family of sets of the form $\bigcup_n H_n$, where $H \in \boldsymbol{R}^N$ 125
$\boldsymbol{R}_\delta$	family of sets of the form $\bigcap_n H_n$, where $H \in \boldsymbol{R}^N$ 125
$\boldsymbol{R}$	σ-algebra generated by the family $\boldsymbol{R}$ 124
$\operatorname{Bor}(\boldsymbol{R})$	least σ-additive and σ-multiplicative family containing $\boldsymbol{R}$ 124, 125, 269
$\prod_{t\in T} F_t$	cartesian product of sets 129
Y^T	cartesian power of the set Y 129
$\mathscr{C}$	Cantor set 135
C_T	the generalized Cantor set 135
$\sim I$	equivalence relation modulo the ideal $\boldsymbol{I}$ 141
$\langle X, Y\rangle$	cut in an ordered set 155

$\leqslant$	order relation between cuts 155
$\mathfrak{P}$	family of all cuts of an ordered set 155
$A \sim B$	"the set A is equipollent to the set B" 164
$\bar{\bar{A}}$	cardinal number or power of the set A 169
$\mathfrak{a}$	cardinal number of infinite countable sets 169
$X \sim_{\mathrm{fin}} Y$	"the sets X and Y are equivalent by finite decomposition" 186
$\mathfrak{c}$	power of the continuum 188
$\bar{A}$	order type of the relational system $\langle A, R \rangle$ 201
$x \prec_R y$, $x \prec y$	the element x precedes y under the relation R 204
$O_R(x)$	initial segment defined by the element x under the ordering relation R 204
ω	order types of sets 210
η	order types of sets 211
λ	order types of sets 214
α^*	inverse of the order type α 218
$W(\alpha) = \{\xi < \alpha\}$	233
$\lim_{\lambda<\gamma} \varphi(\gamma)$	limit of the λ-sequence $\varphi(\gamma)$ for $\gamma < \lambda$ 235
R_α	family of sets of rank at most α 238
$\alpha(+)\beta$	natural sum of the ordinal numbers α and β 253
$\alpha(\cdot)\beta$	natural product of the ordinal numbers α and β 253
VN ordinal	ordinal number in the sense of von Neumann 262
$\bar{\xi}$	cardinal number of the ordinal ξ 267
ω_1	smallest uncountable ordinal number 267
$\aleph_1$	power (cardinal number) of the ordinal number ω_1 267
$\aleph(\mathfrak{m})$	Hartog's aleph function 271
ω_0, ω_1	initial ordinals of the powers $\mathfrak{a}$ and $\aleph_1$ 272
$P(\varphi)$	set of all initial ordinals $\psi < \varphi$ 273
$\iota(\varphi)$	index of the initial ordinal φ 273
ω_α	initial ordinal whose index is equal to α 273
$cf(\alpha)$	the least number ξ such that the limit ordinal α is cofinal with ω_ξ 275
$\aleph_\alpha$	power of an initial number whose index is equal to α 275
$\mathfrak{a}_\xi$	ξth term of the exponential hierarchy of cardinals 285
$O(x) = \{z \in \text{Tree } T: z \leqslant x\}$	317
$x \preccurlyeq y$	317
$\delta(X)$	diameter of the space X 387
$K(p, r)$	r-ball 387
W	sieve 412

$C_\alpha(L)$	αth constituent of L 414, 444
2^X	family of closed non-empty subsets of X 392
$\boldsymbol{K}(X)$	family of compact non-empty subsets of X 393, 398
$\otimes$	inner product 402
$\boldsymbol{S}(X)$	family of analytic subsets of X 434
$\mathscr{A}$	$\mathscr{A}$-operation
$\mathscr{I}$	interval 01
$\mathscr{N} = N^N$	set of irrationals of $\mathscr{I}$
$\mathscr{E}$	set of reals
$\mathscr{C}$	Cantor discontinuum
$\mathscr{Q} = \mathscr{I}^{\aleph_0}$	Hilbert cube
$\mathscr{S}$	family of finite sets of positive integers
$\mathscr{R}_i$	set of binary fractions of $\mathscr{I}$

SUBJECT INDEX

abscissa of a point 62
absorption of a cardinal number 187
accessible cardinal 348
accumulation point 28, 32
aleph 267, 275
α-sequence 231
algebra,
  atomic Boolean 148
  Boolean 33
  Brouwerian 43
  cylindrical 152
algebra of sets 123
ambiguous Borel sets 417
analytic set 434
analytically representable function 236
antecedent of an implication 1
antichain 80, 316
anti-lexicographical ordering 221, 222
antinomies 4, 60
antinomy of Russell 61
$\mathscr{A}$-operation 409
Aronszajn tree 330
associative law 167
  for the addition of cardinal numbers 178
  for the multiplication of cardinal numbers 179
  for the product of cardinal numbers 197
associative laws for operations on sets 10
atom of Boolean algebra 41, 148
atomic Boolean algebra 148
axiom, multiplicative 130
axiom
  of choice 53
axiom
  of difference 6
  of the empty set 52
  of existence 6
  of extensionality 5, 52
  of inaccessible cardinals 357
  of infinity 53
  of pairs 52
  of power sets 53
  of regularity 57
  of relational systems 88
  of replacement 55
  of subsets for a formula 54
  of unions 6, 52
axiom system of set theory
  of Gödel–Bernays (GB system) 58
  of von Neumann 57
  of Zermelo 57
  of Zermelo-Fraenkel (ZFC and ZF) 57
axioms of Boolean algebra 34
axioms of lattice theory 42

back-and-forth method 215
Baire function 236
Baire property 428, 433, 466
Baire space 136
ball 387
base of a space 116, 387
Bernstein formula 282
binary relation 64
Boolean algebra, 33
  atomic 148
  distributive 148
  factor 79

Boolean polynomial 33
Boolean ring (algebra) 33
Borel formula 444, 457
Borel-measurable (*B*-measurable) function 419
Borel sets 126
  of class α 235, 415
boundary of a set 32
boundary set 32
bounded mapping 387
bounded set 387
box 305
branch 84, 316
Brouwerian lattice 43

CA-formula 444
cancellation laws
canonical mapping 79
Cantor–Bernstein theorem 185
Cantor discontinuum 135
Cantor set, 135
  generalized 135
cardinal number 169
cardinal $\mathfrak{c}$ 188
cardinal,
  accessible 348
  hyper-inaccessible 359
  inaccessible 348, 357, 359
  Mahlo 359
  measurable 367, 375
  regular initial 275
  singular initial 275
  strongly inaccessible 348
  strong limit 348
  weakly compact 360
  weakly inaccessible 348
cartesian power of a set 129
cartesian product 129
  of sets 62
  of relations 131
  of spaces 136
  of operations 130
Cauchy sequence 392, 396
chain 80, 255
characteristic function of a set 119
characteristic of a relational system 86
choice function 73, 394, 395
closed base of a topological space 116
closed function 393
closed set 28
closed subbase 116
closure 27, 133
coanalytic (*CA*) set 434
cofinal ordinals 231
cofinal sets 82
coinitial sets 82
commutative diagram 71
commutative law 166
  for addition of cardinal numbers 178
  for multiplication of cardinal numbers 179
  for the product of cardinal numbers 197
commutative laws for operations on sets 10
commutative ring 16
compact topological space 137, 149, 391
compact-open topology 395
comparable elements 80
complement
  of an element of a Boolean algebra 38
  of a set 19
complete lattice 83
complete ordering 83, 201
complete space 396
complex function 132
components of a conjunction 1
composition of relations 64
conjunction 1
connected relation 81
consequent of an implication 1
constituent 21, 414
  determined by a sieve 414

continuous α-sequence 233
continuous function 121
continuous set 206
continuously ordered set 206
continuum hypothesis (CH), 290
  generalized (GCH) 290
convergent sequence of sets 118
coordinate axes (of a product) 62
coset of a function 75
countable (denumerable) set 169
countable reduction property 127
cover of a set 81
critical ordinal of an α-sequence 233
(C)-sets 455
cut 155
cylindric algebras 152

Dedekind infinite set 105
definition by induction 93
degree of a set 303
degree of disjointness 302
δ-lattice 124
δ-multiplicative family of sets 124
dense set 205
dense in itself set 33
densely ordered set 205
denumerable (countable set) 169
derivative of a function 344
derivative of a set 33, 344
derivative of order α 235
diagonalization theorem 175
diameter of a space 387
difference
  of elements of Boolean algebra 34
  of ordinals 240
  of sets 6
direct predecessor 204
direct product of Boolean algebras 131
direct successor 204
directed set 81
Dirichlet's drawer principle 104
Dirichlet's principle for infinite sets 106
disjoint sets 9
disjunction 1
distance
  of a point to a set 388
  of two points 386
  of two sets 393
distributive Boolean algebra 148
distributive lattice 42
distributive laws for operations on sets 10
distributive law for the product of cardinal numbers 197
δ-lattice 416
domain (closed or open) 115
domain
  of a function 69
  of a relation 64
drawer principle of Dirichlet 104

element,
  extendable 183
  maximal (minimal), of ordered set 82, 255
  unit, of ordered set 83
  zero, of ordered set 83
element of a set 5
effective proof 48
elementary formula 143
embeddability of order type 219
embedding preserving suprema (infima) 154
empty set 9
enumeration of a set 343
epsilon-ordinal 247
equipollence 164, 293
equivalence classes of a relation 68
equivalence of sentences 2
equivalence relation 66
η_ξ-set 321
Euclidean algorithm for ordinal numbers 243
Euclidean space 136

even ordinal 236
existential quantifier 47
expansion of an ordinal number for a base, 249
exponential hierarchy of cardinal numbers 284
exponential topology 392
exponentiation of alephs 280
extendable element of a set 183
extension
  of an element 183
  of a function 71
  of a relational system 153

factor Boolean algebra 79
family
  of almost disjoint sets 300
  of Borel sets 126, 269
  of projective sets 456
  of sets closed under a function 121
family of sets,
  δ-multiplicative 124
  independent 297
  inductive 261
  $\mathfrak{m}$-complete 355
  $\mathfrak{m}$-disjoint 300
  monotone 81
  σ-additive 124
Φ-accessible cardinals 348
field
  of sets 34
  of a relation 64
filter 17
final segment 204
finite intersection property 137
finite reduction property 127
finite sequence 92
finite set 102
first category set 428
first distributive law 3
first element of linearly ordered set 202
first graph theorem 404, 454
first monotonic law
  for addition 239
  for ordinal multiplication 241
first separation theorem 451
formula, 46
  Bernstein 282
  elementary 143
  Hausdorff 280
  open 139
  Tarski 281
formula
  of first order 211
  of second order 211
F_σ-set 388
full binary tree 316
function, 69
  analytically representable 236
  Baire 236
  characteristic, of a set 119
  choice 73, 394, 395
  closed set-valued 393
  complex 73
  continuous 77
  Hartogs' aleph 271
  increasing 226
  injective 70
  monotonic, of subsets 187
  nth iteration of 94
  one-to-one 70
  regressive 347
  regular 409
  semi-continuous 393
  universal 423
function
  consistent with a relation 78
  defined on a set 69
  induced by a relation 79
  of two and more variables 72, 73

G_δ-set 388
gap 206
general quantifier 47

generalized associative law (for cardinal numbers) 193
generalized associative laws 110
generalized Cantor set 135
generalized commutative law (for cardinal numbers) 193
generalized commutative laws 110
generalized continuum hypothesis (H) 291
generalized distributive law for multiplication with respect to addition 193
generalized distributive laws 111
generalized Hausdorff formula 282
Gödel–Bernays axiom system of set theory 58
graph
of a function, Gr(f) 397
of a formula 394
greatest lower bound 82

Hamel basis 260
Hartogs' aleph function 271
Hausdorff operations 412
Hausdorff recursion formula 280
Hausdorff space 388
height of a tree 315
higher measurability 400
hereditary set 226
Hilbert cube 137
homeomorphism 77
of class α, β 419
homogeneous set 303
hyper-inaccessible cardinals 359

ideal, 17
prime 158
principal, generated by a lattice 159
ideal of a distributive lattice 158
ideal of a measure 368
image of a set 55
image of a set under relation 74
immediate transformability of polynomials 35
implication 1
inaccessible cardinal 348, 357–359
inclusion relation 7
incomparable elements 80
increasing function 226
independent sets 23, 260
index of an initial ordinal 273
induction, 93
transfinite 226
inductive definition 93
inductive family of sets 261
inductive set 90
inequalities between cardinal numbers 181
infimum of elements 146
infinite set 102
initial ordinal 272, 273, 275
initial segment 204
interior of a set 28
intersection
of sets 7, 107
of sets belonging to a family 60
interval 204
invariance under an isomorphism 86
inverse-function theorem 454
inverse image of a set under relation 74
inverse 64
of an order type 217
of a relation 64
inverse types 217
isolated point 171
isomorphic embedding of relational systems 153
isomorphic relational systems 86

joint continuity 395
ξth beth 239

last element of the linearly ordered set 202
lattice, 42
  complete 83
  distributive 42
  modular 44
lattice generated by a family of sets 123
law,
  associative 167
  associative, for addition of cardinal numbers 178
  associative, for multiplication of cardinal numbers 179
  associative, for the product of cardinal numbers 197
  commutative 166
  commutative, for addition of cardinal numbers 178
  commutative, for multiplication of cardinal numbers 179
  commutative, for the product of cardinal numbers 197
  distributive, for cardinal numbers 197
  distributive, for operations on sets 10
  first distributive 3
  first monotonic, for addition 239
  first monotonic, for ordinal multiplication 241
  full binary tree 84
  generalized associative 193
  generalized commutative 193
  generalized distributive, for multiplication with respect to addition 193
  second distributive 3
  second monotonic, for addition 240
  second monotonic, for ordinal multiplication 241
law
  of associativity of conjunction 3
  of associativity of disjunction 3
law
  of commutativity of conjunction 3
  of commutativity of disjunction 3
  of contradiction 4
  of contraposition 4
  of double complementation 19
  of double negation 4
  of excluded middle 4
  of exponents for the cartesian product 167
  of hypothetical syllogism 4
  of trichotomy 308
laws,
  associative, for operations on sets 10
  commutative 10
  distributive 10
  logical (tautology) 2
  monotonic, for ordinal subtraction 241
  de Morgan's 4, 12, 49
laws
  of absorption 3
  of subtraction 11
  of tautology 3, 11
least upper bound 82
left distributivity of ordinal multiplication with respect to ordinal subtraction 241
lexicographical ordering 221, 319.
level of a tree 315
limit
  of a sequence of points 389
  of a sequence of sets 118
  of λ-sequence 231
limit ordinal 230
limit inferior 118
limit superior 118
linear (total, complete, simple) ordering 81, 201
L-mapping 400
logical product 1
logical sum 1
lower-L mapping 405
lower measurability 400

lower-***R*** partition 463
lower section of a cut 155
lower semi-continuity 393
lower-Souslin mapping 453
Lusin space 427
Mahlo cardinal 359
Mahlo classification of inaccessible cardinals 359
mapping (transformation) 69
mapping of lower class α 462
Marczewski algebra 431
maximal element of an ordered set 82, 255
maximum principle 261
$\mathfrak{m}$-complete family 355
$\mathfrak{m}$-disjoint sets 301
mean-value theorem 186
meager set 428
measurable cardinal 367, 375
measurable mapping (***L***-mapping) 400
measurable selectors 458
measure on a set 366
method of transfinite induction 226
metric space 386
model of axioms 352
modular lattice 44
monotone family 81
monotonic function on subsets 187
monotonic laws for ordinal subtraction 241
minimal element of an ordered set 82, 255
minimal extension of an ordered set 156
de Morgan's laws 4, 12, 49
multiplicative axiom 130

natural addition 252
natural interpretation of axioms 27
natural model of set theory 352
natural multiplication 252
natural numbers 89
natural product 253
natural projection 463
natural sum 253
negation 2
von Neumann's axiom system of set theory 57
von Neumann's ordinals 262
normal form of sets 21
normal function 342
normal set 343
normal space 388
normal tree 334
nowhere dense set 32
*n*th iteration of a function 94

odd ordinal 236
one-to-one function 70
one-to-one sequence with *n* terms 102
open base 116
open formula 139
operation 72
operation $\mathscr{A}$ 409
order
  of a function 77
  of a tree 316
order relation 80
  in Boolean algebra 37
  in a lattice 42
order type 201
order type
  embeddable in an order type 219
  η 211
  λ 214
  ω 210
ordered pair 59
ordered set 80
ordered union of ordered sets 208
ordering,
  anti-lexicographical (by the principle of last differences) 221, 222
  lexicographical (by the principle of first differences) 221
ordinal cofinal with a limit ordinal 231
ordinal exponentiation 245

ordinal number (ordinal) 228
ordinal,
  epsilon- 247
  expansion of 249
  exponent 249
  even 236
  initial 272, 273
  limit 230
  odd 236
  power of 245
  regular initial 275
  singular initial 275
ordinal number in the sense of von Neumann 262
ordinals,
  natural addition of 252
  natural multiplication of 252
  natural sum of 253
  quotient of 243
ordinate of a point 62

partition 81, 67, 463
pair,
  ordered 59
  unordered 58
partition theorems 336
PCA (Σ^1_2)-sets 455
permutation of a set 71
Peano axioms 90
perfect set 259
perfectly normal space 388
point,
  accumulation 28, 32
  isolated 174
points of the space 27
Polish space 396
power set 53
power
  of a set 169
  of an ordinal 245
  of a cardinal number 180
  of the continuum $\mathfrak{c}$ 188
pre-order relation 80
prime ideal 142, 158
prime ideal of a distributive lattice 158
principal ideal generated by a lattice 159
principal ordinals of multiplication 253
principle
  of duality 41
  of induction for ordinals 268
  of transfinite induction 225
  of effectiveness 426
problem of elimination 24
product,
  cartesian, *see* cartesian product
  direct, of Boolean algebras 131
  logical, of sentences 1
  natural **253**
  reduced 141
product-function 73, 132
product
  of cardinal numbers 178, 196
  of order types 218
  of ordinal numbers 239
projection of a relation 64
projective set 456
proof by induction 90
proper cut 206
proper extremum of a function 173
proper subset 8
property invariant under isomorphism 86
property of universality 214
pseudo-complement 44
pseudo-difference 44
pseudo-tree 84, 315

q-additive measure 366, 367
q-$\mathscr{E}$-measure 367
q-2-measure 367
quantifiers 47
quasi-order relation 80
quotient class of sets with respect to a relation 68
quotient of ordinals 243

range
of a relation 64
of a function 69
rank of a set 238
reduced product 141, 376
reduction theorem for Borel sets 418
refinement of a cover 81
regressive function 347
regular closed set 39,115
regular function 49
regular initial ordinal 275
regular open set 115
regular space 388
relation, 64
binary 64
equivalence 66
inclusion 7
inverse 64
"less than", for cardinal numbers 181
n-ary 72
order 80
order, in a Boolean algebra 37
order, in a lattice 42
pre-order 80
quasi-order 80
transitive 8
relation of preceding
for elements of a set 204
for intervals of a set 204
relational system 86
embedded into a relational system 153
relational type 5, 88
remainder
of an ordinal number 251
of a set 231
representation 378
result of the operation $\mathscr{A}$ 409
ring 16
Russell's antinomy 61

saturated set 464
saturation of a set 464
scattered set 205
second distributive law 3
second graph theorem 406, 454
second monotonic law for addition 240
second monotonic law for ordinal multiplication 241
second separation principle 449
selector
for a mapping 458
for a partition 458, 464
semi-continuous function 393
sentence 47
separable space 387
separation theorem 128, 418
sequence,
α- 231
Cauchy 396
convergent 118
finite 92
infinite 92
one-to-one, with *n* terms 102
transfinite, of type α (or: α-sequence) 231
set, 4
analytic 434
Borel 126
Borel, of class α 235, 415
boundary 32
bounded 387
Cantor 135
closed 28, 116
closed ordered 255
continuous 206
continuously ordered 206
countable 169
Dedekind infinite 105
dense 205
dense in itself 33
densely ordered 205
densely ordered in linearly ordered set 206
denumerable 169

set
  directed 81
  empty 9
  η- 321
  finite 102
  finite in the direction of the kth axis 28
  F_σ 388
  generalized Cantor set 135
  hereditary 226
  homogeneous 303
  infinite 102
  independent 260
  inductive 90
  meager 428
  nowhere dense 32
  of the first category 428
  open 28
  open modulo an ideal 408
  ordered 80
  perfect 259
  power 53
  projective 456
  regular closed 39, 115
  saturated 464
  scattered 205
  sifted 412
  successor of 89
  well-ordered 226
set
  of representatives 68
  of type λ 215
sets,
  Borel 415
  (C)- 455
  cofinal 82
  coinitial 82
  congruent modulo ideal 17
  disjoint 9
  $\mathfrak{m}$-disjoint 300
  disjoint modulo ideal 17
  equipollent 164
  equivalent by finite decomposition 185
  independent 23, 260
set-valued function 393
sieve 412, 444
σ-additive family of sets 124
σ-algebra 124, 451
$\bar{\sigma}$-algebra 451
σ-ideal 127
σ-lattice 124, 416
σ-$\mathscr{E}$-measure 367
Σ_2^1-set 455
similar sets 85
similarity of relations 203
singular initial ordinal 275
$\bar{S}$-mapping 452
$\bar{S}$-measurable mapping 452
Souslin formula 444
Souslin problem 332
Souslin set 332
Souslin space 434
Souslin tree 332
space (universe) 18
space,
  Baire 136
  bounded metric 387
  compact topological 137, 139, 391
  complete metric 396
  Hausdorff 388
  Lusin 427
  metric 386
  normal 388
  Polish 396
  regular 388
  separable 387
  Stone 163
  topological 27
  topologically complete 396
stationary set 347
Stone space 163
strong limit cardinals 348
strongly inaccessible cardinal 348
subbase 120
sub-pseudo-tree 84
subset 7

subsystem of a relational system 153
successor of a set 89
sum,
  logical 1
  natural 253
sum
  of cardinal numbers 178, 192
  of elements of a Boolean algebra 34
  of order types 218
  of series of cardinal numbers 192
  of sets 6
supremum of elements 146
symmetric difference
  of sets 7
  of elements of a Boolean algebra 34

Tarski recursion formula 281
tautology (logical law) 2, 11
theorem
  on definitions by transfinite induction 233
  on diagonalization 174
  on the existence of pair 58
  on the existence of union 58
  on the existence of unordered triples, quadruples, etc. 58
Theorems of
  Addison 455, 456, 457
  Alexandrov, P. 396
  Alexandrov, P., Hausdorff 427
  Aronszajn 329
  Baire 399
  Banach 183, 421
  Banach-Kuratowski 366
  Bernstein, F. 260, 282
  Birkhoff, G. 262
  Burali-Forti 230
  Cantor, G. 174, 181, 190, 213, 214, 239–253, 290, 399
  Cantor-Bendixson 270
  Cantor-Bernstein 184
  Castaing 458
  Čoban 458
Theorems of
  Cohen, P. J. 291
  Dauer 458
  Dedekind 156
  Dellachérie 458
  Easton 291
  Engelking 421
  Engelking, Holsztyński, Sikorski 426
  Erdös-Rado 340
  Erdös, Sierpiński 432
  Erdös, Tarski 297, 305, 308, 357
  Fichtenholz-Kantorovitch 297
  Fodor 345
  Fox 395
  Frolik 428
  Gödel 290, 455, 456
  Hamel 261
  Harrington 457
  Hartogs 270, 309
  Hausdorff 280, 282, 325, 421
  Hessenberg 225, 276
  Himmelberg 458
  Huntington 41
  Hurewicz 443
  Jayne 428
  Jónsson 214
  Jónnson, Tarski 163
  Kaniewski 407
  Kaniewski, Pol 463
  Kantorovitch-Livensohn 457
  Keesling 393
  Kleene 457
  Kochen 325
  Kondo 472
  König, D. 326
  König, J. 198
  Kunugui 449
  Kunugui, Novikov 471
  Kuratowski 58, 125, 280, 394, 398, 442
  Kuratowski-Maitra 464
  Kuratowski, v. Neumann 457
  Kuratowski, Ryll-Nardzewski 458

Theorems of
Kuratowski, Tarski 125, 457,
de la Vallée Poussin 421
Lavrentiev 427
Lebesgue 421, 423
Lebesgue, Hausdorff 421
Levi B. 182
Lindelöf 388
Lindenbaum 265, 270, 272, 314
Lindenbaum, Tarski 182, 186
Lusin 295, 412, 417, 425, 438, 448, 449
Lusin, Sierpiński 413, 447
Lusin, Souslin 409
Łoś 140
Mackey 442
Mahlo 359
Maitra-Rao 427, 466
Mansfield 456
Marczewski 119, 305, 431
Mazurkiewicz 443, 457, 466, 468
McKinsey-Tarski 45
Mc Neille 154
Michael 392
Miller 333
Morley-Vaught 214
Mostowski 489
v. Neumann 466
Nikodym 430, 455
Novikov 450, 451, 455
Oxtoby 432
Peano 90
Ramsey 337
Rubin, H.–Rubin, J. 312
Russell-Whitehead 61, 130
Schönflies 209
Schori-West 393
Schröder-Bernstein 184–185
Scott D. 390
Selivanovski 455
Sierpiński 182, 192, 241, 280, 293, 296, 301, 313, 321, 423
Sierpiński, Tarski 354
Sikorski 280

Theorems of
Simpson 338
Skolem 27
Solovay 332
Solovay, Tennenbaum 332
Souslin 437
Specker 437
Stone, A. H. 32
Stone, M. H. 163
Taimanov 428
Tarski 111, 148, 183, 186, 188, 281, 297, 310, 311, 314
Teichmüller 260
Tychonoff 138
Ulam 366, 369, 372
Urysohn 391
Van Vleck 458
Vaught 262
Vitali 432
Wagner 458, 461
Wiener 59
Zermelo 57, 254
third graph theorem 406
topological space 27
topologically complete space 396
transfinite induction 226
transfinite sequence of type α 231
transformation (mapping) 69
transitive relation 8
tree 84, 315
of finite order 316
tree property 329
tree topology 84
Tychonoff topology 133
type of a relational system 88

uniform convergence 389
union of sets 6, 53, 107
over an indexing set 208
unit element 83
of the ring 16
of a Boolean algebra 37

universal function for a family of sets 423
universe (space) 18
universe of a relational system 86
unordered pair 58
upper-***L*** mapping 405
upper section of a cut 155
upper semi-continuity 393
upper-Souslin mapping 453
upper-***R*** partition 463

value
  of a Boolean polynomial 35
  of a function 69
Vietoris topology 392
Vitali set 432
VN ordinal 262

weakly compact cardinal 360
weakly inaccessible cardinal 348
weight of topological space 302
well-ordered set 224
well-ordering theorem (Zermelo's theorem) 254

Zorn maximal principle 256
zero element 34, 83
Zermelo set theory 57
Zermelo-Fraenkel system (ZFC) 57

er
eaf